About Island Press

Island Press is the only nonprofit organization in the United States whose principal purpose is the publication of books on environmental issues and natural resource management. We provide solutions-oriented information to professionals, public officials, business and community leaders, and concerned citizens who are shaping responses to environmental problems.

In 2001, Island Press celebrates its seventeenth anniversary as the leading provider of timely and practical books that take a multidisciplinary approach to critical environmental concerns. Our growing list of titles reflects our commitment to bringing the best of an expanding body of literature to the environmental community throughout North America and the world.

Support for Island Press is provided by The Bullitt Foundation, The Mary Flagler Cary Charitable Trust, The Nathan Cummings Foundation, Geraldine R. Dodge Foundation, Doris Duke Charitable Foundation, The Charles Engelhard Foundation, The Ford Foundation, The George Gund Foundation, The Vira I. Heinz Endowment, The William and Flora Hewlett Foundation, W. Alton Jones Foundation, The John D. and Catherine T. MacArthur Foundation, The Andrew W. Mellon Foundation, The Charles Stewart Mott Foundation, The Curtis and Edith Munson Foundation, National Fish and Wildlife Foundation, The New-Land Foundation, Oak Foundation, The Overbrook Foundation, The David and Lucile Packard Foundation, The Pew Charitable Trusts, Rockefeller Brothers Fund, The Winslow Foundation, and other generous donors.

About Society for Conservation Biology

Our goal is to help develop the scientific and technical means for the protection, maintenance, and restoration of life on this planet—its species, its ecological and evolutionary processes, and its particular and total environment. In the service of this goal, our objectives include (1) the promotion of research and the maintenance of the highest standards of quality and ethics in this activity; (2) the publication and dissemination of scientific, technical, and management information; (3) the encouragement of communication and collaboration between conservation biology and other disciplines (including other biological and physical sciences, the behavioral and social sciences, economics, law, and philosophy); (4) the education, at all levels, preparatory and continuing, of the public, of biologists, and of managers, in the principles of conservation biology; (5) the promotion of all of the above through the provision of adequate funding; and (6) the recognition of outstanding contributions to the field made by individuals and organizations. For more information, visit the SCB Web site at http://conbio.rice.edu/scb/.

CONSERVATION BIOLOGY

Research Priorities for the Next Decade

EDITED BY

MICHAEL E. SOULÉ AND GORDON H. ORIANS

Society for Conservation Biology

ISLAND PRESS

Washington • Covelo • London

Island Press is a trademark of The Center for Resource Economics.

Conservation biology : research priorities for the next decade / edited by Michael E. Soulé and Gordon H. Orians.
  p. cm.
Includes bibliographical references.
  ISBN 1-55963-868-0 (cloth : alk. paper) — ISBN 1-55963-869-9 (pbk. : alk. paper)
  1. Conservation biology—Research. I. Soulé, Michael E. II. Orians, Gordon H.
  QH75 .C6616 2001
  333.95'16—dc21

                                                    2001003355

British Cataloguing-in-Publication Data available.

Printed on recycled, acid-free paper ♻

Manufactured in the United States of America
10 9 8 7 6 5 4 3 2 1

CONSERVATION BIOLOGY

CONTENTS

FOREWORD

AT LEAST ONE BILLION HUMANS, MAINLY IN THE RICHEST countries, currently are practicing an environmentally unsustainable lifestyle. Another billion people, mainly in the poorer countries, lack gainful employment. Both affluence and poverty cause humans to damage terrestrial and aquatic biodiversity, as well as ecosystem functioning, and modify the earth's climate. Human consumption and increasing human numbers are the driving forces behind species extinction and were the impetus for the formation of the Society for Conservation Biology in 1986.

The Society's goal is "to help develop the scientific and technical means for the protection, maintenance, and restoration of life on this planet—its species, its ecological and evolutionary processes, and its particular and total environment." After the formation of the Society, a group of scientists met to develop a research agenda for this new field, and a small book called *Research Priorities for Conservation Biology,* edited by Michael Soulé and Kathryn Kohm, was published in 1989.

Twelve years passed, and in April 2000 the Society for Conservation Biology convened a workshop funded by SCB, the Gilman Foundation, and Conservation International. Our purpose was to update the conservation biology research agenda articulated in 1989. Gordon Orians and Michael Soulé were asked to provide the scientific underpinning for priorities for research in conservation biology. Together with a steering committee, they developed the topics and invited scientists to draft papers for discussion and review. The results and insights of some of the leading conservation biologists, mainly from the United States, are encapsulated in the book before you. They will interest biologists, resource managers, and policy makers.

The open-ended agenda can and should catalyze scientific efforts, provide a meter stick to measure progress, and act as a vehicle for scientific debate. It is not meant to be a comprehensive agenda but reflects the priorities of the editors who assembled in April 2000.

Notably some trends have changed little since 1989, making many of the first book's recommendations still timely. Human uses of the land, water, and flora and fauna continue to accelerate, modifying the life support systems of many species and causing increasing numbers of extinctions. Although workshop participants made no formal assessment of progress over the intervening years, several achievements are recognized. Among the most apparent is that conservation biology is a legitimate interdisciplinary science. Its professional society is among the largest professional biological societies in the world. A more holistic approach in addressing problems and encouraging an interdisciplinary exchange of knowledge is one of the hallmarks that distinguish the field. The maturing of the discipline means that more specific recommendations and a sharper focus are possible than twelve years ago.

As president of the Society for Conservation Biology, I felt that a workshop that produced a book generated by some of the greatest conservation biologists of the twentieth century would act to renew the vigor of practicing conservation biologists and call a new generation of scientists into service for the twenty-first century. Society must increase funding for conservation biologists to gather the scientific information needed if we are to make intelligent decisions on the use of the earth's biodiversity.

Information for conservation problem solving is still in short supply. A major challenge is to identify and learn the distribution, abundance, natural history, and population trends of earth's species. Such a goal is more ambitious and more important than curing cancer, identifying the human genome, going to the moon, or living on Mars. It is a goal on which human society as we know it depends, because the richness of the services life provides—cleaning our air and water, growing our food, and, most important, giving meaning to our lives—depend on biodiversity.

The earth is small, and the creatures that share our planet remain little known. Conservation biologists must provide insights not only into what makes the tapestry of life but also into how we can maintain it, support it, and enrich it while using its services to support human life. Conservation biologists must provide much of the knowledge and help society to use it wisely. If we fail, human life will become more impoverished as species after species is exterminated and the fabric of life unravels. Can we use our knowledge to minimize our damage to the earth while we struggle toward a more sustainable path? Information, in and of itself, is not enough; information must reach the right audience so it can be used, further refined, and developed for maximum impact, much like the headwaters of a stream pick up force and sediment to be deposited as a rich and productive delta. Ideas, like water, should be purified, renewed, recycled, and reused.

The Society has launched a magazine to use stories as a way of conveying the successes and failures of practicing conservation biologists. Stories can portray old ideas in new ways and give new ideas a context and an application not readily apparent at first glance. *Conservation Biology in Practice* is designed to build strategic partnerships to bridge the gaps between conservation practice, theory, and policy and to help humans solve the problems they have created. The two-pronged approach of identifying our scientific needs in a book and providing the vehicle of a magazine to communicate with practitioners and policy makers is one that the Society hopes will increase our effectiveness.

The book before you sets forth a research agenda for conservation scientists of all ages. It should inspire conservation biologists addressing new as well as long-established problems. Orians and Soulé have brought together a group of authors who set forth an agenda for many lifetimes of work.

P. Dee Boersma

PREFACE

CONSERVATION BIOLOGY IS A YOUNG SCIENCE THAT WAS BORN as one response of the scientific community to the current massive environmental changes occurring on Earth. Some of these alterations are threatening the existence of a significant fraction of Earth's millions of species and affecting the evolutionary processes that generate new species. The main tasks of conservation biology are to provide the intellectual and technical tools to enable society to anticipate, prevent, and reduce ecological damage, and to generate the scientific information on which effective conservation policies can be designed and implemented.

1988 Conference

As part of an effort to evaluate the status of the field and to identify major, compelling research priorities in conservation biology, experts from many fields met for three days in Florida in April 1988 (Soulé and Kohm 1989). Their discussions were motivated by the recognition that research by conservation biologists and scientists in related fields had already provided managers and planners with important information and planning guidelines. They already knew that merely setting aside samples of ecosystems often fails to achieve its intended objectives and may have unexpected undesirable outcomes if underlying ecological processes are not understood. Recent scientific insights derived from the study of patch dynamics, edge effects, population viability analyses, ecosystem processes, scale and dynamics of disturbances, invasive exotic species, patterns of endemism, and human-induced environmental changes were providing new guidelines for effective conservation and management policies. Thus, the participants in the workshop were well-aware of the potential utility of well-focused future scientific research.

The 1988 workshop concentrated on identifying priority areas for scientific research for the subsequent decade. Although participants identified many important research topics, they agreed that the following five areas were of highest priority (Soulé and Kohm 1989):

1. *Conduct a crash program of extensive surveys and mapping to identify areas that are critical for the protection of natural and genetic resources because of their high biotic diversity, or high levels of endemism, or because of imminent destruction of critical or unusual habitats and/or biotas.* Although much survey work has been done, this task has not been completed. Rapid environmental assessment methods have been developed, and surveys have been conducted in many poorly known tropical areas (Mittermeier and Forsythe 1997). Most terrestrial areas of unusually high diversity and endemism of vertebrates and vascular plants have been identified, but distributions of terrestrial invertebrates and marine organisms are still poorly known. Enough is known to demonstrate that surveys of limited taxonomic scope are often poor guides to the distribution of species in other taxonomic groups (Heywood and Watson 1995).

2. *Establishment of a small number of research sites in the tropics to develop a coordinated program of comparative research on populations, communities, and ecosystems in relatively undisturbed and secure situations.* Since 1988 the number of tropical field stations has grown, and long-term data sets have been generated at some of them. Comparative results from the four premier neotropical field stations have been summarized in the book *Four Neotropical Forests,* edited by Alwyn Gentry (1990). Volumes synthesizing research results have been published for La Selva, Costa Rica (McDade et al. 1994), and Barro Colorado Island (BCI), Panama (Leigh et al. 1982). A steady stream of publications is emerging from the Station de Recherches des Nouragues in French Guyana (Charles-Dominique 1977, 1986). Nevertheless, coordination of research themes among tropical field stations is still very poor, and no concensus has been reached on a common set of research questions that should be pursued at all stations. Therefore, although cooperation among researchers at La Selva and BCI is increasing and more and more scientists are gaining research experience at both stations, a comparative science of tropical biology has yet to emerge. Long-term research in tropical marine systems is still minimal, the few exceptions being coordinated regional programs, such as CARICOMP in the Caribbean (Ogden and Gladfelter 1986).

3. *Conduct studies at all spatial scales to assess the kinds, mechanisms, and magnitudes of impacts on ecological systems.* Since the 1988 workshop,

there has been increasing recognition of the importance of research at a variety of spatial and temporal scales, especially at spatial scales larger than those that have typified ecological field experimentation (Kareiva and Anderson 1988). Conservation biology concerns have stimulated research into metapopulation models, source-sink population dynamics, and mechanisms and patterns of dispersal, and they have highlighted the importance of long-term studies of contemporary and paleoecological data for detecting and interpreting temporal patterns of change. Thus, this research agenda is as important today as it was in 1988, and it is expected to continue to be an arena of active investigation in both marine and terrestrial systems.

4. *Enhanced support for research that focuses on the physiology, reproduction, behavior, ecological interactions, and viability of individuals, populations, and species, especially with regard to species of critical ecological or economic importance.* Although research into these important issues continues unabated today, enhanced research support has not been forthcoming. For example, the rapidly increasing sophistication of population viability analysis, its penetration into governmental agencies and nongovernmental organizations, and its recognition by the disciplines of ecology and genetics as a legitimate area for scholarly pursuit constitute one of the success stories of conservation biology; yet funding agencies still resist supporting this field, in part because it resides at the interface between basic and applied research. Fortunately, a recent initiative by the U.S. National Science Board may eventually result in increased funding for environmental sciences by the National Science Foundation (NSF 2000).

5. *Increased training for both basic scientists and natural resource managers, particularly in tropical developing countries.* There have been encouraging developments in this area since 1988. The Organization for Tropical Studies has significantly expanded its course offerings for Latin American students, both graduate and undergraduate, and it has initiated courses in Brazil and Peru to supplement its effective educational program in Costa Rica. A consortium of European universities has established an educational program in Africa, and an increasing number of North American universities sponsor tropical field courses. The World Wildlife Fund–U.S. has developed a program of Russell Train Fellowships that supports promising young tropical scientists for varying periods of time at North American universities. Encouraging though these developments are, the need identified in 1988 is far from being met. Indeed, the need is probably greater today than it was a decade ago, in part because the status of parks and similar protected areas is lower today than at any time since the Second World War, and

also because governments can now allow such protected areas to languish without much fear of criticism or adverse consequences. Training of marine conservation biologists and resource managers in tropical developing countries is still inadequate.

Current Needs

While the major tasks identified at the 1988 workshop are being addressed, new needs have arisen because of the major increases that have taken place in human activities during the ensuing twelve years. During that interval the human population has grown from four billion to more than six billion, and it continues to increase. In addition, global commerce has dramatically increased. The budgets of the major international corporations now dwarf the budgets of the governments of most countries. Lowering trade barriers that encourage trade in commodities results in major expanses of land being devoted to the production of agricultural, aquacultural, horticultural, and forest products. Conservation activities now take place in a more crowded and more intensely interconnected planet.

Thus, although it is clear that the top-priority research topics identified by workshop participants in 1988 remain on the high-priority agenda, in 2000 leaders of the Society for Conservation Biology judged that it was time for the conservation biology research community to take a fresh look at research accomplishments and needs and to offer suggestions and guidance to researchers and private and public funders for the first decade of the twenty-first century. Accordingly, twenty-six experts from four countries met at White Oak Conference Center, Florida, April 12–16, to do for the new decade what the previous workshop did for the last decade.

To foster a rapid and comprehensive assessment, ten major topics were identified by a steering committee. They are captured in the chapters of this book. One or more experts on each topic agreed to prepare discussion papers prior to the workshop, and each paper was reviewed by a respondent. Significant contributions by other participants are signaled by their inclusion as coauthors of the respective chapters, but all participants contributed to all chapters of the book.

The audience for this book includes scientists, resource managers, and funding agencies throughout the world. Because threats to biodiversity operate worldwide, success in the protection of nature for its own sake and for the sake of future human generations will require the efforts of dedicated people everywhere. Although this report focuses on the research that is most urgently needed to support wise management decisions, we recognize that research is only one component of the diverse array of activities that must be a part of the successful conservation effort. Indeed, the research we recommend would be of little value in the absence of a cadre of practitioners willing and able to use the research results and insights to further their activities on the ground.

ACKNOWLEDGMENTS

Members of the international steering committee that selected the ten topics for which invited papers were requested and identified authors and respondents were P. Dee Boersma, president of the Society for Conservation Biology; Reed Noss, president-elect of the Society; Sandy Andelman; Julie Denslow; Gustavo Fonseca; Jane Lubchenco; Georgina Mace; Scott Mills; Gordon Orians; Hugh Possingham; John Robinson; and Michael Soulé. We wish to thank P. Dee Boersma, Douglas Bolger, Carla D'Antonio, Jack Ewel, Scott Mills, Joshua Schimel, and John Terborgh for their review of particular chapters.

Funding to support the workshop was provided by the Society for Conservation Biology, the Gilman Foundation, Conservation International, and NASA.

LITERATURE CITED

Charles-Dominique, P. 1977. *Ecology and Behaviour of Nocturnal Primates.* New York: Columbia University Press.

Charles-Dominique, P. 1986. Interrelations between frugivorous vertebrates and pioneer plants: *Cecropia,* birds, and bats in French Guyana. pp. 119–135 in A. Estrada, and T. H. Fleming (eds.), *Frugivores and Seed Dispersal.* Boston: W. Junk.

Gentry, A. H. (ed.). 1990. *Four Neotropical Forests.* New Haven: Yale University Press.

Heywood, V. H., and R. T. Watson (eds.). 1995. *Global Biodiversity Assessment.* Cambridge, England: Cambridge University Press.

Kareiva, P. M., and M. Anderson. 1988. Spatial aspects of species interactions: The wedding of models and experiments. In A. Hastings (ed.), *Community Ecology,* pp. 38–54. New York: Springer-Verlag.

Leigh, E. G. Jr., A. S. Rand, and D. M. Windsor (eds.). 1982. *The Ecology of a Tropical Forest: Seasonal Rhythms and Long-Term Changes.* Washington, D.C.: Smithsonian Institution Press.

McDade, L. A., K. S. Bawa, H. A. Hespenheide, and G. S. Hartshorn (eds.). 1994. *La Selva: Ecology and Natural History of a Neotropical Rain Forest.* Chicago: University of Chicago Press.

Mittermeier, R. A., and A. B. Forsythe. 1997. Setting priorities for biodiversity conservation: One organization's approach. In G. K. Meffe and C. R. Carroll (eds.), *Principles of Conservation Biology,* pp. 141–142. Sunderland, Mass.: Sinauer Associates.

National Science Foundation. 2000. *Environmental Science and Engineering for the 21st Century: The role of the National Science Foundation,* NSB00-02. Washington, D.C.

Ogden, J. C., and E. H. Gladfelter (eds.). 1986. *Caribbean Coastal Marine Productivity.* Paris: UNESCO.

Soulé, M. E., and K. Kohm. 1989. *Research Priorities for Conservation Biology.* Washington, D.C.: Island Press.

I

INTRODUCTION

Gordon H. Orians and Michael E. Soulé

FOR A NUMBER OF YEARS FOLLOWING ITS ESTABLISHMENT, THE Society for Conservation Biology (SCB) was the most rapidly growing professional scientific society in the world. Its expansion reflected the high level of concern among biologists, especially younger ones, about the increasing threats to the survival of millions of species, generated primarily by human activities. The continued rapid growth of the human population, combined with increasing per capita consumption of resources, is generating unprecedented demands on Earth's renewable and nonrenewable resources. Three decades ago, the greatest threat to sustainable use of natural resources was believed to be the high rate of use of nonrenewable resources. The impending exhaustion of those resources was expected to drive up prices and create serious shortages of materials. As it has turned out, technology has been remarkably successful in finding substitutes for many nonrenewable resources, reducing demands on their use and extending the estimates of the lifetimes of their effective supplies.

What has emerged instead as being of most concern are serious problems associated with unsustainable use of Earth's renewable resources. Among these vital resources are physical processes, such as rates of soil formation and the capacities of Earth's oceans, fresh waters, and atmosphere to decompose and dissipate wastes generated by human activities. Humans are exerting substantial influences on the major biogeochemical cycles of carbon, nitrogen, phosphorus, and sulfur, and on renewable biological processes, such as the dynamics of populations of Earth's biota, that is, the species whose activities generate the goods and services on which human society depends.

In addition, the opportunities generated by modern information technology that allow people to live, work, and retire in relatively remote areas are resulting in a sprawl of human residences across landscapes. For example, in many areas of western North America, major sources of income are unrelated to local resources. Thus, for example, brokers conduct their Wall Street businesses from offices in Bozeman, Montana, and internet businesses take orders from rural locations around the world. The spatial pattern of human settlements and their impacts on the environment are changing dramatically.

In addition, conservation biologists typically consider preservation of species to be an ethical responsibility. Conservation biology, together with such disciplines as medicine, is a normative science. Conservation science is employed in the service of an ethical goal, the maintenance of Earth's biodiversity. Indeed, the U.S. Endangered Species Act is fundamentally an ethical statement. It specifies appropriate behavior that is not simply a matter of economic convenience. The combination of this ethical stance with the estimated current high rates of extinction of species gives a sense of urgency to the activities of conservation biologists. This sense of urgency influenced the selection of research priorities identified by participants during the workshop at the White Oak Plantation in Florida, April 12–16, 2000.

The Role of Basic Research

The goal of conservation biology is to provide sound scientific concepts and data to inform the design of effective conservation management actions. Recovery and maintenance plans for species are most likely to be successful if they are based on detailed knowledge of the life history traits and habitat requirements of those species. Ecosystem management should be based on knowledge of successional patterns; responses of vegetation to different types of disturbances; and responses of marine communities to changes in ocean currents, temperatures, and chemical composition. The design of systems of reserves should be based on information about the needs of species with the largest spatial requirements, locations of migration corridors and rare community types, distribution patterns of existing vegetation types, and patterns of dispersal among habitat patches. Restoration of degraded ecosystems requires information on interactions among species, the legacies of past perturbations, and the nature of continuing forces that impinge on the systems.

Thus, conservation practitioners need information on how ecological systems work, how the interactions among species determine the functional properties of the systems, and the spatial and temporal scales at which they operate. They need to know how much of what kinds of perturbations ecological communities can absorb, the consequences of ecosystem fragmentation, and how and why introduced species alter ecosystems.

Unfortunately, conservation is being carried out on a planet that is poorly known. Although about 1.75 million species have been described, the total number of species living on Earth may exceed 50 million. The majority of species of insects, mites, nematodes, and inhabitants of marine sediments have not been described. The geographic ranges and ecological requirements of only a tiny fraction of described species are known. Moreover, we know little about the abundances of species and even less about population trends. Conservation practitioners urgently need much better information on distributions of species, the composition of ecological communities, and the ecological requirements of the species they attempt to preserve. Practitioners could better deal with all of the issues discussed in this report if better taxonomic and ecological information was available.

Given the high rates of environmental changes, conservation biologists are motivated to devote considerable attention to research that can deliver important results quickly. In many parts of the world, opportunities to study large systems that harbor complete biotas at reasonably normal densities are rapidly diminishing or have already been lost. Species are becoming extinct in large numbers before their traits and functional relationships with other species have been determined. Natural areas are disappearing before their potential role as components of reserve systems can be evaluated, much less implemented. Therefore, the research priorities identified during the workshop and reported herein stress activities with the potential of informing decisions to be made within the next decade. Nonetheless, workshop participants fully recognized the importance of long-term data sets and research designed to elucidate ecological processes operating over long temporal scales (Soulé and Terborgh 1999).

Threats to the Generation and Maintenance of Biodiversity

Changes in the number of species on Earth at any moment in time are the result of the rates at which species are being formed and the rate at which they are becoming extinct. Averaged over hundreds of thousands of years, speciation rates have exceeded extinction rates, with the result that Earth's current biota may be richer than it has ever been. However, the general increase in species richness on Earth over evolutionary time has been punctuated by a small number of major extinction events and a larger number of minor extinction events, after which the diversity of life rebounded over the course of several million years.

The time required for speciation—the separation of one evolutionary lineage into two lineages—varies greatly. Among plants, speciation by polyploidy is very rapid; it requires only one generation to double the number of chromosomes. Because more than half of existing species of vascular plants are estimated to be polyploids, the potential for high rates of speciation in small areas probably continues to be high among these organisms. In contrast, speciation by polyploidy is rare among animals. Evidence suggests that most speciation events among animals depend on separation of a formerly continuous geographic range, either by imposition of a barrier to gene flow within the existing range or colonization of a new area by dispersing individuals. Subsequent evolution of differences sufficient to generate reproductive barriers should the populations again become sympatric may require thousands of years.

The current loss of species richness among Earth's existing biota is believed to be driven primarily by a striking increase in extinction rates. Rates of species formation are probably also declining but at a modest rate. For larger animals, current range contractions are probably reducing opportunities for subpopulations to survive long enough to lead to new species.

Increases in extinction rates have stimulated research to determine their causes, because if we wish to maintain Earth's biodiversity and preserve speciation opportunities, our primary task is to find ways to reduce the current high extinction rates. Extinctions are caused by many different agents and combinations of them, but the following few factors are believed currently to be the primary causes.

Habitat Loss and Fragmentation

Conversion of ecosystems to agriculture, forest plantations, rangelands dominated by domestic livestock, and urban and suburban development is greatly reducing the overall extent of many ecosystem types, and the remaining pieces are increasingly small and isolated from one another. Abundant evidence shows that species composition, interactions among species, and climates in small patches differ from those in large patches, and movements of individuals among patches are reduced. The habitats of many species with small ranges are being completely obliterated by conversion to other habitat types in which they cannot survive. Habitat loss and fragmentation are especially pronounced today in tropical forests, but fragmentation is also serious in other terrestrial habitats, especially grasslands. It may be important in some coastal marine regions, as well.

Although many ecosystem types were historically distributed in large, contiguous blocks, many others, such as lakes, reefs, caves, and outcrops of unusual rock types, are typically highly dispersed and fragmented. Thus, organisms that live in them presumably evolved adaptations to survive in and disperse among isolated habitat patches. Species that live in formerly unfragmented habitats, on the other hand, are unlikely to have life history traits that adapt them for survival in small habitat patches. Therefore, conservation attention is appropriately concentrated on those species, but much can be learned by comparing their life history traits with those of species that evolved in patchy habitats.

Overexploitation

Much research has been carried out to assess maximum sustainable exploitation rates of commercially important species and to determine how those rates are influenced by life history traits of species and patterns of environmental variability. However, for primarily political and economic reasons, sustainable yield rates are typically exceeded, with the result that, for example, more than half of Earth's commercial fisheries are currently overexploited. A major trend in modern human societies is the expansion of the number of species that are traded in markets. Especially notable is the explosion of markets for ornamental plants and animals as pets. In addition, many fishing methods capture species that are not commercially important—bycatch—and typically discard them. Some harvesting methods, like dynamiting coral reefs and dredging the ocean floor, destroy the environments of both targeted and associated species.

Exotic Species

Natural processes have always moved individuals of species across substantial environmental barriers. Indeed, the biotas of some environments, such as oceanic islands, have been assembled entirely by such natural long-distance dispersal, followed by in situ speciation. Today, however, Earth's terrestrial and marine biota is being scrambled by an unprecedented movement of individuals of thousands of species by human activity. The biotas of many areas are now dominated by introduced species, and exotics are significant fractions of the biotas in other areas. In addition, many exotic species have become highly invasive and now dominate ecosystems in the areas in which they have become established. Extinction of native species by exotics appears to be relatively rare on continents, but it is common on islands. Nonetheless, major reductions in populations of native species, which are common consequences of invasions of exotics on mainland areas, are of considerable conservation concern.

Pollution

The waste products of human activities are now major components of Earth's atmosphere, soils, fresh waters, and oceans. Some of these materials are highly toxic substances that directly affect the health of organisms. Others are nutrients that cause undesirable effects at unnaturally high concentrations. Nutrient enrichment is especially important in fresh waters and coastal marine areas, but it is also significant in some terrestrial environments, especially those that are naturally nutrient poor. Some substances emitted to the atmosphere are neither nutrients or toxicants, but their increased concentrations result in alterations of the quantity and chemistry of precipitation.

Climate Change

Climates are never constant, and organisms have evolved traits that enable them to adjust to the range of variability that has characterized their environments. However, when climate variability exceeds levels to which organisms are adapted, many species may be unable to maintain viable populations. Considerable evidence exists to suggest that current human activities are perturbing Earth's atmosphere sufficiently to alter the current climate; the forcing factors, particularly greenhouse gases, especially carbon dioxide from burning fossil fuels, continue to increase. Therefore, increased climatic changes are anticipated. Even in the absence of direct anthropogenic effects, climatologists expect future climate variability to exceed that which has characterized the past few centuries. Increasing evidence suggests that extreme events may play powerful roles in determining changes in distribution and abundance of species, suggesting even greater needs for understanding the interaction between climate and biodiversity.

Targets of Conservation Activities—Problem of Baselines

Conservation practitioners work to achieve a variety of goals. They may seek to restore populations of an endangered species or to reintroduce a species into parts of its range from which it has been exterminated. They may attempt to remove an exotic species from a community that it is invading. They may try to reestablish a complex ecological community in an area from which it has been eliminated. They may wish to restore the primary productivity of an area or to reduce rates of soil erosion.

What all of these activities have in common is an attempt to reestablish some ecological state and functioning believed to have characterized a site at some time in the past. To do that, conservationists must make two key decisions. First, they must decide which time in the past should serve as the reference period. Second, they must assess the probable ecological conditions that existed in the area at that time. Neither of these decisions can be made using only scientific criteria. For example, in the United States, should we attempt to recreate ecological conditions as they were when humans first arrived from Asia? When Europeans first colonized North America? When chestnut blight first invaded eastern deciduous forests? When zebra mussels first invaded the Great Lakes? No unambiguous criteria exist for making these decisions. Value judgments may figure prominently and appropriately in deciding on reference periods.

Once a target period has been chosen, a practitioner must assess the probable ecological conditions that existed at that time. Sometimes paleontological or historical data are available to inform those guesses, but such data are always incomplete. The composition of the ecological community prior to the establishment of an exotic invader may be known only within wide limits. Even if the general composition of the former ecological community can be determined fairly precisely, should the goal be to recreate that composition as accurately as possible? Or should the goal be to recreate a community with a physical structure similar to the previous one? Or should a comparable level of biological productivity be sought? Again, multiple goals are possible, and it is not easy to determine which ones are most appropriate for a given situation. Some of the research priorities identified in the chapters to follow deal with methods designed to help make decisions about appropriate baselines, but no advances in conservation biology research will eliminate the need to make subjective judgments about baselines.

Major Themes

In the following eleven chapters, the status of knowledge in an array of fields is summarized, and critical gaps in knowledge that are currently impeding conservation practice in the field are described. For each topic a small number of high-priority research topics are identified. As the White Oak workshop unfolded, a small number of overarching research themes emerged:

- *Find ways to better use satellite imagery and aerial photographs to identify ecological systems and habitats and, if possible, the status of biodiversity within them.* Vast quantities of data are now being gathered by satellites orbiting Earth. For the most part, those data are being used to identify broad categories of land cover (agricultural lands, forests, urban and suburban areas) or to assess primary productivity. The potential of those data to serve as sources of information on habitat quality and species composition has scarcely been explored, especially for the marine environment.

- *Perform research to improve our understanding of the significance of alterations to the structure of food and interaction webs.* Local extinctions (extirpations) of native species, invasions of exotic species, and major changes in the relative abundances of species are likely to precipitate direct and indirect changes in ecological structure and processes, not to mention effects on the capacity of the system to provide goods and services for the human economy. The effective loss of highly interactive species is likely to initiate trophic cascades leading to the loss of species diversity.

- *Perform research to inform the design, development, and maintenance of marine protected areas.* Despite their potentially great importance, relatively little attention has been given to marine reserves, which account for less than 0.25 percent of the oceans. The current rate of alterations of coastal marine environments, especially coral reefs, gives special urgency to this concern.

- *Investigate the large-scale community and ecosystem impacts of eutrophication in marine and fresh waters.* Lakes, streams, rivers, and the oceans are the ultimate repositories of materials introduced into ecosystems. Yet little is currently known about where and how these materials enter aquatic environments, how they interact with the biotic and abiotic components of the ecosystems, what harmful consequences might occur, and what remedial actions should be devised.

- *Perform research to enable us to better predict which exotic species are likely to become invasive and what their effects will be.* In practice it is impossible to completely stop the importation of potentially troublesome exotic species, but significant improvements are possible, including restrictions on importing potentially invasive species and disease vectors. We need to become more adept, however, at identifying potentially harmful species and predicting their public health and ecological hazards.

Organization of the Book

The subsequent eleven chapters in this book are based on the discussion papers written prior to the White Oak workshop by the invited senior authors. They have been modified as a result of contributions by the designated respondents and other workshop participants, and converted to a general format. Each chapter assesses the current state of knowledge about the issue and identifies the gaps in knowledge that most strongly impede the ability of conservation practitioners to carry out their work. These two sections serve to motivate the research priorities that are proposed in the final section of each chapter and highlighted in a box at the end of each chapter. A final synthesis chapter describes some of the changes that have occurred in the world and the field of conservation biology in the last twenty years and places research priorities in a broad political context.

LITERATURE CITED

Soulé, M. E., and J. Terborgh. 1999. *Continental Conservation: Scientific Foundations of Regional Reserve Networks*. Washington, D.C.: Island Press.

2

ASSESSMENT AND MANAGEMENT OF SPECIES AT RISK

Georgina M. Mace, Jonathan E. M. Baillie,
Steven R. Beissinger, and Kent H. Redford

A MASSIVE HUMAN POPULATION, DISTRIBUTED ACROSS MOST AREAS of the Earth's land surface, now competes directly and indirectly with other species for natural resources. As a result of human activities many species have gone extinct, others are restricted to tiny islands of remaining habitat, and still others are committed to extinction in the near future.

For still larger numbers, the future looks extremely uncertain. Recent assessments of the global conservation status of species (IUCN 1996; Oldfield et al. 1998) suggest that more than 10 percent of birds and around 20 percent of mammals and amphibians are now threatened with extinction (table 2.1). Comprehensive assessments of all native U.S. species by The Nature Conservancy show that over 60 percent of molluscs and crayfish, 40 percent of freshwater fishes and amphibians, and 50 percent of freshwater invertebrates are at risk. Extinction risk also is demonstrably nonrandom; species within some phylogenetic clades tend to have higher or lower levels of extinction risk than across taxa as a whole, thereby increasing the loss of evolutionary diversity from species extinctions (Russell et al. 1998; Purvis, Agapow, et al. 2000). Extinctions are also nonrandom with respect to the ecological roles that species play such that loss of certain species will have a disproportionate impact on ecosystems (Jernvall and Wright 1998), as described in chapter 3.

TABLE 2.1. Summary Results of the Application of IUCN Criteria to Various Higher Taxonomic Groupings

Taxon	Approximate number of species assessed	Percent threatened of those assessed	Percent data deficient of total assessed
Mammals	4763 (100%)	23	5
Birds	9946 (100%)	11	1
Reptiles	1480 (20%)	17	5
Amphibians	600 (12%)	21	7
Molluscs	>3000 (4%)	31	18
Trees	10,091 (0.1%)	59	4

Note: The approximate number of species assessed is estimated from IUCN (1996) and Oldfield et al. (1998). The figure in parentheses indicates the proportion assessed of the total species diversity in the group. Since assessors may focus on the most threatened species, and on those that are well known, the threat and data deficiency rates may become more unrepresentative as the proportion of species assessed decreases.

The situation is more serious still for some taxonomic groups in especially vulnerable habitats. Certain dramatic evolutionary radiations have been almost completely eliminated. For example, until the mid-1980s over 120 land snail species of the snail family Partulidae were distributed across the islands of Polynesia; almost all are now either critically endangered or extinct as a result of the introduction of a predatory snail (Mace et al. 1998). The introduction of the Nile perch (*Lates niloticus*) to Lake Victoria in East Africa has now led to the extinction of up to two hundred species of haplochromine cichlid fishes that once formed a unique "species flock" with extraordinary levels of morphological, ecological, and behavioral specializations. These examples illustrate the intensity and potential of the current extinction spasm.

From the perspective of a research agenda, we need to analyze the causes and understand the processes that ultimately lead to species extinction, and thus identify methods and approaches to mitigate the problem. In this chapter we review some key issues for the assessment and management of species at risk.

Habitat Loss, Overexploitation, and Introduced Species

Species have always gone extinct. The fossil record bears testament to the extinctions of both species and entire evolutionary radiations, and extant species probably represent only about 2–4 percent of all species that have ever lived (May et al. 1995). The fragmentary nature of the fossil record makes direct comparison of ancient extinctions with modern extinction problematic (Jablonski 1995). Most paleontological data come from marine invertebrates, usually of abundant and widespread taxa, while modern extinctions have mostly been cataloged for terrestrial vertebrates, many of which were relatively rare and narrowly distributed. Some authors have suggested that, based on the average persistence time of species in the fossil record of between 0.5 and 10 million years, the current extinction rate may be from one hundred to one thousand, and perhaps ten thousand, times higher than background extinction rates (May et al. 1995; Pimm et al. 1995).

Extinction rates have varied throughout the fossil record, with noticeable peaks during periods of mass extinction (Jablonski 1991; Benton 1995; May et al. 1995). The rates that we expect over the coming centuries may begin to approach the most severe periods of extinction in the earth's history (Jablonski 1995). Unlike the mass extinctions of the past, however, the driving processes are not attributable to abiotic geological changes or extraterrestrial impacts. Instead, the direct and indirect causes are almost entirely due to human activities. Our analysis of the nature of extinctions must therefore start with a consideration of these ultimate driving processes.

The major processes leading to extinction are anthropogenic and result from habitat loss, overexploitation, introduced species, and the interactions among these (Diamond 1989). These processes may be regarded as the extrinsic drivers, the ultimate causes of extinction as described by Simberloff (1986), or the agents behind the "declining population paradigm" described by Caughley (1994).

Habitat loss is probably the most common problem, especially for continental species (WCMC 1992; Mace and Balmford 2000). Human population growth and development lead to the appropriation of extensive areas of land for settlement, agriculture, and resource extraction, and the infrastructure to support these activities causes further loss and fragmentation. Habitat loss implies not just clearance but also all kinds of degradation and fragmentation that limit the potential for wild species to persist. Over large continental areas—for example, in northwest Europe and eastern North America—natural habitats are now reduced to a tiny proportion of their historical extent, and many of the land areas set aside for wildlife are degraded and left for conservation management largely because they are unsuitable for development. As the patches of remaining habitat become smaller and more isolated, the species living within them are more likely to go extinct and individuals from neighboring patches are unable to immigrate to repopulate the patch. In eastern North America, species extinctions have been recorded at a rate consistent with simple models concerning the effects of habitat loss (Pimm and

Askins 1995). Other regions, such as the species-rich eastern coastal forests of Brazil, which have endured relatively recent yet extensive clearance, have so far suffered few recorded extinctions. However, large numbers of species restricted to small remaining patches of forest, though not extinct, appear to have very bleak prospects (Brooks and Balmford 1996). Following their isolation in habitat remnants, species may persist for some time before finally succumbing (Brooks et al. 1999; Cowlishaw 1999), perhaps giving us too optimistic a view of the eventual impact of our current activities.

Overexploitation can take a variety of forms, from large-scale commercial extraction of fishes and forests to local, subsistence hunting. There are often limited controls to the access to wildlife resources, providing little incentive to manage the resource sustainably (Hardin 1999). This creates competition that is hard to regulate efficiently, even when there are technical and financial resources to do so (Ludwig et al. 1993). Despite extensive research into methods for sustainably harvesting natural populations, there are still many difficulties in estimating key parameters and implementing and enforcing management plans (Ludwig et al. 1993; Lande et al. 1994). As a result, for example, although the total commercial catches of marine fishes have been relatively constant over the last few decades, data from the Food and Agriculture Organization of the UN (FAO) show a shift in catch records from long-lived, high-trophic-level, piscivorous bottom fish toward short-lived, low-trophic-level invertebrates and planktivorous pelagic fish. These trends signal unsustainable exploitation of marine resources and seem likely to have both direct and indirect impacts on oceanic food webs and communities (Pauly et al. 1998). Indeed, several commercially exploited fish species now show continuing declines in abundance despite international agreements to control their harvest rates (Cook et al. 1997; Myers et al. 1997; Matsuda et al. 1998). Large, long-lived, and slowly reproducing species appear to be the most vulnerable to declines from commercial fishing (Jennings et al. 1998), and one such species, the barndoor skate (*Raja laevis*), now appears to be close to extinction (Casey and Myers 1998).

At more local scales, overharvesting for food, fuelwood, pets, ornaments, and trophies continues to threaten many species. Although international agreements limit the trade in wild species among countries, there are difficulties in controlling exploitation within countries, especially where the wildlife products are of high commercial value (e.g., elephants [Milner-Gulland and Beddington 1993]). Many wild species are also unsustainably harvested in tropical forests as a source of protein (e.g., neotropical vertebrates [Alvard 1997]). Hunting for bushmeat is a serious threat to many wildlife species because the characteristics that make species attractive as a source of protein, such as large body size, also make the species especially vulnerable to overhunting (Robinson and Redford 1991).

Introductions and translocations of species to areas outside their natural ranges are now, and have been, a major driver of anthropogenic extinctions. Many species, especially in areas that have been settled by people from distant lands, have already been driven extinct by introduced species. Pimm et al. (1994) estimate that the Polynesians wiped out five hundred to one thousand bird species in their early explorations across the Pacific, and introductions continue to pose a serious threat to island species (WCMC 1992). Similarly, the Australian mammal fauna has suffered directly and indirectly from introduced species, especially European native species brought by early settlers (Burbidge and McKenzie 1989). The highly diverse endemic plant radiation of the Cape Floristic province of South Africa is threatened by both invasive species and land-use changes. Successful invasive species generally tend to be short lived, with high intrinsic rates of increase, i.e., early successional species. Although they outcompete native species in the short term, over the long term the ability of some invasive species to persist may be limited. In contrast, the species that are most susceptible to the effects of introduced species are those that inhabit isolated regions and occupy habitats that have been heavily modified by anthropogenic influences (Godfray and Crawley 1998), where they may suffer population declines and extinction from competition, predation, or hybridization.

Consequences of Extrinsic, Human-Caused Perturbations

A species' vulnerability to extrinsic threats is often influenced by its ecology, life history, physiology, or distribution. Intrinsic characteristics of extinction-prone species have been more thoroughly investigated by research biologists than have the effects of the external drivers and the interactions between the two (Caughley 1994), leading to some debate about the relative importance of each. However, it is clear that both factors should be considered if we are to improve the reliability of predictions about extinction risk (Hedrick et al. 1996; Beissinger and Westphal 1998).

Several kinds of evidence provide information about the most vulnerable species. First, empirical studies have identified intrinsic characteristics of extinction-prone species. Rates of local extinction have been shown to be higher for species that have restricted ranges or occupy a small number of sites (Hanski 1982; Simberloff and Gotelli 1984; Thomas and Mallorie 1985; Happel et al. 1987; Gaston 1994; Gaston and Blackburn 1996; Gaston and Chown 1999); are local endemics (Terborgh and Winter 1980; Cowling and Bond 1991); or have low abundances, high temporal population variability, and poor dispersal (Karr 1982; Diamond 1984; Pimm et al. 1988; Newmark 1991; Gaston 1994). These studies are all open to the criticism that they may be investigating only correlates of extinction-prone characteristics, since body size, dispersal ability, range size, population variability, and local population density are all interrelated (McArdle et al. 1990; Pimm 1992; Gaston 1994; Gaston and Blackburn 1996). In studies where interrelationships among life history traits and the geographical sampling can be controlled, extinction risk has been shown to be higher for species with low population density, small range size, and habitat and diet specializations (Foufopoulos and Ives 1999; Purvis, Gittleman et al. 2000). The response of a species to a threat is complicated and depends on its life history, the local environmental conditions, and the specific perturbation. For example, the stability of fluctuating populations is reduced by exploitation (Beddington and May 1977); the response of primates to logging is a function of their home-range size and the latitude at which they live (a correlate of habitat variability) (Harcourt 1997); and extinction of carnivores within reserves is higher for those with large home ranges (Woodroffe and Ginsberg 1998).

Second, ecological and life history theory provides some useful insights. It can be shown that small populations are more extinction prone because of their susceptibility to demographic stochasticity (Richter-Dyn and Goel 1972; Goodman 1987); the increased expression of recessive deleterious alleles under inbreeding (Soulé 1980; Soulé and Mills 1998); the change in quantitative characters that allow adaptation; and the accumulation of mildly deleterious mutations (Hedrick and Miller 1992; Frankham 1995a).

Lande (1998) has reviewed these various processes in terms of the minimum viable population sizes that they imply. This has shown that demographic stochasticity is unlikely to be important for any population that has more than one hundred individuals, but random environmental variation or catastrophes are important for populations of all sizes, and they become more significant as environmental variability becomes large in relation to the population growth rate (Lande 1993).

In terms of genetic effects, deleterious recessive alleles become more likely to be expressed when populations are reduced in number. The resulting inbreeding depression may increase short-term effective population size ($N_e s$) up to 50–100 (Mills and Smouse 1994; Allendorf and Ryman in press), translating to effects on actual population size in the range of 200–500 individuals (Soulé et al. 1986; Mace and Lande 1991; Frankham 1995b). To preserve quantitative trait variation, larger populations are needed. One estimate is that to maintain high levels (more than 90 percent) over thousands of years will require minimum effective population sizes of at least 5,000, and to prevent the accumulation of mildly deleterious mutations over tens of thousands of years will require minimum effective population sizes of around 10,000 to 100,000. Because of difficulties in estimating key parameter values for these calculations (Franklin and Frankham 1998; Lynch and Lande 1998; Waples in press), the critical population sizes from these theoretical studies are best interpreted as guides to the relative importance of different characteristics rather than as real thresholds for management (Lande 1998). Moreover, this approach is deficient because it lacks a treatment of the feedbacks between genetics and demography (Soulé and Mills 1998).

Our analysis of the extrinsic and intrinsic factors associated with increased extinction risk has covered much of the field of ecology and evolutionary biology. In order to identify key questions, in the following sections we analyze issues that are of practical significance for conservation practitioners.

Critical Gaps in Knowledge

In discussing the critical gaps in our knowledge about how to protect species at risk, we will examine six major areas. Research priorities for these six areas are given in box 2.1. The research areas are:

- Identifying patterns of species at risk
- Investigating the dynamics of threatening processes
- Investigating the dynamics of population persistence in a changing world
- Assessing the effectiveness of different management strategies
- Incorporating the human dimension in conservation
- Implementing adaptive management and monitoring

Each of these subjects is discussed below, and pertinent research questions are identified. First we discuss research questions, then move on to management actions.

Research Questions

Identifying Patterns of Species at Risk

The first set of research questions requires an analysis of patterns—a simple accounting of where the most species are located and what their relative risks of extinction, endemism, and diversity are. Along with patterns of biodiversity it is important to document the distribution and intensity of dominant processes of threat throughout different regions of the world. The above issues are particularly relevant to the design of networks of protected areas (Soulé and Terborgh 1999). This approach has been widely explored and used for planning at both global and regional levels. Through such studies, high-priority areas, habitats, and ecosystems have been identified and targeted for action (Myers 1988, 1990; Pressey et al. 1993; Beissinger et al. 1996; Williams 1998). Species are not distributed evenly over the earth's surface, and neither are the centers of richness, endemism, or threat particularly well correlated across taxa (WCMC 2000), especially in certain regions, such as the United Kingdom and South Africa (Prendergast et al. 1993; vanJaarsveld et al. 1998). In other areas, there are higher congruence patterns of species' complementarity among sites, such as forest patches in Uganda (Howard et al. 1998). In general, however, the concordance among diversity, endemism, and threat—three different measures of the conservation importance of an area—also varies with scale and location (Balmford and Long 1995; Kershaw et al. 1995) and presents some difficult choices for those setting conservation priorities (Margules et al. 1988; Pressey et al. 1996).

Investigating the Dynamics of Threatening Processes

Analyzing patterns provides only a partial solution. The viability of populations and species ultimately depends upon population and evolutionary dynamics. Within any particular area, species will not share similar future prospects. Those that are highly adaptable (genetically, behaviorally, or both), have good dispersal abilities, or belong to populations whose prospects are more favorable will be expected to have enhanced viabilities. Therefore, we consider the dynamics of threatening processes and their likely impact upon affected species.

Investigating the Dynamics of Population Persistence in a Changing World

A simple consideration of threatening processes reveals that these change in nature and intensity over both time and space—sometimes at alarming rates. This is one reason analyses of patterns do not provide a sufficient basis for management (Smith et al. 1993; Balmford et al. 1998). Our third set of research topics considers studies on how species at risk can cope with dynamic change.

Currently, dynamic changes are accentuated as a result of the rapid human population expansion. Numbers are still increasing rapidly, although at different rates in different areas of the world (Cohen 1995). Globally, human numbers are expected to double over the next sixty years (McNeely et al. 1995). While some regions, such as Europe and North America, are expected to meet their resource needs without further land conversion for agriculture, many other areas are still undergoing rapid development alongside high population growth. Food requirements may double in Central and South America and Asia, and perhaps increase fivefold in Africa (WCMC 2000). Land-use changes will therefore proceed inexorably in some areas of high biodiversity. This demand, and its likely impact on wild species, needs to be understood and managed if conservation activities are to be successful. Even in more stable areas of the world, changing patterns of work and leisure are leading to migrations from urban to rural areas, with significant consequences for land-use planning. Habitat loss and fragmentation will continue but at different rates and in different ways than in the past.

One of the difficulties with the dynamic nature of land changes is that past trends are often not a reliable predictor of the future. Human activities also present a new challenge in the form of global climate change. Current predictions about changes in average temperatures suggest that the impacts will vary regionally and across habitats. There may be profound influences on marine ecosystems, which could experience an increased frequency and intensity of oceanic perturbations with consequences for many marine fish and corals. Global patterns of seawater circulation are expected to alter as a result of increased surface temperatures, which may have consequences for sea levels and for climate in both island and continental areas. Some parts of the land surface, especially temperate and arctic zones, are expected to face large temperature changes, and there is already evidence that European bird and butterfly species ranges are shifting to higher latitudes and animals are altering egg-laying dates in response to current warming trends (Crick and Sparks 1999; Hill et al. 1999; Thomas and Lennon 1999). Under systematic warming, species persistence will depend on a different set of traits than those that offer high fitness in more stable habitats. All this underlines the importance of studying and understanding the spatial, temporal, and adaptational characteristics of species. For example, a study of a forest savannah in Cameroon indicated that, although ecotones have generally been regarded as unimportant areas, natural selection operating on birds in the ecotone may be important for generating evolutionary novelties (Smith et al. 1993).

Management Actions

The final three groups of research priorities discussed below focus on ways to use knowledge gained from research in the development, implementation, and assessment of management activities. Many conservation practices have been tried in the past, some with much more success than others. Unfortunately these experiences are often poorly documented, and knowledge disappears with the people who were responsible for it. A more systematic approach to analyzing and synthesizing alternative approaches is needed.

Assessing the Effectiveness of Different Management Strategies

In this section, we focus on the more biological aspects of management at both the population and ecosystem levels. Fundamental principles from population biology were first developed a couple of decades ago (Soulé 1987). A variety of ever more sophisticated techniques for population management now incorporate environmental and human processes in various forms of population viability analysis (PVA) (Boyce 1992; Burgman et al. 1993). While some empirical evidence suggests that these models are reliable predictors of

BOX 2.1. **Research and Management Priorities**

Research Priorities

Identifying Patterns of Species at Risk

1. Which regions, habitats, taxa, and species are most threatened or soon will be?
2. What are the major processes threatening biodiversity, and where are they currently focused? How do these processes change over time and space?
3. What are the current population trends of species in different locations? Can a valid index of community or ecosystem health based on many species and processes be developed?
4. What do studies of recent and historical correlates among extinction rates, anthropogenic processes, and biological features of species indicate about simple measures that can be used to direct conservation activities? What is the evidence that extinction filters, selective processes of past extinction, have already stripped out vulnerable taxa and communities? Can such knowledge ensure that conservation efforts be directed toward systems that have been less exposed to threats and are therefore more vulnerable to them?
5. Can habitat loss be used as a surrogate to estimate rates of extinction? If so, how will factors such as species richness, habitat type, and latitude influence the calibration?
6. Are indicator, keystone, or flagship species effective conservation monitoring tools, and if so, which species or complex of species best represent the status of a larger system?

Investigating the Dynamics of Threatening Processes

1. How do aspects of species ecology, behavior, genetics, phylogenetic status, and life history affect a species' ability to acclimate and persist in response to the following: habitat loss, fragmentation, and modification; introduction of exotic species; exploitation for sustainable use; introduction of new diseases; and other anthropogenic threats and disturbances?
2. To what degree do extinction lag effects inhibit the identification of species that are at great risk of extinction?
3. What determines when extinction is the endpoint of a decline? Are species that have successfully adapted to the human landscape simply behaviorally preadapted, or are some species better able than others to adapt through learning?
4. What are the best methods for diagnosing population decline? Experimental and comparative methods for diagnosing causes of population decline need to be developed and tested.
5. How important are cumulative threats to the persistence of species? What kinds of interactions among threats are the most severe?
6. What combinations of spatial patterns and biological or physical factors (e.g., nutrient balance, rarity, and abundance) are the most efficient and useful indicators of changing population and ecosystem processes?

7. Can new approaches, such as molecular genetic methods, be used to obtain rapid insight into a species' population dynamics when long-term data are lacking?

Investigating the Dynamics of Population Persistence in a Changing World

1. What determines variation in the resilience of communities or ecosystems? How is it affected by recent isolation, species richness, trophic-level distribution, alien species, humidity, latitude, niche occupation, life history, functional redundancy, and keystone species?

2. How do population dynamics and persistence vary in relation to life history attributes, population structure, environments, and phylogeny? How are habitat types and ecological factors such as ecotones, marginal habitats, niche availability, or the edge of a species range associated with persistence?

3. How much and at what spatial scales does demography vary within a metapopulation and across landscapes? Can sources and sinks be identified given their true definitions? How often do different types of metapopulations and source-sink dynamics occur in nature?

4. What is the relative importance of demographic and environmental stochasticity in affecting extinction compared to biological features of species and anthropogenic factors?

5. To what degree and under what conditions does genetic variation influence demography and persistence? How does inbreeding in small populations affect persistence? How does outbreeding depression affect persistence in admixed populations? What is the relative importance of gene flow versus local adaptation for population persistence? How does genetic architecture of a trait influence its potential for adaptive change?

Management Actions and Priorities

Assess the Effectiveness of Different Management and Implementation Strategies

1. How can the models used in decision making be improved? What is the appropriate context for use of population viability analysis, decision analysis, and other analytical techniques that are intended to guide managers?

2. How should reserves be designed to reflect dynamic landscapes? How large need reserves be so that difficult and expensive management regimes are not required to replace former ecological processes?

3. What conservation strategies maximize both species and community conservation and fulfill specific human needs? When is it best to focus conservation resources on strictly protected areas versus integrated land-use areas? What are the biological and economic costs and benefits of strategies that focus on prevention versus cure? What are the biological and economic costs and benefits of focusing on reversible threats and species that can be saved while rejecting difficult cases, compared to focusing on saving the most threatened first (i.e., triage approaches)?

(continues)

BOX 2.1 *(Continued)*

4. How useful are rules of thumb for conservation planning and population viability assessment? Can they be tested using a combination of modeling techniques and empirical studies?

5. What are the biological and economic costs and benefits of captive breeding, reintroductions, and translocations? How frequently does adaptation to captive propagation prevent successful reintroductions of plants and animals? How can the process of domestication and adaptation be reversed in captive animal and plant populations?

6. What incentives can be used to encourage conservation organizations to share information and pool resources for common projects?

Incorporate the Human Dimension in Conservation

1. What are the most useful approaches to involve local communities in conservation projects? What can we learn from projects that have and have not been successful in accomplishing this? How can we ensure that local support for conservation projects is sustainable over long periods of time?

2. What organizational models lead to improved implementation, coordination, and scientific direction of endangered species recovery programs?

3. How can we plan protected areas and other management units so that they require minimal long-term management, or political and social controls?

Implement Adaptive Management and Monitoring

1. How should managers monitor, assess, and disseminate results of conservation projects? What can be learned from implementation of management recommendations that should feed back into the research agendas and future planning options?

2. What aspects of structure, power, and authority in governmental and nongovernmental organizations enhance or decrease the implementation of adaptive management approaches in decision making?

3. How can management interventions be structured to test theory and collect data to learn from crises?

4. Where should assessment and monitoring be targeted—individual organisms; populations; subspecific units and species; or assemblages of species that are grouped by habitat, ecoregion, phylogeny, or functional criteria? What are the economic and biological consequences of adopting any one of these for a specific instance?

future population trends (Brook et al. 2000), it is unclear how robust they are as management tools when vital rates are well known (Fieberg and Ellner 2000), let alone under uncertainty and catastrophic change (Mangel and Tier 1994; Ludwig 1996); and it is also unclear how best to incorporate many different variables and their interactions into such models (Beissinger and Westphal 1998).

Incorporating the Human Dimension in Conservation

Culture and place affect the choice of conservation tactics. For example, the United States has strong species-based legislation that provides protection for the habitats of listed species in their native areas, whereas in Australia and New Zealand translocations, introductions, and reintroductions have been extensively used for some highly endangered species (Serena and Williams 1994). The role of captive breeding has been widely debated (Hutchins et al. 1995; Balmford et al. 1996; Snyder et al. 1996). The practitioners of many regions focus conservation efforts on the design, implementation, and management of protected areas and reserves (Soulé and Terborgh 1999). In a rapidly changing environment and landscape, it is increasingly difficult to design protected areas that will be efficient in the future. Placing reserves at natural landscape scales with edges at natural boundaries should be efficient (Peres and Terborgh 1995), but the designs and configuration of reserve areas may need to be altered to cope with systematic environmental changes, especially climate change.

We recognize both the challenges and benefits associated with local human communities whose immediate needs and aspirations may conflict with those of biodiversity conservation. While some apparent conflicts are illusory, in other cases strategies and approaches can be taken that will be more effective at achieving multiple short- and long-term goals. It is in this area in particular that conservation biologists must move into more interdisciplinary scientific communities and work with economists, sociologists, land planners, and human demographers to achieve effective results.

Implementing Adaptive Management and Monitoring

A long-term approach to management that includes both monitoring, analysis of actions, and adaptive management is important. Once management plans are in place, the predetermined goals and objectives should be evaluated regularly; if the actions are not proceeding according to plan, then the reasons need to be identified and alterations made (Margoluis and Salafsky 1998). For effective adaptive management, a great deal of planning must be undertaken at the start, with agreement among various stakeholders about what the indicators of altered management might be. Biologists should play an active role in such planning to ensure that the biological principles at the heart of any management plan are not compromised as implementation proceeds.

LITERATURE CITED

Alvard, M., J. Robinson, K. Redford, and H. Kaplan. 1997. The sustainability of subsistence hunting in the neotropics. *Conservation Biology* 11:977–982.

Balmford, A., and A. Long. 1995. Across-country analyses of biodiversity congruence and current conservation effort in the tropics. *Conservation Biology* 9:1539–1547.

Balmford, A., G. Mace, and J. Ginsberg. 1998. Challenges to conservation in a changing world. In G. M. Mace, A. Balmford, and J. Ginsberg (eds.), *Conservation in a Changing World,* pp. 1–28. Cambridge: Cambridge University Press.

Balmford, A., G. M. Mace, and N. Leader-Williams. 1996. Redesigning the ark: Setting priorities for captive breeding. *Conservation Biology* 10:719–727.

Beddington, J. R., and R. M. May. 1977. Harvesting populations in a randomly fluctuating environment. *Science* 197:463–465.

Beissinger, S. R., E. C. Steadman, T. Wohlgenant, G. Blate, and S. Zack. 1996. Null models for assessing ecosystem conservation priorities: Threatened birds as titers of threatened ecosystems in South America. *Conservation Biology* 10:1343–1352.

Beissinger, S. R., and M. I. Westphal. 1998. On the use of demographic models of population viability in endangered species management. *Journal of Wildlife Management* 62:821–841.

Benton, M. J. 1995. Diversification and extinction in the history of life. *Science* 268:52–58.

Boyce, M. S. 1992. Population viability analysis. *Annual Review of Ecology and Systematics* 23:481–506.

Brook, B. W., J. J. O'Gady, A. P. Chapman, M. A. Burgman, H. R. Akcakaya, and R. Frankham. 2000. Predictive accuracy of population viability analysis in conservation biology. *Nature* 404:385–387.

Brooks, T., and A. Balmford. 1996. Atlantic forest extinctions. *Nature* 380:115.

Brooks, T. M., S. L. Pimm, and J. O. Oyugi. 1999. Time lag between deforestation and bird extinction in tropical forest fragments. *Conservation Biology* 13:1140–1150.

Burbidge, A. A., and N. L. McKenzie. 1989. Patterns in the modern decline of western Australia's native fauna: Causes and conservation implications. *Biological Conservation* 50:143–198.

Burgman, M. A., S. Ferson, and H. R. Akcakaya, eds. 1993. *Risk Assessment in Conservation Biology.* London: Chapman & Hall.

Casey, J. M., and R. A. Myers. 1998. Near extinction of a large, widely distributed fish. *Science* 281:690–692.

Caughley, G. 1994. Directions in conservation biology. *Journal of Animal Ecology* 63:215–244.

Cohen, J. E. 1995. *How Many People Can the Earth Support?* New York: W.W. Norton.

Cook, R. M., A. Sinclair, and G. Stefansson. 1997. Potential collapse of North Sea cod stocks. *Nature* 385:521–522.

Cowling, R. M., and W. J. Bond. 1991. How small can reserves be—An empirical approach in Cape Fynbos, South Africa. *Biological Conservation* 58:243–256.

Cowlishaw, G. 1999. Predicting the pattern of decline of African primate diversity: An extinction debt from historical deforestation. *Conservation Biology* 13:1183–1193.

Crick, H. Q. P., and T. H. Sparks. 1999. Climate change related to egg-laying trends. *Nature* 399.

Diamond, J. M. 1984. "Normal" extinctions of isolated populations. In M. H. Nitecki (ed.), *Extinctions*, pp. 191–246. Chicago: University of Chicago Press.

Diamond, J. M. 1989. The present, past and future of human-caused extinctions. *Philosophical Transactions of the Royal Society, London,* Series B 325:469–477.

Fieberg, J., and S. P. Ellner. 2000. When it is meaningful to estimate an extinction probability? *Ecology* 81:2040–2047.

Foufopoulos, J., and A. R. Ives. 1999. Reptile extinctions on land-bridge islands: Life-history attributes and vulnerability to extinction. *American Naturalist* 153: 1–25.

Frankham, R. 1995a. Conservation genetics. *Annual Review of Genetics* 29:305–327.

Frankham, R. 1995b. Effective population-size adult population-size ratios in wildlife—A review. *Genetical Research* 66:95–107.

Franklin, I. R., and R. Frankham. 1998. How large must populations be to retain evolutionary potential? *Animal Conservation* 1:69–70.

Gaston, K. J. 1994. *Rarity.* London: Chapman and Hall.

Gaston, K. J., and T. M. Blackburn. 1996. Global scale macroecology: Interactions between population size, geographic range size and body size in the Anseriformes. *Journal of Animal Ecology* 65:701–714.

Gaston, K. J., and S. L. Chown. 1999. Geographic range size and speciation. In A. E. Magurran and R. M. May (eds.), *Evolution of Biological Diversity,* pp. 236–259. Oxford: Oxford University Press.

Godfray, H. C. J., and M. J. Crawley. 1998. Introductions. In W. J. Sutherland (ed.), *Conservation Science and Action*, pp. 39–65. Oxford: Blackwells.

Goodman, D. 1987. The demography of chance extinction. In M. E. Soulé (ed.), *Viable Populations for Conservation,* pp. 11–34. Cambridge: Cambridge University Press.

Hanski, I. 1982. Dynamics of regional distribution: The core and satellite species hypothesis. *Oikos* 38:210–221.

Happel, R. E., J. F. Noss, and C. W. Marsh. 1987. Distribution abundance and endangerment of primates. In C. W. Marsh and R. A. Mittermeier (eds.), *Primate Conservation in the Tropical Rain Forest,* pp. 63–82. New York: Alan R. Liss.

Harcourt, A. 1997. Ecological indicators of risk for primates as judged by species' susceptibility to logging. In T. Caro (ed.), *Behavioural Ecology and Conservation Biology,* pp. 56–79. New York: Oxford University Press.

Hardin, G. 1999. The tragedy of the commons revisited. *Environment* 41:5+.

Hedrick, P. W., R. C. Lacy, F. W. Allendorf, and M. E. Soulé. 1996. Directions, in

Conservation Biology comments on Caughley. *Conservation Biology* 10:1312–1320.

Hedrick, P. W., and P. S. Miller. 1992. Conservation genetics: Techniques and fundamentals. *Ecological Applications* 2:30–46.

Hill, J. K., C. D. Thomas, and B. Huntley. 1999. Climate and habitat availability determine 20th century changes in a butterfly's range margin. *Proceedings of the Royal Society of London,* Series B 266:1197–1206.

Howard, P. C., P. Viskanic, T. R. B. Davenport, F. W. Kigenyi, M. Baltzer, C. J. Dickinson, J. S. Lwanga, R. A. Matthews, and A. Balmford. 1998. Complementarity and the use of indicator groups for reserve selection in Uganda. *Nature* 394:472–475.

Hutchins, M. K., K. Willis, and R. J. Wiese. 1995. Strategic collection planning: Theory and practice. *Zoo Biology* 14:5–25.

IUCN. 1996. *The 1996 IUCN Red List of Threatened Animals.* Gland, Switzerland: IUCN.

Jablonski, D. 1991. Extinctions: A palaeontological perspective. *Science* 253:754–757.

Jablonski, D. 1995. Extinctions in the fossil record. In J. H. Lawton and R. M. May (eds.), *Extinction Rates,* pp. 389-391. Oxford: Oxford University Press.

Jennings, S., J. Reynolds, and S. Mills. 1998. Life history correlates of responses to fisheries exploitation. *Proceedings of the Royal Society of London,* Series B 265:333–339.

Jernvall, J., and P. C. Wright. 1998. Diversity components of impending primate extinctions. *Proceedings of the National Academy of Sciences of the United States of America* 95:11279–11283.

Karr, J. R. 1982. Population variability and extinction in the avifauna of a tropical land bridge island. *Ecology* 63:1975–1978.

Kershaw, M., G. M. Mace, and P. H. Williams. 1995. Threatened status, rarity, and diversity as alternative selection measures for protected areas: A test using Afrotropical antelopes. *Conservation Biology* 9:324–334.

Lande, R. C. 1993. Risks of population extinction from demographic and environmental stochasticity and random catastrophes. *American Naturalist* 142:911–927.

Lande, R. 1998. Anthropogenic, ecological and genetic factors in extinction. In G. M. Mace, A. Balmford, and J. R. Ginsberg (eds.), *Conservation in a Changing World.* Cambridge: Cambridge University Press.

Lande, R., S. Engen, and B. E. Saether. 1994. Optimal harvesting, economic discounting and extinction risk in fluctuating populations. *Nature* 372:88–90.

Ludwig, D. 1996. Uncertainty and the assessment of extinction probabilities. *Ecological Applications* 6:1067–1076.

Ludwig, D., R. Hilborn, and C. Walters. 1993. Uncertainty, resource exploitation, and conservation—Lessons from history. *Science* 260:17.

Lynch, M., and R. Lande. 1998. The critical effective size for a genetically secure population. *Animal Conservation* 1:70–72.

Mace, G. M., and A. Balmford. 2000. Patterns and processes in contemporary mammalian extinctions. In A. Entwhistle and N. Dunstone (eds.), *Priorities for the Conservation of Mammalian Biodiversity.* Cambridge: Cambridge University Press.

Mace, G. M., and R. Lande. 1991. Assessing extinction threats: Toward a reevaluation of IUCN threatened species categories. *Conservation Biology* 5:148–157, Special Publication No. 2.

Mace, G. M., P. Peace-Kelly, and D. Clarke. 1998. An integrated conservation programme for the tree snails of Polynesia (Partulidae): A review of captive and wild elements. *Journal of Conchology* (special publication) 2:89–96.

Mangel, M., and C. Tier. 1994. Four facts every conservation biologist should know about persistence. *Ecology* 75:607–614.

Margoluis, R. A., and N. N. Salafsky. 1998. *Measures of Success.* Washington, D.C.: Island Press.

Margules, C. R., A. O. Nicholls, and R. L. Pressey. 1988. Selecting networks of reserves to maximise biological diversity. *Biological Conservation* 43:63–76.

Matsuda, H., Y. Takenaka, T. Yahara, and Y. Uozumi. 1998. Extinction risk assessment of declining wild populations: The case of the southern bluefin tuna. *Researches on Population Ecology* 40:271–278.

May, R. M., J. H. Lawton, and N. E. Stork. 1995. Assessing extinction rates. In J. H. Lawton and R. M. May (eds.), *Extinction Rates.* Oxford: Oxford University Press.

McArdle, B. H., K. J. Gaston, and J. H. Lawton. 1990. Variation in the size of animal populations: Patterns, problems and artefacts. *Journal of Animal Ecology* 59:439–454.

McNeely, J. A., M. Gadgil, C. Leveque, C. Padoch, and K. Redford. 1995. Human influences on biodiversity. In V. H. Heywood (ed.), *Global Biodiversity Assessment*, pp. 510–519. Cambridge: Cambridge University Press.

Mills, L. A., and P. E. Smouse. 1994. Demographic consequences of inbreeding in remnant populations. *American Scientist* 144:412–431.

Milner-Gulland, E. J., and J. R. Beddington. 1993. The exploitation of elephants for the ivory trade—An historical perspective. *Proceedings of the Royal Society of London,* Series B 252:29–37.

Myers, N. 1988. Threatened biotas: "hotspots" in tropical forests. *Environmentalist* 8:187–208.

Myers, N. 1990. The biodiversity challenge: Expanded hotspots analysis. *Environmentalist* 10:243–256.

Myers, R. A., J. A. Hutchings, and N. J. Barrowman. 1997. Why do fish stocks collapse? The example of cod in Atlantic Canada. *Ecological Applications* 7:91–106.

Newmark, W. D. 1991. Tropical forest fragmentation and the local extinction of understory birds in the eastern Usambara Mountains, Tanzania. *Conservation Biology* 5:67–78.

Oldfield, S., C. Lusty, and A. MacKinven, eds. 1998. *The World List of Threatened Trees.* Cambridge: World Conservation Press.

Pauly, D., V. Christensen, J. Dalsgaard, R. Froese, and F. Torres. 1998. Fishing down marine food webs. *Science* 279:860–863.

Peres, C. A., and J. W. Terborgh. 1995. Amazonian nature reserves—An analysis of the defensibility status of existing conservation units and design criteria for the future. *Conservation Biology* 9:34–46.

Pimm, S. L. 1992. *The Balance of Nature.* Chicago: University of Chicago Press.

Pimm, S. L., and R. A. Askins. 1995. Forest losses predict bird extinctions in eastern North America. *Proceedings of the National Academy of Science of the United States of America* 92:9343–9347.

Pimm, S. L., H. L. Jones, and J. M. Diamond. 1988. On the risk of extinction. *American Naturalist* 132:757–785.

Pimm, S. L., M. P. Moulton, and L. J. Justice. 1994. Bird extinctions in the central Pacific. *Philosophical Transactions of the Royal Society at London,* Series B 344: 27–33.

Pimm, S. L., G. J. Russell, J. L. Gittleman, and T. M. Brooks. 1995. The future of biodiversity. *Science* 269:347–350.

Prendergast, J. R., R. M. Quinn, J. H. Lawton, B. C. Eversham, and D. W. Gibbons. 1993. Rare species, the coincidence of diversity hotspots and conservation strategies. *Nature* 365:335–337.

Pressey, R. L., C. J. Humphries, C. R. Margules, R. I. Vanewright, and P. H. Williams. 1993. Beyond opportunism—Key principles for systematic reserve selection. *Trends in Ecology & Evolution* 8:124–128.

Pressey, R. L., H. P. Possingham, and C. R. Margules. 1996. Optimality in reserve selection algorithms: When does it matter and how much? *Biological Conservation* 76:259–267.

Purvis, A., P.-M. Agapow, J. L. Gittleman, and G. M. Mace. 2000. Nonrandom extinction and the loss of evolutionary history. *Science* 288:328–330.

Purvis, A., J. L. Gittleman, G. C. Cowlishaw, and G. M. Mace. 2000. Predicting extinction risk in declining species. *Proceedings of the Royal Society of London,* Series B. 267:1947–1952.

Richter-Dyn, N., and N. S. Goel. 1972. On the extinction of a colonising species. *Theoretical Population Biology* 3:406–433.

Robinson, J. G., and K. H. Redford. 1991. Sustainable harvest of neotropical forest mammals. In J. G. Robinson and K. H. Redford (eds.), *Neotropical Wildlife Use and Conservation,* pp. 415–429. Chicago: University of Chicago Press.

Russell, G. J., T. M. Brooks, M. M. McKinney, and C. G. Anderson. 1998. Present and future taxonomic selectivity in bird and mammal extinctions. *Conservation Biology* 12:1365–1376.

Serena, M., and G. A. Williams. 1994. Wildlife conservation and reintroduction: An Australian perspective. In M. Serena (ed.), *Reintroduction Biology of Australian and New Zealand Fauna,* pp. 247–252. Chipping Norton, New South Wales, Australia: Surrey Beatty and Sons.

Simberloff, D. 1986. The proximate causes of extinction. In D. M. Raup and D. Jablonski (eds.), *Patterns and Processes in the History of Life,* pp. 259–276. Berlin: Springer Verlag.

Simberloff, D., and N. Gotelli. 1984. Effects of insularization on plant–species richness in the prairie–forest ecotone. *Biological Conservation* 29:27–46.

Smith, T. B., M. W. Bruford, and R. K. Wayne. 1993. The preservation of processes: The missing element of conservation programmes. *Biodiversity Letters* 1:164–167.

Snyder, N. F. R., S. R. Derrickson, S. R. Beissinger, J. W. Wiley, T. B. Smith, W. D. Toone, and B. Miller. 1996. Limitations of captive breeding in endangered species recovery. *Conservation Biology* 10:338–348.

Soulé, M. E. 1980. Thresholds for survival: Maintaining fitness and evolutionary potential. In M. E. Soulé and B. A. Wilcox (eds.), *Conservation Biology: An Evolutionary-Ecological Perspective,* pp. 151–169. Sunderland, Mass.: Sinauer Associates.

Soulé, M. E. 1987. *Viable Populations for Conservation.* Cambridge: Cambridge University Press.

Soulé, M. E., M. Gilpin, W. Conway, and T. Foose. 1986. The millenium ark: How long a voyage, how many staterooms, how many passengers? *Zoo Biology* 5:101–114.

Soulé, M. E. and L. S. Mills. 1998. Population genetics—No need to isolate genetics science. *Science* 282:1658–1659.

Soulé, M. E., and J. Terborgh. 1999. *Continental Conservation: Scientific Foundations for Regional Conservation Networks.* Washington, D.C.: Island Press.

Terborgh, J., and B. Winter. 1980. Some causes of extinction. In M. E. Soulé and B. A. Wilcox (eds.), *Conservation Biology: An Evolutionary-Ecological Perspective,* pp. 119–133. Sunderland, Mass.: Sinauer Associates.

Thomas, C. D., and J. J. Lennon. 1999. Birds extend their ranges northwards. *Nature* 399:213.

Thomas, C. D., and H. C. Mallorie. 1985. Rarity, species richness and conservation: Butterflies of the Atlas Mountains in Morocco. *Biological Conservation* 33:95–117.

van Jaarsveld, A. S., S. Freitag, S. L. Chown, C. Muller, S. Koch, H. Hull, C. Bellamy, M. Kruger, S. EndrodyYounga, M. W. Mansell, and C. H. Scholtz. 1998. Biodiversity assessment and conservation strategies. *Science* 279:2106–2108.

WCMC. 1992. *Global Diversity: Status of the Earth's Living Resources.* London: Chapman & Hall.

WCMC. 2000. *Global Biodiversity: Earth's Living Resources in the 21st Century.* Cambridge, U.K.: World Conservation Press.

Williams, P. H. 1998. Key sites for conservation: Area-selection methods for biodiversity. In G. M. Mace, A. Balmford, and J. Ginsberg (eds.), *Conservation in a Changing World,* pp. 211–249. Cambridge: Cambridge University Press.

Woodroffe, R., and J. R. Ginsberg. 1998. Edge effects and the extinction of populations inside protected areas. *Science* 280:2126–2128.

3

HUMAN ALTERATION OF FOOD WEBS

Research Priorities for Conservation and Management

Fiorenza Micheli, Gary A. Polis, P. Dee Boersma, Mark A. Hixon, Elliott A. Norse, Paul V. R. Snelgrove, and Michael E. Soulé

MOST NATURAL FOOD WEBS HAVE BEEN ALTERED PROFOUNDLY through human activities (Botsford et al. 1997; Vitousek et al. 1997; Terborgh et al. 1999). Human perturbations of food-web interactions result from overexploitation of species, particularly top predators; introduction of exotics; habitat destruction and fragmentation; and changes in resource availability through alteration of biogeochemical cycles, enhanced loadings of nutrients and organics, and "subsidies" of natural food webs through waste disposal or discarded fishery bycatch (see chapters 4, 5, 7, 9). Alterations of resource availability represent bottom-up perturbations of food-web dynamics, whereas removal or addition of consumers through hunting, fishing, species introductions, and habitat alteration represents a top-down perturbation. Top-down and bottom-up perturbations of natural food webs modify consumer-resource interactions, with subsequent impacts on population abundances, community structure and diversity, and ecosystem processes.

Biological conservation and management in the face of human impacts on whole ecosystems require an understanding of how consumer-resource interactions control populations and shape communities and a shift from focusing on single species to taking into account the complexity of ecological systems and interactions among their components (Lubchenco et al. 1991;

Christensen et al. 1996). This new tendency is exemplified by an increasing recognition among fishery managers that multispecies effects must be considered (NRC 1994; FAO 1997). Traditional, single-species fishery management has often failed to maintain populations of exploited species at sustainable levels (Ludwig et al. 1993). Classic examples include the collapse of the North Sea herring and mackerel stocks (Hempel 1978) and the northern cod (Walters and Maguire 1996). Part of the variability of commercial fish stocks, and the consequent uncertainty about their predicted trajectories, is caused by interactions with other species, which may, in turn, be impacted by harvesting conducted at multiple trophic levels (May et al. 1979; Yodzis 1994; Pauly et al. 1998). The present emphasis on marine reserves as tools for marine conservation and fisheries management exemplifies a shift in focus from an emphasis on single species to more holistic approaches to conservation and management (Roberts and Polunin 1993; Bonsauk 1996; Russ and Alcala 1996; Allison et al. 1998). These new approaches take into account the uncertainty of predictions about the combined effects of environmental variability and human impacts on species assemblages (Clark 1996; Lauck et al. 1998).

Decades of experimental studies have shown dramatic effects of consumer-resource interactions on populations and communities. Bottom-up effects of resources on consumers and top-down effects of consumers on other species in the community include a suite of direct and indirect pathways of interaction. Resource or prey availability controls the rates of population growth of their consumers, whereas predators exert direct effects on their prey abundance, size structure, and spatial distribution (Zaret 1980; Sih et al. 1985). Although even simple predator-prey systems can generate complex dynamics (McCauley et al. 1988), the uncertainty of predictions about resource and predation effects is further increased by responses mediated through other species in the community. Theoretical and empirical studies have shown that predators can influence community structure and diversity through indirect effects (Schoener 1993; Wootton 1993; Menge 1995; Abrams et al. 1996). In theory, there can be an almost unlimited number of different types of indirect effects and resulting dynamics (Hastings and Powell 1991; Abrams 1992), but experimental manipulations have revealed a more limited range of possibilities (Schoener 1993; Menge 1995; Abrams et al. 1996).

Knowledge of the role of consumer-resource interactions in regulating species dynamics and shaping natural communities has important conservation and management implications. Applications include:

- predicting the impacts of enhanced resource availability, predator removal, and species introduction on community structure;
- guiding predator manipulations aimed at decreasing mortality of an endangered or harvested species;

- controlling the abundance of pests and exotics;
- conserving or reintroducing top predators and keystone species with the aim of maintaining diversity or restoring the structure of altered communities through cascading effects;
- controlling the consequences of anthropogenic eutrophication of aquatic ecosystems through manipulation of the food-web structure (biomanipulation); and
- designing reserve networks to conserve whole assemblages and the interactions among their component species.

However, indirect effects and diffuse interactions among multiple species can cause unanticipated changes in community structure and nontarget effects of management interventions. Unraveling the tremendous complexity of the dynamics of multispecies communities is one of the main challenges confronting ecologists, conservation biologists, and environmental managers.

In this chapter, we identify the contributions of ecological studies of food-web interactions to conservation and management, and we highlight promising new research directions. First, we present the ecological theory of consumer-resource interactions and its applications to conservation and management. Second, we review the empirical evidence about the role of top-down and bottom-up forces in influencing the structure and dynamics of ecological communities. Third, we describe the importance of synthesizing experimental and monitoring data to develop generalizations that are more widely applicable, and we provide one example of such synthesis. Then we discuss whether and how to apply what we know about food-web interactions to real-world resource management. Finally, we propose some research priorities for the conservation and management of assemblages of interacting species.

Theory of Consumer-Resource Interactions

Ecological theory has produced a plethora of consumer-resource (i.e., predator-prey) models. One way of classifying these numerous contributions is on the basis of the number of species, or species groups, included in the models. Basic consumer-resource models, such as Lotka-Volterra's, focus on the dynamics of a two-species, enemy-victim system. The predator-prey interaction causes species abundances to cycle, with amplitudes determined by initial population abundances. A suite of biological mechanisms induces stability, including density-dependent (logistic) prey growth, density-dependent predator death rates or predator attack rates (interference among predators), physical refuges for the prey, the presence of an invulnerable life stage for the prey, and external sources of prey or predator recruits (open-system dynamics) (Gurney and Nisbet 1998). In contrast, enrichment is expected to destabilize predator-prey systems (the "paradox of enrichment," Rosenzweig 1971), although there is little empirical evidence that real populations show the predicted instability in nutrient-rich environments (see Murdoch et al. 1998). Enemy-victim models such as Lotka-Volterra's and Nicholson-Bailey's, and later modifications of the original equations, have most commonly been applied to terrestrial systems, particularly in biological pest control (Hassell 1978; Murdoch et al. 1985; Waage and Mills 1992; Murdoch and Briggs 1996).

Predator-prey models of the Lotka-Volterra type typically isolate subsets of interacting species from a complex, multispecies ecosystem. Other species in the community are considered part of the environment and are not mod-eled explicitly. Thus, it is assumed that the dynamics of the target species are determined primarily by strong interactions with its prey or predator, and that links with other species in the community are weaker and less important in determining the system dynamics. However, Lotka-Volterra–type models have also been used to investigate the dynamics of multispecies assemblages. Species that use common resources in similar ways are grouped into trophic levels, which are assumed to act dynamically as populations. This approach was initiated by Hairston, Smith, and Slobodkin (1960) and further devel-oped by Fretwell (1977) and Oksanen et al. (1981).

In Hairston et al.'s classic 1960 paper, food chains are composed of three trophic levels: plants, herbivores, and carnivores. Carnivores control herbi-vore populations, thereby releasing plants from top-down control and allow-ing accumulation of "green" biomass. Fretwell (1977) and Oksanen et al. (1981) expanded this approach to examine food-chain length and interactions along productivity gradients. Increasing productivity supports increasing numbers of trophic levels, which subsequently exert a top-down control on their prey and initiate cascading trophic interactions propagating down the food chain. At equilibrium, top-down control produces a stepped pattern of biomass increase along a productivity gradient, where top trophic levels and those even numbers of steps below are resource limited, whereas trophic lev-els odd numbers of steps below the top are limited by their consumers. Although linear food-chain models were originally developed with terres-trial systems in mind (Hairston et al. 1960; Fretwell 1977; Oksanen et al. 1981), they have been applied most commonly to aquatic systems (Persson et al. 1988, 1992; Crowder et al. 1988; Power 1990; Wootton and Power 1993; Brett and Goldman 1996, 1997).

Similar to linear food-chain models, trophic cascade models (Paine 1980) propose that predator-prey interactions are transmitted through food webs to cause variance in plant biomass and production (Carpenter et al. 1985). Trophic cascades are predation effects across multiple trophic levels resulting in inverse patterns of abundance or biomass across trophic levels of a food web. The trophic cascade hypothesis has been used to explain the ~50 percent of observed variability in primary production of lakes that could not be attributed to variation in nutrient loading (Carpenter et al. 1985). Trophic cascades are also the basis for management manipulations of lake food webs referred to as biomanipulations (Shapiro et al. 1975; Gulati et al. 1990). Biomanipulation typically involves enhancement of piscivorous fish stocks or removal of planktivorous fish with the goal of decreasing predator control on large herbivorous zooplankton and increasing grazing of the phytoplankton. Thus, successful biomanipulation results in the biological control of one of the consequences of anthropogenic eutrophication, namely increased primary production (Carpenter and Kitchell 1992; Kitchell 1992).

Biological complexities that characterize many food webs can dampen trophic cascades and lead to weak or no top-down control. These include: (1) inedible prey or invulnerable life stages of prey (Murdoch 1966; Ehrlich and Birch 1967; McCauley et al. 1988; Leibold 1989; McQueen 1990; Strong 1992; Abrams 1993; Leibold et al. 1997); (2) complex interactions such as cannibalism, ontogenetic diet shifts, feeding on more than one trophic level (omnivory), and feeding on competitors (intraguild predation) (Mittelbach et al. 1988; Arditi and Ginzburg 1989; Strong 1992; Polis and Strong 1996; Holt and Polis 1997; McCann, Hastings, and Huxel 1998; McCann, Hastings, and Strong 1998); and (3) the fact that most food webs are not closed systems but systems that exchange resources and individuals with adjacent systems through nutrient and detritus input or loss, recruitment processes, and organism migration (Polis and Strong 1996; Polis et al. 1997; Huxel and McCann 1998). Inclusion of the above biological complexities in food-web models generates a range of possible community responses to variation in resource or consumer levels (table 3.1).

Basic consumer-resource and food-chain models focus on the dynamics of subsets of real food webs (figure 3.1). These systems are embedded in a more complex web of interactions (figure 3.2), which may affect the dynamics of the focal consumer-resource system (Polis and Strong 1996; Polis et al. 1997; Yodzis 1998, 2000). Attempts to model all trophic interactions within a community result in an overwhelming amount of possible interaction pathways. Can we safely ignore some of these pathways? Yodzis (1998) has tackled this question using the pelagic marine food web of the Benguela ecosystem, off South Africa (figure 3.2). Inclusion of all documented trophic interactions among the twenty-nine species of the food web yielded 203 direct links and millions of possible indirect pathways of interaction. He quantified the ecosystem response to the removal of one of the top carnivores in the system, South African fur seals, by examining predicted changes in the yield of the main commercial fisheries. This analysis shows that 44 percent of the 203 links can be eliminated from the model without altering the response of fish-

TABLE 3.1. Some Models of Top-Down and Bottom-Up Community Regulation

- *Hairston, Smith, and Slobodkin 1960:* Carnivores control herbivore populations, thereby releasing plants from top-down control and allowing accumulation of "green" biomass.
- *Murdoch 1966; McCauley et al. 1988; Leibold 1989; Strong 1992:* Prey defenses (e.g., inedible prey) dampen top-down effects.
- *Fretwell 1977; Oksanen et al. 1981:* Stepped pattern of biomass accrual along productivity gradients. Top trophic levels and those even numbers of steps below them are resource limited; trophic levels odd numbers of steps below the top one are predator limited.
- *Getz 1984; Arditi and Ginzburg 1989:* Interference among predators prevents their efficient exploitation of resources and leads to no top-down control. Biomass increases with increasing resource levels at all trophic levels. [AUTHOR: Please add Getz 1984 to Literature Cited]
- *Mittelbach et al. 1988:* Ontogenetic diet shifts. Predators cannot track resources because resource increases influence only one life stage, leading to weak top-down effects.
- *McQueen 1990:* Bottom-up control is stronger at base of food web; top-down control is stronger at higher trophic levels. Trophic cascades attenuate before reaching plants.
- *Polis and Strong 1996:* Complex interactions among components of food webs dampen top-down control. Multichannel omnivory can promote top-down regulation when top consumers are subsidized by external resources. Thus, omnivory can both dampen and enhance top-down effects.
- *Leibold et al. 1997:* Species replacement through time (predation selects for defended prey). Weak top-down effects in the long-term.
- *McCann, Hastings, and Strong 1998:* Intratrophic interference affecting consumer population growth rates dampen top-down effects. Increased productivity results in increased plant biomass, whereas consumer biomass shows a modest increase.
- *Huxel and McCann 1998:* Allochtonous resource input weakens top-down control.

Source: Modified from M. E. Power, "Top-Down and Bottom-Up Forces in Food Webs: Do Plants Have Primacy?" *Ecology* 73 (1992): 733–746.

ery yields to seal culling, suggesting that some simplification of the system is possible. Even such simplification leaves 112 direct consumer-resource links potentially influencing the system response to perturbation, still a tremendously complex system. Further modeling and empirical work should determine how much detail is needed to describe the dynamics of whole communities and under what circumstances complex systems yield predictable responses to perturbation (e.g., Terborgh et al. 1999).

Empirical Evidence: How Top-Down and Bottom-Up Forces Influence Ecological Communities

Experiments consisting of altering resource levels and adding or removing consumers have shown that consumer-resource interactions can play a critical role in regulating populations and shaping communities and have eluci-

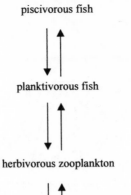

piscivorous fish

planktivorous fish

herbivorous zooplankton

phytoplankton

FIGURE 3.1. A simplified food chain comprising four trophic levels: piscivorous fish (e.g., tuna), planktivorous fish (e.g., anchovy), herbivorous zooplankton (e.g., copepod crustaceans), and phytoplankton.

dated direct and indirect mechanisms underlying predation effects (e.g., Sih et al. 1985; Schoener 1993; Wootton 1993; Menge 1995). Unfortunately, controlled manipulative experiments often cannot be conducted over the spatial and temporal scales relevant to biological conservation and management. Despite the difficulty of interpreting observed patterns in the lack of replication and controlled reference conditions, "natural" experiments provide an opportunity to examine the consequences of food-web alterations at large scales (e.g., Terborgh et al. 1999). The alteration of resource availability through nutrient enrichment or food-web subsidies, the widespread decline or loss of top consumers caused by fishing and hunting, and the introduction of predators to new areas are large-scale manipulations of food-web structure and consumer-resource interactions.

Human activities result in the production of nutrients and organic matter that can fuel natural food webs and alter abundances and interactions of species or whole trophic levels. In particular, industrial and agricultural activities add to terrestrial systems at least as much fixed nitrogen as all natural sources combined (Vitousek et al. 1997). Nitrogenous compounds from industrial activities, agriculture, and sewage, reach rivers, lakes, and oceans through groundwater and atmospheric discharge. Enhanced nitrogen input generally results in increased primary productivity, decreased biological diversity, and changes in plant community composition in both terrestrial and aquatic ecosystems (Schindler 1974; Tilman 1987; Nixon 1995; Jeffries

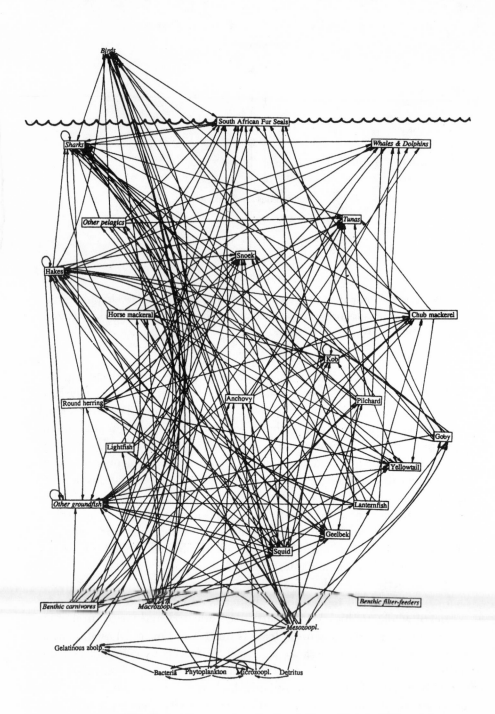

FIGURE 3.2. The Benguela ecosystem food web (from Yodzis 1998, reproduced with author's permission). The species and species groups are embedded in a complex web of trophic and nontrophic interactions.

and Maron 1997). Such increased primary productivity can support greater abundance and biomass of higher trophic levels (Oksanen et al. 1981; McNaughton et al. 1989). However, anthropogenic nutrient enrichment frequently disrupts food-web interactions, causing ungrazed primary production to accumulate in the focal system or in adjacent ecosystems. The consequences of anthropogenic eutrophication, such as algal and microbial blooms (including toxic species), hypoxic/anoxic conditions of the water column, and mass mortality of fish and invertebrates, are particularly widespread and worrisome in coastal marine ecosystems (Hallegraeff 1993; Turner and Rabalais 1994; Nixon 1995; chapter 7).

A range of human activities produces waste that can subsidize natural food webs. For example, trash dumps and the bycatch discarded by commercial fisheries can provide important resources for scavengers and opportunistic predators, including fish, crabs, seabirds, and raptors (Dayton et al. 1995; Garthe et al. 1996; Blanco 1997), which may in turn control populations of competitors or prey and influence community structure. We still know very little about the impacts of different types of resource subsidies on natural ecosystems, how such impacts vary across spatial scales (from the local effects of waste dumping to the global effects of the alteration of the nitrogen and carbon cycles), and how the enhanced productivity supported by the added resources alters food-web dynamics.

The decline and collapse of top carnivores provide ample evidence that predation can structure aquatic and terrestrial communities (Crowder et al. 1996; Terborgh et al. 1999). One of the best-documented case studies is the sea otter–sea urchin–kelp trophic cascade occurring in North American coastal marine habitats (Estes and Palmisano 1974; Estes et al. 1998). Hunting reduced sea otters to a number of widely scattered remnant populations along the northwestern coast of the United States. The absence of sea otters often is associated with an abundance of sea urchins, which overgraze large benthic algae, including kelps (Estes and Palmisano 1974). In recent years, increased predation by killer whales on sea otters, possibly caused by a decline in the pelagic prey of killer whales, may have led to sea otter decline and increased sea urchin abundance and overgrazing of kelp (Estes et al. 1998). In lakes, experimental or fisheries removal of piscivorous fish can cause trophic cascades, leading to increased abundance of planktivorous fish, alterations of the size structure and species composition of the zooplankton, and increased phytoplankton biomass (Carpenter et al. 1985; Kitchell 1992; Carpenter and Kitchell 1993).

McLaren and Peterson (1994) documented a compelling example of a terrestrial trophic cascade on Isle Royale, Michigan, USA. Wolves control moose populations, which in turn influence tree growth rates. When wolves were rare, moose abundances increased and the growth rates of balsam fir were depressed. Hunting, combined with habitat destruction and fragmentation, has drastically reduced the abundance and geographical range of many top carnivores, including wolves, bears, coyotes, cougars, lions, and tigers. In North America, reduction or elimination of top predators has resulted in the "mesopredator release" of other small or midsized predators, such as foxes, skunks, raccoons, and feral and domestic cats (Soulé et al. 1988), thereby altering food-web dynamics and community structure. For example, the decline and disappearance of coyotes in landscapes fragmented by development affects the distribution and abundance of smaller carnivores and the persistence of their avian prey (Crooks and Soulé 1999). In suburban areas and parklands of North America, where hunting eliminated top carnivores in the past but is now prohibited, mammals that were part of the top carnivore prey pool, like deer, have increased in numbers, in some cases becoming road hazards and overbrowsing the vegetation (McShea et al. 1997).

In pelagic marine ecosystems, there are many examples of declines of top predators accompanied by species replacement at the same trophic level or in changes of the prey community at the next trophic level. Hunting of blue whales around Antarctica was followed by increased abundance of other species feeding on krill, including minke whales, crab-eater seals, and penguins (Laws 1984; Brownell et al. 1989). On Georges Bank, off the New England coast, small elasmobranchs (mainly dogfish sharks and skates) have increased in abundance following the collapse of the stocks of large gadid fishes such as haddock and cod. This switch in species dominance is likely due to increased prey availability for the elasmobranchs caused by the decline of the piscivorous gadid and flounder species (Fogarty and Murawski 1998). Extensive removal of large sharks off South Africa led to increased abundances of small sharks, on which the large sharks preyed, and to reductions of commercial fish yields (van der Elst 1979). In the North Sea, decline in the herring and mackerel stocks resulted in increased catch of their prey, mostly smaller fish including sand lance, sprat, and Norway pout (Hempel 1978). In contrast, the decline or depletion of some marine apex predators, including great auks, Steller's sea cows, and gray whales, appears to have caused only subtle changes in populations of their prey and competitors (Boersma and Moore in press). In these case studies, it is not known how other food-web components—for example, plankton or benthos—responded to predator decline.

Marine reserves—portions of the coastline protected from fishing and other human disturbance—constitute large-scale human-exclusion experiments. As such, they present an invaluable opportunity to determine the impacts of food-web alterations on marine communities. In the Las Cruces marine reserve, central Chile, the predatory gastropod *Concholepas concholepas* ("loco") increased in abundance within the reserve compared to adjacent areas where it was harvested. Increased predation by *Concholepas* on its prey, particularly mussels, led to the almost complete elimination of mussel beds, which were replaced by barnacles and algae (Castilla 1999). Increased abundance of predatory fishes within no-take marine reserves in Spain and East Africa has been linked to increased predation on sea urchins, decreased urchin abundance, changes in population size structure, and a proliferation of fleshy algae (Sala and Zabala 1996; McClanahan et al. 1999).

Introductions of exotic predators provide further evidence of the potential for top-down and bottom-up effects to influence communities and ecosystems. Intentional introductions of fishery species to lakes have had devastating consequences for native communities. The introduction of the peacock bass to Gatun Lake, Panama, led to the loss of several native species of fish and to a simplification of the food web (Zaret and Paine 1973). Similarly, the introduction of the Nile perch to Lake Victoria, East Africa, has profoundly altered the native fish community and caused the loss of many endemic species of cyclid fishes (Barel et al. 1985; Kitchell et al. 1997). The accidental introduction of the ctenophore *Mnemiopsis leidyi,* a consumer of zooplankton and juvenile fish, to the Black Sea has caused further decline and loss of several species of crustacean zooplankton and planktivorous fish in a system already heavily impacted by increased nutrient loading, chemical pollution, and fisheries exploitation (Zaitsev 1992).

The introduction of exotic predators to predator-free islands provides evidence for top-down regulation of some terrestrial communities. For example, the accidental introduction of the brown tree snake to Guam led to the extinction of several species of native birds (Savidge 1987). Similarly, the introductions of foxes to the Alaskan Islands and mangoose to tropical Pacific islands have contributed to the collapse of native fauna (King 1984; Bailey 1993). Exotics may also influence ecosystem processes, altering resource availability for other species and the structure and dynamics of the whole ecosystem. One compelling example is the invasion of the nitrogen-fixing tree *Myrica faya* in Hawaii. The establishment of *Myrica faya* in newly formed volcanic habitats can lead to a rapid increase of the biologically available nitrogen in nutrient-poor volcanic soils, causing dramatic changes of the plant and soil communities and favoring subsequent invasions by non-native organisms (Vitousek and Walker 1989).

Synthesis of Experimental and Monitoring Data

Experimental predator or resource manipulations and large-scale "natural" experiments associated with the decline or introduction of predators, nutrient enrichment, and food-web subsidies provide crucial information about the types of community changes caused by food-web alterations and point to urgent research priorities for conservation and management.

Fundamental questions concern how commonly different community responses occur and what conditions are more likely to lead to a particular type of community response. Are all systems unique? Can the changes observed in one system be expected in other systems? Syntheses of experimental and monitoring data can produce generalizations about community responses to food-web alterations and yield predictions about the most likely outcomes of future perturbations and management interventions. Unfortunately, the plethora of empirical data documenting food-web effects in experimental and natural systems have rarely been synthesized to test whether general responses exist. For an example of such quantitative synthesis, we focus on generalizations about trophic cascades as one mechanism leading to community changes following the addition or removal of consumers in aquatic ecosystems.

Trophic cascades (predation effects across multiple trophic levels resulting in inverse patterns of abundance or biomass across a food web) have received much attention for their potential to spread alterations in consumer levels across an entire food web. Thus, loss or introduction of predators may impact the whole community through cascading trophic interactions. In addition, trophic cascades have management applications as a means of controlling the productivity and biomass of the phytoplankton in waters subject to anthropogenic nutrient loading (Shapiro et al. 1975; Gulati et al. 1990; Kitchell 1992).

Trophic cascades have been documented in a variety of terrestrial and aquatic ecosystems, including temperate and tropical reefs, rocky intertidal communities, streams, lakes, and terrestrial insect communities in the tropics (Power 1990; Wootton 1995; Spiller and Schoener 1994; Hixon 1997; Letourneau and Dyer 1998; Pace et al. 1999; Pinnegar et al. 2000; and references in the preceding "Empirical Evidence" section). Most documented cascades occur among a few of the species in the community, particularly in terrestrial systems (Polis 1999; Persson 1999). However, trophic cascades influencing the structure and dynamics of whole communities occur in aquatic ecosystems (e.g., Estes and Palmisano 1974; Power 1990; Carpenter and Kitchell 1993; Shiomoto et al. 1997; Estes et al. 1998). Thus, food-web alterations can initiate cascading trophic interactions influencing the structure of aquatic communities. How commonly do trophic cascades occur, and in what conditions are they more likely to occur?

Syntheses conducted in lakes, marine pelagic ecosystems, and rocky intertidal communities indicate that trophic cascades are uncommon in those systems. In a meta-analysis of predator (fish) manipulations conducted in lakes, Brett and Goldman (1997) found that fish exerted a strong control on their prey, the zooplankton, but phytoplankton biomass exhibited weak responses to fish manipulation. In a more extensive analysis, phytoplankton responded strongly to fish manipulation in approximately one-third of fifty-four enclosure or pond experiments and showed weak responses in the others (Brett and Goldman 1996).

Meta-analyses of data from manipulations of nutrients and consumers conducted in marine mesocosms and long-term monitoring of nutrients, phytoplankton, zooplankton, and zooplanktivorous fish abundance in open pelagic ecosystems indicated that trophic cascades are uncommon in marine pelagic food webs (Micheli 1999). In particular, of the twenty open marine systems included in this analysis, only one exhibited an inverse pattern in biomass across three trophic levels. A ten-year time series of phytoplankton, zooplankton, and a planktivorous predator, the pink salmon *Oncorhynchus gorbuscha,* from the subarctic Pacific Ocean showed that interannual variation in salmon abundance was inversely related to zooplankton biomass and positively related to phytoplankton biomass (Shiomoto et al. 1997).

In general, year-to-year fluctuations in zooplankton biomass were negatively correlated with fish, indicating that fish predation may control zooplankton biomass. In contrast, the zooplankton and the phytoplankton were not significantly correlated, indicating that fish predation does not commonly control the biomass of primary producers in these pelagic ecosystems (Micheli 1999). An important implication is that biomanipulations are unlikely to control phytoplankton production and biomass in the coastal marine environment. Reductions of anthropogenic nutrient loading to coastal waters may be the only means of controlling marine eutrophication.

Whole-lake experiments suggest that trophic cascades may be enhanced under nutrient-rich conditions (Pace et al. 1999). This result has important implications because most aquatic systems are subjected to simultaneous food-web alterations, through fishing and introduction of exotic species, and to increased nutrient loading from agricultural activities and coastal development. Food-web manipulations conducted in marine mesocosms suggest that nutrient enrichment may favor trophic cascades also in marine food webs. Overall, there was a statistically significant increase in phytoplankton biomass following the addition of planktivorous fishes in mesocosms enriched with nitrogen but not in mesocosms with no nutrients added (Micheli 1999). However, patterns from open marine systems do not support this hypothesis: the only system where trophic cascades occurred had the lowest productivity among the twenty systems in the data set (Micheli, unpublished data).

In a survey of direct and indirect effects in experimental manipulations of twenty-three marine rocky intertidal food webs, Menge (1995) found that indirect effects accounted for a large proportion of the community changes caused by manipulations (24 to 61 percent, mean ≈ 40 percent). Nearly half of all indirect effects (total = 565) resulted from manipulations of predators. In these food webs, 6.5 percent of all indirect effects were trophic cascades. The most frequent types of indirect effects were keystone predation (when a predator's consumption of a prey increases the abundance of the prey competitors: 35 percent of all indirect effects) and apparent competition (when species share a common predator and increased abundance of one species enhances predation on the other species: 25 percent).

In conclusion, trophic cascades can have important effects on community structure and species dynamics (see the "Empirical Evidence" section) but appear to be more common in some ecosystem types—such as shallow marine benthic communities, small lakes, and vertebrate terrestrial systems—and in the presence of intense anthropogenic disturbance (Carpenter et al. 1985; Terborgh et al. 1999; Pinnegar et al. 2000). Our present perception may reflect biases in the focus and time frames of the studies conducted to date. For example, shallow benthic marine communities are more accessible and amenable to experimental manipulations than offshore pelagic ecosystems (Pinnegar et al. 2000). In addition, large terrestrial and aquatic predators have been decimated through hunting, fishing, and habitat destruction long before the start of the monitoring programs and ecological studies with which we are attempting to detect the community consequences of predator removal (e.g., Jackson 1997). Determining what conditions lead to trophic cascades, how commonly they occur in different types of ecosystems, and the temporal and spatial scales at which food-web perturbations alter community structure and interactions is an important research priority.

Conclusions

There is ample experimental and correlative evidence that consumer-resource interactions play an important role in regulating population dynamics, community structure, and diversity. Potential applications include:

- predicting the impacts of food-web alterations on population and community dynamics;
- conserving top predators and keystone species, preventing competitive displacement of species, and maintaining diversity;
- manipulating consumers and resources to restore community structure and ecosystem processes;
- conserving species or species groups that effectively control pests and exotic invaders;
- designing reserve networks to conserve whole assemblages and the interactions among their component species; and
- controlling anthropogenic eutrophication of aquatic ecosystems through biomanipulation of food webs.

Do we know enough about bottom-up and top-down regulation of natural communities to predict change and implement these conservation and management strategies successfully?

Theory and data indicate that there is considerable indeterminacy about the type, magnitude, and direction of community changes following food-web alterations. Perturbations caused by the addition or removal of a resource or of a consumer species can cause both direct and indirect effects and result in a large suite of possible outcomes (Bender et al. 1984; Yodzis 1988; Abrams 1993; Wootton 1993). Because of the inherent variability of ecological processes and the fragmentary data available, we will never be able to fully explain and predict species dynamics and community change. Management of species and ecosystems in the face of uncertainty requires implementation of the precautionary principle (FAO 1995; Lauck et al. 1998) and of adaptive management (Walters 1986; Parma et al. 1998). Uncertainty about the consequences of food-web alterations in complex, multispecies communities is an additional reason to use precautionary and adaptive approaches in conservation and management. Nevertheless, a better understanding of processes structuring communities, including food-web interactions, could only help reduce some of the uncertainty.

Perhaps the main challenge in applying food-web research to conservation is how to deal with the multitude of possible direct and indirect pathways of interactions among species. One approach is to focus on subsets of species linked through strong interactions and weakly linked to other species in the system (community "modules," Holt 1996). Simplifying complex communities to smaller subsets of strong interactors that largely drive community dynamics allows a mechanistic understanding of consumer-resource interactions and has a strong theoretical basis in a plethora of predator-prey models. However, we still know very little about the relative frequencies of strong and weak interactions in natural communities (Power et al. 1996). Direct quantification of interaction strength in two marine and one terrestrial community indicated that most interactions were weak, with a few strong interactions (Paine 1992; Power et al. 1996), but broader comparisons are needed before any generalization is possible.

The presence of strong interactors within communities, such as keystone predators, has potentially broad applicability in conservation because it allows focusing monitoring and management on key species that regulate the structure and diversity of the whole community. To date, there are no general guidelines about the traits that characterize keystones or the types of communities where they are more likely to occur (Mills et al. 1993; Power et al. 1996). Identifying the species that are strong interactors in different communities, as well as the context where species are most likely to be strong interactors, is an important research priority for the coming decade. However, ample evidence suggests that the top predators targeted by hunting and fishing tend to exert strong top-down effects and that immediate action should be taken to protect top predators and prevent further disruption of natural food webs.

Attempts to model, manipulate, or measure all possible interactions in a food web quickly exceed data availability and our ability to relate community patterns to the processes that underlie them. Aggregation of species into trophic levels is one means of simplifying complex communities while including multiple species, but it can overlook important biological complexities. For example, intratrophic interference, nontrophic interactions, and long-term species replacement result in community patterns that are not predicted by linear food-chain models (table 3.1). Models of intermediate complexity and determination of what key processes should be included for different ecosystems seem the most promising research directions for producing theory and guidelines for conservation and management. The added realism renders models less general but more applicable to particular systems (Murdoch and Briggs 1996).

The community-wide consequences of fishery collapses, loss of top carnivores from terrestrial systems, introduction of exotic species, and anthropogenic nutrient enrichment of ecosystems indicate that the removal or addition of consumers and resources generally influences prey at the next lower trophic level and the species that directly use the resource. Frequently, species replacement occurs within a trophic level. In some cases, effects can also influence the whole community through trophic and nontrophic interactions cascading through the food web. Establishing the generality of these trends and identifying the key processes underlying community responses to food-web perturbations are urgent research priorities.

Research Priorities

The following priorities and their components are outlined in box 3.1.

Produce Generalizations about Species and Community Responses to Food-Web Alteration

Theory and experiments have shown that food-web alterations influence communities through a range of direct and indirect pathways of interactions, which result in a large suite of possible outcomes (Bender et al. 1984; Yodzis 1988; Abrams 1993; Wootton 1993). Yet recent syntheses of data suggest that some generalization may be possible. In freshwater and marine pelagic food webs, alterations of resource and consumer levels cause similar general patterns of community change (Brett and Goldman 1996, 1997; Micheli 1999). In marine rocky intertidal food webs, experimental manipulations caused several types of indirect effects, but approximately 60 percent of indirect effects were either keystone predation or apparent competition (Menge 1995). "Natural" experiments provided by the variation in space and time of human impacts (e.g., Estes et al. 1998; Crooks and Soulé 1999; Terborgh et al. 1999) represent invaluable opportunities to examine food-web responses to bottom-up and top-down forces over large scales and compare responses across systems. Syntheses of existing data may produce generalizations about community responses to food-web alterations and generate predictions about the likelihood of different effects of future perturbations or management interventions.

Determine Level of Resolution of Community Representation

There is an urgent need to determine how much detail is necessary to detect change and predict community dynamics in the face of environmental variability and human impacts. The difficulty and costs of modeling and monitoring species dynamics and interactions within complex food webs increase quickly with increasing resolution of models and sampling. Some form of simplification is necessary. This is typically achieved by focusing on small groups of interacting species or by lumping species into trophic levels or functional groups. Simulations, resampling of existing data, and investigation of the effects of aggregating species-level data into broader categories are promising avenues for determining how much detail can be omitted from food-web models and for producing guidelines for monitoring multispecies assemblages (e.g., Frost et al. 1995; Cottingham and Carpenter 1998; Yodzis 1998).

Understand the Effects of Nutrient and Organic Enrichment of Ecosystems

The availability of resources to species at all trophic levels of natural food webs is altered through a variety of human activities. Human alteration of the availability of nutrients, detritus, and prey to producers and consumers is a widespread phenomenon, but it includes a variety of types of alteration, occurring over local to global scales. Determining how different types and scales of subsidies influence food-web interactions, community structure, and ecosystem processes is an urgent research priority in the face of the ever increasing human production of pollutants and waste.

Link Pattern and Process to Understand Mechanisms Underlying Food-Web Dynamics

A better understanding of the processes underlying observed community dynamics would greatly improve conservation and management of aquatic

BOX 3.1. Research Priorities for
Food-Web Conservation and Management

Produce Generalizations about Species and Community Responses to Food-Web Alteration

*1. Characterize community responses to food-web alterations and determine which conditions are more likely to lead to a particular type of community response.
2. Determine unique vs. general responses to food-web perturbation.
3. Relate the strength and effects of consumer-resource interactions to environmental gradients in productivity and disturbance.
4. Explore how systems of increasing complexity respond to perturbation and determine under what circumstances complex systems may yield predictable responses to perturbation.
5. Investigate the separate and joint effects of different types of food-web perturbations.

Determine Level of Resolution of Community Representation

*1. Determine how many species and interactions can be omitted from monitoring and models and what level of aggregation still allows accurate representation of community dynamics.
2. Establish the relative frequencies of strong and weak interactions in natural ecosystems.
3. Investigate what traits characterize strong interactors and keystone species.

Understand the Effects of Nutrient and Organic Enrichment of Ecosystems

*1. Determine how different types of subsidies influence food-web interactions, community structure, and ecosystem processes.

ecosystems. One approach is to develop mechanistic models that represent alternative views of the processes driving food-web dynamics, and to compare model predictions to observed patterns. Confidence in alternative models is built through comparisons of model predictions to data, thereby linking observed patterns of community change to specific mechanisms of consumer-resource interactions. For example, Shea et al. (1998) proposed testing alternative process-based models against population time series to improve population management in conservation, fisheries management, and pest control. Similarly, Kendall et al. (1999) combined mechanistic models and time-series statistical models to elucidate the processes underlying population cycles. McCann, Hastings, and Strong (1998) compared predic-

2. Investigate how nutrient and organic matter subsidies influence food-web dynamics in ecosystems already subjected to other perturbations, including overexploitation, habitat degradation, and climate change.
3. Explore how the effects of food-web subsidies vary between natural sources (e.g., oceanic upwelling, litter fall, bird guano) and anthropogenic sources (e.g., sewage and fertilizers, urban waste, discarded fisheries bycatch).
4. Evaluate how the effects of resource alteration vary among local, regional, and global scales.

Link Pattern and Process to Understand Mechanisms Underlying Food-Web Dynamics

1. Link alternative models of consumer-resource interactions to observed patterns of community change.
*2. Develop models that include the influence of human activities on community and ecosystem dynamics and explore alternative conservation and management scenarios before they are implemented.

Determine Scale of Interactions and Food-Web Boundaries

*1. Determine the relevant food-web components influencing community structure and diversity.
*2. Investigate the effects of the movement of energy, matter, and organisms across community and ecosystem boundaries on food-web dynamics.
*3. Establish the temporal and spatial scales over which community change can be detected.
4. Explore how local, regional, and global food-web perturbations influence each other.

Note: Asterisks denote the six highest priorities for the current decade.

tions of food-web models that include different forms of interference among consumers to patterns of biomass change at different trophic levels of lake food webs. Modeling also can be used to simulate scenarios that cannot be simultaneously realized in practice, allowing exploration of the effects of species addition and deletion, of single vs. multiple perturbations, and of different management strategies.

Determine Scale of Interactions and Food-Web Boundaries

A largely unexplored area is that of how delimiting food webs and ecosystems using different criteria and over different spatial and temporal scales may influence conclusions about the community and ecosystem consequences of food-web interactions (e.g., Polis et al. 1997). For example, our understanding of above-ground plant-insect interactions and marine plankton dynamics was dramatically changed by including soil (Strong 1999) and microbial (Pomeroy 1974) communities in these food webs. Similarly, the dynamics of intertidal marine communities and terrestrial communities on small islands are largely influenced by oceanic input of larvae (Roughgarden et al. 1988) and detritus (Bustamante et al. 1995; Polis and Hurd 1996). Historical and paleontological data indicate that coral reef communities have been impacted and modified by fishing and other human activities well before the start of the monitoring programs with which we attempt to detect temporal change (Jackson 1997). Scale and boundary issues are relevant to a suite of conservation and management applications, such as designing networks of reserves, managing watersheds and coastal areas, and determining whether species assemblages are being modified by human disturbances. Tackling scale issues will require conducting research at multiple scales, using a diverse set of perspectives and approaches, and establishing interdisciplinary collaborations.

Acknowledgments

We dedicate this work to the memory of Gary Polis, who died under tragic circumstances before the completion of this book. We thank the workshop participants, in particular A. Covich, G. Mace, J. MacMahon, W. Murdoch, R. Pulliam, K. Redford, and W. Schlessinger, for stimulating discussions and helpful suggestions. Special thanks to G. Orians and M. Soulé for convening the workshop.

Literature Cited

Abrams, P. A. 1992. Predators that benefit prey and prey that harm predators: Unusual effects of interacting foraging adaptations. *American Naturalist* 140: 573–600.

Abrams, P. A. 1993. Effect of increased productivity on the abundances of trophic levels. *American Naturalist* 141: 351–371.

Abrams, P. A., B. A. Menge, G. G. Mittelbach, D. A. Spiller, and P. Yodzis. 1996. The role of indirect effects in food webs. In G. A. Polis and K. O. Winemiller (eds.), *Food Webs—Integration of Patterns and Dynamics*, pp. 371–395. New York: Chapman and Hall.

Allison, G. W., J. Lubchenco, and M. H. Carr. 1998. Marine reserves are necessary but not sufficient for marine conservation. *Ecological Applications* 8: 79–92.

Arditi, R., and L. R. Ginzburg. 1989. Coupling in predator-prey dynamics: Ratio dependence. *Journal of Theoretical Biology* 139: 311–326.

Bailey, E. P. 1993. *Introduction of foxes to Alaskan islands—History, effects on avifauna, and revocation.* Resource publication 193. Washington, D.C.: U.S. Department of Interior, Fish and Wildlife Service.

Barel, C. D. N., R. Dorit, P. H. Greenwood, G. Fryer, N. Hughes, P. B. N. Jackson, H. Kawanabe, R. H. Lowe-McConnell, M. Nagoshi, A. J. Ribbink, E. Trewavas, F. Witte, and K. Yamaoka. 1985. Destruction of fisheries in Africa's lakes. *Nature* 315: 19–20.

Bender, E. A., T. J. Case, and M. E. Gilpin. 1984. Perturbation experiments in community ecology: Theory and practice. *Ecology* 65: 1–13.

Blanco, G. 1997. Role of refuse as food for migrant, floater and breeding black kites (*Milvus migrans*). *Journal of Raptor Research* 31: 71–76.

Boersma, P.D., and S. E. Moore. In press. Seabirds and marine mammals: Political keystones. In P. Kareiva and S. Levin (eds.), *The Importance of Species: Perspectives on Expandibility and Triage.*

Bonsauk, J. A. 1996. Maintenance and recovery of reef fishery productivity. In N. V. C. Polunin and C.M. Roberts (eds.), *Management of Reef Fisheries,* pp. 283–313. London: Chapman and Hall.

Botsford, L.W., J. C. Castilla, and C. H. Peterson. 1997. The management of fisheries and marine ecosystems. *Science* 277: 509–515.

Brett, M. T., and C. R. Goldman. 1996. A meta-analysis of the freshwater trophic cascade. *Proceedings of the National Academy of Science of the USA* 93: 7723–7726.

Brett, M. T., and C. R. Goldman. 1997. Consumer versus resource control in freshwater pelagic food webs. *Science* 275: 384–386.

Brownell, R. L. Jr., K. Ralls, and W. F. Perrin. 1989. The plight of the forgotten whales. *Oceanus* 32: 5–20.

Bustamante, R. H., G. M. Branch, and S. Eekhout. 1995. Maintenance of an exceptional intertidal grazer biomass in South Africa: Subsidy by subtidal kelps. *Ecology* 76: 2314–2329.

Carpenter, S. R., and J. F. Kitchell. 1992. Trophic cascade and biomanipulation: Interface of research and management. *Limnology and Oceanography* 37: 208–213.

Carpenter, S. R., and J. F. Kitchell (eds.). 1993. *The Trophic Cascade in Lakes.* Cambridge: Cambridge University Press.

Carpenter, S. R., J. F. Kitchell, and J.R. Hodgson. 1985. Cascading trophic interactions and lake productivity. *Bioscience* 38: 764–769.

Castilla, J. C. 1999. Coastal marine communities: Trends and perspectives from human-exclusion experiments. *Trends in Ecology and Evolution* 14: 280–283.

Christensen, N.L., A. M. Bartuska, J. H. Brown, et al. 1996. The report of the Ecological Society of America Committee on the scientific basis for ecosystem management. *Ecological Applications* 6: 665–691.

Clark, C. W. 1996. Marine reserves and the precautionary management of fisheries. *Ecological Applications* 6: 369–370.

Cottingham, K. L., and S. R. Carpenter. 1998. Population, community, and ecosystem variates as ecological indicators: Phytoplankton responses to whole-lake enrichment. *Ecological Applications* 8: 508–530.

Crooks, K. R., and M. E. Soulé. 1999. Mesopredator release and avifaunal extinctions in a fragmented system. *Nature* 400: 563–566.

Crowder, L. B., R. W. Drenner, W. C. Kerfoot, D. J. McQueen, E. L. Mills, U. Sommer,

C. Spencer, and M. J. Vanni. 1988. Food web interactions in lakes. In S. R. Carpenter (ed.), *Complex Interactions in Lake Ecosystems,* pp. 141–160. New York: Springer-Verlag.

Crowder, L. B., D. P. Reagan, and D. W. Freckman. 1996. Food web dynamics and applied problems. In G. A. Polis and K. O. Winemiller (eds.), *Food Webs—Integration of Patterns and Dynamics,* pp. 327–336. New York: Chapman and Hall.

Dayton, P. K., S. F. Thrush, M. T. Agardy, and R. J. Hofman. 1995. Environmental effects of marine fishing. *Aquatic Conservation* 5: 205–232.

Ehrlich, P. R., and L. C. Birch. 1967. The "balance of nature" and "population control." *American Naturalist* 101: 97–107.

Estes, J. A., and J. F. Palmisano. 1974. Sea otters: Their role in structuring nearshore communites. *Science* 185: 1058–1060.

Estes, J. A., M. T. Tinker, T. M. Williams, and D. F. Doak. 1998. Killer whale predation on sea otters linking oceanic and nearshore ecosystems. *Science* 282: 473–476.

FAO (Food and Agriculture Organization of the United Nations). 1995. *Draft guidelines on the precautionary approach to capture fisheries.* Report of the technical consultation on the Precautionary Approach to Capture Fisheries, Lysekil, Sweden, June 6–13, 1995. Rome: FAO and Swedish National Board of Fisheries.

FAO (Food and Agriculture Organization of the United Nations). 1997. *Technical guidelines for responsible fisheries.* Rome: FAO.

Fogarty, M. J., and S. A. Murawski. 1998. Large-scale disturbance and the structure of marine systems—Fishery impacts on Georges Bank. *Ecological Applications* 8: S6–S22.

Fretwell, S. D. 1977. The regulation of plant communities by the food chains exploiting them. *Perspectives in Biology and Medicine* 20: 169–185.

Frost, T. M., S. R. Carpenter, A. R. Ives, and T. K. Kratz. 1995. Species compensation and complementarity in ecosystem function. In C. G. Jones and J. H. Lawton (eds.), *Linking Species and Ecosystems,* pp. 224–240. New York: Chapman and Hall.

Garthe, S., C. J. Camphuysen, and R. W. Furness. 1996. Amounts of discards by commercial fisheries and their significance as food for birds in the North Sea. *Marine Ecology Progress Series* 136: 1–11.

Getz, W. M. 1984. Population dynamics: A per capita resource approach. *Journal of Theoretical Biology* 108: 623–643.

Gulati, R. D., E. H. R. R. Lammens, M.-J. Meijer, and E. van Donk (eds.). 1990. *Biomanipulation—Tool for water management.* Dordrecht: Kluwer Press.

Gurney, W. S. C., and R. M. Nisbet. 1998. *Ecological Dynamics.* New York: Oxford University Press.

Hairston, N. G., F. E. Smith, and L. B. Slobodkin. 1960. Community structure, population control, and competition. *American Naturalist* 94: 421–425.

Hallegraeff, G. M. 1993. A review of harmful algal blooms and their apparent global increase. *Phycologia* 32: 79–99.

Hassell, M. P. 1978. *The Dynamics of Arthropod Predator-Prey Systems.* Princeton, N.J.: Princeton University Press.

Hastings, A., and T. Powell. 1991. Chaos in a three-species food chain. *Ecology* 72: 896–903.

Hempel, G. 1978. North Sea fisheries and fish stocks—A review of recent changes. *Rapports et Proces Verbaux des Reunions du Conseil Permanent International pour l'Exploration de la Mer* 173: 145–167.

Hixon, M. A. 1997. The effects of reef fishes on corals and algae. In C. Birkeland (ed.), *Life and Death on Coral Reef,* pp. 230–248. New York: Chapman and Hall.

Holt, R. D. 1996. Community modules. In M. Begon, A. Gange, and V. Brown (eds.), *Multitrophic Interactions,* pp. 333–350. London: Chapman and Hall.

Holt, R. D., and G. A. Polis. 1997. A theoretical framework for intraguild predation. *American Naturalist* 149: 745–764.

Huxel, G. R., and K. McCann. 1998. Food web stability: The influence of trophic flows across habitats. *American Naturalist* 152: 460–469.

Jackson, J. B. C. 1997. Reefs since Columbus. *Coral Reef* 16: S23–S32.

Jeffries, R. L., and J. L. Maron. 1997. The embarrassment of the riches: Atmospheric deposition of nitrogen and community and ecosystem processes. *Trends in Ecology and Evolution* 12: 74–78.

Kendall, B. E., C. J. Briggs, W. W. Murdoch, P. Turchin, S. P. Ellner, E. McCauley, R. M. Nisbet, and S. N. Wood. 1999. Why do populations cycle? A synthesis of statistical and mechanistic modeling approaches. *Ecology* 80: 1789–1805.

King, C. M. 1984. *Immigrant Killers: Introduced Predators and the Conservation of Birds in New Zealand.* Auckland: Oxford University Press.

Kitchell, J. F. (ed.). 1992. *Food Web Management: A Case Study of Lake Mendota.* New York: Springer-Verlag.

Kitchell, J. F., D. F. Schindler, R. Ogutuohwayo, and P. N. Reinthal. 1997. The Nile perch in Lake Victoria—Interactions between predation and fisheries. *Ecological Applications* 7: 653–664.

Lauck, T., C. W. Clark, M. Mangel, and G. R. Munro. 1998. Implementing the precautionary principles in fisheries management through marine reserves. *Ecological Applications* 8: S72–S78.

Laws, R. M. 1984. Seals. *Antarctic Ecology* 2: 621–715.

Leibold, M. A. 1989. Resource edibility and the effect of predators and productivity on the outcome of trophic interactions. *American Naturalist* 134: 922–949.

Leibold, M. A., J. M. Chase, J. B. Shurin, and A. L. Downing. 1997. Species turnover and the regulation of trophic structure. *Annual Review of Ecology and Systematics* 28: 467–494.

Letourneau, D. K., and L. A. Dyer. 1998. Experimental test in a lowland tropical forest shows top-down effects through four trophic levels. *Ecology* 79: 1678–1687.

Lubchenco, J., et al. 1991. The sustainable biosphere initiative: An ecological research agenda. *Ecology* 72: 371–412.

Ludwig, D., R. Hilborn, and C. J. Walters. 1993. Uncertainty, resources exploitation, and conservation: Lessons from history. *Science* 260: 17, 36.

May, R. M., J. R. Beddington, C. W. Clark, S. J. Holt, and R. M. Laws. 1979. Management of multispecies fisheries. *Science* 205: 267–277.

McCann, K. S., A. Hastings, and G. R. Huxel. 1998. Weak trophic interactions and the balance of nature. *Nature* 395: 794–798.

McCann, K. S., A. Hastings, and D. R. Strong. 1998. Trophic cascades and trophic trickles in pelagic food webs. *Proceedings of the Royal Society of London* B 265: 205–209.

McCauley, E., W. W. Murdoch, and S. Watson. 1988. Simple models and variation in plankton densities among lakes. *American Naturalist* 132: 383–403.

McClanahan, T. R., N. A. Muthiga, A. T. Kamukuru, H. Machano, and R. W. Kiambo. 1999. The effects of marine parks and fishing on coral reefs of northern Tanzania. *Biological Conservation* 89: 161–182.

McLaren, B. E., and R. O. Peterson. 1994. Wolves, moose, and tree rings on Isle Royale. *Science* 266: 1555–1558.

McNaughton, S. J., M. Oesterheld, D. A. Frank, K. J. Williams. 1989. Ecosystem-level patterns of primary productivity and herbivory in terrestrial habitats. *Nature* 341: 142–144.

McQueen, D. J. 1990. Manipulating lake community structure: Where do we go from here? *Freshwater Biology* 23: 613–620.

McShea, W. J., H. B. Underwood, and J. H. Rappole. 1997. *The Science of Overabundance: Deer Ecology and Population Management.* Washington, D.C.: Smithsonian Institution Press.

Menge, B. A. 1995. Indirect effects in marine rocky intertidal interaction webs: Patterns and importance. *Ecological Monographs* 65: 21–74.

Micheli, F. 1999. Eutrophication, fisheries, and consumer-resource dynamics in marine pelagic ecosystems. *Science* 285: 1396–1398.

Mills, L. S., M. E. Soulé, and D. F. Doak. 1993. The keystone-species concept in ecology and conservation. *Bioscience* 43: 219–224.

Mittelbach, G. G., C. W. Osenberg, and M. A. Leibold. 1988. Trophic relations and ontogenetic niche shifts in aquatic ecosystems. In B. Ebenman and L. Persson (eds.), *Size-Structured Populations*, pp. 219–233. Berlin: Springer-Verlag.

Murdoch, W. W. 1966. Community structure, population control, and competition—A critique. *American Naturalist* 100: 219–226.

Murdoch, W. W., and C. J. Briggs. 1996. Theory for biological control: Recent developments. *Ecology* 77: 2001–2013.

Murdoch, W. W., J. Chesson, and P. L. Chesson. 1985. Biological control in theory and practice. *American Naturalist* 125: 344–366.

Murdoch, W. W., R. M. Nisbet, F. McCauley, A. M. deRoos, and W. S. C. Gurney. 1998. Plankton abundance and dynamics across nutrient levels: Tests of hypotheses. *Ecology* 79: 1339–1356.

Nixon, S. W. 1995. Coastal marine eutrophication: A definition, social causes, and future concerns. *Ophelia* 41: 199–219.

NRC (National Research Council). 1994. *Improving the Management of U.S. Marine Fisheries.* Washington, D.C.: National Academic Press.

Oksanen, L., S. J. Fretwell, J. Arruda, and P. Niemela. 1981. Exploitation ecosystems in gradients of primary productivity. *American Naturalist* 118: 240–261.

Pace, M. L., J. J. Cole, S. R. Carpenter, and J. F. Kitchell. 1999. Trophic cascades revealed in diverse ecosystems. *Trends in Ecology and Evolution* 14: 483–488.

Paine, R. T. 1980. Food webs, linkage interaction strength, and community infrastructure. *Journal of Animal Ecology* 49: 667–685.

Paine, R. T. 1992. Food-web analysis through field measurement of the per capita interaction strength. *Nature* 355: 73–75.

Parma, A. M., et al. 1998. What can adaptive management do for our fish, forests, food and biodiversity? *Integrative Biology* 1: 16–26.

Pauly, D., V. Christensen, J. Dalsgaard, R. Froese, and F. Torres Jr. 1998. Fishing down marine food webs. *Science* 279: 860.

Persson, L. 1999. Trophic cascades: Abiding heterogeneity and the trophic level concept at the end of the road. *Oikos* 85: 385–397.

Persson, L., G. Andersson, S. F. Hamrin, and L. Johansson. 1988. Predator regulation and primary production along the productivity gradient of temperate lake ecosystems. In S. R. Carpenter (ed.), *Complex Interactions in Lake Ecosystems,* pp. 45–68. New York: Springer-Verlag.

Persson, L., S. Diehl, L. Johansson , G. Andersson, and S. F. Hamrin. 1992. Trophic interactions in temperate lake ecosystems: A test of food chain theory. *American Naturalist* 140: 59–84.

Pinnegar, J. K., N. V. C. Polunin, P. Francour, F. Badalamenti, R. Chemello, M. L. Harmelin-Vivien, B. Hereu, M. Milazzo, M. Zabala, G. D'Anna, and C. Pipitone. 2000. Trophic cascades in benthic marine ecosystems: Lessons for fisheries and protected-area management. *Environmental Conservation* 27: 179–200.

Polis, G. A. 1999. Why are parts of the world green? Multiple factors control productivity and the distribution of biomass. *Oikos* 86: 3–15.

Polis, G. A., W. B. Anderson, and R. D. Holt. 1997. Toward an integration of landscape and food web ecology: The dynamics of spatially subsidized food webs. *Annual Review of Ecology and Systematics* 28: 289–316.

Polis, G.A., and S. D. Hurd. 1996. Linking marine and terrestrial food webs: Allochthonous input from the ocean supports high secondary productivity on small islands and coastal land communities. *American Naturalist* 147: 396–423.

Polis, G. A., and D. R. Strong. 1996. Food web complexity and community dynamics. *American Naturalist* 147: 813–846.

Pomeroy, L. R. 1974. The ocean's food webs: A changing paradigm. *Biological Science* 24: 242–244.

Power, M. E. 1990. Effects of fish in river food webs. *Science* 250: 811–814.

Power, M. E. 1992. Top-down and bottom-up forces in food webs: Do plants have primacy? *Ecology* 73: 733–746.

Power, M. E., D. Tilman, J. E. Estes, B. A. Menge, W. L. Bond, L. S. Mills, G. Daily, J. C. Castilla, J. Lubchenco, and R. T. Paine. 1996. Challenges in the quest for keystones. *Bioscience* 46: 609–620.

Roberts, C. M., and N. V. C. Polunin. 1993. Marine reserves: Simple solutions to managing complex fisheries? *Ambio* 22: 363–368.

Rosenzweig, M. L. 1971. Paradox of enrichment: Destabilization of exploitation ecosystems in ecological time. *Science* 171: 385–387.

Roughgarden, J., S. D. Gaines, and H. Possingham. 1988. Recruitment dynamics in complex life cycles. *Science* 241: 1460–1466.

Russ, G. R., and A. C. Alcala. 1996. Marine reserves: Rates and patterns of recovery and decline of large predatory fish. *Ecological Applications* 6: 947–961.

Sala, E., and M. Zabala. 1996. Fish predation and the structure of the sea urchin *Paracentrotus lividus* populations in the NW Mediterranean. *Marine Ecology Progress Series* 140: 71–81.

Savidge, J. A. 1987. Extinction of an island avifauna by an introduced snake. *Ecology* 68: 660–668.

Schindler, D. W. 1974. Eutrophication and recovery in experimental lakes: Implications for lake management. *Science* 184: 897–899.

Schoener, T. W. 1993. On the relative importance of direct versus indirect effects in ecological communities. In H. Kawanabe, J. E. Cohen, and K. Iwasaki (eds.), *Mutualism and Community Organization: Behavioral, Theoretical, and Food-Web Approaches*, pp. 365–411. New York: Oxford University Press.

Shapiro, J., V. Lamarra, and M. Lynch. 1975. Biomanipulation: An ecosystem approach to lake restoration. In P. L. Brezonik and J. L. Fox (eds.), *Proceedings of a Symposium on Water Quality Management through Biological Control*, pp. 85–96. Gainesville: University of Florida.

Shea, K., et al. 1998. Management of populations in conservation, harvesting and control. *Trends in Ecology and Evolution* 13: 371–375.

Shiomoto, A., K. Tadokoro, K. Nagasawa, and Y. Ishida. 1997. Trophic relations in the subarctic North Pacific ecosystem: Possible feeding effect from pink salmon. *Marine Ecology Progress Series* 150: 75–85.

Sih, A., P. Crowley, M. McPeek, J. Petranka, and K. Strohmeier. 1985. Predation, competition, and prey communities: A review. *Annual Review of Ecology and Systematics* 16: 269–311.

Soulé, M. E., E. T. Bolger, A. C. Alberts, R. A. Sauvajot, J. Wright, M. Sorice, and S. Hill. 1988. Reconstructed dynamics of rapid extinctions of chaparral-requiring birds in urban habitat islands. *Conservation Biology* 2: 75–92.

Spiller, D. A., and T. W. Schoener. 1994. Effects of top and intermediate predators in a terrestrial food web. *Ecology* 75: 182–196.

Strong, D. R. 1992. Are trophic cascades all wet? *Ecology* 73: 747–754.

Strong, D. R. 1999. Predator control in terrestrial ecosystems: The underground food chain of bush lupine. In H. Olff, V. K. Brown, and R. H. Drent (eds.), *Herbivores: Between Plants and Predators,* pp. 577–602. Oxford, U.K.: Blackwell Science.

Terborgh, J., J. A. Estes, P. Paquet, K. Ralls, D. Boyd-Heger, B. J. Miller, and R. F. Noss. 1999. The role of top carnivores in regulating terrestrial ecosystems. In M. E. Soulé and J. Terborgh (eds.), *Continental Conservation,* pp. 39–64. Washington, D.C.: Island Press.

Tilman, D. 1987. Secondary succession and the pattern of plant dominance along experimental nitrogen gradients. *Ecological Monographs* 57: 189–214.

Turner, R. E., and N. N. Rabalais. 1994. Coastal eutrophication near the Mississippi River delta. *Nature* 368: 619–621.

van der Elst, R. P. 1979. A proliferation of small sharks in the shore-based Natal sport fishery. *Environmental Biology of Fish* 4: 349–362.

Vitousek, P. M., H. A. Mooney, J. Lubchenco, and J. M. Melillo. 1997. Human domination of Earth's ecosystems. *Science* 277: 494–499.

Vitousek, P. M., and L. R. Walker. 1989. Biological invasion by *Myrica faya* in Hawaii: Plant demography, nitrogen fixation, and ecosystem effects. *Ecological Monographs* 59: 247–265.

Waage, J. K., and N. J. Mills. 1992. Biological control. In M. J. Crawley (ed.), *Natural Enemies—The Population Biology of Predators, Parasites and Diseases,* pp. 412–430. Oxford: Blackwell Scientific Publications.

Walters, C. J. 1986. *Adaptive Management of Renewable Resources.* New York: McMillan.

Walters, C. J., and J. J. Maguire. 1996. Lessons for stock assessment from the northern cod collapse. *Reviews in Fish Biology and Fisheries* 6: 125–137.

Wootton, J. T. 1993. Indirect effects and habitat use in an intertidal community: Interaction chains and interaction modifications. *American Naturalist* 141: 71–89.

Wootton, J. T. 1995. Effects of birds on sea urchins and algae: A lower-intertidal trophic cascade. *Ecoscience* 2: 321–328.

Wootton, J. T., and M. E. Power. 1993. Productivity, consumers, and the structure of a river food chain. *Proceedings of the National Academy of Science of the USA* 90: 1384–1387.

Yodzis, P. 1988. The indeterminacy of ecological interactions as perceived through perturbation experiments. *Ecology* 69: 508–515.

Yodzis, P. 1994. Predator-prey theory and management of multispecies fisheries. *Ecological Applications* 4: 51–58.

Yodzis, P. 1998. Local trophodynamics and the interaction of marine mammals and fisheries in the Benguela ecosystem. *Journal of Animal Ecology* 67: 635–658.

Yodzis, P. 2000. Diffuse effects in food webs. *Ecology* 81: 261–266.

Zaitsev, Yu. P. 1992. Recent changes in the trophic structure of the Black Sea. *Fisheries Oceanography* 1: 180–189.

Zaret, T. M. 1980. *Predation and Freshwater Communities.* New Haven: Yale University Press.

Zaret, T. M., and R. T. Paine. 1973. Species introduction in a tropical lake. *Science* 182: 449–455.

4

EXOTIC SPECIES AND CONSERVATION

Research Needs

Carla D'Antonio, Laura A. Meyerson, and Julie Denslow

OVER THE PAST TWO DECADES, EXOTIC INVASIVE SPECIES HAVE become recognized as an important cause of species declines and native habitat degradation (Vitousek et al. 1997; Wilcove et al. 1998). Although exotic species may increase species richness temporarily, over the long term they contribute to species extinction and therefore a decline in biological diversity. For years invasive exotic species were thought to be second only to land-use change in causing species extinctions (e.g., Soulé 1990; D'Antonio and Vitousek 1992), a claim now verified for fish (Miller et al. 1989), plants (D'Antonio and Dudley 1995), and threatened and endangered animal species in the United States (Wilcove et al. 1998). Approximately 60 percent of the species listed under the Endangered Species Act are threatened by invasive species (or fire suppression). Outside of the United States the proportion is estimated to reach 80 percent (Armstrong 1995; Wilcove et al. 1998). Crooks and Soulé (1999) predict that invasive species will soon become the leading cause of ecological degradation because of the increasing extent of disturbed lands, many of which are particularly vulnerable to domination by invasive nonindigenous species.

Exotic species threaten the persistence of native species assemblages because they can be predators, disease vectors, and competitors. They may so alter ecosystem processes that sustaining populations of native species or restoring ecosystem structure and function becomes difficult and expensive

(Vitousek et al. 1997). Although fewer than 20 percent of exotic species negatively affect native species or ecosystems (Simberloff 1981; U.S. Congress 1993; Williamson and Fitter 1996; D'Antonio and Haubensak 1998), this 20 percent can cause persistent changes to native biodiversity and ecosystem functioning. Species with this potential should receive the top priority for conservation and management efforts.

Despite increasing attention from conservationists and researchers, ability to predict establishment success and impact of nonindigenous species remains limited. In this chapter we suggest directions for research into the ecology of exotic species as they affect the conservation of native species and ecosystems. We focus on those issues for which quantitative biological research is needed to illuminate conservation challenges. For convenience, we divide research needs into (1) those addressing the ecology of invasive species, including pathways of introduction and factors affecting invasiveness, rates of spread, and impacts; and (2) those associated with their management, although we recognize the overlap. These research priorities are summarized in Box 4.1.

Factors Influencing Establishment and Spread

Trade and Pathways of Introduction

Humans have been a major vector of nonindigenous species from the beginnings of exploration, trade, and human migrations. Both the hulls and the holds of ships carry terrestrial and marine plants, animals, seeds, and disease between continents. Today, ballast water (the seawater a vessel takes on in one port, for ballast, and discharges upon arrival at another port) is an important source of introductions into marine and estuarine systems, carrying everything from cholera and botulism bacteria to invertebrates and fish. Estimates of the number of species carried in ballast water at any given moment range from 3,000 to 7,000 (Carlton 1999). We know little about invasion potentials and likely impacts of most of the species being transported.

Today, global trade, travel, and human migration have greatly increased rates of exotic introductions to all countries (U.S. Congress 1993; Mack et al. 2000). Although major pathways of introduction are well known, and many countries have established inspection and quarantine operations to monitor these routes for undesirable introductions, potentially dangerous organisms often arrive by previously unappreciated routes. We lack a comprehensive understanding of the many and varied routes through which introductions can occur.

Even when we are aware of potential pathways, we often do not have critical epidemiological information necessary to predict probabilities of establishment, such as the minimum numbers of individuals needed to establish a viable population and conditions promoting establishment and movement into native ecosystems. For example, numerous insects and plant pathogens have been found in association with wooden packing material, and the potential damage these species may cause to U.S. forests is well recognized (e.g., Wallner 1996). Much more information is needed on factors that affect establishment and spread of such species in order to develop more effective inspection and monitoring strategies. The risk of introduction is inherent in high trade volume, but we need to better evaluate acceptable levels of risk. Purposefully introduced species also have the potential to be directly problematic themselves or to be vectors of disease. Many are currently screened for associated disease organisms (e.g., White and Waterworth 1996), but the effectiveness of current screening and containment needs more research attention. Research on pathogen testing protocols could improve detection methods.

Undesirable exotic species also may be transported *within* a continent, greatly increasing the rate of spread, the exposure of potentially vulnerable ecosystems, and the magnitude of control problems. We need a better understanding of the relationships among pathway characteristics, frequency and abundance of propagule movements, and rates of invasive spread within continents to best direct policy and control strategies.

Predictors of Invasion and Spread

Species Traits and the Environment

Most exotic species do not become established in their new locales, although the small proportion that do become invasive may cause considerable ecological and economic damage. The attempt to identify common characteristics among those taxa that establish and spread on the one hand and among ecosystems vulnerable to invasions on the other has generated considerable debate. What kinds of species are most likely to invade particular types of ecosystems (Williamson and Fitter 1996; Levine and D'Antonio 1999; Lonsdale 1999; Newsome and Noble 1986; Stohlgren et al. 1999; Symstad 2000)? Because extirpation is close to impossible once an invasive has become established, ecologists are seeking better ways to predict which species are likely to become invasive and which ecosystems are most vulnerable to invasion (Ewel et al. 1999). Reliable predictors of invasibility are not yet available, in part because many models include species invading agricultural lands or other highly disturbed landscapes as well as those that invade natural areas (Parker and Reichard 1998). While establishment of many weedy plants is facilitated by disturbance, the process may differ for species that are invasive in native ecosystems (Hobbs and Huenneke 1992; D'Antonio et al. 1999). Likewise, the relative importance of ecosystem characteristics and disturbance in stimulating the establishment of animals is poorly known.

The most reliable determinant of potential invasibility into natural areas is whether a species has been invasive elsewhere. Reichard and Hamilton (1997) found that the best characteristic for predicting invasiveness among woody taxa in North America was whether the species had been reported to be invasive elsewhere. In the absence of a prior history of introductions, invasive potential must be evaluated based on a species' natural history or on the natural history of better-known congeners. However, congeneric species can differ greatly in their invasiveness, suggesting that the predictive capability of current models is not reliable. We need more information on traits correlated with successful establishment and spread and the circumstances under which they are likely to be important, particularly among closely related or ecologically similar species.

The history of intentional introductions of birds and biological control agents has shown that the number of times a species is introduced and the number of individuals per introduction attempt are the best predictors of successful establishment within the region (Newsome and Noble 1986; Hopper and Roush 1993; Crawley 1989; Veltman et al. 1996; Duncan 1997; Green 1997). Such intentional introductions generally have been made into highly modified ecosystems, like agricultural or urban landscapes. We have less information on correlates of success for exotic species entering native ecosystems. Nonetheless, a recent survey found that the rate of propagule arrival was an important determinant for the spread of exotic species into natural areas (Lonsdale 1999). A regular source of propagules greatly increases the likelihood of spread for many reasons. Small populations of colonists are at risk due to Allee effects, genetic bottlenecks, and the high likelihood of stochastic population extirpation (Williamson 1996). In addition to overcoming genetic or breeding problems, a high rate of propagule rain increases the chance of propagules encountering favorable habitats. D'Antonio et al. (in press) suggest that given high propagule supply rates and suitable climate conditions, virtually any ecosystem may be vulnerable to the establishment of exotic species. Research is needed to determine the circumstances under which factors contributing to population or community resistance can be overwhelmed by high rates of propagule arrival.

Although stochastic natural events can be responsible for establishment failure, they can also contribute to the success of an invader. For example, during the 1989–92 drought in England, river flow dropped sufficiently to allow the Asian mitten crab (*Eriocheir sinensis*), which had long been established at the river mouth, to migrate upriver and become established there. Ironically, even beneficial environmental changes, such as reduced pollution, can facilitate invasion by nonindigenous species that are present in low numbers. Populations of the wood-boring gribble (*Limnoria tripunctata*) exploded in the Long Beach–Los Angeles Harbor in the late 1960s because of pollution reduction, even though the isopod had been present in the harbor prior to the 1900s (Crooks and Soulé 1999 and references therein). Given widespread effects of chemical pollution, climate change, and alteration of nutrient cycles, it is essential that we develop models to anticipate movement of nonnative species across the landscape as a consequence of global change. This may be particularly important for disease organisms whose hosts (presumably native species) may be weakened by environmental stresses.

Until recently, low-diversity simple systems were thought to be most vulnerable to biological invasions (Elton 1958; Levine and D'Antonio 1999; Lonsdale 1999; Stohlgren et al. 1999). Recent analyses suggest, however, that invasions are more likely in regions where species diversity is high (Levine and D'Antonio 1999; Stohlgren et al. 1999; Levine 2000; Stadler et al. 2000). These results are contrary to theory, which holds that high species diversity confers resistance to invasion because resources are more fully used (see Levine and D'Antonio 1999 for a review). However, the same environmental conditions that favor diversity of native species (e.g., high resource availability) may also favor exotic species (Stohlgren et al. 1999; Levine and D'Antonio 1999; Levine 2000; Stadler et al. 2000).

As a result, two relatively new hypotheses have emerged: (1) The most diverse communities may in fact be the most vulnerable to invasion, and (2) species losses from diverse communities may lower their resistance to invasion. The correlation between high diversity and high invasibility suggests also that relatively resource-poor ecosystems are at less risk of invasion. Richardson et al. (2000) suggest that in resource-poor environments invasions may be facilitated by mutualisms either between native species and invaders or between previously established and newly arriving invaders. In parts of the tropics, low rates of invasion have been attributed to resource exhaustion and lack of mutualisms, rather than to competition or community resistance (Richardson et al. 2000; Stadler et al. 2000), but experimental evidence in support of these hypotheses is lacking.

The importance of diversity (and species interactions) to invasibility can vary with spatial scale (Levine and D'Antonio 1999; Planty-Tabacchi et al. 1996; Stohlgren et al. 1999; Symstad 2000; Levine 2000; Wiser et al. 1998). For example, Stohlgren et al. (1999) and Levine (2000) found that only at their subplot and plot levels were the most diverse ecosystems resistant to plant invasions. The relationship between neighborhood mechanisms and community patterns relative to invasions rarely has been examined.

Time Lags during Invasion

The detection of a biological invasion may be delayed considerably after the event because the initial population size is small and difficult to detect. In addition, some populations may exhibit an extended lag period before population growth becomes explosive (Crooks and Soulé 1999; Hobbs and Humphries 1995; Schmitz and Brown 1997; Mack et al. 2000). Despite the apparent commonness of this phenomenon, its causes are not well understood. Among the potential mechanisms contributing to such lags in population growth are slow intrinsic growth rates of the invader, the occurrence of environmental changes promoting more rapid growth rates, the occurrence of genetic changes leading to higher rates of reproduction (Crooks and Soulé 1999), and the continued introduction of new colonists.

For example, the explosive growth of *Melaleuca* into new Florida habitats may have occurred following a genetic mutation allowing broader environmental tolerance (Ewel 1986; Crooks and Soulé 1999). Richardson et al. (2000) suggested that a long lag phase may reflect the absence and subsequent establishment of a necessary symbiont. For example, pines did not become invasive in the southern hemisphere until human activities distributed spores of ectomycorrhizal fungi, which facilitated the naturalization of the pines (Richardson et al. 2000). Because many mutualisms are nonspecific, an invader may benefit from symbionts not found in its native habitat (Richardson et al. 2000). Other cases of seemingly sudden explosive growth can occur when grazing pressure is reduced by removal of herbivores. On Santa Cruz Island in Southern California, fennel (*Foeniculum vulgare*), an introduced European perennial species, was present but not widespread until introduced cattle and feral sheep were removed as part of a conservation effort. Upon removal of the grazers, fennel grew explosively and is now dominant on more than 10 percent of the island (Crooks and Soulé 1999). Despite the enormous potential impacts of introduced disease organisms on native species and ecosystems, we know almost nothing about rates of population increase after introduction or factors influencing apparent lags or the onset of exponential growth. Because control is more likely when populations are small, it is essential that we understand the nature of the lag phase and factors that affect its duration.

Importance of Genetic Diversity to Invasion

Since the classic publications of Baker and others (*Genetics of Colonizing Species,* Baker and Stebbins, 1965), there has been long-standing interest in genetic changes occurring in populations during expansion stages and the importance of genetic diversity to invasion. Several investigators have demonstrated that species with low genetic diversity can become widespread invaders (e.g., Raybould et al. 1991; Stiven and Arnold 1995), while others have found that high genetic diversity occurs in many widespread invaders (e.g., Novak et al. 1993; Demelo and Herbert 1994). More information is needed on the circumstances under which intraspecific genetic diversity promotes or restricts invasion and the relative importance of phenotypic plasticity in allowing widespread invasion in species populations with low genetic variability. Hybridization between invaders and closely related congeners can also accelerate invasion (Vila et al. in press; Daehler and Carino in press).

Impacts of Exotic Species on Native Species, Communities, and Ecosystems

Because exotic species can affect many properties of native populations, communities, and ecosystems, it is difficult to suggest simple and/or comparable measures of impact suitable for use in different ecosystems. Parker et al. (1999) argue that we need a common currency by which to assess species impacts so that investigators can more easily compare their findings and concerns, and species can be more easily prioritized for control. Here we consider effects of exotic species on the likely persistence of native species to be the primary currency of concern. Very few countries recognize ecological pests as species whose movements should be controlled, and species can pass unimpeded into many countries as long as they are not known to be agricultural pests. Quantitative information is badly needed on the potential for ecological damage of many species, particularly those that are being actively traded in the horticultural or pet trade and those that come along as hitchhikers.

Genetic Effects

Invasive, non-native species can affect native species populations through hybridization. Introgression of genes from non-native species can lead to the almost complete loss of the native gene pool. This kind of "genetic swamping" can lead to extinction of a rare species (Levin et al. 1996; Rhymer and Simberloff 1996). Fisheries studies provide a clear example of how this can occur at the subspecific level. Both exogenous and artificially reared Atlantic salmon (*Salmo salar*) have been released for more than a century (Hindar et al. 1991). These releases overwhelmingly have reduced fitness of native populations regardless of whether the introduced salmon were wild or cultured. Salmonid populations are considered to be genetically adapted to their local environments, and hybridization with cultured salmon has disrupted local gene pools (Hindar et al. 1991).

Numerous examples have been documented of this mechanism of loss or change within a native taxon, particularly in animal species (Miller et al. 1989; Rhymer and Simberloff 1996; Vila et al. 2000). Despite published information documenting such impacts, interspecific hybridization or hybridization among subspecies is generally ignored as a serious threat to native species. So-called native species from nonlocal gene pools are commonly used in restoration, revegetation, and landscaping projects. Yet we know little about the degree to which these represent a genetic threat to the persistence of native genotypes or subspecies in the surrounding landscape or the degree to which local adaptation of native species even occurs or can be disrupted. Knapp and Rice (1996, 1998) found that the two native grasses most commonly planted in restoration projects in California, *Nassella pulchra* and *Elymus glaucus,* show high among-site genetic heterogeneity and local adaptation. They also found that restoration practitioners were likely responsible for the movement of genotypes around the state to locales distant from their place of origin (E. Knapp, personal communication) and caution that seed collection zones may need to be very restricted for some species to avoid genetic contamination (Knapp and Rice 1996).

The extent to which these exotic genotypes will decrease the fitness of local genotypes is not known. In at least one instance (*Phragmites australis* in the eastern United States), the introduction of an exotic genotype has led to a native species becoming more invasive, with potentially undesired impacts on wildlife habitat (Crooks and Soulé 1999; Chambers et al. 1999). In many cases interspecific hybridization has been demonstrated to contribute to the formation of more invasive genotypes, which then replace both native species and the original invader (Thompson 1991; Ayres et al. 1999).

Population, Community, and Ecosystem Impacts

Invasive exotic species can cause the decline of native species directly through competition, predation, or disease, or by altering ecosystem processes such that native species begin to die out. Predicting species impacts depends on understanding how traits of the invading species operate under the circumstances of the recipient ecosystem. Impact quality and quantity are affected by the abundance of the invasive species, its particular traits (e.g., rates of food consumption or resource uptake), and characteristics of species in the invaded community. Very few studies quantify impacts of plant invaders, and no systematic review of species traits and their population or community-level impact has been carried out.

Most extinctions caused by species invasions have been due to the introduction of predators of a size (e.g., Nile perch in Lake Victoria) or a feeding type (e.g., snakes in Guam) with no historical precedence in the invaded ecosystem. Pathogenic organisms also have caused extinctions (e.g., avian malaria in Hawaii), presumably when native species lack appropriate resistance. By contrast, exotic species that merely compete with native species contribute to species declines (e.g., Daehler and Carino 1999) but appear less likely to cause extinctions by themselves (Frankel and Soulé 1981).

Vitousek (1990) suggested a conceptual framework for predicting when the addition or deletion of a species is likely to have an ecosystem impact. He predicted that species causing ecosystem impacts are those that (1) introduce a new trophic level to the system, (2) alter the rate of resource supply to the system, or (3) alter the disturbance regime. Similarly, Chapin et al. (1996) suggest that species affecting ecosystem processes have traits that are qualitatively different from the native species in the invaded sites. They and others (Mack and D'Antonio 2001) predict that invaders whose traits differ only quantitatively from native species will affect native species largely through competitive interactions. Some effects on ecosystem processes and community structure may take many years to be manifested. Selective comparisons—e.g., of the impacts of an invasive through a range of invaded habitats, or of the impacts of different kinds of invaders—will be particularly useful to elucidate these ideas.

In addition, we have an urgent need to understand how impacts at one trophic level will translate to other levels. Pathogens such as plant diseases and animal viruses can reduce the abundance of their hosts quickly (see chapter 8). If these species interact strongly with other species—e.g., are important sources of food or shelter—their elimination could be catastrophic for entire communities.

While there has been much debate about the relationship between species diversity and community susceptibility to invasion, there has been little discussion of how native diversity moderates the *impacts* of exotic species. On the basis of his observation that introduced species had large impacts on island ecosystems, Elton (1958) suggested that more simple (and therefore presumably less diverse) ecosystems were more likely to be affected strongly by nonnative species. Simberloff (1995) and D'Antonio and Dudley (1995) likewise found that extinctions caused by invasive exotic species were more common on island ecosystems originally lacking some functional groups. However, there have been few studies of the relationship between diversity and the impact of invaders in continental ecosystems, particularly where diversity within and among functional groups or trophic levels is manipulated.

If invading species interact with natives primarily through competition for limited resources, removal of the exotic should produce a compensatory increase in native populations if propagule supply is not limited. If so, then impacts of exotics should be reversible in the course of a restoration program. However, if invaders alter ecosystem processes such as disturbance regimes or soil processes, impacts may not be readily reversed. For example, the introduction of fire-enhancing grasses to semi-arid ecosystems has greatly increased fire frequency in many ecosystems (D'Antonio and Vitousek 1992). Grass populations are not controlled easily nor are grass-fire cycles easily interrupted. In addition, the changes caused by the altered disturbance regimes are not easily reversed. More research is needed to determine the types of impacts that are reversible and the relationship among traits of the invaders, likelihood of control, and reversibility of impacts.

Effects and Invasiveness of Genetically Modified Organisms

Genetically modified organisms (GMOs) are produced by the insertion of genes into or the removal of genes from a target organism to confer more desirable or delete undesirable traits. This technology has been used in place of traditional plant breeding programs, for example, to significantly decrease the development time for commercial varieties of crops or to increase resistance to herbivores (Paoletti and Pimentel 1996). Several ecologists have pointed out that the release of GMOs is analogous to the introduction of exotic species (Levin 1988; Parker and Kareiva 1996). An important question in terms of conservation biology is whether GMOs will invade natural ecosystems, hybridize with related species, and/or in some way threaten native populations and communities (Raybould and Gray 1994; Parker and Kareiva 1996; Beringer 2000; Hails 2000). Concern has also been raised that wild animal populations will be attracted to pollen produced by a genetically modified crop but will be killed by compounds such as insecticides that are produced by the crop (Poppy 2000). There are almost no ecological data to bring to bear on the controversy.

There has been considerable alarm among the public over the introduction of genetically modified organisms, particularly in Europe. While it is generally assumed that most intentional introductions of GMOs will be benign, it is also acknowledged that some risks exist, since novel genotypes will be created and introduced to environments that are new to them (Levin 1988). In the United States, companies that want to commercialize genetically modified crops are required to show that their product will not become more of a pest species than its unaltered counterpart. However, these requirements are not considered to be rigorous enough by some ecologists, since it can be very difficult to anticipate the invasiveness of particular species or genotypes (Purrington and Bergelson 1995) or to anticipate the rate of pollen movement between GMOs and wild relatives. Considerable doubt surrounds the successful prediction of invasiveness based on the examination of DNA sequences, especially because thus far progress has been limited in predicting invasiveness based on factors better understood than DNA, such as character traits (Purrington and Bergelson 1995; Bergelson 1994; but see Rejmanek 1996).

Experimental and observational work to date does suggest that GMO pollen will spread to wild populations (e.g., Timmons et al. 1995; Lefol et al. 1996), but the extent to which this will occur or is a threat to native populations is debated (Salisbury 2000; Wilkinson et al. 2000). The potential "weediness" of GMOs themselves is also controversial. In a set of experiments with *Arabidopsis thaliana,* Bergelson (1994) found that the genetically altered genotypes bred for herbicide resistance produced fewer seeds and thus had reduced fitness relative to wild-type genotypes. They were nonetheless equally as weedy as their wild relatives. She interpreted this outcome as being the result of population size being limited by something other than seed number. A broad conclusion from this work is that field experiments under a variety of conditions are necessary to ascertain the true ecological risks associated with GMOs (Bergelson 1994).

From an ecological point of view, a significant flaw in the testing of GMOs for commercial release has been the failure to determine whether these species will become invasive if they escape to natural areas. Such testing would require experimentation under a range of realistic field conditions, and this is rarely done. In addition, weeds that are already present in a system are equally likely to benefit from the new traits such as herbicide resistance, insect resistance, stress tolerance, and the ability to fix nitrogen. For example, there is the potential introgression of herbicide resistance to closely related species that are already considered "weedy," making management even more difficult (Hails 2000). Further, reciprocal hybrids of crops and wild species must be studied even if they do not exist in the country where the GMO originated, since some countries that receive GMO products may be home to the wild relatives, and many countries may lack sufficient regulations and/or resources for proper screening (Purrington and Bergelson 1995).

While genetically modified crops might have some environmental benefits because they reduce the need for the use of herbicides and pesticides (Beringer 2000), the topic remains highly controversial among ecologists and the public alike. At a minimum, more rigorous experimental testing, as well as limiting use of GMO crops to areas where no wild relatives are found, will help to reduce the risk of genetically engineered plant invaders (Barrett 2000).

Control of Invasive Species and Restoration of Communities and Ecosystems

Limited resources and lack of consensus by land managers, conservationists, and the general public mandate that we develop a better understanding of the biological and economic consequences of prevention, mitigation, and control measures against invasive species. Potential control mechanisms include import restrictions, mechanical and chemical control programs, and the use of other exotic species to reduce target species and restore environments and communities. Ultimately, the allocation of resources for invasive species management and the protocols through which such management is implemented are matters for public debate and resolution. In many cases, however, we lack both the factual and the theoretical basis with which to inform those discussions. We will need a broad understanding of the consequences of both action and inaction in the application of control protocols on which to base appropriate risk analyses.

Fundamental ecological information is necessary for developing priorities for exotic species removal, control, or use (see above). For example, development of reliable risk analyses will require more extensive research into the combination of species characteristics, community composition, and ecosystem processes that lead to explosive population growth; a better understanding of the impacts invasive species have on community and ecosystem characteristics; and research into the relative roles of propagule availability, resource supply, and community structure in determining establishment rates of both exotic and native species.

Such information will form the necessary foundation on which to develop priorities for exotic species management in natural ecosystems. Many naturalized exotic species may not threaten native species or compromise management goals (Williams 1996). In Hawaii, of more than nine hundred exotic species naturalized in native communities, less than 10 percent are seen as presenting a current threat to native species or ecosystem processes (Wester 1992). However, invasive species may persist at low population levels for many years before explosive population growth brings them to the attention of managers (Crooks and Soulé 1999), yet control is more easily effected while populations and ranges are small.

Even where ecological impacts of invasive species are substantial, their removal or control is likely to be only part of a long-term management program. Reestablishment of native communities is rarely accomplished by elimination of an invasive exotic. Seed stocks may be lost, disturbance regimes may be altered, and high resource availability after the disturbance associated with removal may favor establishment of other exotics. A substantial part of community structure may depend on timing and opportunity for establishment, and biotic structuring forces may be overwhelmed by seed inputs from exotic species. Removal or reduction of a dominant exotic may not reestablish critical ecosystem processes such as disturbance regimes (Adler et al. 1998; Cabin et al. 2000; Mack and D'Antonio 1998), soil chemistry (Macdonald and Richardson 1986; Vivrette and Muller 1977; D'Antonio 1990; Vitousek et al. 1987), and hydrology (Lacey et al. 1989). Experimental studies of the removal or reduction of exotic species would contribute to understanding of their impacts and of processes affecting subsequent community structure and ecosystem processes (Morrison 1997). Priorities for exotic control should include projections of the composition and structure of the post-removal community, yet the consequences of control are poorly understood and are often not anticipated. Under what circumstances does control or removal of an invasive species reverse its impact or lead to a less desirable condition?

Although exotic species are often introduced intentionally for habitat restoration and invasive species control, our understanding of the risks involved is limited. Cover crops are used as part of large-scale restoration efforts to slow environmental degradation, reduce erosion, improve soil structure and nutrient supply, moderate microclimates, and alter disturbance regimes (NRC 1993; Richardson 1999). Selected species are often exotics, because cultural information on native species may be wanting, seed supply may be insufficient, or native species may not possess the appropriate characteristics for the intended purposes. Although native species may establish under such cover crops (e.g., Parotta 1992; Lugo 1992), exotics also may retard early seedling establishment, persist as a semipermanent component of the landscape, and in some cases become invasive as well. *Leucaena leucocephala,* a nitrogen-fixing legume, was broadcast aerially over Guam and Saipan to control erosion following the devastating World War II bombardment; fifty years later, the northern portion of the island is dominated by this early successional species, and native forests are much restricted in distribution. Moreover, Richardson (1998, and references cited therein) offers evidence that tree species that cause the greatest problems as invasives are those that have been most widely planted for the longest period of time, suggesting that under appropriate circumstances most widely planted trees in alien environments are likely to become problematic.

Predators, herbivores, and pathogens from their native ranges may be introduced as control agents for invasive species. Biological control often offers the only viable option for reducing populations of widespread exotics, for which it provides a low-cost method with relatively low impact on native ecosystems (Odour 1999). Although the risk of adverse impact on nontarget species has been reduced from the early days of biological control, when introductions were accomplished with less research and the value of nontarget indigenous species was less appreciated (Simberloff and Stiling 1996; Howarth 1991), appropriate pre- and post-release research on impacts to native species and communities is relatively recent. The unique combinations of species, environment, and community that characterize any introduction of an exotic species make predicting the consequences difficult. Higher-order interactions (parasite-host interactions, pollination systems, trophic structures) are particularly difficult to predict in advance (Memmott 1999). A better understanding of the consequences of such introductions based on field and lab experiments in habitats of origin and introduction would provide a better basis for assessing risk.

BOX 4.1. **Research Priorities for Invasive, Non-Native Species and Their Potential Impacts on Natural Populations and Communities of Ecosystems**

Investigate Pathways of Introduction:

†1. What are the critical pathways of introduction of new species, and how do they differ in contributing harmful nonindigenous species? For example:

  †• Introduced plant pathogens can have devastating consequences for entire ecosystems. What are the most important pathways for their arrival, and how do they subsequently spread?

  †• Introduced insects may also strongly affect forested ecosystems and may carry pathogens. What are the most common pathways for harmful nonindigenous insect species to arrive in new locales, and how does the likelihood of their successful establishment scale with volume of trade?

†*2. What are acceptable levels of risk of entry of known potential invaders, how well do protocols established to prevent accidental introductions really work, and how can protocols be improved?

3. How do minimum viable population sizes of invaders vary among species, ecosystems, and establishment circumstances? Are there useful generalizations to be made here that might help development of monitoring and screening strategies?

†*4. Under what circumstances do intentional introductions for commercial purposes contribute to the introduction and spread of harmful invasive species? Can we develop reliable risk assessment protocols to screen intentional introductions for potential invaders, particularly harmful ones ?

Investigate the Processes of Invasion and Spread

1. What traits characterize species with high potential for rapid spread beyond their site of introduction?

2. What are the characteristics of natural communities that affect their resistance to invasion? How does propagule pressure interact with resistance, and under what circumstances can we expect propagule pressure to overwhelm resistance?

*3. How will the spread of non-native species be affected by other global changes, such as chemical pollution, climate change, altered disturbance regimes, and alteration of biogeochemical cycles? For example:

  • Will nitrogen deposition increase rates of plant invasions by favoring fast-growing non-native species?

  • Will changes in storm frequencies and intensities affect the persistence of native populations and potentially favor disturbance-loving exotic species?

  • Will increasing environmental stresses such as air- or water-borne pollutants make native species more susceptible to introduced diseases?

†4. Why are tropical ecosystems less invaded by nonindigenous species than their temperate counterparts? Will increasing fragmentation of tropical habitats and propagule pressure from exotic species alter this pattern?

5. What is the relationship between neighborhood-scale species interactions that affect invader success and landscape-level patterns of invasion and impact?

*6. Why are there often long time lags between establishment and the explosive growth and spread of introduced populations? Are there commonalities among species in their invasion patterns relative to the occurrence of time lags?

7. How does genetic diversity influence rates or patterns of invasion?

†8. How do human activities and cultural patterns—e.g., road construction, land-use patterns, traditional uses of plants, and visitation to reserves—affect the introduction and spread of invasive species?

Assess Impacts

†*1. What is the potential for introgression of introduced genes to native species, and under what circumstances is this likely to cause a change (either positive or negative) in fitness (and hence ecological performance)? How does the likelihood of such introgression vary among mating systems and life history characteristics of introduced taxa?

2. What traits of exotic species increase the danger of genetic threats to native species? What ecosystem characteristics are associated with high rates of genetic introgression?

3. Which species traits (or combinations thereof) are most likely to threaten local persistence of native species or create difficult-to-reverse impacts on ecosystem processes?

4. Does the arrival and establishment of one or a few non-native species influence the establishment of further alien species?

*5. How can knowledge of species traits be overlain or interfaced with ecosystem traits to predict species impact?

6. What kinds of higher-order effects—e.g., on other trophic levels or on community processes—are associated with interspecific interactions involving introduced species?

7. How do species richness, functional diversity, and trophic complexity influence the impact of an invader?

*8. Under what circumstances are impacts of an invasive species likely to be reversible? Are ecosystem effects longer lasting or farther reaching than competition or predation effects? Are impacts due to competition or predation more likely to cause population declines or extinction among native species?

(continues)

BOX 4.1. *(Continued)*

Consider Genetically Modified Organisms

†1. Under what circumstances might GMOs or their genes be able to spread beyond points of introduction?

†2. Under what circumstances might the spread of GMOs or their genes into wildland habitat pose a threat for native species or ecosystem structure and function?

†3. What criteria are needed to develop protocols for release and risk assessment associated with GMOs?

Study Control, Restoration, and Their Interactions

†1. How do we develop priorities for exotic species removal, control, or use?

2. Under what circumstances does control or removal of an invasive species lead to a less desirable condition?

3. Under what circumstances is the introduction of exotic species warranted for restoration or for biological control of invasive non-native species?

Note: An asterisk indicates a top research question. A dagger indicates a research priority that needs an answer within next ten years, or it will be too late for many species or natural communities.

Concluding Remarks

Native biological diversity is valued for many reasons. Its preservation is one of the major challenges of this century. Invasive nonindigenous species contribute to this challenge because a small but significant fraction of them interfere with native species and contribute to their demise on a local or even global scale. Some alter ecosystem processes such that there is a complete alteration of community composition, structure, and function. To reduce the risk of introduction and spread of harmful invaders and better manage existing nonindigenous species for the benefit of native biological diversity, we need greater access to basic information on their ecology and the ecology of the potential systems they might enter. In addition, we need greater coordination among agencies and institutions conducting that research and coordinated development of databases that can be made available to managers.

LITERATURE CITED

Adler, P., C. M. D'Antonio, and J. T. Tunison. 1998. Understory succession following a dieback of *Myrica faya* in Hawaii Volcanoes National Park. *Pacific Science* 52:69–78.

Armstrong, S. 1995. Rare plant protects Cape's water supplies. *New Scientist* (Feb. 11): 8.

Ayres, D. R., D. Garcia-Rossi, H. G. Davis, and D. R. Strong. 1999. Extent and degree of hybridization between exotic (*Spartina alterniflora*) and native (*S. foliosa*) cordgrass (Poaceae) in California, USA, determined by random amplified polymorphic DNA (RAPDs). *Molecular Ecology* 8:1179–1186.

Baker, H. G., and G. L. Stebbins. 1965. *The Genetics of Colonizing Species.* New York: Academic Press.

Barrett, S. C. H. 2000. Microevolutionary influences of global changes on plant invasions. In *Invasive Species in a Changing World,* edited by H. A. Mooney and R. J. Hobbs, pp. 115–139. Washington, D.C.: Island Press.

Bergelson, J. 1994. Changes in fecundity do not predict invasiveness: A model study of transgenic plants. *Ecology* 75:249–252.

Beringer, J. E. 2000. Releasing genetically modified organisms: Will any harm outweigh the advantage? *Journal of Applied Ecology* 37:207–214.

Cabin, R. J., S. Weller, D. Lorence, T. Flynn, D. A. Sakai, D. Sandquist, and L. Hadway. 2000. Effects of long-term ungulate exclusion and recent alien species control on the preservation and restoration of a Hawaiian tropical dry forest. *Conservation Biology* 14:439–453.

Carlton, J. T. 1999. The scale and ecological consequences of biological invasions in the world's oceans. In *Invasive Species and Biodiversity Management,* edited by O. T. Sandlund, P. J. Schei, and Å. Viken, pp. 195–212. New York: Kluwer Academic Publishers.

Chambers, R. M., L. A. Meyerson, and K. Saltonstall. 1999. Expansion of *Phragmites australis* into tidal wetlands of North America. *Aquatic Botany* 64:261–273.

Chapin, F. S. III, H. Reynolds, C. M. D'Antonio, and V. Eckhart. 1996. The functional role of species in terrestrial ecosystems. In *Global Change in Terrestrial Ecosystems,* edited by B. Walker and W. Steffan, pp. 403–430. Cambridge, U.K.: Cambridge University Press.

Crawley, M. J. 1989. Chance and timing in biological invasions. In *Biological Invasions: A Global Perspective,* edited by J. A. Drake et al., pp. 407–423. New York: John Wiley.

Crooks, J. A., and M. E. Soulé. 1999. Lag times in population explosions of invasive species: Causes and implications. In *Invasive Species and Biodiversity Management,* edited by O. T. Sandlund, P. J. Schei, and Å. Viken, pp. 103–125. New York: Kluwer Academic Publishers.

Daehler, C. C., and D. A. Carino. 1999. Threats of invasive plants to the conservation of biodiversity. In *Biodiversity and Allelopathy: From Organisms to Ecosystems in the Pacific,* edited by C. H. Chow et al., pp. 21–27. Taipei: Academia Sinica.

Daehler, C. C. and D. A. Carino. 2001. Hybridization between native and alien plants and its consequences. In *Biotic Homogenization: The Loss of Biodiversity through Extinction and Invasion,* edited by J. L. Lockwood and M. L. McKinney, pp. 85–106. New York: Kluwer Academic Publishers.

D'Antonio, C. M. 1990. *Invasion of Coastal Plant Communities by the Introduced Succulent,* Carpobrotus edulis *(Aizoaceae),* Ph.D. dissertation, University of California, Santa Barbara.

D'Antonio, C. M., and Dudley, T. L. 1995. Biological invasions as agents of change on islands versus mainlands. In *Islands: Biological Diversity and Ecosystem Function,* edited by P. M. Vitousek, L. Loope, and H. Adsersen, pp. 103–121. Berlin: Springer-Verlag.

D'Antonio, C. M., T. L. Dudley, and M. C. Mack. 1999. Disturbance and biological invasions: Direct effects and feedbacks. In *Ecosystems of Disturbed Ground,* edited by L. Walker, pp. 413–452. Oxford, U.K.: Elsevier.

D'Antonio, C. M., and K. A. Haubensak. 1998. Community and ecosystem impacts of introduced species. *Fremontia* 26:13–18.

D'Antonio, C. M., J. Levine, and M. Thomsen. In press. Ecosystem resistance to invasion and the role of propagule supply: A California perspective. *Journal of Mediterranean Environments* 1.

D'Antonio, C. M., and Vitousek, P. M. 1992. Biological invasions by exotic grasses, the grass/fire cycle, and global change. *Annual Review of Ecology and Systematics* 23:63–87.

Demelo, R., and P. D. M. Herbert. 1994. Founder effects and geographical variation in the invading cladoceran *Bosmina* (Eubosmina) *coregoni* (Baird 1857) in North America. *Heredity* 73:490–499.

Duncan, R. P. 1997. The role of competition and introduction effort in the success of passeriform birds introduced into New Zealand. *American Naturalist* 149:903–915.

Elton, C. S. 1958. *The Ecology of Invasions by Animals and Plants.* London: Methuen.

Ewel, J. 1986. Invasiblity: Lessons from south Florida. In *Ecology of Biological Invasions of North America and Hawai'i,* edited by H. A. Mooney and J. Drake, pp. 214–230. New York: Springer-Verlag.

Ewel, J. J., D. J. O'Dowd, and 19 others. 1999. Deliberate introductions of species: research needs. *Bioscience* 49:619–630.

Frankel, O. H., and M. E. Soulé. 1981. *Conservation and Evolution.* New York: Cambridge University Press.

Green, R. E. 1997. The influence of numbers released on the outcome of attempts to introduce exotic bird species to New Zealand. *Journal of Animal Ecology* 66:25–35.

Hails, R. S. 2000. Genetically modified plants—The debate continues. *Trends in Ecology and Evolution* 15: 14–18.

Hindar, K., N. Ryman, and F. Utter. 1991. Genetic effects of cultured fish on natural populations. *Canadian Journal of Fisheries and Aquatic Science* 48:945–957.

Hobbs, R., and L. F. Huenneke. 1992. Disturbance, diversity and invasion: Implications for conservation. *Conservation Biology* 6:324–337.

Hobbs, R., and S. E. Humphries. 1995. An integrated approach to the ecology and management of plant invasions. *Conservation Biology* 9:761–770.

Hopper, K. R., and R. T. Roush. 1993. Mate finding, dispersal, number released and the success of biological control introductions. *Ecological Entomology* 18:321–333.

Howarth, F. G. 1991. Environmental impacts of classical biological control. *Annual Review of Entomology* 36:485–509.

Knapp, E., and K. J. Rice. 1996. Genetic structure and gene flow in *Elymus glaucus* (blue wildrye): Implications for native grassland restoration. *Restoration Ecology* 4:1–10.

Knapp, E., and K. J. Rice. 1998. Comparison of isozymes and quantitative traits for evaluating patterns of genetic variation in purple needlegrass (*Nassella pulchra*). *Conservation Biology* 12:1031–1041.

Lacey, J. R., C. B. Marlow, and J. R. Lane. 1989. Influence of spotted knapweed (*Centaurea maculosa*) on surface runoff and sediment yield. *Weed Technology* 3:627–631.

Lefol, E., A. Fleury, and H. Darmency. 1996. Gene dispersal from transgenic crops: II. Hybridization between oilseed rape and the wild hoary mustard. *Sexual Plant Reproduction* 9:189–196.

Levin, D. J., J. Francisco-Ortega, and R. K. Jansen. 1996. Hybridization and the extinction of rare plant species. *Conservation Biology* 10:10–16.

Levin, S. A. 1988. Safety standards for the environmental release of genetically engineered organisms. *Trends in Ecology and Evolution* 3:S47–S49.

Levine, J. M. 2000. Species diversity and biological invasions: Relating local process to community pattern. *Science* 288:761–763.

Levine, J. M., and C. M. D'Antonio. 1999. Elton revisited: A review of evidence linking diversity and invasibility. *Oikos* 87:15–26.

Lonsdale, W. M. 1999. Global patterns of plant invasions and the concept of invasibility. *Ecology* 80:1522–1536.

Lugo, A. E. 1992. Tree plantations for rehabilitating damaged forest lands in the tropics. In *Ecosystem Rehabilitation: Preamble to Sustainable Development, Volume 2,* edited by M. K. Wali, pp. 247–255. The Hague, Netherlands: Academic Press.

Lugo, A. E. 1988. The future of the forest: Ecosystem rehabilitation in the tropics. *Environment* 30:41–45.

Macdonald, I. A. W., and D. M. Richardson. 1986. Alien species in terrestrial ecosystems of the fynbos biome. In *The Ecology and Management of Biological Invasions in Southern Africa,* edited by I. A. W. Macdonald et al., pp. 77–91, Cape Town, Republic of South Africa: Oxford University Press.

Mack, M., and C. M. D'Antonio. 1998. Impacts of biological invasions on disturbance regimes. *Trends in Ecology and Evolution* 13:195–198.

Mack, M., C. M. D'Antonio, and R. Ley. 2001. Pathways through which exotic grasses alter ecosystem nitrogen cycling in seasonally dry Hawaiian woodlands. *Ecological Application,* in press.

Mack, R. N., D. Simberloff, W. M. Lonsdale, H. Evans, and M. Clout. 2000. Biotic invasions: Causes, epidemiology, global consequences and control. *Ecological*

Applications 10:689–710.

Memmott, J. 1999. Food webs as a tool for studying nontarget effects in biological control. In *Non-Target Effects of Biocontrol,* edited by P. A. Follett and J. J. Duan, pp. 147–280. New York: Kluwer Academic Publishers.

Miller, R. R., J. D. Williams, and J. E. Williams. 1989. Extinctions of North American fishes during the past century. *Fisheries* 14: 22–38.

Morrison, M. L. 1997. Experimental design for plant removal and restoration. In *Assessment and Management of Plant Invasions,* edited by J. O. Luken and J. W. Thieret, pp. 104–117. New York: Springer.

Newsome, A. E., and I. R. Noble. 1986. Ecological and physiological characters of invading species. In *Ecology of Biological Invasions: An Australian Perspective,* edited by R. H. Groves and J. J. Burdon, pp. 1–20. New York: Cambridge University Press.

Novak, S. J., R. N. Mack, and P. S. Soltis. 1993. Genetic variation in *Bromus tectorum* (Poaceae): Introduction dynamics in North America. *Canadian Journal of Botany* 71:1441–1448.

NRC (National Research Council). 1993. *Vetiver Grass: A Thin Green Line Against Erosion.* Washington, D.C.: National Academy Press.

Odour, G. I. 1999. Biological pest control for alien invasive species. In *Invasive Species and Biodiversity Management,* edited by O. T. Sandlund, P. J. Schei, and Å. Viken, pp. 305–321. New York: Kluwer Academic Publishers.

Paoletti, M. G., and D. Pimentel. 1996. Genetic engineering in agriculture and the environment. *Bioscience* 46:665–673.

Parker, I. M., and P. Kareiva. 1996. Assessing the risks of invasion for genetically engineered plants: Acceptable evidence and reasonable doubt. *Biological Conservation* 78:193–203.

Parker, I. M., and S. H. Reichard. 1998. Critical issues in invasion biology for conservation science. In *Conservation Biology for the Coming Decade,* edited by P. L. Fielder and P. M. Kareiva, pp. 283–305. New York: Chapman and Hall.

Parker, I. M., D. Simberloff, W. M. Lonsdale, K. Goodell, M. Wonham, P. Kareiva, M. H. Williamson, B. Von Holle, P. Moyle, J. E. Byers, and L. Goldwasser. 1999. Impact: Towards a framework for understanding the ecological effects of invaders. *Biological Invasions* 1:3–19.

Parotta, J. A. 1992. The role of plantation forests in rehabilitating degraded tropical ecosystems. *Agricultural Ecosystems and the Environment* 41:115–133.

Planty-Tabacchi, A., E. Tabacchi, R. J. Naiman, C. DeFerrari, and H. DeCamps. 1996. Invasibility of species rich communities in riparian zones. *Conservation Biology* 10:598–607.

Poppy, G. 2000. GM crops: Environmental risks and non-target effects. *Trends in Plant Science* 5:5–9.

Purrington, C. B., and J. Bergelson. 1995. Assessing weediness of transgenic crops: Industry plays plant ecologist. *Trends in Ecology and Evolution* 10:340–342.

Raybould, A. F., and A. J. Gray. 1994. Will hybrids of genetically modified crops invade natural communities? *Trends in Ecology and Evolution* 9:85–89.

Raybould, A. F., A. J. Gray, M. J. Lawrence, and D. F. Marshall. 1991. The evolution of *Spartina anglica* CE Hubard (Gramineae): Genetic variation and status of the parental species in Britain. *Biological Journal of the Linnean Society* 44:369–380.

Reichard, S. H., and C. W. Hamilton. 1997. Predicting invasions of woody plants introduced into North America. *Conservation Biology* 11:193–203.

Rejmanek, M. 1996. What attributes make some plant species more invasive? *Ecology* 77:1655–1661.

Rhymer, J., and D. Simberloff. 1996. Extinction by hybridization and introgression. *Annual Review of Ecology and Systematics* 27: 83–109.

Richardson, D. M. 1999. Commercial forestry and agroforestry as sources of alien trees and shrubs. In *Invasive Species and Biodiversity Management,* edited by O. T. Sandlund, P. J. Schei, and Å. Viken, pp. 237–257. New York: Kluwer Academic Publishers.

Richardson, D. M. 1998. Pines as invaders in the southern hemisphere. In *Ecology Biogeography of Pines,* edited by D. M. Richardson, pp. 450–473. New York: Cambridge University Press.

Richardson, D. M., N. Allsopp, C. M. D'Antonio, S. Milton, and M. Rejmanek. 2000. Plant invasions—The role of mutualisms. *Biological Reviews* 75:65–93.

Salisbury, P. A. 2000. The myths of gene transfer: A canola case study. *Plant Protection Quarterly* 15:71–76.

Schmitz, C., and T. C. Brown. 1997. *Strangers in Paradise.* Washington D.C.: Island Press.

Simberloff, D. 1995. Why do introduced species appear to devastate islands more than mainland areas? *Pacific Science* 49:87–97.

Simberloff, D. 1981. Community effects of introduced species. In *Biotic Crises in Ecological and Evolutionary Time,* edited by M. H. Nitecki, pp. 53–71. New York: Academic Press.

Simberloff, D., and P. Stiling. 1996. Risks of species introduced for biological control. *Biological Conservation* 78:185–192.

Simberloff, D., and B. Von Holle. 1999. Positive interactions of non-indigenous species: Invasional meltdown. *Biological Invasions* 1: 21–32.

Soulé, M. E. 1990. The onslaught of alien species, and other challenges in the coming decades. *Conservation Biology* 4:233–239.

Stadler, J., A. Trefflich, S. Klotz, and R. Brandl. 2000. Exotic plant species invade diversity hot spots: The alien flora of northwestern Kenya. *Ecography* 23: 169–176.

Stiven, A. E., and G. L. Arnold. 1995. Phenotypic differentiation among four North Carolina populations of the exotic mussel *Corbicula fluminea* (Muller 1974). *Journal of the Elisha Mitchell Scientific Society* 111:103–115.

Stohlgren, T., D. Binkley, G. W. Chong, et al. 1999. Exotic plant species invade hot spots of native plant diversity. *Ecological Monographs* 69:25–46.

Symstad, A. J. 2000. A test of the effects of functional group richness and composition on grassland invasibility. *Ecology* 81: 99–109

Thompson, J. D. 1991. The biology of an invasive plant. *Bioscience* 41:393–401.

Timmons, A. M., E. T. O'Brien, Y. M. Charters, S. J. Dubbels, and M. J. Wilkinson. 1995. Assessing the risks of wind pollination from fields of genetically modified

Brassica napus spp. oleifera. *Euphytica* 85:417–423.

U.S. Congress, OTA (Office of Technology Assessment). 1993. *Harmful Non-Indigenous Species in the United States.* Washington, D.C.: U.S. Govt. Printing Office.

Veltman, C. J., S. Nee, and M. J. Crawley. 1996. Correlates of introduction success in exotic New Zealand birds." *American Naturalist* 147:542–557.

Vila, M., E. Weber, and C.M. D'Antonio. 2000. Conservation implications of invasion by plant hybridization. *Biological Invasions* 2:207–217.

Vitousek, P. M. 1990. Biological invasions and ecosystem processes: Towards an integration of population biology and ecosystem studies. *Oikos* 57:7013.

Vitousek, P., C. D'Antonio, L. Loope, M. Rejmanek, and R. Westbrooks. 1997. Introduced species: A significant component of human-caused global change. *New Zealand Journal of Ecology* 21:1–16.

Vitousek, P. M., L. Walker, L. Whiteaker, D. Mueller-Dombois, and P. Matson. 1987. Biological invasion by *Myrica faya* alters ecosystem development in Hawaii. *Science* 238:802–804.

Vivrette, N. J., and C. H. Muller. 1977. Mechanism of invasion and dominance of coastal grassland by *Mesembryanthemum crystallinum. Ecological Monographs* 47:301–318.

Wallner, W. E. 1996. Invasive pests ("biological pollutants") and US forests: Whose problem, who pays? *OEPP/EPPO Bulletin* 26:167–180.

Wester, L. 1992. Origin and distribution of adventive alien flowering plants in Hawaii. In *Alien Plant Invasions in Native Ecosystems of Hawai'i,* edited by C. P. Stone, C. W. Smith, and J. T. Tunison, pp. 99–155. Honolulu: University of Hawaii Cooperative National Park Resources Study Unit.

White, G. A. and H. E. Waterworth. 1996. International exchange of horticultural crop germplasm. *HortScience* 31:315–321.

Wilcove, D. S., D. Rothstein, J. Dubow, A. Phillips, and E. Losos. 1998. Quantifying threats to imperiled species in the United States. *Bioscience* 48:607–617.

Wilkinson, M. J., I. J. Davenport, Y. M. Charters, A. E. Jones, J. Allainguillaume, H. T. Butler, D. C. Mason, and A. F. Raybould. 2000. A direct regional scale estimate of transgene movement from genetically modified oilseed rape to its wild progenitors. *Molecular Ecology* 9:983–991.

Williams, C. E. 1996. Potential valuable ecological functions of non-indigenous plants. In *Assessment and Management of Plant Invasions,* edited by J. O. Luken and J. W. Thieret, pp. 26–36, New York: Springer.

Williamson, M. 1996. *Biological Invasions.* London: Chapman and Hall.

Williamson, M., and A. Fitter. 1996. The varying success of invaders. *Ecology* 77: 1661–1666.

Wiser, S. K., R. Allen, P. W. Clinton, and K. H. Platt. 1998. Community structure and forest invasion by an exotic herb over 23 years. *Ecology* 79:2071–2081.

Zedler, P. H., and G. A. Scheid. 1988. Invasion of *Carpobrotus edulis* and *Salix lasiolepis* after fire in a coastal chaparral site in Santa Barbara County, California. *Madroño* 35: 196–201.

5

HABITAT FRAGMENTATION

Consequences, Management, and Future Research Priorities

Kendi F. Davies, Claude Gascon, and Chris R. Margules

Change in land use and land cover, and the associated fragmentation of habitat, is one of the most pervasive effects of human activities on the face of the globe. Habitat destruction and fragmentation are the likely primary causes of the increase in the rate of extinction over recent decades (Henle et al. 1996). All measures of habitat destruction and fragmentation in all areas on earth today indicate a severe and accelerating problem (e.g., Whitmore 1997). Even large wilderness areas like the Amazon are becoming fragmented. In the Amazon, forest clearing increased exponentially during the 1970s and 1980s (Fearnside 1987) and continues at an alarming rate. This is significant because the tropics are highly diverse and relatively unknown. In Peninsular Malaysia, for example, there are over three thousand tree species of over 30 cm diameter, compared to fifty species indigenous to continental Europe north of the Alps and west of the Urals (Whitmore 1997).

Many temperate regions are already severely fragmented. The western Australian wheatbelt retains just 7 percent of the original woodland, heath, shrubland, and mallee as fragments spread throughout the landscape (Hobbs 1998). The wheatbelt roughly corresponds with the southwest botanical province of western Australia, an area of 300,000 square kilometers that is a recognized hot spot for floristic biodiversity, with about eight thousand species and 75 percent endemism (Myers et al. 2000; Hobbs 1998; Groombridge 1992).

Habitat fragmentation interrupts all ecological patterns and processes, from ecosystem functions and species interactions (Saunders et al. 1991; Kapos 1989) through species behavior and dispersal patterns (Hanski et al. 1995; Redpath 1995) to the genetic makeup of populations (Oostermeijer et al. 1996; Sarre 1995). The science of ecology is only beginning to come to terms with the impacts of this wide array of phenomena on the persistence or extinction of species. This review is an attempt to summarize what we do know and discuss the implications of that knowledge for the management of fragmented landscapes, and to identify the most pressing needs for future research in habitat fragmentation.

Consequences of Habitat Fragmentation

In this section we present the known effects of fragmentation, first on processes operating at the within-fragment spatial scale, and second on processes operating at the between-fragment spatial scale. Then we extend our discussion by recognizing that fragmented landscapes are dynamic and these known effects are not one-time events. On the contrary, they may be the start of changes that continue over longer time scales and have more far-reaching consequences than is first apparent. In the last part of this section we note that habitat remnants are all different and therefore contribute differentially to regional biodiversity goals.

Processes Operating at the Within-Fragment Scale

Isolation

Fragmentation can increase the risk of local extinction because remnant populations within fragments may be smaller and more isolated than they were in the unfragmented landscape. The factors that contribute to the extinction risk of small populations are: (1) environmental stochasticity, particularly the effect of extreme events or disturbance (e.g., drought, fire); and (2) deterministic threats (e.g., habitat degradation). Two other often-cited effects are probably of less importance but may help to "finish off" a declining population. These are: (3) demographic stochasticity, which affects only very small populations, and (4) loss of genetic variation, which acts relatively slowly in all but the smallest populations (Harrison and Taylor 1997).

Size and Edge Effects

Mounting empirical evidence suggests that habitat modification (a deterministic threat) may be one of the most significant consequences of habitat fragmentation for the persistence of species. Habitat is modified as a result of changes in physical conditions including fluxes of wind, water, solar radiation, fire, and disturbance (Saunders et al. 1991). This in turn can cause changes to vegetation structure (Laurance et al., *Effects,* 1998; Malcolm 1994), microclimate (Kapos 1989), and ground cover (Didham et al. 1998). However, not all species are affected by a given change in the same way. Often a change that negatively affects one species benefits another (e.g., Davies et al. 2000; Didham et al. 1998; Gascon 1993).

These physical changes to habitat are usually greatest at fragment edges, particularly in forests (e.g., Laurance et al., *Rain Forest,* 1998; Murcia 1995; Malcolm 1994). For example, the large-scale tropical forest fragmentation experiment at Manaus in Brazil recorded an overall reduction in forest biomass, a reduction of the overstory vegetation, and an increase in the understory vegetation of fragments (Laurance et al., *Effects,* 1998; Laurance et al. 1997; Malcolm 1994). Fragment edges were hotter, drier, and windier (Kapos et al. 1997; Kapos 1989), resulting in higher tree mortality at edges (Laurance et al., *Effects,* 1998, Ferreira and Laurance 1997) and increased rates of leaf fall near recent edges (Gascon and Lovejoy 1998; Lovejoy et al. 1986). Air temperature, litter biomass and moisture, and percentage ground cover of twigs were all correlated with distance from edges (Didham et al. 1998). In a similar large-scale forest fragmentation experiment in temperate eucalypt forest at Wog Wog in southeastern Australia, significantly higher counts of fallen logs were recorded near fragment edges, and higher levels of solar radiation were found to be reaching the forest floor. In addition, litter depth and percentage ground cover of leaves and grass were lower in fragments, while cover of bare ground and bark were higher in fragments than in continuous forest (Davies, unpublished data).

Changes in animal distribution follow these edge-induced physical changes to fragments. For example, in fragments of coastal scrub in Southern California, water runoff from surrounding developed areas and reductions in the cover of native vegetation near edges have both facilitated the invasion of the exotic Argentine ant. All three changes have significantly reduced the richness of the native ant fauna, particularly near edges (Saurez et al. 1998). At Manaus in Brazil, beetle species composition changed with both decreasing distance from fragment edges and decreasing fragment area. Environmental variables that changed with distance from fragment edges helped to explain the change in species composition. These variables included canopy height, air temperature, litter biomass, percentage ground cover of twigs, and the moisture content of litter (Didham et al. 1998). At Wog Wog, beetle species composition at small fragment edges was most different post-fragmentation from beetle composition in continuous forest, while beetle species composition at large fragment interiors was not different from that of continuous forest (Davies et al. 2001). These changes in composition were not related directly to the recorded physical changes to fragments, but changes in the abundance of some individual species were (Davies, unpublished data). Many other studies have hypothesized that physical changes to fragments at least partly explain species responses to fragmentation (e.g., Margules et al. 1994; Stouffer and Bierregaard 1995).

An important point that has been recognized about edge effects is that the edge effect at a site within a fragment is the sum of the influences from all points on all edges (Malcolm 1994; Laurance and Yensen 1991). This means that the size and shape of a fragment are important because, in fragments with large perimeter-to-area ratios, edge effects penetrate a larger proportion of the fragment. However, given a deeply penetrating edge effect, even very large fragments can be entirely physically modified by edge effects. It also means that populations on small fragments can be at risk, not only because populations become small but also because they are subjected to more signif- icant habitat modification. This makes management challenging, because the effects of habitat modification through edge effects are landscape specific, and guidelines must be developed for each situation. In contrast, a metapop- ulation model of the landscape that focuses on isolation and dispersal allows general guidelines to be developed, but if habitat modification is important, such guidelines may be inappropriate.

Processes Operating at the Between-Fragment Scale

Dispersal

Processes operating at the between-patch scale also affect the persistence of species. The core concept of both metapopulation theory (Levins 1969) and the equilibrium theory of island biogeography (MacArthur and Wilson 1967) is that within-patch dynamics are influenced not only by local extinc- tion but also by colonization from elsewhere in the landscape. As the land- scape becomes more fragmented, colonization rates are reduced. As a result, the number of species within a patch declines (MacArthur and Wilson 1967), and extinction risk at the scale of the entire landscape or metapopulation increases (Hanski 1994; Levins 1969).

The focus on metapopulation theory in fragmentation studies has cen- tered conservation ecologists' attention on the importance of dispersal and between-patch dynamics for persistence. However, although theoretical models demonstrate the importance of dispersal, empirical evidence from real landscapes suggests that local dynamics are at least as important over the short term. In a recent review of fragmentation studies, Harrison and Bruna (1999) concluded that the role of spatial dynamics in determining species responses to fragmentation might have been overemphasized. The studies reviewed suggest that local processes, like edge effects and physical changes, are the main cause of changes to the distribution and abundance of organisms in fragmented landscapes.

However, we must bear in mind that the effects of local processes may have been documented more often simply because those effects are easier to detect than are extinction, colonization, and dispersal dynamics. In addition, it is likely that local processes are important in the short term, but over longer time scales spatial dynamics play a greater role. Many of the studies in real landscapes have documented the effects of fragmentation on particular plant or animal species over only a few generations since fragmentation, whereas the role of spatial dynamics might come into play on time scales of many generations.

Few studies have tried to separate these two classes of mechanism: the importance of local deterministic changes like edge effects and the importance of spatial population dynamics. To address this issue, we need to ask: what kinds of species in what kinds of landscapes lead to important effects of spatial dynamics? An important consideration here is the correspondence between the spatial scale of a local population and the scale at which the landscape is fragmented. We use the term *local population* to refer to the spatial scale over which individuals are well mixed such that individuals can be considered to share the same location, and we use the term *metapopulation* to refer to a collection of local populations (Hanski and Simberloff 1997). At one extreme, the scale of a local population is larger than a single fragment. Thus, individual fragments support parts of a local population but not an entire local population (i.e., a "patchy population," Harrison 1991; e.g., Diffendorfer et al.1995; Andreassen et al. 1998; Andreassen and Ims 1998). An intermediate match of scales is where one fragment supports a well-mixed local population, and persistence at the landscape scale depends on dispersal among fragments. This is the scale at which metapopulation theory is most often applied to fragmented landscapes (e.g., Hanski 1994; Hanski et al. 1995). At the other extreme, if the scale of a local population is smaller than a fragment, then fragments may be internally patchy, supporting many populations (a metapopulation within fragments). Then, recolonization from within fragments increases persistence at both fragment and landscape scales, even when between-fragment dispersal is low (Holt 1992).

The Matrix

A feature that distinguishes fragmented landscapes from simple island-sea models is that the suitability of the matrix of disturbed or converted lands, in which habitat fragments are embedded, varies among species. The matrix, therefore, has a strong influence on between-patch processes (Holt and Gaines 1993) and has at least four potential roles:

1. The matrix can alter dispersal and colonization rates, which may be reduced or enhanced. Movement through the matrix is more likely when the vegetation structure of the matrix is similar to the fragments (e.g., Stouffer and Bierregaard 1995; Pearson 1993).

2. The matrix can provide alternative habitat for the original species (e.g., Whitmore 1997; Foster and Gaines 1992) and generalist species in particular (Harrison and Bruna 1999; Holt 1997).

Putting roles 1 and 2 together, empirical evidence shows that those species that use matrix habitat surrounding fragments for dispersal or as additional habitat are less vulnerable in fragmented landscapes than those restricted to remnants of wildlands. Gascon et al. (1999) found those species of birds, small mammals, and frogs that were able to tolerate or exploit the matrix were less vulnerable to fragmentation, while species that avoided the matrix tended to decline or disappear. Davies et al. (2000) also found that the group of beetle species that were never detected in the matrix declined in abundance in fragments relative to continuous forest controls, while the group of species that was abundant in matrix habitat increased in abundance in fragments.

3. The matrix may provide suitable habitat for exotic (non-native) species and therefore act as a source of species invading habitat fragments (Gascon et al. 1999; Saurez et al. 1998; Whitmore 1997; Fahrig and Merriam 1994; Halme and Niemelä 1993). Invasions by exotics have frequently been implicated as the cause of ecological change in fragments. There are many examples of highly competitive weed species displacing the original vegetation (e.g., Viana et al. 1997; McIntyre and Lavoral 1994; Parsons 1972) and of generalist predators causing declines in birds and mammals (reviewed in Tewksbury et al. 1998; Bolger et al. 1997; Soulé et al. 1988).

4. The matrix vegetation type and its structure can determine the severity of edge effects. For example, Amazon rain forest fragments surrounded by pasture had higher tree mortality rates near edges than those surrounded by regrowth forest, and this edge effect penetrated farther in those fragments surrounded by pasture (Mesquita et al. 1999).

Fragmented Landscapes Are Dynamic

To summarize this discussion of consequences so far, we now recognize that fragments are not simply cutout pieces of the original continuous habitat. Physical conditions within fragments are likely to have been altered. We know that for many species, these local alterations may be as important in determining persistence in fragmented landscapes as are processes operating at the between-patch scale, such as dispersal, at least in the short term. However, it is time we recognized that fragmented landscapes are dynamic and that habitat and physical conditions in fragments are likely to change over time scales longer than the initial fragmentation process (Soulé et al. 1988). We have identified three classes of mechanism that can cause fragments to change in time: loops of positive feedback, cumulative impacts, and cascades of events.

Mechanisms

POSITIVE FEEDBACK LOOPS

Feedback loops place fragments at risk of becoming significantly altered or vanishing altogether. For example, in the Amazon rain forests of Brazil, fragments are embedded in a matrix of sugar cane and eucalyptus plantations. Fire, which is used for management in the matrix, is becoming an important element in the dynamics of these landscapes. Fires encroaching into rain forest fragments create a positive feedback loop (Cochrane et al. 1999). Low-intensity fires kill most contacted rain forest trees because of their characteristically thin bark. This increases the fuel load on the forest floor, which in turn dries out because the canopy becomes more open. Thus, after an initial low-intensity fire, fragments become more susceptible to new fires. In this way, primary forest is readily replaced by secondary species that are in turn eliminated and replaced by exotics that are resistant to fire (Laurance et al., *Effects,* 1998; Viana et al. 1997). In addition, edges can act as a conduit to carry fire into the margins of the primary forest, because invading grasses and weedy vines add highly combustible live fuels to the already high fuel loads (Cochrane et al. 1999). The increasing incidence of fire in this landscape may be sufficient to cause degeneration of forest at fragment edges, leading to receding edges of primary rain forest and the potential for fragments to vanish altogether (Gascon et al. 2000).

In another example, in the wheatbelt of western Australian, the water table is rising because of widespread clearing of the native woodland, heath, shrubland, and mallee for planting to wheat (Hobbs 1993). This region has one of the highest levels of endemism in the world, but this diversity now persists in remnants scattered across the agricultural landscape. The rising water table causes waterlogging but also brings with it salt stored at depth, causing salinization at the soil surface. Even given the best intentions of conservation managers, the rising water table ultimately threatens those remnants in the lower parts of the landscape (Hobbs 1993). As the vegetation on those remnants is killed, the water table may rise even further, threatening, in turn, a new set of remnants. Without active tree planting in key parts of the landscape, the unique flora and fauna of this region is at great risk.

CUMULATIVE IMPACTS

A cumulative impact can degrade a fragment over time. In the western Australian wheatbelt, the addition of fertilizer in combination with physical disturbance (from sheep grazing) increases both the rate of establishment and the growth of weed species at the edges of remnants (Hobbs and Atkins 1988). The woodland soils are naturally nutrient poor, but fertilizer is applied to wheat crops and nutrient is readily transferred across remnant edges by wind and by defecating livestock (Hobbs 1993; Scougall et al. 1993). Thus, the cumulative addition of nutrients may result in increasingly severe weed invasion, reducing the recruitment of native species (Hobbs 1991) and resulting in the gradual degradation of woodland fragments. In southeastern Australia, the buildup of nutrients in woodland fragments has led to increases in the abundance of defoliating insects, which can eventually kill trees, resulting in the gradual degradation and loss of woodland fragments in those agricultural landscapes (Yates and Hobbs 1997; Landsberg et al. 1990).

CASCADES OF EVENTS

Some of the clearest examples of cascades of events due to fragmentation are trophic cascades (see chapter 3). In a trophic cascade, changes in the presence or abundance of a species at one trophic level in turn affect species at other trophic levels (Carpenter et al. 1985). For example, in coastal Southern California, sage-scrub habitat remains in steep-sided canyons embedded in an urban matrix. The decline and disappearance of the top predator, the coyote, from smaller habitat fragments have resulted in elevated numbers and activity of mesopredators that exert stronger predation pressure on native bird species. As a result, those fragments with coyotes have retained more bird species (Crooks and Soulé 1999).

Trophic cascades can also occur when a species is added to a fragmented landscape or increases in abundance. For example, in the midwestern United States, nest predation and parasitism by cowbirds is greater in forest landscapes that have been more severely fragmented. This has affected the reproductive rates of some neotropical migrant birds so severely that their populations have become sinks, dependent on immigration from populations in landscapes with more extensive forest cover (Robinson et al. 1995).

Another example in which fragmentation has resulted in changes to ecological relationships among species is the reduced rates of pollination observed in fragments of dry Chaco forest in Argentina (Aizen and

Feinsinger 1994) and of rain forest in Manaus (Powell and Powell 1987). Over time, the altered pollination regime could have a cascading effect that leads to the loss of plant species and in turn to their associated biota. Since many tree species live for centuries, we might not see such cascading effects for some time (Whitmore 1997).

Apart from trophic cascades, there is also potential for cascades to be set off when biophysical processes are altered. For example, in the wheatbelt of western Australia, there is evidence that the extent of land clearing and the resulting change in albedo has changed the rainfall pattern: there is now less rain over areas dominated by the agricultural matrix and more rain over areas that are still largely covered by native vegetation (Hobbs 1998). With reduced annual rainfall in remnants, it is easy to imagine that recruitment of some plant species may decline, eventually resulting in the loss of those species. In turn, animals that rely on them as food sources (e.g., herbivores, frugivores, seed eaters) may be threatened, and so on.

There are now many other examples in which ecological function has been impaired by fragmentation. It is easy to imagine that these losses of function could set off cascades of events in time. For example, Klein (1989) showed that fragmentation reduced numbers of dung beetles, resulting in slower decomposition rates of organic matter in small fragments compared to continuous forest. Accumulating organic matter on the forest floor could have cascading effects that place those fragments on a trajectory of change. For example, fires may become more severe from the increased fuel load, resulting in higher tree mortality and loss of species dependent on trees.

Representativeness and Complementarity

As a final point, we recognize that original continuous habitat is always spatially heterogeneous, as plant and animal communities turn over in the landscape in response to environmental gradients or patchiness. This means that within a fragmented landscape, the biota of remnants will be different from one another to some extent, not just miniature versions of a previous continuous habitat (Laurance et al. 1999). Thus, each habitat remnant will contribute different species to make up the full complement of species of the fragmented landscape under consideration (Brooker and Margules 1996). The same principle applies more generally to biodiversity attributes, such as genetic or ecosystem diversity. The contribution of remnants should be measured and priorities then set for protection and the allocation of scarce management resources, according to the magnitude of that contribution (Margules and Nicholls 1987; see also, e.g., Margules 1999; Pressey et al. 1993).

BOX 5.1. **Research Priorities on Habitat Fragmentation**

Investigate Within-Fragment Effects

*1. What is the relative importance, for the local persistence of species in frag-
ments, of threats due to fragment size caused by either habitat modification
(edge effects and other deterministic threats) or small population size (i.e., local
extinction due to environmental and demographic stochasticity, genetic deterioration)?

 2. How does the relative importance of these threats (habitat modification versus
small population size) vary given:
   • the type of ecosystem (e.g., forest, grasslands)?
   • the type of matrix (e.g., is the structure of the matrix similar to or different
     from that of the fragments)?
   • the degree of fragmentation (e.g., does one or the other play a greater role in
     landscapes that are highly fragmented)?
   • the size of a fragment relative to the spatial scale of a local, focal species
     population?
   • time since fragmentation?

 3. In cases where both habitat modification and small population size affect persistence, are their effects additive or multiplicative (synergistic)? Does this
depend on the items listed above?

Investigate Between-Fragment Effects

*1. What is the relative importance of either within-fragment effects (habitat
modification and small population size) or between-fragment effects (dispersal) in determining species persistence over the short term (a couple of generations) and over the long term (many generations)?

 2. How does the importance of within-fragment effects and between-fragment
effects differ in different types of organisms (with respect to traits such as
trophic group, rarity, dispersal strategy, habitat specialization) and in different
types of landscapes?

 3. How important are metapopulation dynamics in populations that have been
anthropogenically fragmented?

 4. Are metapopulation dynamics more important for species persistence in fragmented landscapes, in landscapes that have been fragmented for long periods?

*5. How important in determining species persistence in fragmented landscapes
is the match of the spatial scale at which populations of a given species are
structured and the spatial scale at which the landscape is fragmented?

 6. To understand how populations of a given species are structured, we need to
know the scale of a local population and the scale of dispersal. To answer 5, we
need to study this for many species in many different types of landscapes.

7. How general is the finding that species that can persist in the matrix habitat are more likely to persist, or even increase in abundance, on fragments? (As far as we are aware, this has been found only for forest.)

*8. Do those species that can persist in the matrix persist or increase in abundance on fragments at the detriment of other species, through interactions?

9. If yes, how important (strong) are those interactions, and does it depend on the type of landscape and/or how similar the matrix habitat is in structure to the fragment habitat?

10. In which landscapes does the matrix facilitate invasions into fragments: When the matrix habitat is similar or different from fragment habitat? When the matrix is intensively managed or not?

Investigate the Dynamics of Species Loss

*1. Are there dynamic processes that are causing loss of biodiversity (i.e., positive feedback loops, cumulative impacts, cascades)?

*2. Can we learn anything from synthesizing the findings of existing studies that describe dynamics in fragmented landscapes? Are fragments more likely to deteriorate in particular types of landscapes? Does it depend on the harshness of the matrix, on how intensively the matrix is managed?

3. Are there useful landscape-level indicators for detecting the sustainability or deterioration of fragments within landscapes (e.g., persistence of a range of successional stages or vegetation classes, degree of connectivity)?

4. In particular, determine whether any of these landscape-level indicators have a relationship with trends in populations of individual species in the landscape. If so, for which types of species (what traits do they have in common)? And which, if any, indicator is most effective in which types of landscapes?

Investigate Landscape-Level Strategies

1. What is the best method for making the tradeoff between, on the one hand, preserving important components of biodiversity represented only in small fragments that will need to be intensively managed to prevent them from deteriorating, and, on the other hand, investing scarce management resources in large fragments that may not be representative of regional biotas but are more secure?

2. Where managing small fragments is considered cost effective, what are the best management options?
   • Should we make them larger, or surround them by a buffer zone, and if the latter, what should the physical characteristics of the buffer zone be?
   • How does the particular landscape affect strategies?

Note: Asterisks denote the highest priorities for the current decade.

Research Priorities for the Conservation of Biodiversity

In a research program designed to promote the conservation of biodiversity in fragmented landscapes, we need to develop an understanding of within-fragment, or local, effects because they seem to manifest first. However, understanding those effects is not necessarily of general interest to ecologists because that knowledge may not contribute to general ecological principles. Rather, those effects tend to be system specific and require research that increases local knowledge of natural history. For example, in woodland remnants in the western Australian wheatbelt, there is little or no regeneration of the dominant tree species, *Eucalyptus salmonophloia*. Only detailed experiments are helping to determine which factors, of many candidates, are responsible (e.g., soil compaction from livestock grazing, competition with weeds) (Yates et al. 2000). However, this knowledge probably will not help to build generalizations that are transferable to other systems. Nonetheless, this type of research is critical. In box 5.1 we have tended to emphasize research priorities that build general principles because it is not possible to prescribe system-specific research. However, system-specific research is also essential.

How Should We Proceed?

We advocate the use of well-designed field experiments. We also encourage the intelligent use of natural experiments and comparative studies to increase their explanatory power and extend their inference to greater temporal and spatial scales than are approachable through experimental methods. To date, experiments have provided us with the clearest answers, but their spatial and temporal scales can be limited.

It is important to be able to approach the unplanned or natural "experiments" in such a way that clear conclusions can be drawn. In unplanned fragmented landscapes, the opportunities to incorporate key elements of experimental design, such as controls and prefragmentation data, are usually limited. Then, the way forward is through a combination of observation and fitting of specific models (Steinberg and Kareiva 1997). Thus, perhaps the most important issue for studies in unplanned fragmented landscapes is the formulation of adequate models, which should consider effects of the matrix, population structure within and between fragments, and habitat modification.

ACKNOWLEDGMENTS

The manuscript was greatly improved by the comments of Doug Bolger, Brett Melbourne, Michael Soulé, and Richard Hobbs. We also thank Brett Melbourne and Richard Hobbs for discussions, which helped to clarify our ideas.

LITERATURE CITED

Aizen, M.A., and P. Feinsinger. 1994. Forest fragmentation, pollination and plant reproduction in a Chaco dry forest, Argentina. *Ecology* 75: 330–351.

Andreassen, H.P., K. Hertzberg, and R.A. Ims. 1998. Space use responses to habitat fragmentation and connectivity in the root vole *Microtus oeconomus*. *Ecology* 79: 1223–1235.

Andreassen, H.P., and R. A. Ims. 1998. The effects of experimental habitat destruction and patch isolation on space use and fitness parameters in female root vole *Microtus oeconomus*. *Journal of Animal Ecology* 67: 941–952.

Bolger, D.T., A.C. Alberts, R.M. Sauvajot, P. Potenza, C. McCalvin, D. Tran, S. Mazzoni, and M.E. Soulé. 1997. Response of rodents to habitat fragmentation in coastal Southern California. *Ecological Applications* 7: 552–563.

Brooker, M.G., and C.R. Margules. 1996. The relative conservation value of remnant patches of native vegetation in the wheatbelt of western Australia: I. Plant diversity. *Pacific Conservation Biology* 2: 268–278.

Carpenter, S.R., J.F. Kitchell, and J.R. Hogson. 1985. Cascading trophic interactions and lake productivity. *Bioscience* 35: 634–639.

Cochrane, M.A., A. Alencar, M.D. Schulze, C.M. Souza, D.C. Nepstad, P. Lefebvre, and E.A. Davidson. 1999. Positive feedbacks in the fire dynamic of closed canopy tropical forests. *Science* 284: 1832–1835.

Crooks, K.R., and M.E. Soulé. 1999. Mesopredator release and avifaunal extinctions in a fragmented system. *Science* 400: 563–566.

Davies, K.F., and C.R. Margules. 1998. Effects of habitat fragmentation on carabid beetles: Experimental evidence. *Journal of Animal Ecology* 67: 460–471.

Davies, K.F., C.R. Margules, and J.F. Lawrence. 2000. Which traits of species predict population declines in experimental forest fragments? *Ecology* 81: 1450–1461.

Davies, K.F., B.A. Melbourne, and C.R. Margules. 2001. Effects of within and between-patch processes on beetle-community dynamics in experimentally fragmented forest. *Ecology:* in press.

Didham, R.K., P.M. Hammond, J.H. Lawton, P. Eggleton, and N.E. Stork. 1998. Beetle species responses to tropical forest fragmentation. *Ecological Monographs* 68: 295–323.

Diffendorfer, J.E., M.S. Gaines, and R.D. Holt. 1995. Habitat fragmentation and movements of three small mammals (*Sigmodon, Microtus,* and *Peromyscus*). *Ecology* 76: 827–839.

Fahrig, L., and G. Merriam. 1994. Conservation of fragmented populations. *Conservation Biology* 8: 50–59.

Fearnside, P.M. 1987. Deforestation and international economic development projects in Brazilian Amazon. *Conservation Biology* 1: 214.

Ferreira, L.V., and W.F. Laurance. 1997. Effects of forest fragmentation on mortality and damage of selected trees in central Amazonia. *Conservation Biology* 11: 797–801.

Foster, J., and M.S. Gaines. 1992. The effects of a successional habitat mosaic on a small mammal community. *Ecology* 72: 1358–1373.

Gascon, C. 1993. Breeding-habitat use by five Amazonian frogs at forest edges. *Biodiversity and Conservation* 2: 438–444.

Gascon, C., and T.E. Lovejoy. 1998. Ecological impacts of forest fragmentation in central Amazonia. *Zoology, Analysis of Complex Systems* 101: 273–280.

Gascon, C., T.E. Lovejoy, R.O.J. Bierregaard, J.R. Malcolm, P.C. Stouffer, H. Vasconcelos, W.F. Laurance, B. Zimmerman, M. Tocher, and S. Borges. 1999. Matrix habitat and species persistence in tropical forest remnants. *Biological Conservation* 91: 223–229.

Gascon, C., G.B. Williamson, and G.A.B. da Fonseca. 2000. Receding forest edges and vanishing reserves. *Science* 288: 1356–1358.

Groombridge, B. 1992. *Global Biodiversity Status of the Earth's Living Resources.* London: Chapman and Hall.

Halme, E., and J. Niemelä. 1993. Carabid beetles in fragments of coniferous forest. *Annales Zoologica Fennici* 30: 17–30.

Hanski, I. 1994. Patch-occupancy dynamics in fragmented landscapes. *Trends in Ecology and Evolution* 9: 131–135.

Hanski, I., T. Pakkala, M. Kuussaari, and G. Lei. 1995. Metapopulation persistence of an endangered butterfly in a fragmented landscape. *Oikos* 72: 21–28.

Hanski, I.A., and D. Simberloff. 1997. The metapopulation approach, its history, conceptual domain, and application to conservation. In *Metapopulation Biology: Ecology, Genetics and Evolution,* eds. I. A. Hanski and M. E. Gilpin, pp. 5–26. San Diego: Academic Press.

Harrison, S. 1991. Local extinction in a metapopulation context: An empirical evaluation. *Biological Journal of the Linnean Society* 42: 73–88.

Harrison, S., and E. Bruna. 1999. Habitat fragmentation and large-scale conservation: What do we know for sure? *Ecography* 22: 225–232.

Harrison, S., and A.D. Taylor. 1997. Empirical evidence for metapopulation dynamics. In *Metapopulation Biology: Ecology, Genetics and Evolution,* eds. I.A. Hanski and M.E. Gilpin, pp. 27–42. San Diego: Academic Press.

Henle, K., P. Poschlod, C. Margules, and J. Settele. 1996. Species survival in relation to habitat quality, size and isolation: Summary conclusions and future directions. In *Species Survival in Fragmented Landscapes,* edited by J. Settele, C. Margules, and P. Poschlod, pp. 373–381. Dordrecht: Kluwer Academic Publishers,

Hobbs, R.J. 1991. Disturbance a precursor to weed invasion in native vegetation. *Plant Protection Quarterly* 6: 99–104.

Hobbs, R.J. 1993. Effects of landscape fragmentation on ecosystem processes in the western Australian wheatbelt. *Biological Conservation* 64: 193–201.

Hobbs, R. 1998. Impacts of land use on biodiversity in southwestern Australia. In *Landscape Disturbance and Biodiversity in Mediterranean-Type Ecosystems,* eds. P.W. Rundel, G. Montenegro, and F.M. Jaksic, pp. 81–106. Berlin: Springer-Verlag.

Hobbs, R.J., and L. Atkins. 1988. The effect of disturbance and nutrient addition on native and introduced annuals in the western Australian wheatbelt. *Australian Journal of Ecology* 13: 171–179.

Holt, R.D. 1992. A neglected facet of island biogeography: The role of internal spatial dynamics in area effects. *Theoretical Population Biology* 41: 354–371.

Holt, R.D. 1997. From metapopulation dynamics to community structure: Some consequences of spatial heterogeneity. In *Metapopulation Biology: Ecology, Genetics and Evolution,* eds. I.A. Hanski and M.E. Gilpin, pp. 149–164. San Diego: Academic Press.

Holt, R.D, and M.S. Gaines. 1993. The influence of regional processes on local communities: Examples from an experimentally fragmented landscape. In *Patch Dynamics,* eds. S.A. Levin, T.M. Powell, and J.H. Steele, pp. 260–276. New York: Springer-Verlag.

Kapos, V. 1989. Effects of isolation on the water status of forest patches in the Brazilian Amazon. *Journal of Tropical Ecology* 5: 173–185.

Kapos, V., E. Wandelli, J.L. Camargo, and G. Ganade. 1997. Edge-related changes in environment and plant responses due to forest fragmentation in central Amazonia. In *Tropical Forest Remnants: Ecology, Management, and Conservation of Fragmented Communities,* eds. W.F. Laurence and R.O. Bierregaard, pp. 33–44. Chicago: University of Chicago Press.

Klein, B.C. 1989. Effects of forest fragmentation on dung and carrion beetle communities in central Amazonia. *Ecology* 70: 1715–1725.

Landsberg, J., J. Morse, and P. Khanna. 1990. Tree dieback and insect dynamics in remnants of native woodlands on farms. *Proceedings of the Ecological Society of Australia* 16: 149–165.

Laurance, W.F., L.V. Ferreira, J.M. Rankin–de Merona, and S.G. Laurance. 1998. Rain forest fragmentation and the dynamics of Amazonian tree communities. *Ecology* 79: 2032–2040.

Laurance, W.F., L.V. Ferreira, J.M. Rankin–de Merona, S.G. Laurance, R. Hutchings, and T. Lovejoy. 1998. Effects of forest fragmentation on recruitment patterns in Amazonian tree communities. *Conservation Biology* 12: 460–464.

Laurance, W.F., and C. Gascon. 1997. How to creatively fragment a landscape. *Conservation Biology* 11: 577–579.

Laurance, W.F., C. Gascon, and J.M. Rankin–de Merona. 1999. Predicting the effects of habitat destruction on plant communities: A test of a model using Amazonian trees. *Ecological Applications* 9: 548–554.

Laurance, W.F., S.G. Laurance, L.V. Ferreira, J.M. Rankindemerona, C. Gascon, and T.E. Lovejoy. 1997. Biomass collapse in Amazonian forest fragments. *Science* 278: 1117–1118.

Laurance, W.F., and E. Yensen. 1991. Predicting the impact of edge effects in fragmented habitats. *Biological Conservation* 55: 77–92.

Levins, R. 1969. Some demographic and genetic consequences of environmental heterogeneity for biological control. *Bulletin of the Entomological Society of America* 15: 237–240.

Lovejoy, T.E., R.O. Bierregaard, A.B. Rylands, J.R. Malcolm, C.E. Quintela, L.H. Harper, K.S. Brown, A.H. Powell, G.V.N. Powell, H.O.R. Schubart, and M.B. Hays. 1986. Edge and other effects of isolation on Amazon forest fragments. In *Conservation Biology: The Science of Scarcity and Diversity,* ed. M.E. Soulé, pp. 257–285. Sunderland, Mass.: Sinauer.

MacArthur, R.H., and E.O. Wilson. 1967. *The Theory of Island Biogeography,* Princeton, N.J.: Princeton University Press.

Malcolm, J. R. 1991. *The Small Mammals of Amazonian Forest Fragments: Pattern and Process.* Thesis, University of Florida.

Malcolm, J.R. 1994. Edge effects in central Amazonian forest fragments. *Ecology* 75: 2439.

Margules, C.R. 1999. Conservation planning at the landscape scale. In *Issues in Landscape Ecology,* eds. J.A. Wiens and M.R. Moss, pp. 83–87. Boulder, Colo.: International Association for Landscape Ecology, Pioneer Press.

Margules, C.R., G.A. Milkovits, and G.T. Smith. 1994. Contrasting effects of habitat fragmentation on the scorpion *Cercophonius squama* and an amphipod. *Ecology* 75: 2033–2042.

Margules, C.R., and A.O. Nicholls. 1987. Assessing the conservation value of remnant habitat "islands": Mallee patches on the western Eyre Peninsula, South Australia. In *Nature Conservation: The Role of Remnants of Native Vegetation,* eds. D.A. Saunders, G.W. Arnold, A.A. Burbidge, and A.J.M. Hopkins, pp. 89–102. Chipping Norton, Sydney, Australia: Surrey Beatty and Sons.

McIntyre, S., and S. Lavoral. 1994. Predicting richness of native, rare and exotic plants in response to habitat disturbance variables across a variegated landscape. *Conservation Biology* 8: 521–531.

Mesquita, R.C.G., P. Delamonica, and W.F. Laurance. 1999. Effect of matrix type on edge-related tree mortality in Amazonian forest fragments. *Biological Conservation* 91: 129–134.

Murcia, C. 1995. Edge effects in fragmented forests: Implications for conservation. *Trends in Ecology and Evolution* 10: 58–62.

Myers, N., R. Mittermeier, C.G. Mittermeier, G.A.B. da Fonseca, and J. Keats. 2000. Biodiversity hotspots for conservation priorities. *Nature* 403: 853–858.

Oostermeijer, J.G.B., A. Berholz, and P. Poschlod. 1996. Genetical aspects of fragmented populations: A review. In *Species Survival in Fragmented Landscapes,* eds. J. Settele, C. Margules, P. Poschlod, and K. Henle, pp. 93–110. Dordrecht: Kluwer Academic Publishers.

Parsons, J.J. 1972. Spread of African pasture grasses to the American tropics. *Journal of Range Management* 25: 12–24.

Pearson, S.M. 1993. The spatial extent and relative influence of landscape-level factors on wintering bird populations. *Landscape Ecology* 8: 3–18.

Powell, A.H., and G.V.N. Powell. 1987. Population dynamics of euglossine bees in Amazonian forest fragments. *Biotropica* 19: 176–179.

Pressey, R.L., C.J. Humphries, C.R. Margules, R.I. Vane-Wright, and P.H. Williams. 1993. Beyond opportunism: Key principles for systematic reserve selection. *Trends in Ecology and Evolution* 8: 124–128.

Redpath, S.M. 1995. Habitat fragmentation and the individual: Tawny owls, *Strix aluco,* in woodland patches. *Journal of Animal Ecology* 64: 652–661.

Robinson, S. K., F. R. Thompson, T.M. Donovan, D.R. Whitehead, and J. Faaborg. 1995. Regional forest fragmentation and the nesting success of migratory birds. *Science* 267: 1987–1990.

Sarre, S. 1995. Mitochondrial DNA variation among populations of *Oedura reticulata* (Reptilia: gekkonidae) in remnant vegetation: Implications for metapopulation structure and population decline. *Molecular Ecology* 4: 395–405.

Saunders, D.A., R.J. Hobbs, and C.R. Margules. 1991. Biological consequences of ecosystem fragmentation: A review. *Conservation Biology* 5: 18–32.

Saurez, A.V., D.T. Bolger, and T.J. Case. 1998. Effects of fragmentation and invasion on native ant communities in coastal Southern California. *Ecology* 79: 2041–2056.

Scougall, S.A., J.D. Majer, and R.J. Hobbs. 1993. Edge effects in grazed and ungrazed western Australian wheatbelt remnants in relation to ecosystem reconstruction. In *Nature Conservation 3: Reconstruction of Fragmented Ecosystems, Global and Regional Perspectives,* eds. D.A. Saunders, R.J. Hobbs, and P.R. Ehrlich, pp. 163–178. Sydney, Australia: Surrey Beatty and Sons.

Soulé, M.E., D.T. Bolger, A.C. Alberts, R.S. Sauvajot, J. Wright, M. Sorice, and S. Hill. 1988. Reconstructed dynamics of rapid extinctions of chaparral-requiring birds in urban habitat islands. *Conservation Biology* 2: 75–92.

Steinberg, E. K., and P. Kareiva. 1997. Challenges and opportunities for empirical evaluation of "spatial theory." In *Spatial Ecology: The Role of Space in Population Dynamics and Interspecific Interactions,* eds. D. Tilman and P. Kareiva, pp. 318–332. Princeton, N.J.: Princeton University Press.

Stouffer, P.C., and R.O. Bierregaard. 1995. Use of Amazonian forest fragments by understory insectivorous birds. *Ecology* 76: 2429–2445.

Tewksbury, J.J., S.J. Heil, and T.E. Martin. 1998. Breeding productivity does not decline with increasing fragmentation in a western landscape. *Ecology* 79: 2890–2903.

Viana, V., A.A.J. Tabanez, and J.L.F. Batista. 1997. Dynamics and restoration of forest remnants in the Brazilian Atlantic moist forest. In *Tropical Forest Remnants: Ecology, Management, and Conservation of Fragmented Communities,* eds. W.F. Laurance and R.O. Bierregaard, pp. 351–365. Chicago: University of Chicago Press.

Whitmore, T.C. 1997. Tropical forest disturbance, disappearance, and species loss. In *Tropical Forest Remnants: Ecology, Management, and Conservation of Fragmented Communities,* eds. W.F. Laurance and R.O. Bierregaard, pp. 3–12. Chicago: University of Chicago Press.

Yates, C.J., and R.J. Hobbs. 1997. Temperate eucalypt woodlands: A review of their status, processes threatening their persistence and techniques for restoration. *Australian Journal of Botany* 45: 949–973.

Yates, C.J., R.J. Hobbs, and L. Atkins. 2000. Establishment of perennial shrub and tree species in degraded *Eucalyptus salmonophloia* (Salmon gum) remnant woodlands: Effects of restoration treatments. *Restoration Ecology* 8: 135–143.

6

CONSERVATION PRIORITIES FOR SOIL AND SEDIMENT INVERTEBRATES

Diana H. Wall, Paul V. R. Snelgrove, and Alan P. Covich

Recent syntheses compare the soil and sediment domains of terrestrial, freshwater, and marine ecosystems in terms of biodiversity and their role in ecosystem functioning (Brussaard et al. 1997; Palmer et al. 1997; Snelgrove et al. 1997; Wall Freckman et al. 1997; Covich et al. 1999; Groffman and Bohlen 1999; Snelgrove 1999; Wall and Moore 1999). These reviews suggest that much of the biodiversity in soils and sediments is yet to be described. Although soils and sediments are critical components of almost every habitat on Earth, and numerous species contribute to many ecosystem processes (e.g., nutrient cycling, carbon storage, etc.), little is known of the role that biodiversity plays in ecosystem processes. We first provide a brief description of soils and sediments, the associated biodiversity, and what is known about the relationships of biodiversity to ecosystem processes. This description summarizes evidence of the important role that soil and sediment biota play in sustaining ecosystem processes and, hence, human welfare. The remainder of the chapter discusses how these biota are affected by global change (human activities and accelerating rates of disturbances). We then provide our views on the priorities for conservation and protection of the organisms found in Earth's least explored ecosystems, its soils and sediments.

The Domains

Soils and sediments are finite resources. Hundreds of years may be required to form a centimeter of soil or sediment. Through geologic time, climate, parent material, topography, and biota have contributed to the physical and chemical characteristics that are the key elements of soil and sediment habitats (Anderson 1975; Swift et al. 1979; Wall Freckman et al. 1997; Clark et al. 1999; Kump et al. 2000). In freshwater and marine environments, the direction, magnitude, and variability of water flow, which are influenced by topography and climate, are also important regulators of smaller-scale patterns in sedimentary habitats. The same variables that act over geologic time to develop soils and sediments also act locally on very short time scales. For example, these living organisms have an immediate impact on the characteristics of the soil and sediment in which they reside, through bioturbation, repackaging sediment and soil particles as fecal pellets, and, in death, by contributing organic matter to the ecosystem (Levinton 1995).

The habitats of soils and sediments are amazingly diverse at small and large spatial scales. A few cubic millimeters can contain hundreds of species of microscopic organisms (bacteria, fungi, protozoa, rotifers, tardigrades, nematodes, mites, and many other small invertebrates). At scales of meters and kilometers, larger invertebrates and vertebrates add to the species richness of these habitats. Similarly, the quality and quantity of organic matter, sediment/soil composition and geochemistry, rate of disturbance, climate, and geological history are all factors that contribute to local- and regional-scale patterns of species diversity and abundance among soil and benthic (aquatic, bottom-living) organisms (Snelgrove and Butman 1994; Brussaard et al. 1997; Covich et al. 1999; Wall and Virginia 2000). On a global scale, variability in geologic parent materials and climates has created highly productive soils that are used intensively for agriculture (1.44 billion hectares of a total 13.02 billion hectares) (WRI 1994).

The remaining soils, some 89 percent of terrestrial land, are either not productive due to high salinity, permafrost, or other conditions, or are marginal land that is being modified for short-term use in agriculture. Moreover, approximately 65 percent of the earth's surface is covered in sediments at ocean depths at which no photosynthetic production occurs (Snelgrove 1999).

These soils and sediments cover most of the earth's surface and support a tremendous diversity of species that are a source of ecosystem services needed to maintain life on Earth (Wardle 1995; Brussaard et al. 1997; Costanza et al. 1997; Daily et al. 1997; Palmer et al. 1997; Snelgrove et al. 1997; Wall Freckman et al. 1997; Edinger et al. 1998). In addition to providing a diversity of habitats for below-surface organisms, soils and sediments support food chains that ultimately yield the food (fisheries, crops), fuel, and fiber upon which civilization depends. The biota of soil and sedimentary environments helps maintain processes in above-ground habitats that have high recreational and aesthetic value, such as forests, deserts, wetlands, rivers, and coastal waterways. Protection and restoration of these subsurface communities are essential if ecosystems are to continue to function and provide ecosystem services.

Unfortunately, many human activities are rapidly accelerating the degradation of soils, sediments, and their biota. Globally, the disruption of soils (e.g., through agricultural practices and construction) has rapidly increased erosion and transport of particles into the atmosphere, rivers, and coastal marine ecosystems, with adverse effects on biota and ecosystem functioning. Coral reefs and their accompanying biota are being rapidly destroyed by increased nutrients from terrestrial ecosystems (Edinger et al. 1998; Naylor et al. 1998; Wagener et al. 1998; Harvell et al. 1999; Hixon et al. 2001; Ewel et al. in press). As productive soils become nutrient depleted or lost through erosion, other natural ecosystems (e.g., forests and wetlands) are claimed for agriculture.

The high demand for coastline property, for development of housing, industry, and aquaculture, has accelerated the loss of coastal wetlands as natural habitats (Levin et al. in press; Ewel et al. in press). Fishing trawls and dredges sweep across the ocean floor creating swaths of highly disturbed bottom habitat (Dayton et al. 1995; Watling and Norse 1998), and dredging of coastal waterways to maintain harbors degrades benthic communities. Some sedimentary taxa are more resilient to disturbance than others (e.g., Prena et al. 1999). For example, organisms in dynamic sandy bottoms frequently moved around by currents may be less vulnerable than species living where physical disturbance is rare (e.g., Watling and Norse 1998). Deep-sea sedimentary communities are more vulnerable to fishing impact than their shallow-water counterparts because deep-sea sedimentary species typically produce few offspring, attain reproductive maturity at a relatively late age, and grow relatively slowly (Merrett and Haedrich 1997). These short-term actions of disturbance have long-term effects on soil and sediment biodiversity and ecosystem processes that affect ecosystem services at local and global scales.

Ecosystem Functioning and Services

Soil and sediment biota contribute many services that are beneficial to humans beyond the value of food production (tables 6.1–6.4; Costanza et al. 1997; Daily et al. 1997; Wall and Virginia 2000). Organisms directly and indirectly drive processes such as nutrient mineralization, nitrogen fixation, hydrologic processes, oxygenation, soil and sediment structure and stabilization, remediation, burial and/or remobilization of pollutants, and redistribution and decomposition of organic matter (Wall Freckman et al. 1997; Bardgett and Chan 1999; Snelgrove et al. 2000). ·

Soils and sediments (including deep marine sediments) are the world's largest reservoir of carbon. Therefore, decomposition of organic matter into inorganic forms is a major ecosystem process involving a diversity and succession of organisms (Anderson 1975; Schlesinger 1997). In terrestrial ecosystems, with the exception of grazed, agricultural, and logged lands, most net primary production is returned to soil in the form of organic matter (roots and plant litter). In deep freshwater and marine ecosystems, much of the primary production is provided by phytoplankton that eventually reaches the bottom. In many shallow freshwater ecosystems and nearshore marine areas, macrophytes within the ecosystems, and detrital plant material transported in from adjacent domains, are primary sources of organic matter for detritivores and decomposers.

In all of these subsurface ecosystems, groups such as nematodes and protozoans influence bacterial production and affect rates of decomposition and mineralization (Anderson et al. 1981; Denton et al. 1999). Many groups of invertebrates, such as earthworms, millipedes, isopods, nematodes, and Collembola in soils; polychaetes, molluscs, and crustaceans in marine sediments; and insects, crustaceans, oligochaetes and nematodes in freshwater sediments, move organic matter and microbes through the soil and sediments. These bioturbators that move materials and fluids also provide pathways for diffusion of gases and oxygenate the soils and sediments, contributing to the microbial degradation process through favoring the more efficient aerobic over anaerobic decomposition (e.g., Aller 1983; Kristensen et al. 1992; Coleman and Crossley 1996; Gilbert et al. 1998). Thus, microbial productivity is enhanced by the feeding activity of organisms that obtain their nutrition by ingesting sediments and attached microbes, a phenomenon called microbial gardening (Hylleberg 1975; Grossman and Reichardt 1991).

Biodiversity

The majority (potentially 95 percent) of the species in soil and sediment domains are thought to be undescribed, particularly in marine ecosystems (Wall Freckman et al. 1997). Invertebrates are the major component of known soil and sediment species richness, with larger macroinvertebrates (e.g., arthropods, annelids) contributing the largest numbers of described species. However, it is the smallest organisms (e.g., fungi, bacteria, protozoa, nematodes, mites) with the least numbers of described species that are estimated to have the greatest species diversity, based on recent analyses in habitats using both molecular and morphological techniques (Behan-Pelletier and Newton 1999; Wall and Virginia 2000).

TABLE 6.1. Terrestrial Soil Habitats, Services, and Vulnerability

Habitat	Diversity[a]	Ecosystem Services	Key Taxa	Threats	Global Area	Habitat Degradation
Deserts	Very low to low	Decay, water filtering, stabilizing	Algae, termites ants, bacteria	Exotics, salinization, erosion	Large	Medium
Tropical forests	Medium to very high	Soil fertility, decay, water filtering, soil structure, stabilization	Mycorrhizae, termites, ants earthworms, mites, fungi	Agriculture, cutting, exotics	Large	High
Temperate forests	Low	Decay, water filtering, soil structure, stabilization, sediment trapping	Earthworms, mites, fungi, ants	Acidification, agriculture, cutting, erosion, exotics	Medium	High
Grasslands	High	Herbivory, soil fertility, water filtering, carbon storage, pest control	Earthworms enchytraeids, termites, nematodes, bacteria	Agriculture, overgrazing, urbanization, desertification	Medium	Low
Agriculture	Low	Soil quality	Mycorrhizae, plant pathogens	Reforestation	Low	High

[a]Known diversity. Diversity comparisons are within rather than across domains.

TABLE 6.2. Marine Sedimentary Habitats, Services, and Vulnerability

Habitat	Diversity[a]	Ecosystem Services	Key Taxa	Threats	Global Area	Habitat Degradation
Estuaries	Low	Nutrient cycling, production, bioremediation, water filtering	Bacteria, fish, macroinverts	Pollution, exotics, fisheries	Low	High
Mangels	Low	Nutrient cycling, production, shore stability, sediment trapping, water filtering	Bacteria, plants, fish, macroinverts	Habitat loss, pollution, exotics	Low	Very high
Salt Marshes	Low	Nutrient cycling, production, shore stability, bioremediation, sediment trapping water filtering	Bacteria, plants, fish, macroinverts	Habitat loss, exotics, pollution,	Low	Very high
Shelf	Low–high	Nutrient cycling, production,	Bacteria, fish, macroinverts	Fishing, habitat loss	Large	Medium
Deep Sea	Very high	Nutrient cycling	Bacteria, macroinverts	—	Very large	Low

[a]Known diversity. Diversity comparisons are within rather than across domains.

TABLE 6.3. Freshwater Sedimentary Habitats, Services, and Vulnerability

Habitat	Diversity[a]	Ecosystem Services	Key Taxa	Threats	Global Area	Habitat Degradation
Deep lakes	High	Production, sediment storage	Fish, amphipods, gastropods	Pollution, overfishing, exotics,	Low	Low
Shallow lakes	Medium	Production, nutrient cycling, sediment trapping	Fish, microbes, macrophytes	Acidification, overfishing, exotics	High	Medium
Rivers	High	Production, flood control, nutrient cycling, sediment transport	Fish, mussels, microbes	Dams, levees, exotics, pollution	Low	High
Reservoirs	Medium	Production, flood control, nutrient cycling, sediment transport	Fish, mussels, microbes, decapods	Diversion, pollution, exotics, overfishing	Medium	Low–high
Springs	High	Water supply	Microbes, macrofauna	Pollution, diversion	High	Low–high
Ponds and wetlands	Low	Nutrient cycling, flood control	Macrophytes, invertebrates	Land use, exotics	Medium	Medium–high
Groundwater	High	Water supply	Microbes, invertebrates	Diversion	High	Low

[a]Known diversity. Diversity comparisons are within rather than across domains.

TABLE 6.4. Ecosystem Services Provided by Soil and Sediment Biota

Regulation of major biogeochemical cycles

Retention and delivery of nutrients to plants and algae

Generation and renewal of soil and sediment structure and soil fertility

Bioremediation of wastes and pollutants

Provision of clean drinking water

Modification of the hydrological cycle, including mitigation of floods and droughts, and
   erosion control

Translocation of nutrients, particles, and gases

Regulation of atmospheric trace gases (e.g., CO_2, NO_x)(production and consumption)

Modification of anthropogenically driven global change (e.g., carbon sequestration,
   modifiers of plant and algae responses)

Regulation of animal and plant (including algae, macrophyte) populations

Control of potential agricultural pests and vertebrate pathogens

Contribution to plant production for food, fuel, and fiber and to fisheries for food

Contribution to landscape heterogeneity

Vital component of habitats important for recreation and natural history

Source: Modified from G. C. Daily, P. A. Matson, P. M. Vitousek. "Ecosystem Services Supplied by Soil," in *Nature's Services: Societal Dependence on Natural Ecosystems,* G. C. Daily,
ed., pp. 113–132 (Washington, D.C.: Island Press, 1997); D. H. Wall, G. Adams, and
A. N. Parsons, "Soil Biodiversity under Global Change Scenarios," in *Scenarios of Future
Biodiversity,* O. Sala, F. S. Chapin III, and E. Huber-Sannwald, eds. (New York: Springer-
Verlag, in press); D. H. Wall and R. A. Virginia, "The World Beneath Our Feet: Soil Biodi-
versity and Ecosystem Functioning," in *Nature and Human Society: The Quest for a
Sustainable World,* P. R. Raven and T. Williams, eds., pp. 225–241 (Washington, D.C.:
National Academy of Sciences and National Research Council, 2000).

There are large differences among soils, freshwater sediments, and
marine sediments in terms of numbers of species within the different phyla
(figure 6.1). Species richness is extremely high in all three domains, and diversity patterns are determined by several factors (Bardgett et al. in press). Most
major groups of freshwater species are much less diverse than their marine
counterparts. This difference is illustrated in the subphylum Crustacea (phy-
lum Arthropoda): most of the 780 species of mysids are marine; worldwide,
only 25 species occur in freshwater, and 18 additional species live in freshwater caves. Generally, only about 10 percent of the nearly 40,000 extant crustacean species occur in the inland pools, lakes, and streams of the world
(Thorp and Covich 1991). Some groups such as gastropods, bivalves, and
amphipods are exceptionally diverse in inland waters, especially in deep,
ancient lakes, caves, and rivers (Allan and Flecker 1993; Martens 1997; Ricciardi and Rasmussen 1999; Culver et al. 2000; Master et al. 2000).

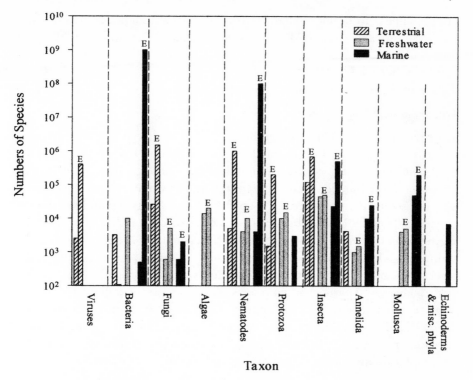

FIGURE 6.1.   Estimated numbers of different taxa from terrestrial, freshwater, and marine soils and sediments. Bars denote global estimates of numbers of species already described; those with "E" above them denote extrapolated numbers. Sources for data are Brussaard et al. 1997; Palmer et al. 1997; Snelgrove et al. 1997; and Wall and Virginia 2000.

How does above-surface diversity influence soil and sediment biodiversity? There is insufficient evidence at present to answer this question fully. It appears, however, that (1) above-surface disturbances generally reduce local soil and sediment biodiversity (Freckman and Ettema 1993; Wardle 1995; Covich et al. 1996; Ettema et al. 1999; Wall Freckman et al. 1997); (2) soil and freshwater sediment biodiversity generally is not correlated with latitudinal gradients of plant diversity (Hooper et al. 2000); and (3) marine sediment diversity shows a latitudinal gradient (Rex et al. 1993; Roy et al. 1998). The mechanistic linkages are just being explored for the above- and below-ground (air-soil, water-sediment) components of ecosystems (Hooper et al. 2000; Palmer et al.2000; Snelgrove et al. 2000). Communities in soils resemble above-surface diversity in that a few species are common and the majority of species are rare (Denton et al. 1999; Wall and Virginia 2000); however, rare species appear to be more characteristic of marine sediments than of the water column above them (McGowan and Walker 1985; Carney 1997).

Global Change Threats to Ecosystem Functioning and Ecosystem Services

Overview on Species Loss

Here we examine the evidence for the effect of some drivers of global change (invasive species, pollution, habitat destruction, and elevated CO_2) on soil and sediment species and ecosystem functioning. Although little knowledge exists about the functional attributes of the majority of individual species in soil and sediment communities or how their loss affects soil and sediment processes (Schimel 1995; Brussaard et al. 1997; Covich et al. 1999; Bardgett et al. in press), there is information on the effects of larger species on soils and sediments. Macrofaunal species have significant consequences on physical structure and hydrology of soils and sediments over larger spatial (e.g., forest floor, lake, coastal harbor) or temporal scales. Examples of well-documented recent extinctions and endangerment of species in soil or sediment ecosystems are rare, although disappearances of individual species from parts of their pre-human ranges are known, especially for endemic species with limited geographical ranges such as freshwater bivalves (Master et al. 2000).

Invasive Species

Some of the best examples of the potential impacts of shifts in biodiversity on ecosystem functioning are from research on invasive species. In these studies, the focus has often been on the impact of a single species rather than effects of changes in biodiversity per se (Trudgill et al. 1994; Boag et al. 1998; Boag et al. 1999). The global movement of goods over greater distances and at faster speeds than ever before has greatly increased the number of species invasions into soils and sediments, with resulting impacts on biodiversity and ecosystem functioning, and on economics and ecosystem services (Vitousek, Mooney, et al. 1997; Strayer et al. 1999; McMahon 2000).

Although marine environments would appear at first glance to have few barriers to prevent the transport and natural spreading of species, the open ocean is a major barrier to the dispersal of many coastal species. Coastal waters, which have the most numerous and well studied exotic species, are the major focus of most research on introduced species in marine ecosystems (Nichols et al. 1986; Hixon et al. 2001). The predominant mechanism of colonization of coastal benthic species is through transport in the ballast water of ships (Carlton 1985, 2000). Vessels take on ballast water in one port, along with the reproductive propagules, juveniles, and even adults of coastal sedimentary species, and dump it when they arrive in another port. This mechanism has been a major means of inadvertent introductions of exotic species in the Great Lakes and marine coastal waters (Mills et al. 1993; Mackie 2000). As with terrestrial species, physical and chemical barriers as well as biological barriers (such as the presence of predators or strong competitors) limit range expansions of species into inland waters. Species with diapausing stages can be dispersed passively by wildlife and people, as microscopic larvae are transported in bait buckets by fishermen and releases occur from the aquarium trade.

There are numerous examples of accidental or deliberate species introductions that have changed some aspect of the below-surface ecosystem (e.g., biogeochemistry, species diversity, soil or sediment structure), and these disruptions are magnified throughout the whole ecosystem. In most cases, deliberate introductions have been ecologically catastrophic; they have rarely been beneficial.

A number of freshwater and marine species have deliberately been introduced into developing countries for aquaculture, and, although these species often have increased protein availability and economic markets, they are causing environmental degradation (Naylor et al. 1998; Naylor et al. 2000). Significant impacts on the sedimentary fauna of these and neighboring ecosystems occur as a result of the activities of the industries themselves and through biological interactions between native and non-native species. Unfortunately, little research effort is being directed toward this question, which affects many developing countries, even as the scale of aquaculture increases.

Introductions of animals and plants have impacts on the physical and chemical properties of soil and sediment habitats, and indirectly on other soil and sediment biota. For example, an introduced earthworm species in New York forests changed organic matter composition, water infiltration patterns, and soil chemistry (Burtelow et al. 1998). Even the introduction of two exotic plant species, *Berberis thunbergii* and *Microstegium vimineum,* into North American forests has affected soil pH and organic matter and increased earthworm population densities and, subsequently, soil porosity (Kourtev et al. 1998). The periwinkle *Littorina littorea,* introduced from Europe in the early 1800s (Carlton 1982) has displaced the mudsnail *Ilyanassa obsoleta* from New England salt marshes and seagrass beds. Mudsnails are now confined to sand and mudflats that periwinkles cannot utilize, with potential impacts on other species (Brenchley and Carlton 1983). In the soils of the United Kingdom and northern Europe, the invasion of the Australasian planarian flatworm *Artioposthia triangulata,* a predator of earthworms, is causing great concern (Boag et al. 1998). The flatworm has no known beneficial impact on soils, while earthworms are a source of food for birds and other animals, as well as maintaining soil structure and improving water infiltration. In North America, all soil flatworms are introduced (Ogren and Kawakatsu 1998). Amazonian forest converted to pasture resulted in a 68 percent decrease in the original soil macrofaunal community and a 90 percent dominance in biomass of the colonizing earthworm, *Pontoscolex corethrurus*. This species has an identifiable type of burrowing and casting, which explained the reduction in soil macroporosity from 3 to 1.5 $cm^3/100g$ soil and increased bare ground in pasture areas (Chauvel et al. 1999).

Invasive species in freshwater ecosystems represent some of the clearest examples of how dramatically ecosystem functioning can be altered by accidental introduction of a single non-native species. One of the best-known examples is that of the zebra mussel, *Dreissina polymorpha,* a suspension-feeding bivalve native to the Caspian Sea that was inadvertently introduced into the Great Lakes sometime around 1988. Within the Great Lakes it has attained densities that greatly exceed those achieved by native bivalves in rivers and lakes, and it has altered entire aquatic ecosystems as it spreads through watersheds in the United States (Strayer et al. 1999; McMahon 2000). Because zebra mussels attain extraordinarily high densities and filter feed very efficiently, they can effectively strip rivers and lakes of phytoplankton, with significant impacts on food webs, frequency and composition of phytoplankton blooms, nutrient cycling, and water flow (Strayer et al. 1999).

The best-studied examples of inadvertent marine introductions are in San Francisco Bay, where hundreds of non-native species now flourish (Carlton 2000), compressing the habitat of native species and substantially changing the biology of the bay. The explosion of the filter-feeding Asian clam *Potamocorbula amurensis* has greatly increased the capacity of ocean organisms to remove suspended particles, including phytoplankton that is the base of much of the food chain. As a result, the seasonal phytoplankton blooms that characterized the bay have been heavily dampened (Alpine and Cloern 1992), along with other impacts on the pelagic biota (Kimmerer et al. 1994).

Pollution

Pollution affects soil and sediment biodiversity, as well as soil chemical and physical properties, and can have long-term effects on the functioning of ecosystems. As human populations increase, levels of many pollutants have continued to rise (Vitousek, Aber, et al. 1997; Downing et al. 1999). Increased awareness has reduced the rate of input of many compounds, such as pesticides and heavy metals, but persistence of compounds in soils and sediments continues to create problems locally and globally. Agricultural fertilizers entering surface and ground waters result in the decline of many soil and benthic species. Estuaries are affected by coastal pollution and eutrophication (inputs of phosphorous and nitrogen) because human populations along coasts are rapidly increasing (Ewel et al. in press).

Nutrient enrichment via sewage discharge, storm drainage, and over-application of fertilizers on land and subsequent runoff into freshwater and coastal areas is a major problem. Increased nutrients in aquatic ecosystems favor algal blooms that sink to the bottom, where their decomposition renders the water anoxic. These events are increasingly occurring across domains from midwestern farm soils to the Mississippi drainage and Gulf of Mexico (Turner and Rabalais 1994; Justiz et al. 1997; Malakoff 1998; Downing et al. 1999).

Land-Use Change, Habitat Destruction, and Harvesting

Anything (e.g., tillage, erosion, plant removal) that changes the soil structure, microclimate, physical and chemical properties, and carbon/nitrogen food base of soil food webs (e.g., plant chemical composition, quality and quantity of detritus, oil spills, manure) can have a direct impact on soil diversity and food web structure. Land-use change, the conversion of natural systems to managed systems (as in agriculture, forestry, grazing, urbanization, industrialization), has the greatest effect on soil biota, because it generally decreases the diversity of all soil taxa and dramatically changes the primary determinants of soil biodiversity, e.g., soil structure, vegetation, and microclimate (Wall et al. in press). For example, Lawton et al. (1998) found that species richness of nematodes and termites, as well as six other above-ground animal groups, declined with increasing disturbance in a Cameroon forest. The species decline was most obvious in active slash-and-burn areas and cleared tropical forest.

Intensive tillage can result in decoupling of nutrient release from plant uptake and a need for greater amounts of fertilizer. Efforts to conserve soils by methods such as reduced tillage and no tillage have beneficial effects of increasing soil carbon storage, improving soil moisture, and decreasing disruptions of soil food webs. However, reduced tillage methods may require increased inputs of herbicides and result in increased densities of soil pathogens (bacteria, fungi, nematodes, some insects). The banning of many nematicides and fungicides, such as methyl bromide and DBCP, due to their adverse environmental effects, has reduced the arsenal of pesticides used to manage soil pathogens for the past fifty years (Duncan and Noling 1998). The demand for new control measures for soil pathogens that allow maintenance of plant productivity under reduced tillage systems presents a new research and management challenge.

The wetlands of the world, both freshwater and marine, have suffered major destruction in the last few centuries. In marine ecosystems, the greatest habitat loss is occurring in coastal environments, particularly transition zones, or regions where soil and freshwater domains meet marine habitats. Transition zones are critical habitats for many species, and the sedimentary organisms that reside within them provide an important food resource. Destruction of mangrove habitats for shrimp farming and other forms of coastal development result in the loss of habitat and food for many species (Beveridge et al. 1994; Naylor et al. 2000; Ewel et al. in press).

Climate and Elevated CO_2

Increased frequency of extreme climatic events predicted as a result of global change can have varying effects on soil and sediment biota (Boag et al. 1991; Covich et al. 1997; Schindler 1997; Hogg et al. 1998; Sala et al. 2000; Smith et al. in press; Wall et al. in press; Wolters et al. in press). Climate change, increasing elevated atmospheric CO_2, and the resulting changes in rainfall patterns, soil temperatures, and plant performance are predicted to alter the geographical distribution of soil species indirectly through changes in vegetation distribution (Groffman and Bohlen 1999; Wall et al. in press; Wolters et al. in press). Frequency and amount of rainfall can increase populations of soil pathogenic fungi by reducing soil aeration, resulting in shifts in species composition and richness.

Ecosystem-level effects of climate change on soil biodiversity are just beginning to be analyzed through manipulative experiments that increase elevated CO_2 and vary rainfall and temperature. Direct effects of elevated CO_2 on soil organisms are considered to be less important than indirect effects, because most soils naturally have higher than ambient CO_2 concentrations due to root respiration. Indirect effects on soil biota have occurred through changes in plant metabolism, resulting in reallocation of carbon and nitrogen. Increased below-ground plant resources affect both decomposition and herbivory (Zak et al. 1996).

There are many projections of the indirect effect of global-change drivers on freshwater ecosystems (Covich 1993; Kolar and Lodge 2000). For example, in increasingly arid regions increasing human demands will result in construction of more dams and irrigation canals. These structures will increase habitat fragmentation and alter disturbance regimes, refugia, and microhabitats of freshwater species. Alteration of rates and direction of water flow for irrigation has a major impact on terrestrial, freshwater, and coastal marine ecosystems; hydrological changes impact soil and sediment composition, pH of water and soils, and salinity, among other variables (Strange et al. 1999). Dam construction in the southeastern United States constitutes one of the largest threats to persistence of freshwater species, especially bivalves

(Bogan 1993; Master et al. 2000). There are about 334 species of freshwater bivalves in North America. The recent loss of many of these bivalve species and the invasion of inland waters by non-native bivalves represent major challenges for ecosystem management.

Little direct evidence suggests that biodiversity changes among marine sedimentary organisms are attributed to global climate change, but studies of pelagic and intertidal communities (Southward et al. 1995) and seagrass communities (Marba and Duarte 1997) suggest that effects may be expected. One of the most serious projections in some global-warming scenarios for marine sedimentary fauna is the possible shifting of ocean currents (Manabe et al. 1994), which influence food supply and transport of reproductive propagules from bottom-dwelling species.

Conservation Research Priorities

Policy makers, the public, and national and international trade and economic organizations have yet to realize that soil and sediment biota are interlinked with the above surface we see, or to realize that the biodiversity in soils and sediments contributes to ecosystem services and goods that help drive economics and benefit humanity. Much of this wealth is as yet undescribed, mapped, or recognized as a new world for exploration. Yet evidence suggests that the destruction of soil and sediment habitats changes biodiversity with cascading effects on ecosystem processes and endangers the ability of the biota to maintain goods and services. These habitats and their biodiversity cannot be easily restored.

Protection and conservation of soil and sediment biodiversity are dependent on an increase in education, research, and publicity to address the gaps in the scientific knowledge of species distributions, impacts of global-change drivers, and consequences for goods and services. Education at all levels must incorporate more opportunities for learning about the role of biodiversity in soils and sediments and how these subsurface habitats are linked to each other and to above-ground habitats. Soils, sediments, and invertebrate biodiversity, particularly biodiversity at the microscopic level, can become more relevant to students if the traditionally taught biology curriculum in elementary schools, high schools, and colleges is changed to include more comprehensive approaches to understanding how ecosystems function and the importance of nutrient cycling, decomposition, and the complex physical and chemical environment. Such an approach would increase understanding of our dependence on multiple species in soil and sediments. This would incorporate new technologies and a sense of discovery into classrooms and field trips. The separation of chemistry and biology as different subjects and the isolation of related topics on atmospheric and earth sciences prevents most students from a comprehensive understanding of the important roles that many different microbial and macroscopic species have in regulating rates of ecosystem processes.

Most people of any age recognize their "own" ecosystem (backyard, garden, soil, city, stream, park, field), but, in many classrooms, the local environment and its interactions, functions, and linkages to regional or global ecosystems are given limited time. Nor is there discussion of how individuals or human activities contribute to the conservation, sustainable management, and restoration of an essential natural resource: soils, including freshwater and marine sediments. Most high school and college curricula in biology today are focused on premedical perspectives, and that important information dominates discussion of the overall relationships of emerging diseases and sustainable food production and availability of clean waters.

New technologies (e.g., GIS, molecular analysis) are rapidly advancing the study of biodiversity in soils and sediments when, unfortunately, the number of systematists globally has declined. It is now possible to geo-reference soils with global positioning systems as they are sampled; extract different invertebrate groups; then rapidly sort the numerous individuals from many families using microscopes with enhanced computer image analysis; and integrate the data with other related GIS data on soil physical and chemical characteristics, vegetation, and disturbance factors (Oliver et al. 2000). The study of species in soils and sediments is further advanced with the use of molecular tools, which contribute to both taxonomic and ecological research such as faster assessments of geographic distribution patterns, phylogenetic relationships, and consequences of disturbance (e.g., effect of climate change on below-surface organisms and ecosystem processes). Incorporating these new techniques for estimating soil and sediment biodiversity into existing research efforts at aquatic and terrestrial sites will enhance and broaden the quantitative information on ecosystem functioning and lead to better recommendations for conservation, restoration, and sustainability.

We offer the following statements as a basis for the research priorities summarized in box 6.1:

- Although we know very little about the diversity and composition of most soil and sediment communities, we do have detailed studies of some species.

- We do know that soil and sediment communities perform functions that are critical for the functioning of the ecosystem as a whole, although the role of biodiversity in those processes is poorly understood.

- The number of taxonomists specializing in below-surface species has declined when our abilities to learn about these species have increased with new technologies.

- Soils and sediments are being altered by acceleration of rates of inputs of nutrients and organic matter and by physical disruption or destruction of habitats.

- The public is generally unaware of the essential ecosystem services provided by subsurface organisms.

BOX 6.1. **Summary of Research Priorities for Soil and Sediment Ecosystems**

1. Develop a strategy to increase the numbers of systematists working with below-surface invertebrates, as there are presently very few of these specialists globally.
2. Develop a strategic approach to better represent and understand the functions of poorly sampled taxa (e.g., bacteria, protozoa, fungi, nematodes) and geographic areas (e.g., tropical soils, tropical shallow- and deep-sea sediments) to generate species estimates for these habitats with a greater degree of certainty.
3. Identify which taxa and habitats are most vulnerable to biodiversity loss, including those areas that represent hot spots of biodiversity and ecosystem services within soils and sediments.
4. Determine which soil and sediment habitats are most amenable to restoration once degraded, and the mechanisms and time scales by which they may be most effectively restored.
5. Document linkages between different soil and sediment domains, between communities above and below the soil/sediment-water interface, and between natural versus human-dominated systems, and how changes in one system may influence another.
6. Identify aspects of the natural history of key taxa as they compare to other species that may fill similar ecological niches, with the goal of evaluating whether functional redundancy actually exists within different soil and sediment ecosystems and whether species loss will have significant functional consequences to the ecosystem.
7. Using natural history information, predict which species are likely to be invasive, thus offering potential strategies to reduce their spread and evaluate which species and services may be most at risk to invasive species.
8. Improved budgets for global processes, such as carbon and nitrogen cycling, where the traditional approach of treating species as black boxes is replaced with an effort to determine the role played by individual species in these key processes.

Linking Conservation and Research Priorities

Many soil and sedimentary ecosystems have been degraded by human activities, both in terms of the biodiversity within these systems and the ecosystem services that they provide. As scientists, we are just beginning to communicate to the general public and policy makers regarding the potential gravity of these threats to global sustainability. Experiments are needed to provide essential data as to the degree of linkages and dependency of ecosystem processes on soil and sediment species under current environmental conditions, and the role biological diversity plays in ensuring continued ecosystem functioning under scenarios of global change. New research can provide a detailed analysis of soil and sedimentary organisms, their value, and their contribution to ecosystem services.

We do know that there are certain processes in which a diversity of soil and sedimentary organisms are integral players, and we know in some instances that certain species are critical to those processes. However, our level of scientific knowledge on soil and sediment organisms is insufficient. The majority of data on linkages between soil and sediment species is for a single species and a single effect, and there are few data on multiple species across diverse taxonomic groups within a community, their interactions and effects on ecosystem processes. Although biodiversity is very likely being lost, we, at present, cannot predict where and when the loss will have minor or catastrophic consequences. In some instances, scientists may expect that policy implementation will follow a precautionary principle, but in the majority of instances, it will be necessary to provide irrefutable data to demonstrate that as human activities affect soil and sedimentary systems at an accelerating rate, conservation principles are in the best interests of individuals, communities, and nations.

ACKNOWLEDGMENTS

The authors would like to thank G. Adams, D. Krumm, N. DeCrappeo, and D. Porazinska for helpful suggestions on this manuscript, and R. Pulliam, M. Soulé, and G. Orians for their thoughtful contributions. D. Wall acknowledges the support of NSF DEB 97 08596, DEB 96 26813, and NSF OPP 9211773 and OPP 9624743; P. Snelgrove acknowledges the support of NSERC of Canada; and A. Covich acknowledges the support of NSF DEB 9705814.

LITERATURE CITED

Allan, J. D., and A. S. Flecker. 1993. Biodiversity conservation in running waters: Identifying the major factors that threaten destruction of riverine species and ecosystems. *BioScience* 43:32–43.

Aller, R. C. 1983. The importance of the diffusive permeability of animal burrow linings in determining marine sediment chemistry. *Journal of Marine Research* 41:299–322.

Alpine, A. E., and J. E. Cloern. 1992. Trophic interactions and direct physical effects control phytoplankton biomass and production in an estuary. *Limnology and Oceanography* 37:946–955.

Anderson, J. M. 1975. Succession, diversity and trophic relationships of some soil animals in decomposing leaf litter. *Journal of Animal Ecology* 44:475–494.

Anderson, R. V., D. C. Coleman, C. V. Cole, and E. T. Elliott. 1981. Effect of the nematodes *Acrobeloides* sp. and *Mesodiplogaster iheritieri* on substrate utilization and nitrogen and phosphorus mineralization in soil. *Ecology* 62:549–555.

Bardgett, R. D., J. M. Anderson, V. Behan-Pelletier, L. Brussaard, D. C. Coleman, C. Ettema, A. Moldenke, J. P. Schimel, and D. H. Wall. In press. The influence of soil biodiversity on hydrologic pathways and the transfer of materials between terrestrial and aquatic ecosystems. *Ecosystems.*

Bardgett, R. D., and K. F. Chan. 1999. Experimental evidence that soil fauna enhance nutrient mineralization and plant nutrient uptake in montane grassland ecosystems. *Soil Biology & Biochemistry* 31:1007–1014.

Behan-Pelletier, V., and G. Newton. 1999. Linking soil biodiversity and ecosystem function: The taxonomic dilemma. *Bioscience* 49:149–152.

Beveridge, M. C. M., L. G. Ross, and L. A. Kelly. 1994. Aquaculture and biodiversity. *Ambio* 23:497–502.

Boag, B., J. W. Crawford, and R. Neilson. 1991. The effect of potential climatic changes on the geographical distribution of the plant-parasitic nematode *Xiphinema* and *Longidorus* in Europe. *Nematologica* 37:312–323.

Boag, B., H. D. Jones, R. Neilson, and G. Santoro. 1999. Spatial distribution and relationship between the New Zealand flatworm *Arthurdendyus triangulata* and earthworms in a grass field in Scotland. *Pedobiologia* 43:340–344.

Boag, B., G. W. Yeates, and P. M. Johns. 1998. Limitations to the distribution and spread of terrestrial flatworms with special reference to the New Zealand flatworm (*Artioposthia triangulata*). *Pedobiologia* 42:495–503.

Bogan, A. E. 1993. Freshwater bivalve extinctions (Mollusca, Unionoida)—A search for causes. *American Zoologist* 33:599–609.

Brenchley, G. A., and J. T. Carlton. 1983. Competitive displacement of native mud snails by introduced periwinkles in the New England intertidal zone. *Biological Bulletin* 165:543–558.

Brussaard, L., V. M. Behan-Pelletier, D. E. Bignell, V. K. Brown, W. Didden, P. Folgarait, C. Fragoso, D. W. Freckman, V. Gupta, T. Hattori, D. L. Hawksworth, C. Klopatek, P. Lavelle, D. W. Malloch, J. Rusek, B. Soderstrom, J. M. Tiedje, and R. A. Virginia. 1997. Biodiversity and ecosystem functioning in soil. *Ambio* 26:563–570.

Burtelow, A., P. J. Bohlen, and P. M. Groffman. 1998. Influence of exotic earthworm invasion on soil organic matter, microbial biomass and denitrification potential in forest soils of the northeastern U.S. *Applied Soil Ecology* 9:197–202.

Carlton, J. T. 1982. The historical biogeography of *Littornia littorea* on the Atlantic coast of North America, and implications for the interpretation of the structure of New England intertidal communities. *Malacological Review* 15:146.

Carlton, J. T. 1985. Transoceanic and interoceanic dispersal of coastal marine organisms: The biology of ballast water. *Oceanography and Marine Biology: An Annual Review* 23:313–371.

Carlton, J. T. 2000. Global change and biological invasions. In *Invasive Species in a Changing World,* eds. H. A. Mooney and R. J. Hobbs, pp. 31–53. Washington, D.C.: Island Press.

Carney, R. S. 1997. Basing conservation policies for the deep-sea floor on current-diversity concepts: A consideration of rarity. *Biodiversity and Conservation* 6:1463–1485.

Chauvel, A., M. Grimaldi, E. Barros, E. Blanchart, T. Desjardins, M. Sarrazin, and P. Lavelle. 1999. Pasture damage by an Amazonian earthworm. *Nature* 398:32–33.

Clark, P. U., R. B. Alley, and D. Pollard. 1999. Climatology—Northern hemisphere ice-sheet influences on global climate change. *Science* 286:1104–1111.

Coleman, D. C., and D. A. Crossley. 1996. *Fundamentals of Soil Ecology.* San Diego: Academic Press.

Costanza, R., R. dArge, R. deGroot, S. Farber, M. Grasso, B. Hannon, K. Limburg, S. Naeem, R. V. Oneill, J. Paruelo, R. G. Raskin, P. Sutton, and M. vandenBelt. 1997. The value of the world's ecosystem services and natural capital. *Nature* 387:253–260.

Covich, A. P. 1993. Water and ecosystems. In *Water in Crisis: A Guide to the World's Fresh Water Resources,* ed. P. H. Gleick, pp. 40–55. Oxford: Oxford University Press.

Covich, A. P., T. A. Crowl, S. J. Johnson, and M. Pyron. 1996. Distribution and abundance of tropical freshwater shrimp along a stream corridor: Response to disturbance. *Biotropica* 28:484–492.

Covich, A. P., S. C. Fritz, P. J. Lamb, R. D. Marzolf, W. J. Matthews, K. A. Poiani, E. E. Prepas, M. B. Richman, and T. C. Winter. 1997. Potential effects of climate change on aquatic ecosystems of the Great Plains of North America. *Hydrological Processes* 11:993–1021.

Covich, A. P., M. A. Palmer, and T. A. Crowl. 1999. The role of benthic invertebrate species in freshwater ecosystems: Zoobenthic species influence energy flows and nutrient cycling. *Bioscience* 49:119–127.

Culver, D. C., L. L. Master, M. C. Christman, and H. H. Hobbs, III. 2000. Obligate cave fauna of the 49 contiguous United States. *Conservation Biology* 14:386–397.

Daily, G. C., P. A. Matson, and P. M. Vitousek. 1997. Ecosystem services supplied by soil. In *Nature's Services: Societal Dependence on Natural Ecosystems,* ed. G. C. Daily, pp. 113–132. Washington, D.C.: Island Press.

Dayton, P. K., S. F. Thrush, M. T. Agardy, and R. J. Hofman. 1995. Environmental effects of marine fishing. *Aquatic Conservation: Marine and Freshwater Ecosystems* 5:205–232.

Denton, C. S., R. D. Bardgett, R. Cook, and P. J. Hobbs. 1999. Low amounts of root herbivory positively influence the rhizosphere microbial community in a temperate grassland soil. *Soil Biology & Biochemistry* 31:155–165.

Downing, J. A., M. McClain, R. Twilley, J. M. Melack, J. Elser, N. N. Rabalais, W. M. Lewis, R. E. Turner, J. Corredor, D. Soto, A. Yanez-Arancibia, J. A. Kopaska, and R. W. Howarth. 1999. The impact of accelerating land-use change on the N-Cycle of tropical aquatic ecosystems: Current conditions and projected changes. *Biogeochemistry* (Dordrecht) 46:109–148.

Duncan, L. W., and J. W. Noling. 1998. Agricultural sustainability and nematode integrated pest management. In *Agronomy Monograph 36,* eds. K. R. Barker, G. A. Pederson, and G. L. Windham, pp. 251–287. Madison, WI: American Society of Agronomy, Crop Science Society of America, Soil Science Society of America.

Edinger, E. N., J. Jompa, G. V. Limmon, W. Widjatmoko, and M. J. Risk. 1998. Reef degradation and coral biodiversity in Indonesia: Effects of land-based pollution, destructive fishing practices and changes over time. *Marine Pollution Bulletin* 36:617–630.

Ettema, C. H., R. Lowrance, and D. C. Coleman. 1999. Riparian soil response to surface nitrogen input: Temporal changes in denitrification, labile and microbial C and N pools, and bacterial and fungal respiration. *Soil Biology & Biochemistry* 31:1609–1624.

Ewel, K. C., C. Cressa, R. T. Kneib, P. S. Lake, L. A. Levin, M. Palmer, P. Snelgrove, and D. H. Wall. In press. Managing critical transition zones. *Ecosystems.*

Freckman, D. W., and C. E. Ettema. 1993. Assessing nematode communities in agroecosystems of varying human intervention. *Agriculture, Ecosystems and Environment* 45:239–261.

Gilbert, F., G. Stora, and P. Bonin. 1998. Influence of bioturbation on denitrification activity in Mediterranean coastal sediments: An in situ experimental approach. *Marine Ecology Progress Series* 163:99–107.

Groffman, P. M., and P. J. Bohlen. 1999. Soil and sediment biodiversity: Cross-system comparisons and large-scale effects. *Bioscience* 49:139–148.

Grossman, S., and W. Reichardt. 1991. Impact of *Arenicola marina* on bacteria in intertidal sediments. *Marine Ecology Progress Series* 77:85–93.

Harvell, C. D., K. Kim, J. M. Burkholder, R. R. Colwell, P. R. Epstein, D. J. Grimes, E. E. Hofman, E. K. Lipp, A. D. M. E. Osterhaus, R. M. Overstreet, J. W. Porter, G. W. Smith, and G. R. Vasta. 1999. Emerging marine diseases—Climate and anthropogenic factors. *Science* 285:1505–1510.

Hixon, M. A., E. A. Norse, M. L. Hunter Jr., P. D. Boersma, F. Micheli, H. P. Possingham, and P. V. R. Snelgrove. 2001. Oceans at risk: Research priorities in marine conservation biology. In *Conservation Biology: Research Priorities for the Next Decade,* eds. M. E. Soulé and G. H. Orians, pp. 125–154.

Hogg, I. D., J. M. Eadie, and Y. de Lafontaine. 1998. Atmospheric change and the diversity of aquatic invertebrates: Are we missing the boat? *Environmental Monitoring and Assessment* 49:291–301.

Hooper, D. U., D. E. Bignell, V. K. Brown, L. Brussaard, J. M. Dangerfield, D. H. Wall, D. A. Wardle, D. C. Coleman, K. E. Giller, P. Lavelle, W. H. van der Putten, P. C. de Ruiter, J. Rusek, W. Silver, J. Tiedje, V. Wolters. 2000. Interactions between above and belowground biodiversity in terrestrial ecosystems: Patterns, mechanisms, and feedbacks. *Bioscience* 50:1049–1061.

Hylleberg, J. 1975. Selective feeding by *Abarenicola pacifica* with notes on *Abarenicola vagabunda* and a concept of gardening in lugworms. *Ophelia* 14:113–137.

Justiz, D., N. N. Rabalais, and R. E. Turner. 1997. Impact of climate change on net productivity of coastal waters: Implications for carbon budgets and hypoxia. *Climate Change* 8:225–237.

Kimmerer, W. J., E. Gartside, and J. J. Orsi. 1994. Predation by an introduced clam as the likely cause of substantial declines in zooplankton of San Francisco Bay. *Marine Ecology Progress Series* 113:81–93.

Kolar, C., and D. M. Lodge. 2000. Freshwater nonindigenous species: Interactions with other global changes. In *Invasive Species in a Changing World,* eds. H. Mooney and R. Hobbs, pp. 3–30. Washington, D.C.: Island Press.

Kourtev, P. S., J. G. Ehrenfeld, and W. Z. Huang. 1998. Effects of exotic plant species on soil properties in hardwood forests of New Jersey. *Water Air and Soil Pollution* 105:493–501.

Kristensen, E., F. O. Andersen, and T. H. Blackburn. 1992. The effects of benthic macrofauna and temperature on degradation of macroalgal detritus: The fate of organic carbon. *Limnology and Oceanography* 37:1404–1419.

Kump, L. R., S. L. Brantley, and M. A. Arthur. 2000. Chemical weathering, atmospheric CO_2, and climate. *Annual Review of Earth and Planetary Sciences* 28:611–667.

Lawton, J. H., D. E. Bignell, B. Bolton, G. F. Bloemers, P. Eggleton, P. M. Hammond, M. Hodda, R. D. Holt, T. B. Larsen, N. A. Mawdsley, N. E. Stork, D. S. Srivastava, and A. D. Watt. 1998. Biodiversity inventories, indicator taxa and effects of habitat modification in tropical forest. *Nature* (London) 391:72–76.

Levin, L. A., D. F. Boesch, A. P. Covich, C. Dahm, C. Erséus, K. C. Ewel, R. T. Kneib, M. Moldenke, M. A. Palmer, P. Snelgrove, D. Strayer, and J. M. Weslawski. In press. The function of marine critical transition zones and the importance of sediment biodiversity. *Ecosystems*.

Levinton, J. 1995. Bioturbators as ecosystem engineers: Control of the sediment fabric, inter-individual interactions, and material fluxes. In *Linking Species and Ecosystems,* eds. C. G. Jones and J. H. Lawton, pp. 29–36. London: Chapman and Hall.

Mackie, G. L. 2000. Ballast water introductions of Mollusca. In *Nonindigenous Freshwater Organisms,* eds. R. Claudi and J. H. Leach, pp. 219–254. Boca Raton, Fla.: Lewis Publishers.

Malakoff, D. 1998. Death by suffocation in the Gulf of Mexico. *Science* 281:190–192.

Manabe, S., R. J. Stouffer, and M. J. Spelman. 1994. Response of a coupled ocean-atmosphere model to increasing atmospheric carbon dioxide. *Ambio* 23:44–49.

Marba, N., and C. M. Duarte. 1997. Interannual changes in seagrass (*Posidonia oceanica*) growth and environmental change in the Spanish Mediterranean littoral zone. *Limnology and Oceanography* 42:800–810.

Martens, K. 1997. Speciation in ancient lakes. *Trends in Ecology and Evolution* 12:177–182.

Master, L. L., B. A. Stein, L. S. Kutner, and G. A. Hammerson. 2000. Vanishing assets: The conservation status of U.S. species. In *Precious Heritage: The Status of Biodiversity in the United States,* eds. B. A. Stein, L. S. Kutner, and J. S. Adams, pp. 93–118. New York: Oxford University Press.

McGowan, J. A., and P. W. Walker. 1985. Dominance and diversity maintenance in an oceanic ecosystem. *Ecological Monographs* 55:105–118.

McMahon, R. F. 2000. Invasive characteristics of the freshwater bivalve *Corbicula fluminea.* In *Nonindigineous Freshwater Organisms,* eds. R. Claudi and J. H. Leach, pp. 315–343. Boca Raton, Fla.: Lewis Publishers.

Merrett, N. R., and R. L. Haedrich. 1997. *Deep-Sea Demersal Fish and Fisheries.* London: Chapman and Hall.

Mills, E. L., J. H. Leach, J. T. Carlton, and C. L. Secor. 1993. Exotic species in the Great Lakes: A history of biotic crises and anthropogenic introductions. *Journal of Great Lakes Research* 19:1–54.

Naylor, R. L., R. J. Goldberg, H. Mooney, M. Beveridge, J. Clay, C. Folke, N. Kautsky, J. Lubchenco, J. Primavera, and M. Williams. 1998. Nature's subsidies to shrimp and salmon farming. *Science* 282:883–884.

Naylor, R. L., R. J. Goldberg, J. H. Primavera, N. Kautsky, M. C. M. Beveridge, J. Clay, C. Folke, J. Lubchenco, H. Mooney, and M. Troell. 2000. Effect of aquaculture on world fish supplies. *Nature* 405:1017–1024.

Nichols, F. H., J. E. Cloern, S. N. Luoma, and D. H. Peterson. 1986. The modification of an estuary. *Science* 231:567–573.

Ogren, R. E., and M. Kawakatsu. 1998. American nearctic and neotropical land planarians (Tricladida: Terricola) faunas. *Pedobiologia* 42:441–451.

Oliver, I., S. Pik, D. Britton, J. M. Dangerfield, R. K. Colwell, and A. J. Beattie. 2000. Virtual biodiversity assessment systems. *BioScience* 50:441–450.

Palmer, M. A., A. P. Covich, B. J. Finlay, J. Gibert, K. D. Hyde, R. K. Johnson, T. Kairesalo, P. S. Lake, C. R. Lovell, R. J. Naiman, C. Ricci, F. F. Sabater, and D. L. Strayer. 1997. Biodiversity and ecosystem processes in freshwater sediments. *Ambio* 26:571–577.

Palmer, M. A., A. P. Covich, S. Lake, P. Biro, J. J. Brooks, J. Cole, C. Dahm, J. Gibert, W. Goedkoop, K. Martens, J. Verhoeven, and W. J. van de Bund. 2000. Linkages between aquatic sediment biota and life above sediments as potential drivers of biodiversity and ecological processes? *Bioscience* 50:1062–1075.

Prena, J., P. Schwinghamer, T. W. Rowell, D. C. Gordon, K. D. Gilkinson, W. P. Vass, and D. L. McKeown. 1999. Experimental otter trawling on a sandy bottom ecosystem of the Grand Banks of Newfoundland: Analysis of trawl bycatch and effects on epifauna. *Marine Ecology—Progress Series* 181:107–124.

Rex, M. A., C. T. Stuart, R. R. Hessler, J. A. Allen, H. L. Sanders, and G. D. F. Wilson. 1993. Global-scale latitudinal patterns of species diversity in the deep-sea benthos. *Nature* (London) 365:636–639.

Ricciardi, A., and J. B. Rasmussen. 1999. Extinction rates of North American freshwater faunas. *Conservation Biology* 13:1220–1222.

Roy, K., D. Jablonski, J. W. Valentine, and G. Rosenberg. 1998. Marine latitudinal diversity gradients: Tests of causal hypotheses. *Proceedings of the National Academy of Sciences of the United States of America* 95:3699–3702.

Sala, O. E., F. S. Chapin III, J. J. Armesto, E. Berlow, J. Bloomfield, R. Dirzo, E. Huber-Sanwald, L. F. Huenneke, R. Jackson, A. Kinzig, R. Leemans, D. Lodge, H. A. Mooney, M. Oesterheld, N. L. Poff, M. T. Sykes, B. H. Walker, M. Walkder, and D. H. Wall. 2000. Global biodiversity scenarios for the year 2100. *Science* 287:1770–1774.

Schimel, J. 1995. Ecosystem consequences of microbial diversity and community structure. *Ecological Studies* 113:239–254.

Schindler, D. W. 1997. Widespread effects of climatic warming on freshwater ecosystems in North America. *Hydrological Processes* 11:1043–1067.

Schlesinger, W. H. 1997. *Biogeochemistry: An Analysis of Global Change*. San Diego: Academic Press.

Smith, C. R., M. C. Austen, G. Boucher, C. Heip, P. A. Hutchings, G. M. King, I. Koike, P. J. D. Lambshead, and P. Snelgrove. 2000. Global change and biodiversity linkages across the sediment-water interface. *Bioscience* 50:1108–1120.

Snelgrove, P. V. R. 1999. Getting to the bottom of marine biodiversity: Sedimentary habitats: Ocean bottoms are the most widespread habitat on earth and support high biodiversity and key ecosystem services. *Bioscience* 49:129–138.

Snelgrove, P. V. R., M. C. Austen, G. Boucher, C. Heip, P. A. Hutchings, G. M. King, I. Koike, P. J. D. Lambshead, and C. R. Smith. 2000. Linking biodiversity above and below the marine sediment-water interface. *Bioscience* 50:1076–1088.

Snelgrove, P. V. R., T. H. Blackburn, P. A. Hutchings, D. M. Alongi, J. F. Grassle, H. Hummel, G. King, I. Koike, P. J. D. Lambshead, N. B. Ramsing, and V. Solis-Weiss. 1997. The importance of marine sediment biodiversity in ecosystem processes. *Ambio* 26:578–583.

Snelgrove, P. V. R., and C. A. Butman. 1994. Animal-sediment relationships revisited: Causes versus effect. *Oceanography and Marine Biology: An Annual Review* 32:111–177.

Southward, A. J., S. J. Hawkins, and M. T. Burrows. 1995. Seventy years' observations of changes in distribution and abundance of zooplankton and intertidal organisms in the western English Channel in relation to rising sea temperature. *Journal of Thermal Biology* 20:127–155.

Strange, E. M., K. D. Fausch, and A. P. Covich. 1999. Sustaining ecosystem services in human-dominated watersheds: Biohydrology and ecosystem processes in the South Platte River basin. *Environmental Management* 24:39–54.

Strayer, D. L., N. F. Caraco, J. J. Cole, S. Findlay, and M. L. Pace. 1999. Transformation of freshwater ecosystems by bivalves: A case study of zebra mussels in the Hudson River. *Bioscience* 49:19–27.

Swift, M. J., O. W. Heal, and J. M. Anderson. 1979. *Decomposition in Terrestrial Ecosystems*. Oxford: Blackwell.

Thorp, J. H., and A. P. Covich. 1991. *Ecology and Classification of North American Freshwater Invertebrates*. San Diego: Academic Press.

Trudgill, D., V. Blok, M. Phillips, S. Gowen, and M. Hahn. 1994. The importance of understanding the centres of origins of nematode pests. *Aspects of Applied Biology* 39:133–138.

Turner, R. E., and N. N. Rabalais. 1994. Coastal eutrophication near the Mississippi River delta. *Nature* 368:619–621.

Vitousek, P. M., J. D. Aber, R. W. Howarth, G. E. Likens, P. A. Matson, D. W. Schindler, W. H. Schlesinger, and D. G. Tilman. 1997. Human alteration of the global nitrogen cycle: Sources and consequences. *Ecological Applications* 7:737–750.

Vitousek, P. M., H. A. Mooney, J. Lubchenco, and J. M. Melillo. 1997. Human domination of Earth's ecosystems. *Science* 277:494–499.

Wagener, S. M., M. W. Oswood, and J. P. Schimel. 1998. Rivers and soils: Parallels in carbon and nutrient processing. *BioScience* 48:104–108.

Wall, D. H., G. Adams, and A. N. Parsons. In press. Soil biodiversity. In *Future Scenarios of Global Biodiversity,* eds. F. S. Chapin III, O. E. Sala, and E. Huber-Sannwald. New York: Springer-Verlag.

Wall, D. H., and J. C. Moore. 1999. Interactions underground: Soil biodiversity, mutualism, and ecosystem processes. *BioScience* 49:109–117.

Wall, D. H., and R. A. Virginia. 2000. The world beneath our feet: Soil biodiversity and ecosystem functioning. In *Nature and Human Society: The Quest for a Sustainable World,* eds. P. R. Raven and T. Williams, pp. 225–241. Washington, D.C.: National Academy of Sciences and National Research Council.

Wall Freckman, D., T. H. Blackburn, L. Brussaard, P. Hutchings, M. A. Palmer, and P. V. R. Snelgrove. 1997. Linking biodiversity and ecosystem functioning of soils and sediments. *Ambio* 26:556–562.

Wardle, D. A. 1995. Impacts of disturbance on detritus food webs in agro-ecosystems of contrasting tillage and weed management practices. *Advances in Ecological Research* 26:105–183.

Watling, L., and E. A. Norse. 1998. Disturbance of the seabed by mobile fishing gear: A comparison to forest clearcutting. *Conservation Biology* 12:1180–1197.

Wolters, V., W. L. Silver, D. E. Bignell, D. C. Coleman, P. Lavelle, W. van der Putten, P. de Ruiter, J. Rusek, D. H. Wall, D. A. Wardle, L. Brussaard, J. M. Dangerfield, V. K. Brown, K. E, Giller, D. U. Hooper, O. Sala, J. Tiedje, and J. A. van Veen. 2000. Global change effects on above- and belowground biodiversity in terrestrial ecosystems: Interactions and implications for ecosystem functioning. *BioScience* 50:1089–1098.

WRI (World Resources Institute). 1994. *World Resources 1994–1995.* New York: Oxford University Press.

Zak, D. R., D. B. Ringelberg, K. S. Pregitzer, D. L. Randlett, D. C. White, and P. S. Curtis. 1996. Soil microbial communities beneath *Populus grandidentata* grown under elevated atmospheric CO_2. *Ecological Applications* 6:257–262.

7

OCEANS AT RISK

Research Priorities in Marine Conservation Biology

Mark A. Hixon, P. Dee Boersma,
Malcolm L. Hunter Jr., Fiorenza Micheli, Elliott A. Norse,
Hugh P. Possingham, and Paul V. R. Snelgrove

In the end, we will conserve only what we love.
We will love only what we understand.
—Baba Dioum, Senegalese naturalist and poet

The marine environment encompasses a broad array of ecosystems, ranging from spectacular coral reefs and kelp forests to coastal mangroves, seagrass beds, and salt marshes; to expansive deep-sea plains interspersed with trenches, seamounts, ridges, and hydrothermal vents; to the vast open water column up to thousands of meters deep. Although the oceans cover 70.8 percent of the earth's surface, we know little about the three levels of biodiversity in the sea: genetic, species, and ecosystem (Norse 1993; NRC 1995; McAllister 1996; Ormond et al. 1997). However, we do know that marine biodiversity is extremely valuable to humankind, accounting for over 60 percent of the economic value of the biosphere (Costanza et al. 1997). Sea life provides five basic services to humans (Norse 1993; Daily 1997; Costanza 1999; Moberg and Folke 1999):

- *Ecosystem services.* Examples range from the global "biological pump" that sequesters atmospheric carbon dioxide and transports carbon to the deep sea, to the regional role of coral reefs and salt marshes in moderating coastal erosion.

- *Food.* About 20 percent of the animal protein consumed by humans comes from marine fisheries.

- *Medicines.* Marine organisms are increasingly found to contain biomedically active compounds, including antitumor agents.

- *Minerals and chemicals.* Examples include abiotic resources (fossil fuels, manganese, table salt, etc.), as well as chemicals derived from organisms (such as alginate from seaweeds and chitin from crustaceans, both used in a broad variety of food, medical, and technological applications).

- *Recreation and ecotourism.* Marine life, especially the charismatic megafauna of the seas (marine mammals, etc.), has inspired humankind since time immemorial. Recreational use of coral reefs supports many regional economies.

Despite their immense value, marine ecosystems are deteriorating rapidly due to human activities, especially physical alteration of habitat, overexploitation, species introductions, global climate change, and marine pollution (reviews by GESAMP 1990; Norse 1993; NRC 1995; Peterson and Estes 2000; Steneck and Carlton 2000). The most threatened systems are coastal, especially wetlands (including estuaries, salt marshes, and mangroves), coral reefs, and communities associated with the seafloor of the continental shelves.

Unfortunately, because detailed exploration of the oceans is a recent endeavor, we have little knowledge of what species are being lost (Irish and Norse 1996). It is nonetheless becoming increasingly clear that human-induced extinction in the sea is a sad and threatening reality (Carlton et al. 1991; Carlton 1993; Norse 1993; Vermeij 1993; Culotta 1994; Vincent and Hall 1996; Malakoff 1997; Casey and Myers 1998; Carlton et al. 1999; Roberts and Hawkins 1999). To date, it has been documented that only four marine snails, five seabirds, and three marine mammals have gone extinct in recent history due to human activities, but these cases are undoubtedly the mere tip of the iceberg of ongoing marine extinctions. Conservative estimates suggest that over 50,000 species of coral-reef organisms have already been lost (Carlton et al. 1999).

The goals of this chapter are twofold. First, we provide a brief review of the major differences between marine and terrestrial systems relevant to conservation biology, emphasizing the present limits of our knowledge. Second, we propose nine broad scientific research priorities in marine conservation biology, emphasizing six of these as particularly urgent. These priorities were derived from the literature, from canvassing experts (the thirty-five respondents listed in the acknowledgments), and from the workshop participants. Our list extends related research priorities previously proposed by Grassle et al. (1991) (table 7.1) and by the NRC (1995) (table 7.2), as well as the general research thrust described in the Sustainable Biosphere Initiative (Lubchenco et al. 1991). The focus is on natural science rather than socioeconomic issues, although the latter are also crucially important (e.g., Myers and Kent 1998; Costanza et al. 1999; chapter 10). Additionally, we focus more on urgent empirical priorities than on theoretical research questions, even though theory clearly has a role in all conservation efforts.

Addressing research priorities in marine conservation biology will require a broadly based interdisciplinary approach, including oceanography, toxicology, physiology, genetics, paleontology, taxonomy, systematics, and ecology.

TABLE 7.1. Research Questions Regarding Marine Biodiversity Previously Reported by Grassle et al. (1991).

1. Is the spectrum of environmental variation in marine and terrestrial ecosystems fundamentally different?
2. Are biogeographic patterns of biodiversity and ecosystem function determined by a combination of environmental patterns (i.e., are single-factor theories not viable)?
3. Are offshore primary production and nutrient cycling dominated by pelagic processes that determine biogeographic differences in biodiversity?
4. Do increases in environmental heterogeneity in space and time, including disturbance, increase biodiversity, especially in the coastal zones?
5. Do keystone species play a more important role in marine than in terrestrial ecosystems, and is that role more important in the lower latitudes?
6. Do species introductions have major consequences for marine ecosystem function?
7. Are extinctions less likely to occur in marine than in terrestrial systems?
8. Are increases in airborne and waterborne pollutants (including terrestrially derived disease species) and/or overfishing currently resulting in widespread changes in marine systems?
9. Have marine ecosystems and organisms developed less robust internal processes to respond to the low-magnitude short-term variations, and would this result in a reduced ability to respond to large-scale environmental changes?
10. Is redundancy of genes and species necessary for the long-term survival of marine ecosystems?
11. Is there greater genetic variation at the molecular level within species in marine environments than in terrestrial ones?

Source: F. Grassle, P. Lasserre, A. D. McIntyre, and G. C. Ray, "Marine Biodiversity and Ecosystem Function: A Proposal for an International Programme of Research." *Biology International* 23 (Special Issue, 1991): 1–19.

TABLE 7.2. Research Questions Regarding Marine Conservation Biology
Previously Reported by the National Research Council (1995).

Natural Variation in Biodiversity Pattern and Why Biodiversity Matters

1. How do genetic, species, and ecosystem diversity vary in space and time at different
   regional scales and within habitats within those regions? Examples of specific research
   questions are:
   - To what extent does the maintenance of local biodiversity (genetic or species)
     depend on linkages between distant populations, the dispersal between them, and
     the availability of suitable habitat?
   - How does genetic diversity within a species influence reproduction and population
     growth or susceptibility to epidemic disease?
   - To what extent do changes in biodiversity at one site within a region—or between
     regions—affect the biodiversity at another site or in another region?
   - What specific characteristics of a habitat directly or indirectly influence genetic and
     species diversity? For example, are there parallels in the origin and maintenance of
     coral reef and deep-sea biodiversity?
2. What is the functional significance of biodiversity at the genetic, species, and ecosystem
   levels? Are species within a functional group interchangeable? What might be learned
   from comparing and contrasting systems in terms of the functional significance of biodiversity? (For example, are there parallels between the ecological significance of
   microbial diversity as coral reef symbionts [zooxanthellae] and as open-ocean primary
   producers [picoplankton]?)
3. To what extent does the diversity of a community determine (a) "stability," (b) productivity, (c) resistance to invasion or disease, and (d) ability to recover from natural and
   human impacts? Equally important, how do these factors interact? Do high-diversity
   systems have higher or lower production than systems whose diversity has been
   impaired? What is the role of biological invasions in altering system production or
   energy flow?
4. How good are the estimates of genetic, species, and ecosystem biodiversity, and how do
   the limitations (i.e., understanding of the scale of error) influence an understanding of
   biodiversity patterns and of ecosystem structure and function?

Given the scope of challenges in conserving sea life, we cite key publications
to facilitate entry into the marine biodiversity and conservation literature.
For recent accounts in the popular press regarding the plight of the seas, we
suggest Marx (1999), Thorne-Miller (1999), and Woodard (2000).

Differences between Marine and Terrestrial Conservation Biology

There are substantial ecological differences between the sea and the land
(reviews by Steele 1974, 1985; Norse 1993; Cohen 1994; May 1994; NRC
1995; Field et al. 1998), several of which pose contrasts regarding challenges
in conservation research:

Human Impact on Processes Responsible for Biodiversity Change

1. What are the direct impacts on biodiversity of human-altered systems? That is, what is the variation in biodiversity over spatial and temporal scales relevant to the critical environmental issues? Examples of specific research questions are:
   - How do human influences on biodiversity differ from those caused by natural processes?
   - To what extent do human effects alter the probability of ecosystem collapse in different systems?
   - To what extent are particular changes in biodiversity due to human activities reversible?
   - Given the often direct impacts on certain target species within a region, are species within functional groups interchangeable within a system?
   - How does the addition or loss of species due to human activities affect community structure and resilience?

2. What are the indirect impacts on biodiversity of human-altered systems? Examples of specific research questions are:
   - What characteristics of species enhance susceptibility or provide immunity to precipitous declines?
   - In what types of habitats are alternative ecological communities stable?
   - Are threshold processes involved in precipitous declines (and the persistence of those declines) in biodiversity, and ultimately, in the risk of extinction of individual species?
   - Does genetic or species diversity provide a buffer against irreversible or massive perturbations?
   - What are the long-term effects of species replacements (e.g., exotic species) on ecosystem function?

Source: National Research Council, *Understanding Marine Biodiversity: A Research Agenda for the Nation* (Washington, D.C.: National Academy Press, 1995).

- *Oceans are huge and difficult to study.* Relative to that of air, the high density of seawater provides greater buoyancy and support for organisms, thereby distributing planktonic (drifting) and nektonic (swimming) sea life over a vast three-dimensional environment. Assuming that the terrestrial biosphere effectively averages 50 m thick, Childress (1983) estimates that the oceans constitute approximately 99.5 percent of the volume of the biosphere (21.0 percent less than 1,000 m deep and 78.5 percent greater than 1,000 m deep). Averaging some 4,000 m in depth and very difficult to access by humans, the abyssal plain is the largest continuous environment on the earth, accounting for about 42 percent of the area of the oceans and about 30 percent of the planet.

Marine biodiversity is high but largely unknown. As the probable original source of life on the earth, the sea has a greater diversity of animal phyla (fifteen phyla are solely marine vs. one solely terrestrial) than the land (Norse 1993; NRC 1995). Newly discovered marine ecosystems have revealed a variety of taxa previously unknown to science. For example, over twenty new families, one hundred new genera, and two hundred new species (with high endemism) have been discovered on and near deep-sea hydrothermal vents (Tunnicliffe 1991; Van Dover 2000). Hundreds of newly discovered species of invertebrates and fishes are described annually (WCMC 1992).

The immense volume of the oceans, combined with the small size of most marine plankton, offers particular challenges in documenting and understanding the diversity of tiny marine organisms. Entire species groups, such as the prochlorophyte picoplankton, have been discovered only recently (Chisholm et al. 1988; Olson et al. 1990). Undiscovered marine microorganisms may comprise at least thirty-four phyla and eighty-three classes (Corliss 1994).

As a consequence of the huge size, the variety of life and habitats, and the inaccessibility of the oceans, estimates of species diversity in the sea are very rough (reviews by WCMC 1992; Norse 1993; NRC 1995). Grassle and Maciolek (1992) estimate that ten million macroscopic animal species inhabit the deep sea. Snelgrove et al. (1997) report that there are some one hundred thousand species described from marine sediments in general, and perhaps one hundred million undescribed species. Reaka-Kudla (1997) estimates that up to nine million species, including about 30 percent of marine fish species (Roberts et al. 2000), inhabit coral reefs. Accounting for less than 1 percent of the ocean surface, coral reefs can be considered the tropical rain forests of the seas in terms of both high biodiversity and threatened status, with 50–70 percent of reefs under direct threat from human activities (Wilkinson 1999).

Genetic diversity among and within marine species is also high, including many cryptic and sibling species, but is very poorly understood (Palumbi 1992, 1994; Knowlton 1993; Avise 1998). The genetic diversity of marine viruses is particularly immense, with typically fifteen to forty visibly distinct genome sizes in a given sample of seawater, including high spatial and temporal variation (Fuhrman 1999). Genetic diversity of marine bacteria is also huge (Giovannoni et al. 1990). Given that less than 15 percent of the approximately 1.7–1.8 million described species worldwide are marine (WCMC 1992; May 1994), we are sorely in need of a comprehensive assessment of marine species richness and biodiversity. However, immediate conservation action to conserve this diversity is needed before we can afford the time for such an assessment.

- *Marine food webs are extremely complex.* The small size of marine phytoplankton strongly influences the structure of oceanic food webs. Most of the primary production in the sea is consumed by herbivores, unlike the situation on land, where most plant material dies and enters the decomposer food web (detritus nonetheless being the major food input to the deep sea). Indeed, the turnover of phytoplankton can be so high that there can be inverted pyramids of biomass, in which the standing crop of herbivorous zooplankton actually exceeds that of the phytoplankton. Thus, "ecological efficiency," the percentage of energy stored in one trophic level (say, plants) that becomes incorporated in the next higher trophic level (say, herbivores), tends to be greater in the sea than on land (about 20 percent vs. about 10 percent). This fact, combined with the virtual absence of size constraints in marine animals due to the support provided by water, results in marine food webs often having more trophic levels than those on land. Moreover, the relative morphological and evolutionary complexity of marine life results in many cases of omnivory (organisms that consume more than one trophic level below them) and mixotrophy (organisms that both photosynthesize and consume other organisms). These phenomena multiply the number of trophic linkages in marine food webs, creating ecologically complex communities. This complexity, including a variety of direct and indirect interactions, offers challenges in understanding, managing, and conserving marine life at the scale of entire ecosystems (see chapter 3).

- *Oceanic currents transport both larvae and pollutants.* The common reproductive mode of multicellular organisms in the sea (with the exception of sharks, marine mammals, and a few other groups) is broadcast spawning of small gametes, with dispersal occurring during a pelagic (open-water) larval stage. Especially in species that are associated with the seafloor as adults (benthos and demersal fishes), this life history pattern may result in local populations that are largely demographically and genetically open, linked by larval dispersal (a population of populations typically called a metapopulation). However, at some larger spatial scale, all marine populations are reproductively closed (Jones et al. 1999; Swearer et al. 1999). Poorly known is the extent of larval retention within local populations as well as the level of connectivity among populations (Cowen et al. 2000), both of which have important ramifications for conservation of marine species. Larval dispersal may in fact reduce the risk of extinction in some marine organisms. However, this hypothesis has not been adequately tested, and severe declines in populations of large predatory fishes due to overfishing suggest that reliance on this hypothesis is dangerous. In fact, increasing recognition of large numbers of sibling species in the oceans suggests that there may be substantial subdivision of species and populations (Knowlton 1993). Patterns of larval dispersal between spatially isolated local populations may result in genetically identifiable subpopulations or "stocks" delineated by ocean circulation patterns. At the same time, larval transport creates fluid boundaries between oceanic ecosystems and biogeographic regions. Consequently, geographical ranges of species tend to be larger in the sea than on land (reviews by Norse 1993; NRC 1995; Ormond et al. 1997). The spatial scales of resulting "large marine ecosystems" (Sherman et al. 1990) are typically greater than those of political boundaries, creating management problems for straddling fish stocks and highly migratory fishery species.

Finally, oceanic currents transport chemicals as well as organisms, so that it may be more difficult to contain the spread of both pollutants and exotic species in the sea vs. on land (excluding atmospheric dispersal). The oceans are biogeochemical sinks located downstream from land. Coastal marine ecosystems, particularly near the mouths of major rivers, receive tremendous inputs of terrestrially generated pollutants, especially via agricultural runoff.

Overall, compared to that of its terrestrial counterpart, the science of marine conservation biology is in its infancy for three related reasons: (1) the relative size and inaccessibility of the seas (e.g., SCUBA and research submersibles have been widely used by biologists only since the 1970s); (2) the relatively scant funding of marine conservation (e.g., in FY 1999, the U.S. National Parks Service received US$1.7 billion in funding, whereas the U.S. National Marine Sanctuary Program received US$14.3 million); and (3) the relatively sparse literature on this subject (e.g., until recently, terrestrial papers in *Conservation Biology* outnumbered marine papers thirteen to one). Scant funding and sparse literature may be a consequence of the general focus of marine science on issues other than conservation biology. All told, only 0.25 percent of the oceans are afforded some level of conservation protection, compared to over 5 percent of the land (McAllister 1996).

Research Priorities in Marine Conservation Biology

Given our general lack of detailed knowledge relevant to marine conservation biology, research priorities for the next decade (indeed, the next century) include the nature of marine biodiversity, the threats to that biodiversity, and the tools for conserving life in the sea. Nine research priorities are discussed in sequence below and listed with specific priority actions in box 7.1.

Understanding the Nature of Marine Biodiversity

It is difficult to conserve something that is unknown. The key questions underlying this section are: What are we trying to conserve, and how does it function? As is true on land, substantial efforts at documenting marine biodiversity at the level of ecosystems, species, and genes are important. There is also an urgent need for long-term monitoring of marine ecosystems at multiple spatial scales (e.g., CARICOMP in the Caribbean, Ogden and Gladfelter 1986). (Note, however, that biodiversity assessment and monitoring are not necessarily related, as the latter can focus on only a few indicator species.) Marine taxonomy and systematics must also be revitalized (Winston 1992; Feldmann and Manning 1992; Vecchione and Collette 1996). Less formal but equally important, local cultural knowledge can be a major source of information on marine species and their ecology, as exemplified by Johannes's (1981) studies of Pacific coral reef fishes and fisheries.

However, while immediate estimation of undescribed species richness is an important tool, pursuing long-term monitoring and alpha taxonomy should not delay resolution of urgent conservation issues. Therefore, we advocate a major initiative to organize the spatial and temporal information on marine biodiversity that is already available, constructing databases that will be of immediate use for conservation efforts. Geographic information systems (GIS) mapping patterns of marine habitats and biodiversity will be especially valuable for the establishment of substantial networks of no-take marine reserves. At the same time, we see the importance of intensive ecological studies of several key marine ecosystems. We thus propose three research priorities concerning the nature of marine biodiversity:

- *Map the distribution of and threats to biodiversity (ecosystems, species, genes).* Cataloging the distribution of marine life among ecosystem types is the traditional realm of marine biogeography (Briggs 1974). However, conservation efforts require detailed geographic information systems that overlay different aspects of biodiversity (as well as threats) at nested spatial scales, including species-area relationships (Johnston 1998). This endeavor is the focus of seascape ecology (Bartlett and Carter 1991; Ray 1991; Jones and Andrew 1992). GIS assessments of species richness are already being developed and are proving effective for some taxa, such as coral reef fishes (Roberts et al. 2000). An important issue is whether habitat-generating species (such as corals, large seaweeds) and visually dominant groups (such as fishes, large invertebrates) can provide accurate indices of total diversity within highly species-rich communities or those that are otherwise difficult to study (Ward et al. 1999).

We emphasize that much of the needed mapping will require not new research but assembly of existing data. Related research is required to determine the most effective ways to map marine habitats and associated biodiversity. Some approaches may be biophysically based, while others may be based on actual species composition at different locations. For select groups of organisms, indirect mapping tools can be ground-truthed by detailed field assessments.

Regarding genetic diversity, marine conservation genetics is in its infancy (Knowlton 1993; Palumbi 1992, 1994; Avise 1998). Expanded use of modern molecular methods will provide insight on population genetics, identify cryptic and sibling species, elucidate levels of endemism, and generate novel measures of genetic biodiversity. Of course, such analyses take time that we can ill afford, so immediate conservation action remains a higher priority than detailed assessments of genetic diversity.

GIS assessments should also incorporate ecosystem diversity, including coastal zones integrated with associated terrestrial systems (Ray and Hayden 1992; Ray 1996). Overall, these tools will identify regions of high species richness and high endemism, spawning and nursery habitats, migration routes, unique or special environments (such as hydrothermal vents), and particularly sensitive ecosystems. Overlying data on the nature, intensity, and urgency of threats to these regions will enable policy makers and managers to choose and rank areas for immediate conservation efforts, as well as identify sites for long-term study and monitoring.

- *Document temporal changes in biodiversity (ecosystems, species) over historical and geological time scales.* Because long-term knowledge of marine communities is lacking for most systems, the status of marine ecosystems and especially our perceptions of their pre-impact baseline conditions are both shifting rapidly (Pauly 1995; Sheppard 1995; Jackson 1997; Dayton et al. 1998, Steneck and Carlton 2000). Increasing evidence suggests that human impacts on marine ecosystems occurred long before the latter half of the twentieth century (reviews by Pauly 1995; Steneck and Carlton 2000). Examples have been documented in Alaska (Simenstad et al. 1978), the Caribbean (Jackson 1997), California (Dayton et al. 1998), the Gulf of Maine (Steneck and Carlton 2000), and elsewhere (Aronson 1990). Continued examination of historical accounts, as well as intensified monitoring, will provide better estimates of recent changes in marine communities and patterns of biodiversity. Over longer time scales, the fossil record can provide insight on the relative stability of species assemblages (e.g., Sepkoski 1992; Jackson 1995). Further exploration of existing community-level paleontological data will provide estimates of background rates of extinction in the sea. Such baselines will provide a basis of comparison for modern trends that may help to convince governments of the urgency of marine conservation issues.

- *Explore the ecological mechanisms driving population dynamics, structuring communities, and affecting biodiversity in several key ecosystems.* Despite advances in our knowledge of rocky intertidal communities and other reasonably well-studied systems, we know little of basic population and community ecology as they relate to marine conservation biology. Certainly, such information will come only with time and substantial effort, but this knowledge is crucial for understanding what naturally regulates marine populations and maintains species diversity. We advocate intensive ecological study of several ecologically important and representative systems, including: (1) small open-ocean fishes and krill, which constitute the major trophic links between plankton and high-seas fishery species; (2) coastal bottom-oriented fishes (and the seafloor communities of which they are a part), which are often severely overexploited and their habitats physically altered by trawling (see below); and (3) coral reefs, the most species-rich and among the most threatened of all marine ecosystems (see above).

At the population level, increasing use of genetic methods (Avise 1998), otolith (fish ear-stone) microchemistry (Swearer et al. 1999), larval tagging (Jones et al. 1999), and physical oceanography (Cowen et al. 2000) will answer questions regarding population connectivity, metapopulation structure, and stock boundaries. These issues are particularly important for designing and implementing marine protected areas (see below). The mechanisms driving and regulating population fluctuations in the sea are also largely unknown, although hypotheses abound (Rothschild 1986; Sale 1991; Cushing 1995; Caley et al. 1996). Increasing use of controlled field experiments, especially at larger spatial scales, will be especially informative, but not always possible (Hixon and Webster 2001). Central topics in community ecology relevant to conservation include the roles of habitat complexity, disturbance and succession, webs of direct and indirect interactions, and diversity-stability and diversity-function relationships (conceptual reviews by Huston 1994; Pickett et al. 1998).

Ultimately, such knowledge will allow us to answer crucial questions regarding extinction and conservation in the sea. Is our knowledge of terrestrial species relevant for conserving marine species (reviews by WCMC 1992; Carlton et al. 1999; Roberts and Hawkins 1999)? Does larval dispersal render marine species less prone to extinction than terrestrial species (Grassle et al. 1991)? What are minimal viable population sizes (review by Soulé 1987)—from both a demographic and a genetic perspective—in marine species with relatively open vs. closed populations? How (if at all) does the Allee effect (a decreasing population growth rate at low population sizes, reviewed by Courchamp et al. 1999) operate in the sea? Are increases in population sizes of protected marine species hastening the decline of other threatened species, such as sea otters contributing to the demise of white abalone in California (Tegner et al. 1996) or orcas causing the decline in sea otters off Alaska (Estes et al. 1998)?

Understanding the Threats to Marine Biodiversity

The ultimate threat to biodiversity in the sea, as on land, is human overpopulation and overconsumption spurred by technological developments, global commerce, ignorance, greed, and inadequate conservation programs. Proximally, there are five major categories of human-induced threats in the oceans (general reviews by GESAMP 1990; Norse 1993; NRC 1995; Botsford et al. 1997; Peterson and Estes 2000; Steneck and Carlton 2000):

- *Physical alteration of habitat.* Marine habitats are physically degraded by (1) coastal development (including mariculture practices that destroy or alter estuaries, salt marshes, and mangroves), dredging, shoreline erosion, and resulting sedimentation; (2) ocean mining and seafloor drilling; and (3) destructive fishing practices, including bottom trawling and dynamite fishing.
- *Overexploitation.* Widespread overfishing of the seas has resulted in stock collapses and alterations of population and community structure.
- *Species introductions.* Transported across the seas attached to ship hulls or contained in ballast water, introduced exotic species have potential competitive, predatory, and biological disturbance impacts on native species and communities.
- *Marine pollution.* Located downstream from land, the sea carries the burden of human-generated excess nutrients via runoff of sewage and fertilizers (as well as atmospheric deposition of nitrogenous compounds), petroleum spills, halogenated hydrocarbons (pesticides, PCBs, dioxins, etc.), heavy metals, radioactive waste, plastics, etc.
- *Global climate change.* Global warming is altering oceanic currents and patterns of marine productivity, is associated with increasing coral

bleaching, and threatens coastal marine ecosystems via sea level rise, while stratospheric ozone depletion is increasing UV-B penetration in the sea, with documented negative effects on marine organisms.

All these threats are important. We see global climate change as one of the most dangerous threats to marine biodiversity, but due to its universal nature, we defer to the chapter on that topic (chapter 9). Therefore, we advocate focused research priorities within each of the four remaining categories of threats. Despite this separation of impacts into distinct categories, it is important to keep in mind that anthropogenic threats to marine biodiversity are ubiquitous, cumulative, and synergistic, as exemplified by the sorry states of the Baltic Sea (Elmgren 1989), the Black Sea (Zaitsev 1992), and the Caribbean Sea (Hughes 1994). The four remaining categories of threats are as follows:

- *Document how physical alteration and fragmentation of habitats affect seafloor ecosystems.* Seafloor habitats are structured physically by geological and oceanographic processes (resulting in different seafloor types, such as sand vs. rock), and biologically by biogenic habitat builders (species such as mangroves, salt marsh grasses, seagrasses, seaweeds, and corals that create living space for other species), as well as by agents of biological disturbance (such as stingrays digging for buried prey and infaunal clams turning over sediments).

  Degradation and outright destruction of the physical structure of habitat is the most direct negative impact of humanity on the seas. Studies to date indicate that nearshore marine populations and communities are negatively affected by human-induced coastal erosion and sedimentation, artificial beach nourishment, and dredging (reviews by Nelson 1993; Maragos et al. 1996; Peterson et al. 2000). About half of all mangroves are already lost, increasingly due to construction of mariculture ponds, mostly for shrimp and prawn production (Fortes 1988; Kaly and Jones 1998). Offshore, where we have fewer data, trawl and dredge fisheries are clearly degrading crucial seafloor environments and communities (reviews by Jones 1992; Dayton et al. 1995; Watling and Norse 1998; Auster and Langton 1999). Between 1976 and 1991, Georges Bank off New England was trawled and dredged an average of 200 to 400 percent of its area annually. In impoverished developing nations, both demand and greed have led to degradation of coral reefs via dynamite, cyanide, and bleach fishing, all devastating forms of Malthusian overfishing (Pauly 1988). This tragedy is exacerbated by a developing "live fish" restaurant and aquarium trade (Johannes and Riepen 1995).

We need more accurate information on the extent of these impacts, the effects they have on biodiversity over different spatial and temporal scales, and how reversible they are. How much more marine habitat—including salt marshes, mangroves, coral reefs, seagrass beds, and kelp forests—can be lost without the substantial loss of associated species? If human impacts are ameliorated, how long will it take these ecosystems to recover (if they ever can)? Regarding habitat fragmentation, we need more information on population dynamics and movements of bottom-living fish and invertebrates among patches of seafloor habitat created by human activities (Butman et al. 1995; Irlandi and Crawford 1997; Micheli and Peterson 1999).

• *Document how overfishing alters marine food webs.* Compared to terrestrial systems, food extracted by humans from the sea comprises wild (rather than domestic) species at high (rather than low) trophic levels (Pauly et al. 1998). There is increasing evidence that the seas are severely overfished (e.g., Dayton et al. 1995; Pauly and Christensen 1995; Safina 1995; Botsford et al. 1997; FAO 1997; Jennings and Kaiser 1998; Pauly et al. 1998; Hall 1999; NRC 1999). The general question is: What are the direct (i.e., demographic and genetic) and indirect (i.e., community and ecosystem) ecological impacts of fishing on marine populations and communities? Direct effects have been the conceptual realm of fishery science, and we have no intention of presumptuously setting research priorities for fisheries management agencies. We are encouraged by ongoing fisheries research designed to predict sustainable levels of exploitation more realistically by increasing the accuracy of stock assessments (NRC 1998). Also encouraging is a trend for fisheries scientists to examine deeper questions regarding the effects of exploitation on both the demography (e.g., Musick, Berkeley, et al. 2000; Musick, Burgess, et al. 2000; Coleman et al. 2000; Parker et al. 2000) and the genetics (e.g., Ryman 1991; Smith et al. 1991) of targeted species.

However, the indirect effects of overfishing on entire marine communities are still seldom addressed due to the historical single-species focus of fishery biology and general lack of funding. Importantly, fished species are parts of larger systems, so altering their abundance often has ramifications for the entire system. Decreases in population sizes of both targeted and bycatch species can result in "biomass dominance shifts," due to alterations of competitive and predatory interactions (e.g., May et al. 1979; Fogarty and Murawski 1998). In extreme cases, such shifts can lead to systems switching between "alternate stable states" (e.g., Simenstad et al. 1978; Estes et al. 1998). An area of special concern is bycatch, which is the capture of nontargeted species by fisheries, and which accounts for about a quarter of the entire marine catch (reviews by Alverson et al. 1994; Hall 1996; Crowder and Murawski 1998). Most bycatch is discarded at sea, which provides a supplemental food source for seabirds, sharks, crabs, and other scavengers, with unknown community-wide consequences. These and other indirect effects of overfishing are examined in detail in chapter 3, but few are well documented and all clearly pose important research priorities.

- *Document how species introductions affect native species and alter community structure.* The largely inadvertent translocation of marine species, both as sessile adults attached to ship hulls and as larvae carried in ballast water, is increasingly well documented, but the ecological impacts of most introduced exotics are poorly known (reviews by Carlton 1985, 1999; Carlton and Geller 1993; Ruiz et al. 1997, 1999). Unfortunately, in many systems we do not always know which species are truly native vs. introduced vs. cryptogenic (of unknown history; Carlton 1996). Indeed, more than one thousand common intertidal and subtidal species may have been introduced by ships worldwide between 1500 and 1800 (Carlton 1999), and an estimated three thousand species are in transit daily in the ballast water of ships (Carlton and Geller 1993). Mariculture programs may introduce not only cultured species, but also their close associates and their diseases (Naylor et al. 2000). The rate at which marine diseases are spread by human activities and especially the consequences (such as the Caribbean-wide pandemic affecting long-spined urchins in the early 1980s; Lessios 1988) are areas of special concern (Harvell et al. 1999).

  There is need for the geographic origin of marine species to be elucidated by paleontological data (the recent fossil record; e.g., Pandolfi and Minchin 1995; Pandolfi and Jackson 2001), anthropological information (such as human middens; e.g., Simenstad et al. 1978; Borque 1996), and genetic analyses (Avise 1998). Ultimately, there is need to understand the determinants of whether and how introduced species survive and spread in their new habitats, and whether a successfully invasive introduced species comes to dominate its new habitat via competition, predation, or biological disturbance. Of particular concern are the ecological impacts of genetically modified mariculture species, such as the effects of cultured salmon on wild stocks (Naylor et al. 2000).

- *Document how the increasing scale of human-induced eutrophication alters ecosystems.* Given that the oceans are downstream from land, there are numerous pollutants in the seas. These include excess nutrients via runoff of sewage and fertilizers (as well as atmospheric deposition of nitrogenous compounds), petroleum spills, halogenated hydrocarbons (pesticides, PCBs, dioxins, etc.), heavy metals, plastics (including abandoned fishing nets), and radioactive waste (GESAMP 1990; Kennish 1998; Sindermann 1996). GESAMP estimates that the sources of marine pollutants are runoff and land-based discharge (44 percent), atmospheric deposition (33 percent), maritime transportation (12 percent), ocean dumping (10 percent), and offshore oil production (1 percent). Most coastal pollutants are deposited directly from the land (runoff), whereas most open-ocean pollutants are deposited from the atmosphere.

While all these pollutants pose substantial threats, we believe that human-induced eutrophication of the seas is the most pressing problem and the highest research priority regarding marine pollution. Human activities now add at least as much fixed nitrogen to terrestrial ecosystems as do all natural sources combined (Schlesinger 1997; Vitousek et al. 1997), and the oceans receive this nitrogenous pollution from both coastal runoff and via atmospheric deposition. The resulting eutrophication induces macroalgal and microbial blooms, some of which are highly toxic (Hallegraeff 1993), produce expanding "dead zones" of decomposing primary producers (Turner and Rabalais 1994), and alter associated marine communities (Diaz and Rosenberg 1995; Burkholder 1998; Micheli 1999). Ample data show how individual sites respond to organic enrichment, but we are less able to predict the consequences of the increasing scale of eutrophication in terms of specific threats to marine species and general effects on ecosystem services.

Taking Action to Conserve Marine Biodiversity

We emphasize that, although we know relatively little about the nature of and threats to marine biodiversity, enough is known to justify immediate conservation action (Ludwig et al. 1993; Costanza et al. 1998). By the time we understand enough about the oceans to implement conservation policies that are strongly empirically based, it may well be too late (see also Johannes 1998). Scientific uncertainty (Ludwig et al. 1993) and the precautionary principle (Earll 1992) dictate that we take substantial conservation action now, preferably in the experimental context of true adaptive management (Walters and Hilborn 1978; Holling 1978). We see the implementation of marine protected areas as the most immediate and effective conservation action, and the development of marine restoration ecology as an important general priority:

- *Implement and evaluate networks of marine protected areas.* Given that less than 0.25 percent of the oceans are now offered some level of protection (McAllister 1996), the immediate implementation of a substantial network of no-take marine reserves is crucial (Murray et al. 1999). Protecting regions of the sea from all direct human impacts prevents physical alteration of habitat, overexploitation, and point-source pollution (but not global climate change or widespread pollution, and perhaps not species introductions). Such ecosystem-based management applies the precautionary principle by conserving entire systems that we do not fully understand (Earll 1992; Ludwig et al. 1993; Griffis and Kimball 1996; Agardy 1997; Murray et al. 1999; Palumbi 2001).

The first step in implementing reserves is using available data to map regional patterns of habitat and biodiversity, then selecting initial sites for protection (Leslie et al. in press). Identification of both representative and crucial ecosystems for protection (both relatively pristine and degraded) requires detailed GIS assessments (see above). Once provisional networks of no-take marine reserves are in place, we can begin to learn how they function ecologically and, politics permitting, implement adaptive management in terms of evaluating and optimizing the number, size, and spacing of reserves. At present, the rule-of-thumb recommendation is that at least 20 percent of the oceans be protected such that all ecosystems are represented proportionally within no-take reserves (Murray et al. 1999). Currently, the only scientific basis for this percentage is that fishery biologists believe that at least 20 percent of a spawning stock must be conserved to ensure population viability (Bohnsack et al. manuscript). Of course, the siting of reserves is also crucial. For example, conservation of Pacific salmon and other species that migrate between fresh and saltwater will require integrated land-and-sea reserves (see Lichatowich et al. 2000).

Evaluating and optimizing the effectiveness of existing marine reserve networks will require substantial research in several areas *after* reserves are already implemented in an experimental framework. First, knowledge of the level of connectivity of open populations via larval dispersal is essential for understanding whether populations within reserves replenish those outside (the "seeding effect"). Second, knowledge of the movement patterns of animals into and out of reserves is required to understand whether individuals that settle and grow within reserves eventually move to adjacent unprotected areas, thereby augmenting fisheries (the "spillover effect"). Documenting this phenomenon requires knowledge of home ranges, erratic movements, dispersal,

BOX 7.1. **General Research Priorities and Specific Action Items in Marine Conservation Biology**

Understanding the Nature of Marine Biodiversity

*1. Map the distribution of and threats to biodiversity (ecosystems, species, genes).
   - Use geographic information systems to assemble existing data.
   - Determine whether habitat-generating and visually dominant species can provide accurate indices of total biodiversity.
   - Identify regions of high species richness, crucial habitats, special environments, and sensitive ecosystems.
2. Document temporal changes in biodiversity (ecosystems, species) over historical and geological time scales.
   - Examine historical records to document recent changes in ecosystems and biodiversity.
   - Examine paleontological data to estimate background rates of extinction.
3. Explore the ecological mechanisms driving population dynamics, structuring communities, and affecting biodiversity in several key ecosystems.
   - Focus on understanding the ecology of small open-ocean fishes and krill, coastal bottom-oriented fishes, and coral reefs.
   - Determine population boundaries and connectivity, as well as natural mechanisms of population regulation.
   - Determine community-level mechanisms that naturally maintain biodiversity.

Understanding the Threats to Marine Biodiversity

*1. Document how physical alteration and fragmentation of habitats affect seafloor ecosystems.
   - Determine relationship between habitat loss and species loss.
   - Determine effects of habitat fragmentation on population viability.

and migrations of mobile juveniles and adults, and thus innovations in tagging and telemetry. Third, simultaneous socioeconomic studies are necessary to document feedback between exploited marine populations and human society. Again, such knowledge should be used in the framework of adaptive management after immediate implementation of provisional reserves based on existing data.

- *Develop the science of marine restoration ecology.* Given the high level of degradation already suffered by many coastal ecosystems—especially estuarine salt marshes and mangroves, coral reefs, and seagrass beds—

*2. Document how overfishing alters marine food webs.
  • Document cascading effects of declining large vertebrates (predatory fishes, sea turtles, seabirds, and marine mammals) on ecosystem function and stability.
  • Explore effects of biomass dominance shifts on ecosystem function and stability.
  • Document effects of discarded bycatch on ecosystem function and stability.
*3. Document how species introductions affect native species and alter community structure.
  • Determine mechanisms by which introduced exotic species become established and displace native species.
  • Document impacts of exotic disease organisms introduced by human activities.
  • Document effects of genetically modified mariculture species on native species.
*4. Document how the increasing scale of human-induced eutrophication alters ecosystems.
  • Determine effects of algal blooms and resulting dead zones on biodiversity and ecosystems.

Taking Action to Conserve Marine Biodiversity

*1. Implement and evaluate networks of marine protected areas.
  • Implement adaptive management to evaluate and optimize siting, number, size, and spacing of reserves.
  • Document ecological changes inside vs. outside reserves and test whether and how populations inside reserves replenish those outside.
 2. Develop the science of marine restoration ecology.
  • Develop methods to enhance recovery of degraded ecosystems.

  Note: Asterisks denote the six highest priorities for the decade.

research on how to augment recovery of these systems will be useful if and when environmental assaults are ameliorated. The science of marine restoration ecology is in its infancy (Thayer 1992; NRC 1994) and focuses mostly on estuarine systems (Kennish 2000; Zedler 2000). Besides controlling deleterious inputs, case studies to date on enhancing recovery include cleanup following oil spills (Doerffer 1992; see also Paine et al. 1996; Peterson 2000), rehabilitation of mangroves (Day et al. 1999), and transplants of corals and seagrass (Rinkevich 1995; Bortone 2000). General issues and research priorities in restoration ecology are reviewed in chapter 11.

Conclusions

Because funding constraints often force prioritization of priorities, box 7.1 indicates what we believe to be the six most important research thrusts needed to conserve marine biodiversity. Importantly, so little is known about all aspects of marine biodiversity and its demise that precautionary action is needed immediately, even before new research initiatives (Ludwig et al. 1993; Costanza et al. 1998). The most prudent precautionary measure is to set aside areas of the oceans for protection from all direct human activities. Establishing substantial networks of no-take marine reserves in an adaptive management framework will allow us to separate the effects of direct and local human impacts (physical alteration of habitat, overexploitation, local species introductions, and point-source pollutants) from indirect and ubiquitous effects (wide-ranging species introductions, global climate change, widespread pollutants), as well as to examine experimentally the most effective size, shape, spacing, and location of reserves (Murray et al. 1999). Additional precautionary action would be to reverse the burden of proof (Dayton 1998), whereby the instigators of potentially deleterious activities would have to demonstrate that their proposed actions are not a threat (rather than the present situation in which environmental regulators have to document that a deleterious effect has occurred). Success of these measures will require environmental scientists to educate the public and policy makers concerning the importance of action despite scientific uncertainty. Fundamental change is needed regarding the role of the scientist-citizen in society so that scientific advocacy is no longer considered an oxymoron (Hixon 2000). Once precautionary policies are implemented, major research efforts are required to effect adaptive management and enlightened conservation efforts that balance the necessity of conserving life in the sea with society's need to exploit marine resources. The risks of continued research without immediate conservation action are too great to ignore.

ACKNOWLEDGMENTS

We thank G. H. Orians and M. E. Soulé for organizing the workshop, and M. E. Soulé for reviewing the manuscript. Although not all suggestions were incorporated, this compilation benefited immensely from ideas generously shared by dedicated colleagues: A. C. Alcala, P. J. Auster, J. C. Avise, R. T. Barber, J. A. Bohnsack, J. T. Carlton, J. J. Childress, A. N. Cohen, D. P. Costa, P. K. Dayton, J. A. Estes, T. J. Goreau, M. E. Hay, R. Hilborn, R. E. Johannes, L. S. Kaufman, J. Lindholm, T. R. McClanahan, J. C. Ogden, R. T. Paine, S. R. Palumbi, W. G. Pearcy, C. H. Peterson, S. L. Pimm, N. N. Rabalais, G. C. Ray, G. R. Russ, M. Ruth, D. R. Schiel, M. P. Sissenwine, J. A. Sobel, R. S. Steneck, M. J. Tegner, L. E. Watling, and S. L. Williams. Special thanks to J. T. Carlton, J. A. Estes, L. S. Kaufman, S. R. Palumbi, C. H. Peterson, and R. S. Steneck for sharing drafts of upcoming book chapters. MAH dedicates this work to his children, Chelsea and Sean, in hopes that they, their children, and all future generations will benefit from healthy seas. Manuscript preparation partially supported by NSF grant OCE-96-17483 (Hixon).

LITERATURE CITED

Agardy, M. T. 1997. *Marine Protected Areas and Ocean Conservation.* San Diego: Academic Press.

Alverson, D. L., M. H. Freeberg, J. G. Pope, and S. A. Murawski. 1994. *A global assessment of fisheries bycatch and discards.* FAO Fisheries Technical Paper 339. Rome: FAO.

Aronson, R. B. 1990. Onshore-offshore patterns of human fishing activity. *Palaios* 5: 88–93.

Auster, P. J., and R. W. Langton. 1999. The effects of fishing on fish habitat. *American Fisheries Society Symposium* 22: 150–187.

Avise, J. C. 1998. Conservation genetics in the marine realm. *Journal of Heredity* 89: 377–382.

Bartlett, D., and R. W. G. Carter. 1991. Seascape ecology: The landscape ecology of the coastal zone. *Ecology* (CSSR) 10: 43–53.

Bohnsack, J., B. Causey, M. Crosby, R. Griffis, M. Hixon, T. Hourigan, K. Koltes, J. Maragos, A. Simons, and J. Tilmant. In review. *A Rationale for a Minimum of 20% No-Take Protection for Coral Reefs.*

Borque, B. J. 1996. *Diversity and Complexity in Prehistoric Maritime Societies: A Gulf of Maine Perspective.* New York: Plenum Press.

Bortone, S. A., ed. 2000. *Seagrasses: Monitoring, Ecology, Physiology, and Management.* Boca Raton, Fla: CRC Press.

Botsford, L. W., J. C. Castilla, and C. H. Peterson. 1997. The management of fisheries and marine ecosystems. *Science* 277: 509–515.

Briggs, J. C. 1974. *Marine Zoogeography.* New York: McGraw-Hill.

Burkholder, J. M. 1998. Implications of harmful microalgae and heterotrophic dinoflagellates in management of sustainable marine fisheries. *Ecological Applications,* 8(1) Supplement: S37–S62.

Butman, C. A., J. T. Carlton, and S. R. Palumbi. 1995. Whaling effects on deep-sea biodiversity. *Conservation Biology* 9: 462–464.

Caley, M. J., M. H. Carr, M. A. Hixon, T. P. Hughes, G. P. Jones, and B. A. Menge. 1996. Recruitment and the local dynamics of open marine populations. *Annual Review of Ecology and Systematics* 27: 477–500.

Carlton, J. T. 1985. Transoceanic and interoceanic dispersal of coastal marine organisms: The biology of ballast water. *Oceanography and Marine Biology Annual Review* 23: 313–371.

Carlton, J. T. 1993. Neoextinctions of marine invertebrates. *American Zoologist* 33: 499–509.

Carlton, J. T. 1996. Biological invasions and cryptogenic species. *Ecology* 77: 1653–1655.

Carlton, J. T. 1999. The scale and ecological consequences of biological invasions in the world's oceans. Pages 195–212 in O. T. Sandlund, P. J. Schei, and Å. Viken, eds., *Invasive Species and Biodiversity Management.* Dordrecht, Netherlands: Kluwer Academic Publishers.

Carlton, J. T., and J. B. Geller. 1993. Ecological roulette: The global transport of non-indigenous marine organisms. *Science* 261: 78–82.

Carlton, J. T., J. B. Geller, M. L. Reaka-Kudla, and E. A. Norse. 1999. Historical extinctions in the sea. *Annual Review of Ecology and Systematics* 30: 515–538.

Carlton, J. T., G. J. Vermeij, D. R. Lindberg, D. A. Carlton, and E. Dudley. 1991. The first historical extinction of a marine invertebrate in an ocean basin: The demise of the eelgrass limpet *Lottia alveus*. *Biological Bulletin* 180: 72–80.

Casey, J. M., and R. A. Myers. 1998. Near extinction of a large, widely distributed fish. *Science* 281: 690–692.

Childress, J. J. 1983. Oceanic biology: Lost in space? Pages 127–135 in P. Brewer, ed., *Oceanography: The Present and Future.* New York: Springer-Verlag.

Chisholm, S. W., R. J. Olson, E. R. Zettler, R. Goericke, J. B. Waterbury, and N. A. Welschmeyer. 1988. A novel free-living prochlorophyte abundant in the oceanic euphotic zone. *Nature* 334: 340–343.

Cohen, J. E. 1994. Marine and continental food webs: Three paradoxes? *Philosophical Transactions of the Royal Society,* Series B 343: 57–69.

Coleman, F. C., C. C. Koenig, and C. B. Grimes. 2000. Long-lived reef fishes: The grouper-snapper complex. *Fisheries* 25(3): 14–21.

Corliss, J. O. 1994. An interim utilitarian ("user friendly") hierarchical classification and characterization of the protists. *Acta Protozoologica* 33: 1–51.

Costanza, R. 1999. The ecological, economic, and social importance of the oceans. *Ecological Economics* 31: 199–213.

Costanza, R., F. Andrade, P. Antunes, M. van den Belt, D. Boersma, D. F. Boesch, F. Catarino, S. Hanna, L. Limburg, B. Low, M. Molitor, J. G. Pereira, S. Rayner, R. Santos, J. Wilson, and M. Young. 1998. Principles for sustainable governance of the oceans. *Science* 281: 198–199.

Costanza, R., F. Andrade, P. Antunes, M. van den Belt, D. Boesch, D. Boersma, F. Catarino, S. Hanna, L. Limburg, B. Low, M. Molitor, J. G. Pereira, S. Rayner, R. Santos, J. Wilson, and M. Young. 1999. Ecological economics and sustainable governance of the oceans. *Ecological Economics* 31: 171–187.

Costanza, R., R. d'Arge, R. d. Groot, S. Farber, M. Grasso, B. Hannon, K. Limburg, S. Naeem, R. V. O'Neill, J. Paruelo, R. G. Raskin, P. Sutton, and M. van den Belt. 1997. The value of the world's ecosystem services and natural capital. *Nature* 387: 253–260.

Courchamp, F., T. Clutton-Brock, and B. Grenfell. 1999. Inverse density dependence and the Allee effect. *Trends in Ecology and Evolution* 14: 405–410.

Cowen, R. K., K. M. M. Lwiza, S. Sponaugle, C. B. Paris, and D. B. Olson. 2000. Connectivity of marine populations: Open or closed? *Science* 287: 857–859.

Crowder, L. B., and S. A. Murawski. 1998. Fisheries bycatch: Implications for management. *Fisheries* 23(6): 8–17.

Culotta, E. 1994. Is marine biodiversity at risk? *Science* 263: 918–920.

Cushing, D. 1995. *Population production and regulation in the sea: A fisheries perspective.* Cambridge, U.K.: Cambridge University Press.

Daily, G. C., ed. 1997. *Nature's Services.* Washington, D.C.: Island Press.

Day, S., W. J. Streever, and J. J. Watts. 1999. An experimental assessment of slag as a substrate for mangrove rehabilitation. *Restoration Ecology* 7: 139–144.

Dayton, P. K. 1998. Reversal of the burden of proof in fisheries management. *Science* 279: 821–822.

Dayton, P. K., M. J. Tegner, P. B. Edwards, and K. L. Riser. 1998. Sliding baselines, ghosts, and reduced expectations in kelp forest communities. *Ecological Applications* 8: 309–322.

Dayton, P. K., S. F. Thrush, M. T. Agardy, and R. J. Hofman. 1995. Environmental effects of marine fishing. *Aquatic Conservation: Marine and Freshwater Ecosystems* 5: 205–232.

Diaz, R. J., and R. Rosenberg. 1995. Marine benthic hypoxia: A review of its ecological effects and the behavioural responses of benthic macrofauna. *Oceanography and Marine Biology Annual Review* 33: 245–303.

Doerffer, J. W., ed. 1992. *Oil Spill Response in the Marine Environment.* Oxford, U.K.: Pergamon Press.

Earll, R. C. 1992. Commonsense and the precautionary principle: An environmentalist's perspective. *Marine Pollution Bulletin* 24: 182–186.

Elmgren, R. 1989. Man's impact on the ecosystem of the Baltic Sea: Energy flows today and at the turn of the century. *Ambio* 18: 326–332.

Estes, J. A., M. T. Tinker, T. M. Williams, and D. F. Doak. 1998. Killer whale predation on sea otters linking coastal with oceanic ecosystems. *Science* 282: 473–476.

FAO. 1997. *Review of the State of World Fishery Resources: Marine Fisheries.* FAO Fisheries Circular 920. Rome: FAO.

Feldmann, R. M., and R. B. Manning. 1992. Crisis in systematic biology in the "Age of Biodiversity." *Journal of Paleontology* 66: 157–158.

Field, C. B., M. J. Behrenfeld, J. T. Randerson, and P. Falkowski. 1998. Primary production of the biosphere: Integrating terrestrial and oceanic components. *Science* 281: 237–240.

Fogarty, M. J., and S. A. Murawski. 1998. Large-scale disturbance and the structure of marine ecosystems: Fishery impacts on Georges Bank. *Ecological Applications* 8(1) Supplement: S6–S22.

Fortes, M. D. 1988. Mangrove and seagrass beds of East Asia: Habitats under stress. *Ambio* 17: 207–213.

Fuhrman, J. A. 1999. Marine viruses and their biogeochemical and ecological effects. *Nature* 399: 541–548.

GESAMP. 1990. *The State of the Marine Environment.* Oxford, U.K.: Blackwell Scientific Publications.

Giovannoni, S. J., T. B. Britschgi, C. L. Moyer, and K. G. Field. 1990. Genetic diversity in Sargasso Sea bacterioplankton. *Nature* 345: 60–63.

Grassle, J. F., P. Lasserre, A. D. McIntyre, and G. C. Ray. 1991. Marine biodiversity and ecosystem function: A proposal for an international programme of research. *Biology International* (Special Issue) 23: 1–19.

Grassle, J. F., and N. J. Maciolek. 1992. Deep-sea species richness: Regional and local diversity estimates from quantitative bottom samples. *American Naturalist* 139: 313–341.

Griffis, R. B., and K. W. Kimball. 1996. Ecosystem approaches to coastal and ocean stewardship. *Ecological Applications* 6: 708–712.

Hall, M. A. 1996. On bycatches. *Reviews in Fish Biology and Fisheries* 6: 319–352.

Hall, S. J. 1999. *The Effects of Fishing on Ecosystems and Communities.* Oxford, U.K.: Blackwell Science.

Hallegraeff, G. M. 1993. A review of harmful algal blooms and their apparent global increase. *Phycologia* 32: 79–99.

Harvell, C. D., K. Kim, J. M. Burkholder, R. R. Colwell, P. R. Epstein, D. J. Grimes, E. E. Hofmann, E. K. Lipp, A. D. M. E. Osterhaus, R. M. Overstreet, J. W. Porter, G. W. Smith, and G. R. Vasta. 1999. Emerging marine diseases—Climate links and anthropogenic factors. *Science* 285: 1505–1510.

Hixon, M. A. 2000. Environmental advocacy: Dilemma of the citizen-scientist. *Reflections* Special Issue 4: 13.

Hixon, M. A., and M. S. Webster. 2001. Density dependence in marine fishes: Coral-reef populations as model systems. In P. F. Sale, ed., *Advances in the Ecology of Fishes on Coral Reefs.* San Diego: Academic Press.

Holling, C. S. 1978. *Adaptive Environmental Assessment and Management.* New York: John Wiley and Sons.

Hughes, T. P. 1994. Catastrophes, phase-shifts, and large-scale degradation of a Caribbean coral reef. *Science* 265: 1547–1551.

Huston, M. A., ed. 1994. *Species Diversity: The Coexistence of Species on Changing Landscapes.* Cambridge, U.K.: Cambridge University Press.

Irish, K. E., and E. A. Norse. 1996. Scant emphasis on marine biodiversity. *Conservation Biology* 10: 680.

Irlandi, E. A., and M. K. Crawford. 1997. Habitat linkages: The effect of subtidal saltmarshes and adjacent subtidal habitats on abundance, movement, and growth of an estuarine fish. *Oecologia* 110: 231–236.

Jackson, J. B. C. 1995. Constancy and change of life in the sea. Pages 45–54 in J. H. Lawton and R. M. May, eds., *Extinction Rates.* Oxford, U.K.: Oxford University Press.

Jackson, J. B. C. 1997. Reefs since Columbus. *Coral Reefs* 16 (supplement): S23–S32.

Jennings, S., and M. J. Kaiser. 1998. The effects of fishing on marine ecosystems. *Advances in Marine Biology* 34: 201–352.

Johannes, R. E. 1981. *Words of the Lagoon: Fishing and Marine Lore in the Palau District of Micronesia.* Berkeley: University of California Press.

Johannes, R. E. 1998. The case for data-less marine resource management: Examples from tropical nearshore finfisheries. *Trends in Ecology and Evolution* 13: 243–246.

Johannes, R. E., and M. Riepen. 1995. *Environmental, Economic and Social Implications of the Live Reef Fish Trade in Asia and the Western Pacific.* Jakarta, Indonesia: The Nature Conservancy.

Johnston, C. A. 1998. *GIS: In ecology.* Oxford, U.K.: Blackwell.

Jones, J. B. 1992. Environmental impact of trawling on the seabed: A review. *New Zealand Journal of Marine and Freshwater Research* 26: 59–67.

Jones, G. P., and N. L. Andrew. 1993. Temperate reefs and the scope of seascape ecology. Pages 63–76 in C. N. Battershill, D. R. Schiel, G. P. Jones, R. G. Creese, and A. B. MacDiarmid, eds., *Proceedings of the Second International Temperate Reef Symposium.* NIWA Marine, Wellington, New Zealand.

Jones, G. P., M. J. Milicich, M. J. Emslie, and C. Lunow. 1999. Self-recruitment in a coral reef fish population. *Nature* 402: 802–804.

Kaly, U. L., and G. P. Jones. 1998. Mangrove restoration: A potential tool for coastal management in tropical developing countries. *Ambio* 27: 656–661.

Kennish, M. J. 1998. *Pollution Impacts on Marine Biotic Communities.* Boca Raton, Fla.: CRC Press.

Kennish, M. J., ed. 2000. *Estuary Restoration and Maintenance: The National Estuary Program.* Boca Raton, Fla.: CRC Press.

Knowlton, N. 1993. Sibling species in the sea. *Annual Review of Ecology and Systematics* 24: 189–216.

Leslie, H., M. Ruckelshaus, I. Ball, H. Possingham, and S. Andelman. In press. *Using Siting Algorithms in the Design of Marine Reserve Networks.*

Lessios, H. A. 1988. Mass mortality of *Diadema antillarum* in the Caribbean: What have we learned? *Annual Review of Ecology and Systematics* 19: 371–393.

Lichatowich, J. A., G. R. Guido, S. M. Whidden, and C. R. Steward. 2000. Sanctuaries for Pacific salmon. Pages 675–686 in E. E. Knudsen, C. R. Steward, D. D. MacDonald, J. E. Williams, and D. W. Reiser, eds., *Sustainable Fisheries Management: Pacific Salmon.* Boca Raton, Fla.: Lewis Publishers.

Lubchenco, J., A. M. Olson, L. B. Brubaker, S. R. Carpenter, M. M. Holland, S. P. Hubbell, S. A. Levin, J. A. MacMahon, P. A. Matson, J. M. Melillo, H. A. Mooney, C. H. Peterson, H. R. Pulliam, L. A. Real, P. J. Regal, and P. G. Risser. 1991. The sustainable biosphere initiative: An ecological research agenda. *Ecology* 72: 371–412.

Ludwig, D., R. Hilborn, and C. Walters. 1993. Uncertainty, resource exploitation, and conservation: Lessons from history. *Science* 260: 17, 36.

Malakoff, D. 1997. Extinction on the high seas. *Science* 277: 486–488.

Maragos, J. E., M. P. Crosby, and J. W. McManus. 1996. Coral reefs and biodiversity: A critical and threatened relationship. *Oceanography* 9: 83–99.

Marx, W. 1999. *The Frail Ocean.* Point Roberts, Wash.: Hartley and Marks Publishers.

May, R. M. 1994. Biological diversity: Differences between land and sea. *Philosophical Transactions of the Royal Society,* Series B 343: 105–111.

May, R. M., J. R. Beddington, C. W. Clark, S. J. Holt, and R. M. Laws. 1979. Management of multispecies fisheries. *Science* 205: 267–277.

McAllister, D. E. 1996. *The Status of the World Ocean and Its Biodiversity.* Ottawa, Canada: Ocean Voice International.

Micheli, F. 1999. Eutrophication, fisheries, and consumer-resource dynamics in marine pelagic systems. *Science* 285: 1396–1398.

Micheli, F., and C. H. Peterson. 1999. Estuarine vegetated habitats as corridors for predator movements. *Conservation Biology* 13: 869–881.

Moberg, F., and C. Folke. 1999. Ecological goods and services of coral reef ecosystems. *Ecological Economics* 29: 215–233.

Murray, S. N., R. F. Ambrose, J. A. Bohnsack, L. W. Botsford, M. H. Carr, G. E. Davis, P. K. Dayton, D. Gotshall, D. R. Gunderson, M. A. Hixon, J. Lubchenco, M. Mangel, A. MacCall, D. A. McArdle, J. C. Ogden, J. Roughgarden, R. M. Starr, M. J. Tegner, and M. M. Yoklavich. 1999. No-take reserve networks: Protection for fishery populations and marine ecosystems. *Fisheries* 24: 11–25.

Musick, J. A., S. A. Berkeley, G. M. Cailliet, M. Camhi, G. Huntsman, M. Nammack, and M. L. Warren Jr. 2000. Protection of marine fish stocks at risk of extinction. *Fisheries* 25(3): 6–8.

Musick, J. A., G. Burgess, G. Cailliet, M. Camhi, and S. Fordham. 2000. Management of sharks and their relatives (Elasmobranchii). *Fisheries* 25(3): 9–13.

Myers, N., and J. Kent. 1998. *Perverse Subsidies: Tax $s Undercutting Our Economies and Environments Alike.* Winnipeg, Canada: International Institute for Sustainable Development.

Naylor, R. L., R. J. Goldberg, J. H. Primavera, N. Kautsky, M. C. M. Beveridge, J. Clay, C. Folke, J. Lubchenco, H. Mooney, and M. Troell. 2000. Effect of aquaculture on world fish supplies. *Nature* 405: 1071–1024.

Nelson, W. G. 1993. Beach restoration in the southeastern U.S.: Environmental effects and biological monitoring. *Ocean Coastal Management* 19: 157–182.

Norse, E. A., ed. 1993. *Global Marine Biological Diversity: A Strategy for Building Conservation into Decision Making.* Washington, D.C.: Island Press.

NRC (National Research Council). 1994. *Restoring and Protecting Marine Habitat: The Role of Engineering and Technology.* Washington, D.C.: National Academy Press.

NRC (National Research Council). 1995. *Understanding Marine Biodiversity: A Research Agenda for the Nation.* Washington, D.C.: National Academy Press.

NRC (National Research Council). 1998. *Improving Stock Assessments.* Washington, D.C.: National Academy Press.

NRC (National Research Council). 1999. *Sustaining Marine Fisheries.* Washington, D.C.: National Academy Press.

Ogden, J. C., and E. H. Gladfelter, eds. 1986. *Caribbean Coastal Marine Productivity.* Paris: UNESCO.

Olson, R. J., S. W. Chisholm, E. R. Zettler, M. Altabet, and J. Dusenberry. 1990. Spatial and temporal distributions of prochlorophyte picoplankton in the North Atlantic Ocean. *Deep-Sea Research II* 37: 1033–1051.

Ormond, R. F. G., J. D. Gage, and M. V. Angel, eds. 1997. *Marine Biodiversity: Patterns and Processes.* Cambridge, U.K.: Cambridge University Press.

Paine, R. T., J. L. Ruesink, A. Sun, E. L. Soulanille, M. J. Wonham, C. D. G. Harley, D. R. Brumbaugh, and D. L. Secord. 1996. Trouble on oiled waters: Lessons from the *Exxon Valdez* oil spill. *Annual Review of Ecology and Systematics* 27: 197–235.

Palumbi, S. R. 1992. Marine speciation on a small planet. *Trends in Ecology and Evolution* 7: 114–118.

Palumbi, S. R. 1994. Genetic divergence, reproductive isolation, and marine speciation. *Annual Review of Ecology and Systematics* 25: 547–572.

Palumbi, S. R. 2001. The ecology of marine protected areas. In M. D. Bertness, S. D. Gaines, and M. E. Hay, eds., *Marine Community Ecology,* pp. 509–530. Sunderland, Mass.: Sinauer Associates.

Pandolfi, J. M., and P. R. Minchin. 1995. A comparison of taxonomic composition and diversity between reef coral life and dead assemblages in Madang Lagoon, Papua New Guinea. *Palaeogeography Palaeoclimatology Palaeoecology* 119: 321–341.

Pandolfi, J. M., and J. B. C. Jackson. 2001. Community structure of Pleistocene coral reefs of Curaçao, Netherlands Antilles. *Ecological Monographs* 71: 49–67.

Parker, S. J., S. A. Berkeley, J. T. Golden, D. R. Gunderson, J. Heifetz, M. A. Hixon, R. Larson, B. M. Leaman, M. S. Love, J. A. Musick, V. M. O'Connell, S. Ralston, H. J. Weeks, and M. M. Yoklavich. 2000. Management of Pacific rockfish. *Fisheries* 25(3): 22–30.

Pauly, D. 1988. Some definitions of overfishing relevant to coastal zone management in southeast Asia. *Tropical Coastal Area Management* 3: 14–15.

Pauly, D. 1995. Anecdotes and the shifting baseline syndrome of fisheries. *Trends in Ecology and Evolution* 10: 430.

Pauly, D., and V. Christensen. 1995. Primary production required to sustain global fisheries. *Nature* 374: 255–257.

Pauly, D., V. Christensen, J. Dalsgaard, R. Froese, and F. Torres Jr. 1998. Fishing down marine food webs. *Science* 279: 860–863.

Peterson, C. H. 2001. The *Exxon Valdez* oil spill in Alaska: Acute, indirect, and chronic effects on the ecosystem. *Advances in Marine Biology* 39: 1–103.

Peterson, C. H., and J. A. Estes. 2001. Conservation and management of marine communities. Pages 469–507 in M. D. Bertness, S. D. Gaines, and M. E. Hay, eds., *Marine Community Ecology.* Sunderland, Mass.: Sinauer Associates.

Peterson, C. H., D. H. M. Hickerson, and G. G. Johnson. 2000. Short-term consequences of nourishment and bulldozing on the dominant large invertebrates of a sandy beach. *Journal of Coastal Research* 16: 368–378.

Peterson, M. N. A., ed. 1992. *Diversity of Oceanic Life: An Evaluative Review.* Washington, D.C.: Center for Strategic and International Studies.

Pickett, S., R. S. Ostfeld, M. Shachak, and G. E. Likens, eds. 1998. *The Ecological Basis of Conservation: Heterogeneity, Ecosystems, and Biodiversity.* Hingham, Mass.: Kluwer Academic Publishers.

Ray, G. C. 1991. Coastal-zone biodiversity patterns. *BioScience* 41: 490–498.

Ray, G. C. 1996. Coastal-marine discontinuities and synergisms: Implications for biodiversity conservation. *Biodiversity and Conservation* 5: 1095–1108.

Ray, G. C. 1999. Coastal-marine protected areas: Agonies of choice. *Aquatic Conservation: Marine and Freshwater Ecosystems* 9: 607–614.

Ray, G. C., and B. P. Hayden. 1992. Coastal zone ecotones. Pages 403–420 in A. J. Hansen and F. D. Castri, eds., *Landscape Boundaries: Consequences for Biotic Diversity and Ecological Flows.* New York: Springer-Verlag.

Reaka-Kudla, M. L. 1997. The global biodiversity of coral reefs: A comparison with rainforests. Pages 83–108 in M. L. Reaka-Kudla, D. E. Wilson, and E. O. Wilson, eds., *Biodiversity II: Understanding and Protecting Our Biological Resources.* Washington, D.C.: National Academy Press.

Rinkevich, B. 1995. Restoration strategies for coral reefs damaged by recreational activities: Use of sexual and asexual recruits. *Restoration Ecology* 3: 241–251.

Roberts, C. M., and J. P. Hawkins. 1999. Extinction risk in the sea. *Trends in Ecology and Evolution* 14: 241–246.

Roberts, C. M., J. P. Hawkins, D. E. McAllister, and F. W. Schueler. 2001. Hotspots, endemism and the conservation of coral reef fish biodiversity. *Coral Reefs,* in press.

Rothschild, B. J. 1986. *Dynamics of Marine Fish Populations.* Cambridge, Mass.: Harvard University Press.

Ruiz, G. M., J. T. Carlton, E. D. Grosholz, and A. H. Hines. 1997. Global invasions of marine and estuarine habitats by non-indigenous species: Mechanisms, extent, and consequences. *American Zoologist* 37: 621–632.

Ruiz, G. M., P. Fofonoff, and A. H. Hines. 1999. Non-indigenous species as stressors in estuarine and marine communities: Assessing invasion impacts and interactions. *Limnology and Oceanography* 44: 950–972.

Ryman, N. 1991. Conservation genetics considerations in fishery management. *Journal of Fish Biology* 39: 211–234.

Safina, C. 1995. The world's imperiled fish. *Scientific American* 273: 46–53.

Sale, P. F., ed. 1991. *The Ecology of Fishes on Coral Reefs.* San Diego: Academic Press.

Schlesinger, W. H. 1997. *Biogeochemistry: An Analysis of Global Change.* San Diego: Academic Press.

Sepkoski, J. J. 1992. Phylogenetic and ecologic patterns in the Phanerozoic history of marine biodiversity. Pages 77–100 in N. Eldredge, ed., *Systematics, Ecology, and the Biodiversity Crisis.* New York: Columbia University Press.

Sheppard, C. 1995. The shifting baseline syndrome. *Marine Pollution Bulletin* 30: 766–767.

Sherman, K., L. M. Alexander, and B. D. Gold, eds. 1990. *Large Marine Ecosystems: Patterns, Processes, and Yields.* Washington, D.C.: American Association for the Advancement of Science.

Simenstad, C. A., J. A. Estes, and K. W. Kenyon. 1978. Aleuts, sea otters, and alternate stable-state communities. *Science* 200: 403–410.

Sindermann, C. J. 1996. *Ocean Pollution: Effects on Living Resources and Humans.* Boca Raton, Fla.: CRC Press.

Smith, P. J., R. I. C. C. Francis, and M. McVeagh. 1991. Loss of genetic diversity due to fishing pressure. *Fisheries Research* 10: 309–316.

Snelgrove, V. R., T. H. Blackburn, P. A. Hutchings, D. M. Alongi, J. F. Grassle, H. Hummel, G. King, I. Koike, P. J. D. Lambshead, N. B. Ramsing, and V. Solis-

Weiss. 1997. The importance of marine sediment biodiversity in ecosystem processes. *Ambio* 26: 578–583.

Soulé, M. E., ed. 1987. *Viable Populations for Conservation.* Cambridge, U.K.: Cambridge University Press.

Steele, J. H. 1974. *The Structure of Marine Ecosystems.* Cambridge, Mass.: Harvard University Press,

Steele, J. H. 1985. A comparison of terrestrial and marine ecological systems. *Nature* 313: 355–358.

Steneck, R. S., and J. T. Carlton. 2001. Human alterations of marine communities: Students beware! *Caveat studium.* Pages 445–468 in M. D. Bertness, S. D. Gaines, and M. E. Hay, eds., *Marine Community Ecology.* Sunderland, Mass.: Sinauer Associates.

Swearer, S. E., J. E. Caselle, D. W. Lea, and R. R. Warner. 1999. Larval retention and recruitment in an island population of a coral-reef fish. *Nature* 402: 799–802.

Tegner, M. J., L. V. Basch, and P. K. Dayton. 1996. Near extinction of an exploited marine invertebrate. *Trends in Ecology and Evolution* 11: 278–280.

Thayer, G. W., ed. 1992. *Restoring the Nation's Marine Environment.* College Park, Md.: Maryland Sea Grant.

Thorne-Miller, B. 1999. *The Living Ocean: Understanding and Protecting Marine Biodiversity.* Washington, D.C.: Island Press.

Tunnicliffe, V. 1991. The biology of hydrothermal vents: Ecology and evolution. *Oceangraphy and Marine Biology Annual Review* 29: 319–407.

Turner, R. E., and N. N. Rabalais. 1994. Coastal eutrophication near the Mississippi River delta. *Nature* 368: 619–621.

Van Dover, C. L. 2000. *The Ecology of Deep-Sea Hydrothermal Vents.* Princeton, N.J.: Princeton University Press.

Vecchione, M., and B. B. Collette. 1996. The central role of systematics in marine biodiversity problems. *Oceanography* 9: 44–45.

Vermeij, G. J. 1993. Biogeography of recently extinct marine species: Implications for conservation. *Conservation Biology* 7: 391–397.

Vincent, A. C. J., and H. J. Hall. 1996. The threatened status of marine fishes. *Trends in Ecology and Evolution* 11: 360–361.

Vitousek, P. M., H. A. Moones, J. Lubchenco, and J. M. Melillo. 1997. Human domination of the earth's ecosystems. *Science* 277: 494–499.

Walters, C. J., and R. Hilborn. 1978. Ecological optimization and adaptive management. *Annual Review of Ecology and Systematics* 9: 157–188.

Ward, T. J., M. A. Vanderklift, A. O. Nicholls, and R. A. Kenchington. 1999. Selecting marine reserves using habitats and species assemblages as surrogates for biological diversity. *Ecological Applications* 9: 691–698.

Watling, L., and E. A. Norse. 1998. Disturbance of the seabed by mobile fishing gear: A comparison to forest clearcutting. *Conservation Biology* 12: 1180–1197.

WCMC. 1992. *Global Biodiversity: Status of the Earth's Living Resources.* London: Chapman and Hall.

Wilkinson, C. R. 1999. Global and local threats to coral reef functioning and existence: Review and predictions. *Marine and Freshwater Research* 50: 867–878.

Winston, J. E. 1992. Systematics and marine conservation. Pages 144–168 in N. Eldredge, ed., *Systematics, Ecology, and the Biodiversity Crisis*. New York: Columbia University Press.

Woodard, C. 2000. *Ocean's End: Travels through Endangered Seas*. New York: Basic Books.

Zaitsev, Y. 1992. Recent changes in the trophic structure of the Black Sea. *Oceanography* 1: 180–189.

Zedler, J. B., ed. 2000. *Handbook for Restoring Tidal Wetlands*. Boca Raton, Fla.: CRC Press.

8

CONSERVATION BIOLOGY AND THE HEALTH SCIENCES
Defining the Research Priorities of Conservation Medicine

Gary M. Tabor, Richard S. Ostfeld, Mary Poss,
Andrew P. Dobson, and A. Alonso Aguirre

Human impacts on the health of the planet can be characterized within four areas of environmental concern (Colborn et al. 1993; Colborn et al. 1996; Rees 1996; Wilson 1988; Woodwell and Houghton 1990):

1. increasing biological impoverishment, which includes the loss of biodiversity and modification of ecological processes;

2. increasing global "toxification," which includes the spread of hazardous wastes and toxic substances and the impact of endocrine-disrupting hormones;

3. global climate change and ozone depletion; and

4. an increasing human "ecological footprint" as a result of our exponentially growing population and use of resources (an ecological footprint is an area-based calculation used in environmental economics to determine human impacts on natural resources).

The discreet and cumulative impacts of these human-induced global changes not only have diminished the environmental capital of the planet but also have yielded an array of health concerns, including the increasing spread of infectious (e.g., malaria) and noninfectious (e.g., melanoma) diseases, and the growing physiological impacts on species' reproductive health, developmental biology, and immune system response (e.g., endocrine disruptors) (Blaustein et al. 1994; Blaustein and Wake 1995; Colborn et al. 1996; Laurance et al. 1996; Epstein 1995; McMichael and Haines 1997; Patz et al. 1996).

The ecological impacts of humans can ripple throughout ecological communities. The demise of one species, or the rise of one species at the expense of another, may establish a trophic cascade of ecological responses (see chapter 3). When predator-prey or species competition relationships are disrupted, ecological impacts may extend beyond the predator and prey or the competitors (Estes et al. 1998; Epstein et al. 1997; Pace et al. 1999; Ostfeld and Keesing 2000b). The spread of Lyme disease as a result of the changing ecology of white-tailed deer (*Odocoileus virginianus*) and white-footed mice (*Peromyscus leucopus*) in a landscape devoid of large predators and diminished biodiversity is a good example of ecological magnification of disease (Ostfeld and Keesing 2000a).

Certainly global climate change adds to the complexity of disease impacts in conservation. This issue is one of the most pressing ecological health concerns. For instance, global climate change may impact health by promoting the spread of infectious disease from more tropical ranges to temperate areas, expanding beyond latitudinal and altitudinal zones. This phenomenon may be apparent in diseases such as malaria and dengue fever (Epstein 1995, 1997; Patz et al. 1996; Rogers and Randolph 2000). The potential impact of climate change on ecological processes may be profound: increased precipitation in some regions and drought in others; increased erosion of the coastal zone with rising sea levels; and the inability of many species to adapt to the relatively rapid changes in climatic regimes, potentially resulting in mass extinctions (Epstein 1999; McMichael 1997; McMichael at al. 1999). Changes in atmospheric volatilization and deposition of pollutants also are of concern as warmer temperatures and greater precipitation enhance pollutant transport (Blais et al.1998; Schindler 1999).

Since its origin, conservation biology has understood the implications of health and disease in management of species and habitat. The integration of theory into practice is only recently emerging with the rise of the new field of conservation medicine (Frankel and Soulé 1981; Dobson and May 1986; Grenfell and Dobson 1995; Meffe 1999; Pokras et al. 1999). Understanding the ecological aspects of health and disease is a future challenge for conservation biologists, who are increasingly expected to provide management remedies for ecological health concerns.

Conservation Medicine: A Brief Description

Conservation biology and conservation medicine have the common aim of trying to achieve ecological health. Conservation medicine studies the two-way interactions between pathogens and disease on the one hand, and species and ecosystems on the other. It focuses on the study of the ecological context of health and the remediation of ecological health problems. In response to the growing health implications of environmental degradation, conservation medicine has emerged as a new interdisciplinary field to address the complex interrelationship between health and ecological concerns. As such, conservation medicine's purview includes examining the linkages among (1) changes in habitat structure and land use; (2) emergence and reemergence of infectious agents, parasites, and environmental contaminants; and (3) maintenance of biodiversity and ecosystem functions as they sustain the health of plant and animal communities including humans. For example, conservation medicine is concerned with the effects of disease on rare or endangered species and on the functioning of ecosystems. It is also concerned with the impacts of changes in species diversity or rarity on disease maintenance and transmission.

The dynamic balance that we term health is conceptualized using a set of widely varying spatial scales by many disciplines including human medicine, public health, epidemiology, veterinary medicine, toxicology, ecology, and conservation biology. Conservation medicine represents an approach that bridges those disciplines to examine the health of individuals, groups of individuals, and the landscapes in which they live as an indivisible continuum (Meffe 1999; Pokras et al. 1999).

Conservation medicine, like other interdisciplinary approaches, brings together scientific disciplines "long separated by academic and practical tradition" (Wilson 1992). By reaching out to multiple disciplines, conservation medicine provides new skills, tools, and vision to the fields of both conservation biology and medicine. This includes bringing biomedical research and diagnostic resources to address conservation problems such as development of new noninvasive health monitoring techniques; training health professionals and conservation biologists in the promotion and practice of ecological health; and establishing interdisciplinary teams of health and ecological professionals to assess and redress ecological health problems.

Why Should Conservation Biologists Care about Disease?

Pathogens and infectious diseases create special problems for both the protection of endangered species and the maintenance of biodiversity. In several cases the introduction of a pathogen has produced a significant further decline in the numbers of an already endangered species: canine distemper in black-footed ferrets (*Mustela nigripes*) (Thorne and Williams 1988; Williams et al. 1988); rabies and canine distemper in hunting dogs (*Lycaon pictus*) in Serengeti (Dye 1996; Ginsberg et al. 1995; Heinsohn 1992); and avian malaria in endemic Hawaiian birds (Warner 1968; van Riper III et al. 1986). In all of these examples, the pathogens involved utilized common or domestic species as reservoir hosts. There is increasing evidence that species diversity may provide an important buffer that minimizes the impact of a pathogen on a population. Paradoxically, pathogens may also play a role in both maintaining biodiversity and driving processes that lead to increases in local diversity (Augspurger 1984; Gilbert 1995; Gilbert et al. 1995; Van der Putten and Van der Stoel 1998; Van der Putten et al. 1993).

While increased species diversity may affect the ability of a pathogen to "find" a susceptible host (especially if the pathogen utilizes multiple hosts), many pathogens are so species specific that only the density of a population affects their ability to cause disease. And thus, any sudden change in population demographics and proportion of susceptible hosts within a population can alter pathogen distribution (Aguirre, Hansen, et al. 1995). This density-dependent phenomenon can be demonstrated in various management practices, such as translocations; reintroductions; the creation of feeding stations in winter that bring animals into close contact, such as the National Elk Refuge in Wyoming; and the alteration of traditional migratory corridors that may bring wildlife into closer contact with domestic animals—all of which can facilitate disease spread (Aguirre, Starkey, et al. 1995). Many activities used to manage wildlife populations have the potential to alter pathogen transmission patterns, and conservation biologists need to be mindful of the consequences. To understand disease threats to wildlife populations, some level of basic disease ecology is required. Conservation biologists need to know what pathogens are endemic and use that information to assess and remedy the introduction of exotic pathogens into a population (Aguirre and Starkey, 1994).

Three topics are central to our understanding of the role pathogens play in conservation biology:

1. What roles do relative abundance, susceptibility, and overlap in spatial distribution play in determining rates of pathogen transmission in populations of vulnerable or endangered species? In pathogens that utilize a range of host species, how do these features contribute to cross-species (inter-specific) transmission? More precisely, does ecological information on the behavior and spatial distribution of potential host species complement epidemiological data and allow us to identify species that are most likely to act as reservoir hosts for pathogens that provide a significant threat to rare or endangered species?

2. What are the dynamics of a recently introduced pathogen into a population? In particular, how do deterministic and stochastic factors interact to determine the persistence and extinction of pathogens in populations? Significantly, how are the dynamics affected if the population is small? How do different types of

pathogens contribute to further declines in the abundance of potentially endangered host species?

3. Does the local diversity of species used by a pathogen tend to buffer or increase either its persistence or rate of spread? Can a pathogen alter the diversity of species that coexist in any community?

Central to all of these problems is the development of a quantitative understanding of how pathogens affect the dynamics of communities that contain one or more potential host species. Here we need to develop empirical and theoretical analyses that examine the role that pathogens play in mediating the coexistence of potentially competing species or in maintaining genetic diversity within a single host species. *We then need to examine whether the diversity of host species tends to either buffer or amplify disease outbreaks.* This is probably the central question in understanding the role that pathogens play in conservation biology. It has very practical applications in conservation, because many proposed management strategies are based on the rationale that biological diversity leads to greater resilience (Pimm 1984; Schulze and Mooney 1994; Hollings et al. 1995).

Determining the answers will require the exploration of the ecological interactions between mode of transmission, host diversity, and pathogen persistence. This can be performed only by examination of empirical data for a number of systems in which we can dissect the relative roles played by host abundance, host resource utilization, and host spatial distribution in determining rates of intra- and interspecific transmission.

In conjunction with these largely deterministic and analytical exercises, we also need to understand the population dynamics of epidemic outbreaks in small (endangered) populations. These analyses will eventually allow the development of stochastic models that determine the impact of pathogens on populations that are too small to sustain a long-term persistent source of infection. Here we also need to focus on examining the costs and benefits of intervention to prevent further spread of the disease, particularly in the limiting case where small host population size always leads to the eventual extinction of the pathogen.

Understanding the Ecology of Disease: An Essential Guide for Conservation Biologists

Pathogens, also known as infectious agents, include viruses, bacteria, fungi, protozoans, and helminths. They undergo population fluctuations within hosts; they compete with one another for access to resources, such as cells and nutrients within hosts; they are preyed upon by host immune agents; they select habitats (tissues); and they disperse within and among hosts. For those

pathogens that are zoonotic (i.e., those that reside within nonhuman vertebrate animals and are transmitted to humans), and those that are transmitted by vectors, ecological interactions are even more complex. Such
pathogens must contend with differing environmental conditions within
human hosts, nonhuman hosts, and vectors. In addition, the vertebrate hosts
and arthropod vectors are involved in complex ecological interactions of
their own that will influence the maintenance and transmission of
pathogens.

An example of an emerging infectious disease with strong implications for
conservation medicine is the 1999–2000 "outbreak" of West Nile virus
(WNV) in the northeastern United States (Lanciotti et al. 1999). This mosquito-borne disease, endemic to Europe, Africa, and the Middle East,
appeared suddenly in birds and humans in and near New York City and
elicited a response by public officials that included extensive aerial spraying
of insecticides in densely populated areas. This remedial action, which may
be more dangerous to public health and wildlife than the disease itself, *and*
ineffective in reducing disease incidence, was implemented despite our ignorance about the ecology of the mosquito vectors, bird reservoirs, and the
pathogen.

For a disease such as this, that has achieved almost daily attention from the
mass media, our level of ignorance of the risk and its ecological causes is
astonishing. Questions to which no thorough answer exists include: How did
WNV disperse to the Western Hemisphere, or was it here all along? Which
species function as reservoir hosts (i.e., those from which the mosquito vectors obtain the pathogen)? In which species is the virus pathogenic or lethal,
and how does infection influence population dynamics of those species? Over
what area does WNV occur, and how does it get there? How do patterns of
temperature and rainfall influence breeding performance and populations of
container-breeding mosquitoes in urban and suburban ecosystems? Can
WNV exist in mosquitoes other than *Culex pipiens* and *Aedes japonicus*? How
does the pathogen overwinter? Does it migrate south or stay dormant during
the cold winter months (Agulric et al. 1992)?

Several general principles relevant to conservation medicine are represented by the WNV example. First, in the absence of information on the ecological determinants of disease risk, as well as on the magnitude of that risk,
environmentally destructive but largely ineffective measures (e.g., adulticide
spraying) are unfortunately employed. Ecological knowledge about zoonoses
can help prevent uninformed decision making. Second, environmentally
more benign solutions to the WNV problem may require a solid scientific
understanding of the roles played by specific bird or mammal reservoirs,
diversity in the community of reservoirs, and landscapes that influence diver-

sity. Third, a likely cause of the transoceanic dispersal of the virus is either via movements by people or by domesticated or wild animals (Rappole et al. 2000). Long-distance movements by pathogens such as WNV may represent instances of invasion by exotic species and, therefore, are relevant to both conservation biologists and health scientists. Fourth, some evidence suggests that the container-breeding mosquitoes that carry arthropod-borne viruses become particularly abundant when hot, dry weather prevents flushing of urban pools of water, such as storm drains. Climatic factors may influence disease transmission in unexpected ways (Reisen et al. 1992; Hubalek and Halouzka 1999).

Ecological research can help us anticipate and ameliorate the adverse effects of infectious diseases on humans, domestic animals, and wildlife. From an ecological framework, these infectious diseases occur when a pathogen population disperses to a host, invades, and, via infection, triggers a host immune response resulting in clinical signs. The key elements to which ecological research can contribute is an understanding of where pathogens are maintained in nature, how they disperse to the focal population, and what environmental processes are involved in the maintenance of the pathogen within populations and in transmission among them. Armed with such knowledge, conservation managers may be able to intervene in natural systems to reduce the likelihood of exposure. In box 8.1, we have listed potential research priorities, posed as questions, together with a brief rationale, that link disease, ecology, and conservation biology.

Connecting Pathobiology to Ecology

Investigation of disease and disease pathogenesis is based primarily on studies in domestic animals. However, it may not be accurate to assume that principles developed by studies in domesticated animals will apply to wildlife. Even within the same species, extrapolation across animals from a range of different habitats may be problematic. This is because many infectious agents endemic to a population are adapted to the population structure of their hosts. Dynamics and structure of a host population determine the nature of its microbiologic fauna. Isolated or small populations may maintain different types of pathogens than large and contiguous populations that freely exchange individuals. Because both the genetics of the pathogen and the host environment determine the outcome of infection, it is plausible that in isolated populations, genetic bottlenecking of both host and pathogen may lead to a unique outcome of infection. Thus, although detailed pathogen-host studies provide suitable information, and frequently the only baseline information available, the potential for unique pathogenic outcomes in wildlife species should be considered.

A primary objective should be to determine the extent to which disease caused by endemic microbes can influence survival or fecundity of individuals in a population. Endemic diseases are infrequently considered as factors affecting population survival. Yet infections with endemic organisms may more commonly lead to chronic conditions affecting an animal's ability to forage or escape predators than directly to high mortality. In addition, disease induced by infectious organisms common to a population may have a more profound effect on specific age-classes such as neonates or old adults.

Whereas many pathogens are fastidiously host specific, others, such as canine distemper (family Paramyxoviridae, genus *Morbillivirus*), may be transmitted across species boundaries. Wildlife may also acquire a foreign pathogen from their domesticated counterparts. A significant level of mortality can occur if an exotic pathogenic microorganism is introduced into a naive susceptible population (Anderson 1995; Roelke-Parker et al. 1996). Mathematical models based on dynamics and structure of the susceptible species, and on pathogenic processes and transmission properties of the pathogen, are invaluable in assessing risk of introduced pathogens to host population survival.

In general, infectious agents are relatively easy to identify genetically and serologically. Many diagnostic laboratory techniques are available for assisting conservation biologists with disease study. In vitro investigation is perhaps one of the most useful tools in ascertaining pathogen-host interaction. In vitro studies must, however, be complemented by observations on the natural course of infection in the species of interest to provide meaningful results. Data collection that assists field researchers in determining the health profile of a species or community should be incorporated in the research methodology. Much of this information can be gathered through direct field observation of behavior and mortality. Where possible, blood and other body fluid samples of all animals that are immobilized for any reason should be obtained. Collection and detailed evaluation of tissues from mortalities are also critical.

By learning the pathobiology of infectious agents, new areas of conservation application can be gained. The story of bovine brucellosis in bison (*Bison bison*) in Yellowstone National Park represents one example of the need for disease ecological research in wildlife management. Over one thousand bison have been culled in northern Yellowstone Park in order to prevent transmission of brucellosis from bison to livestock. In North America, brucellosis is a disease that causes infertility and abortion in cattle, elk (*Cervus elaphus*), and bison. Cattle and elk more often show these clinical signs; the disease does not usually affect bison. Initially introduced into bison from cattle, brucellosis has been virtually eradicated in Montana's livestock, leaving bison in Yellowstone as a potential reservoir of the disease. Around 50 percent of Yellowstone's bison herd may be infected. Management practices to date have focused on adhering to government protocols on brucellosis-free certification within livestock. This provides a substantial economic benefit to livestock growers.

Nevertheless, there is still a relative lack of knowledge about the transmissibility of the bacterial infectious agent *Brucella abortus* between domestic cattle and wildlife in free-ranging situations (Zaugg et al. 1993; Williams et al. 1993; Meagher and Meyer 1994; Dobson and Meagher 1996; Roffe et al. 1999). The conservation implications are profound because the bison story encapsulates many of the issues conservation biologists must grapple with today: introduced alien species (in this case a pathogen); growing conflicts between humans, their domesticated animals, and wildlife; and increased population impacts of disease through cross-species transmission (brucellosis may be shared between resident Yellowstone bison and elk populations). With elk winter feeding stations concentrating animals, the management of one species obviously impacts another. Brucella is also a zoonotic organism capable of causing infection in people—although rare. More so, the brucellosis story is a shining example of the increasingly complex public policy arena facing conservation managers as they grapple with disease and health issues.

Putting Pathogens to Conservation Work: Potential Biomarkers

Beyond the health ramifications of disease, pathogens can also serve as potential tools for studying the ecology of populations and individuals. Understanding of disease-causing organisms and the pathogenesis of disease has markedly increased due to the availability of molecular tools. For instance, information gained from studies of the Retroviridae, the family of RNA viruses that include the human immunodeficiency virus (HIV), feline leukemia virus (FeLV), feline immunodeficiency virus (FIV), simian immunodeficiency virus (SIV), and others show how the specific biology of a pathogen can provide a specific signature of infection and evolution (Levy 1993; Leitner et al. 1996; Gao et al. 1999; Poss and Overbaugh 1999; Hahn et al. 2000).

Of specific interest are efforts relating viral genetic diversity with population structure of the host. Many of the retroviruses, like other viruses, have potential evolutionary rates approximately one million times higher than those of their vertebrate hosts (Ricchetti and Buc 1990). This suggests that viruses circulating in one population will be genetically distinct from those circulating in a different population (which has been well documented with HIV in humans) (Bachmann et al. 1994; Louwagie et al 1995). Furthermore, adjacent populations that exchange infected individuals can be distinguished from those that do not by examining the genetics of the resident viruses. In the former case the virus population will be randomly distributed, while in the latter phylogenetic reconstruction of virus sequences will indicate distinct viral subpopulations (Poss and Overbaugh 1999). There is great potential benefit to applying these principles to wildlife species, as very recent changes in population dynamics are likely to be discernible due to rapid evolution of the viral genome.

By measuring viral genetic changes, we have the potential to use information that is used to determine contact and transmission of diseases between individuals and populations to delineate metapopulation dynamics of certain species such as mountain lions and lynx. This information can assist conservation managers in the determination of wildlife corridors and linkage areas and zones of wildlife and domestic animal contact.

Research Priorities in Conservation Medicine: Looking 10 Years Ahead

Disease affects every issue of concern to conservation biologists, including biodiversity, small or isolated population persistence, alien species introduction, and trophic cascades. Virtually any ecological perturbation that affects wildlife species can also alter the balance and composition of their pathogen communities. Future efforts in conservation need to incorporate the influence of disease on wildlife and of wildlife on disease. In this chapter, we provide key elements of a framework for inquiry that may assist conservation biologists in moving forward.

There are four broad areas for enhancement of the links between conservation biology and diseases of wildlife and humans:

1. *Expansion of interdisciplinary interaction.* Engage the talents of new disciplines to address conservation concerns. This can include the integration of conservation biology within existing disciplines such as veterinary and human medicine and social sciences such as economics (Rapport 1995). Medical and veterinary practitioners and researchers will certainly enrich their discipline by increasing their knowledge of how conservation issues such as biodiversity loss, climate and land use change, and pollution influence human and animal health. Ecology and conservation biology become enriched when pathogens can be detected and their effects monitored.

2. *Integrated health and ecological assessment and monitoring.* Develop and refine environmental and physiological health assessment and monitoring protocols. The first step in conducting a current health assessment of an ecological system is to characterize the baseline health parameters for selected species or group of species (e.g., health indicator or sentinel species) and determine their responses to naturally fluctuating environmental variables. This includes monitoring variables that contribute to the physiological responses to stress within a population or in an individual (Hofer and East 1998). It will be neces-

sary to establish normal values for bioindicator species based on (1) existing archived data; (2) data from processing of stored, unanalyzed samples where appropriate; (3) opportunistic collections of material from both pristine and disrupted ecosystems; (4) targeted collections of appropriate materials from healthy specimens reflecting that specific ecosystem; and (5) new monitoring efforts. The development of testing protocols, identification of suitable diagnostic procedures and tests, identification of laboratory support, development of a quality-control-assurance plan for laboratory results, and development of data management and data analysis plans are all required to conduct a thorough assessment.

Sentinel species may assist in increasing monitoring efficiency at the ecosystem level. They can be utilized during rapid risk assessments to provide information on the environmental conditions of an area. Sentinel species or health indicator species can be selected for their ability to reflect environmental perturbations (Caro and O'Doherty 1999). Based on their life history and physiological attributes, selected species can provide insightful information about environmental changes at various spatial, temporal, and trophic scales. Given the complexity of ecosystems, sentinel species should be thought of as being specific to particular environmental conditions. In some cases, an assemblage of species may be suitable for providing an umbrella effect in monitoring the cumulative impacts of multiple environmental variables.

Beyond species- and population-oriented monitoring, ecological health information needs to be collected through ecosystem monitoring efforts, including long-term ecological monitoring. New research developments in determining ecological integrity should incorporate health assessment indicators (Karr and Chu 1995, 1999).

3. *Expanded diagnostic capability.* Present diagnostic efforts are focused on clinical testing of human or domestic animals. The potential for applying existing diagnostic capacity to health monitoring of wildlife species or those of conservation concern is relatively untapped. The limitation of financial resources has been a major barrier. Laboratory support for performing diagnostic testing in species other than humans and domesticated animals is minimal. Cross-species diagnostic testing, in which one laboratory can do comparative assessment of disease between species, is extremely rare and is usually found within a couple of national disease testing centers (e.g., USDA Research Labs, Centers for Disease Control, National Wildlife Health Center, and the Southeastern Cooperative Wildlife Disease Study).

Diagnostic laboratory testing can play an important role in monitoring the health of wildlife populations. During health evaluations, testing is necessary to detect exposure to an agent (serology), to identify the agents that are endemic in a population (bacteriology or virology) or to determine the cause of a catastrophic event such as a die-off or disease outbreak (clinical pathology, gross

BOX 8.1. **Potential Research Priorities That Link Disease, Ecology, and Conservation Biology**

1. Investigate vector-borne zoonotic diseases (such as mosquito-borne West Nile virus and malaria, sand fly–borne leishmaniasis, tick-borne Lyme disease, and bug-borne Chagas' disease).

   • What biotic and abiotic factors regulate vector populations? *Rationale for investigation:* Predators, parasites, and competitors may influence the population density and growth rates of arthropod vectors. Similarly, vectors are often limited by extremes in temperature and humidity, and those limitations may vary with life stage of the vector.

   • How do changes in climate influence vector populations, through either biotic or abiotic pathways? *Rationale for investigation:* Global climate change may influence both abundance and large- and small-scale distribution of vectors.

   • How do changes in land use influence vector populations, through either biotic or abiotic pathways? *Rationale for investigation:* Habitat destruction, conversion, and fragmentation may influence vectors directly or via their hosts or natural enemies, thereby affecting population size and variability.

   • What abiotic and biotic factors influence the proportion of vectors that are infected by the pathogen? *Rationale for investigation:* A key risk factor for many vector-borne diseases is the infection prevalence in the vector population. This can be influenced by abiotic factors that affect pathogen development and by ecological factors that affect the relative abundance of host species, which often vary in their competence as reservoirs for pathogens.

   • How does biodiversity affect disease risk? *Rationale for investigation:* High genetic or species diversity in the (reservoir) host community has recently been shown to reduce disease incidence in both human and crop systems.

2. Investigate directly transmitted zoonotic diseases (such as hantavirus pulmonary syndrome, Lassa fever and other arenaviral diseases, *Bartonella*).

   • What factors regulate reservoir populations? *Rationale for investigation:* For many directly transmitted zoonoses, rodents are the primary reservoirs. Rodent populations may be regulated by resource levels, predators, competitors, or pathogens (although most zoonotic pathogens appear relatively benign in reservoir hosts).

   • How do reservoir populations respond to changing climate, land use, and biodiversity? *Rationale for investigation:* Climate, land use, and diversity may influence reservoirs either directly, by allowing expansion or contraction of geographic ranges or population sizes, or indirectly, by influencing biota that regulate reservoir populations.

- How does biodiversity in vertebrate communities affect disease risk? *Rationale for investigation:* High genetic or species diversity within vertebrate communities may regulate populations of reservoir species or otherwise reduce the probability of pathogen transmission.
- How do local and global patterns of transportation influence the distribution of pathogens? *Rationale for investigation:* Travel by people and movements of ships and planes worldwide may introduce pathogens and their reservoirs or vectors to new areas.

3. Investigate water-borne and food-borne zoonotic diseases (such as *Vibrio* cholera, marine biotoxins and cyanobacteria, *Salmonella, Campylobacter*).
   - How do changes in global temperature influence populations of pathogens, either free-living or commensal with other aquatic organisms? *Rationale for investigation:* Population growth of many water-borne pathogens is sensitive to water temperature, as is population growth of planktonic organisms with which pathogens may associate (e.g., *Vibrio cholerae*).
   - How do influxes of nutrients to lakes and estuaries influence populations of pathogens, either free-living or commensal with other aquatic organisms? *Rationale for investigation:* Population growth of many water-borne pathogens is sensitive to nutrient concentrations (e.g., nitrogen, phosphorus), as is population growth of planktonic organisms with which pathogens may associate. Human activities, such as livestock production, agriculture, and septic leakage, are often responsible for production and release of these nutrients.

4. Investigate noninfectious diseases (such as chronic and acute toxic effects of environmental contaminants; asthma; malignant tumors).
   - To what degree are wildlife populations, especially those of conservation concern, affected by pollutants that decrease viability or reproductive success? *Rationale for investigation:* Evidence is mounting that sublethal effects of pollutants (e.g., endocrine mimics) can reduce population viability.
   - How do noninfectious diseases interact in wildlife populations? *Rationale for investigation:* Noninfectious agents may reduce effectiveness of the immune system, thereby influencing the ability of organisms to resist other environmental insults.

Sources: Ostfield and Keesing 2000a; Zhu et al 2000.

pathology, histopathology). Tests can be used for epizootic and epidemiologic purposes and can estimate prevalence, incidence, and geographic distribution of selected disease agents; determine infection in subpopulations; and delineate intra- and interspecific disease transmission. The results of diagnostic testing are most useful in complementing thorough field-based ecological research. Ecological studies provide the necessary context for any results concerning disease dynamics and can even discern certain physiological responses due to stress imposed on populations by environmental change (Hofer and East 1998).

4. *Resolution of human and wildlife conflict—implementing conservation medicine practice in conservation reserve design and management.* With increasing fragmentation of habitat, the consequences of enhanced human and animal conflicts are more apparent. Disease is only one manifestation of this global phenomenon. Nevertheless, conservation managers need to be savvy in developing management scenarios that minimize disease impacts. Buffer zones and wildlife corridors may require new designs to mitigate or diminish disease threats. Resolving human and wildlife conflicts is the challenge ahead in conservation. Those conflicts include the interaction between domesticated animals and wildlife. With growing intrusions of people and their animals into wildland areas, such as livestock grazing on public lands, salmon aquaculture pens in coastal marine areas, primate ecotourism in the tropics, and the influx of backpackers with their dogs into designated wilderness areas, disease mitigation will be a major factor in conservation management. The design and management of conservation reserves need to be open to the consequences of disease threats—not as a marginal issue of concern, which is the case at this time, but as an issue more central to achieving long-term conservation success.

Conclusions

For many of the research questions presented in this chapter, answers will come from a combination of monitoring, modeling, experimentation, and comparison. Monitoring is crucial for detecting changes in disease distribution or incidence as a result of human-caused changes in climate, land use, or pollutant concentration. For example, surveillance for diseases using sentinel animals can provide advance warning of changes in distribution or virulence of pathogens, and suggest possible causes. Modeling allows exceedingly complex parts of nature to be abstracted; relationships among populations, species, habitats, and ecosystems to be specified; the logic of interactions to be clarified; and specific hypotheses to be derived. Experimentation is important for testing specific hypotheses on relatively small scales, and for determining mechanisms for observed patterns at larger scales or within more complex, inclusive systems. For instance, effects of biodiversity or habitat destruction on the maintenance and transmission of pathogens can be examined in a laboratory setting by experimentally altering diversity or landscape patterns and measuring responses.

Comparative methods are analogous to experiments but consist of monitoring effects of conditions that have not been experimentally manipulated and differ between systems in specific ways. For example, population dynamics of an imperiled species can be monitored before and after a toxic spill, or inside and outside an area where a pathogen has invaded. Effects of diminishing biodiversity on vector-borne vs. directly transmitted diseases may elucidate the influence of transmission mode on ecology of infectious diseases.

Ecological studies of infectious and noninfectious diseases are therefore crucial to understanding, predicting, and reducing risk to the health of humans, domesticated animals, and wildlife and the environment that sustains them. Conservation medicine can provide new skill sets to address disease impacts in conservation. These include the incorporation of biomedical research and resources in conservation problem solving; the development of diagnostic methods that examine ecological and health parameters; and the design and implementation of integrated health and ecological assessment models.

LITERATURE CITED

Aguirre, A. A., D. E. Hansen, E. E. Starkey, and R. G. McLean. 1995. Serologic survey of wild cervids for potential disease agents in selected national parks in the United States. *Preventive Veterinary Medicine* 21:313–322.

Aguirre, A. A., R. G McLean, R. S. Cook, and T. J. Quan. 1992. Serologic survey for selected arboviruses and other potential pathogens in wildlife from Mexico. *Journal of Wildlife Diseases* 28:435–442.

Aguirre, A. A., and E. E. Starkey. 1994. Wildlife disease in U.S. national parks: Historical and coevolutionary perspectives. *Conservation Biology* 8:654–661.

Aguirre, A. A., E. E. Starkey, and D. E. Hansen. 1995. Wildlife diseases in national park ecosystems. *Wildlife Society Bulletin* 23:415–419.

Anderson, E. C. 1995. Morbillivirus infections in wildlife (in relation to their population biology and disease control in domestic animals). *Veterinary Microbiology* 44(2–4): 319–332.

Augspurger, C. K. 1984. Seedling survival of tropical tree species: Interactions of dispersal distance, light gaps, and pathogens. *Ecology* 65:1705–1712.

Bachmann, M. H., E. L. Delwart, E. G. Shpaer, P. Lingenfelter, R. Singal, J. I. Mullins. 1994. Rapid genetic characterization of HIV type 1 strains from four World Health Organization–sponsored vaccine evaluation sites using a heteroduplex mobility assay. *AIDS Research and Human Retroviruses* 10(11):1345–1353.

Blais, J. M., D. W. Schindler, D. C. G. Muir, D. B. Donald, and B. Rosenberg. 1998. Accumulation of persistent organochlorine compounds in mountains of western Canada. *Nature* 395: 585–588.

Blaustein, A. R., P. D. Hoffman, D. G. Hokit, and J. M. Kiesecker. 1994. UV repair and resistance to solar UV-B in amphibian eggs: A link to population declines. *Proceedings of the National Academy of Sciences* 91:1791–1795.

Blaustein, A. W., and D. B. Wake. 1995. The puzzle of declining amphibian populations. *Scientific American* (April): 52–57.

Caro, T. M., and G. O'Doherty. 1999. On the use of surrogate species in conservation biology. *Conservation Biology* 13(4):805–814.

Colborn, T., D. Dumanoski, and J. P. Meyers. 1996. *Our Stolen Future*. New York: Dutton.

Colborn, T., R. S. vom Saal, and A. M. Soto. 1993. Developmental effects of endocrine-disrupting chemicals in wildlife and humans. *Environmental Health Perspectives* 101:378–384.

Dobson, A. P., and R. M. May. 1986. Disease and conservation. In M. Soulé (ed.), *Conservation Biology: The Science of Scarcity and Diversity*, pp. 345–365. Sunderland, Mass.: Sinauer Associates.

Dobson, A. P., and M. Meagher. 1996. The population dynamics of brucellosis in the Yellowstone National Park. *Ecology* 74(4):1026–1036.

Dye, C. 1996. Serengeti wild dogs: What really happened? *Trends in Ecology and Evolution* 11:188–189.

Epstein, P. R. 1995. Emerging diseases and ecosystem instabilities: New threats to public health. *American Journal of Public Health* 85:168–172.

Epstein, P. R. 1997. Climate, ecology, and human health. *Consequences* 3(2):2–19.

Epstein, P. R. 1999. Climate and health. *Science* 285:347–348.

Epstein, P. R., A. Dobson, and J. Vandermeer. 1997. Biodiversity and emerging infectious diseases: Integrating health and ecosystem monitoring. F. Grifo and J. Rosenthal (eds.), *Biodiversity and Human Health*. Washington, D.C.: Island Press.

Estes, J.A., M. T. Tinker, T. M. Williams, and D. F. Doak. 1998. Killer whale predation on sea otters linking oceanic and nearshore ecosystems. *Science* 282:473–476.

Frankel, O. H., and M. E. Soulé. 1981. *Conservation and Evolution*. Cambridge, U.K.: Cambridge University Press.

Gao, F., E. Bailes, D. L. Robertson, Y. Chen, C. M. Rodenburg, S. F. Michael, L. B. Cummins, L. O. Arthur, M. Peeters, G. M. Shaw, P. M. Sharp, and B. H. Hahn. 1999. Origin of HIV-1 in the chimpanzee *Pan troglodytes troglodytes*. *Nature* 397(6718):436–441.

Gilbert, G. S. 1995. Rain forest plant diseases: The canopy-understory connection. *Selbyana* 16:75–77.

Ginsberg, J. R., et al. 1995. Handling and survivorship of African wild dog (*Lycaon pictus*) in five ecosystems. *Conservation Biology* 9:665–674.

Grenfell, B. T., and Dobson, A. P. 1995. *Population Dynamics of Infectious Diseases in Natural Populations*, Cambridge, U.K.: Cambridge University Press.

Hahn, B. H., G. M. Shaw, K. M. De Cock, and P. M. Sharp. 2000. AIDS as a zoonosis: Scientific and public health implications. *Science* 287(5453):607–614.

Heinsohn, R. 1992. When conservation goes to the dogs. *Trends in Ecology and Evolution* 7:214–215.

Hofer, H., and M. L. East. 1998. Biological conservation and stress. *Advances in the Study of Behavior* 27:405–525.

Hollings, C. S., D. W. Schindler, B. W. Walker, and J. Roughgarden 1995. Biodiversity in the functioning of ecosystems: An ecological synthesis. In C. Perrings, K.-G. Maler, C. Folke, C. S. Hollings, and B.-O. Jansson (eds.), *Biodiversity Loss: Economic and Ecological Issues,* pp. 44–83. New York: Cambridge University Press.

Hubalek, Z., and J. Halouzka. 1999. West Nile fever—A reemerging mosquito-borne viral disease in Europe. *Emerging Infectious Diseases* 5:643–650.

Karr, J. R., and E. W. Chu. 1995. Ecological integrity: Reclaiming lost connections. In L. Westra and J. Lemons (eds), *Perspectives on Ecological Integrity,* pp. 34–48. Dordrecht, Netherlands: Kluwer.

Karr, J. R., and E. W. Chu. 1999. *Restoring Life in Running Waters: Better Biological Monitoring.* Washington, D.C.: Island Press.

Lanciotti, R. S., J. T. Roehrig, V. Deubel, et al. 1999. Origin of the West Nile virus responsible for an outbreak of encephalitis in the northeastern United States. *Science* 286:2333–2337.

Laurance, W. F., K. R. McDonald, and R. Speare. 1996. Epidemic disease and the catastrophic decline of Australian rain forest frogs. *Conservation Biology* 10(2): 406–413.

Leitner, T., D. Escanilla, C. Franzen, M. Uhlen, and J. Albert. 1996. Accurate reconstruction of a known HIV-1 transmission history by phylogenetic tree analysis. *Proceedings of the National Academy of Sciences USA* 93(20): 10864–10869.

Levy, J. A. 1993. Pathogenesis of human immunodeficiency virus infection. *Microbiological Reviews* 57(1): 183–289.

Louwagie, J., W. Janssens, J. Mascola, L. Heyndrickx, P. Hegerich, G. van der Groen, F. E. McCutchan, and D. S. Burke. 1995. Genetic diversity of the envelope glycoprotein from human immunodeficiency virus type 1 isolates of African origin. *Journal of Virology* 69(1): 263–271.

McMichael, A. J. 1997. Global environmental change and human health: Impact assessment, population vulnerability, and research priorities. *Ecosystem Health* 3:200–210.

McMichael, A. J., B. Bolin, R. Costanza, G. C. Daily, C. Folke, K. Lindahl-Kiessling, B. Lindgren, and E. Niklasson. 1999. Globalization and the sustainability of human health: An ecological perspective. *Bioscience* 49(3):205–210.

McMichael, A. J., and A. Haines. 1997. Global climate change: The potential effects on health. *British Medical Journal* 315:805–809.

Meagher, M., and M. E. Meyer. 1994. On the origin of brucellosis in bison in Yellowstone National Park: A review. *Conservation Biology* 8:645–653.

Meffe, G. 1999. Conservation medicine. *Conservation Biology* 13: 953–954.

Ostfeld, R. S., and F. Keesing. 2000a. Biodiversity and disease risk: The case of Lyme disease. *Conservation Biology* 14:722–728.

Ostfeld, R. S., and F. Keesing. 2000b. Pulsed resources and community dynamics of consumers in terrestrial ecosystems. *Trends in Ecology and Evolution* 15:232–237.

Pace, M. L., J. J. Cole, S. R. Carpenter, and J. F. Kitchell. 1999. Trophic cascades revealed in diverse ecosystems. *TREE* 14(12): 483–488.

Patz, J. A., P. R. Epstein, T. A. Burke, and J. M. Balbus. 1996. Global climate change and emerging infectious diseases. *JAMA* 275:217–223.

Pimm, S. L. 1984. The complexity and stability of ecosystems. *Nature* 307:321–326.

Pokras, M., G. M. Tabor, M. Pearl, D. Sherman, and P. Epstein. 1999. Conservation medicine: An emerging field. In P. K. Raven (ed.), *Nature and Human Society: The Quest for a Sustainable World,* pp. 551–556. Washington, D.C.: National Academy Press.

Poss, M., and J. Overbaugh. 1999. Variants from the diverse virus population identified at seroconversion of a clade A human immunodeficiency virus type 1-infected woman have distinct biological properties. *Journal of Virology* 73(7): 5255–5264.

Rappole, J. H., S. R. Derrickson, and Z. Hubalek. 2000. Migratory birds and spread of West Nile virus in the Western Hemisphere. *Emerging Infectious Diseases* 6:319–328.

Rapport, D. J. 1995. Ecosystem health: An emerging transdisciplinary science. In D. J. Rapport, C. Gaudet, and P. Calow (eds.), *Evaluating and Monitoring the Health of Large-Scale Ecosystems,* pp. 5–31. Heidelberg: Springer-Verlag.

Rees, W. E. 1996. Revisiting carrying capacity: Area-based indicators of sustainability. *Population and Environment* 7:195–215.

Reisen, W. K., R. P. Meyer, M. M. Milby, et al. 1992. Ecological observations on the 1989 outbreak of St. Louis encephalitis virus in the southern San Joaquin Valley of California. *Journal of Medical Entomology* 29:472–482.

Ricchetti, M., and H. Buc. 1990. Reverse transcriptases and genomic variability: The accuracy of DNA replication is enzyme specific and sequence dependent. *European Molecular Biology Organization Journal* 9(5): 1583–1593.

Roelke-Parker, M. E., L. Munson, C. Packer, R. Kock, S. Cleaveland, M. Carpenter, S. J. O'Brien, A. Pospischil, R. Hofmann-Lehmann, H. Lutz, et al. 1996. A canine distemper virus epidemic in Serengeti lions (*Panthera leo*). *Nature* 379(6564): 441–445.

Roffe, T. J., J. C. Rhyan, K. Aune, L. M. Philo, D. R. Ewalt, T. Gidlewski, and S. G. Hennager. 1999. Brucellosis in Yellowstone National Park bison: Quantitative serology and infection. *Journal of Wildlife Management* 63(4):1132–1137.

Rogers, D. J., and S. E. Randolph. 2000. The global spread of malaria in a future, warmer world. *Science* 289:1763–1766.

Schindler, D. W. 1999. From acid rain to toxic snow. *Ambio* 28:350–355.

Schulze, E.-D., and H. A. Mooney (eds.). 1994. *Biodiversity and Ecosystem Function*. Berlin: Springer-Verlag.

Thorne, E. T., and E. S. Williams. 1988. Disease and endangered species: The black-footed ferret as a recent example. *Conservation Biology* 2:66–74.

Van der Putten, W. H., and C. D. Van der Stoel. 1998. Plant parasitic nematodes and spatio-temporal variation in natural vegetation. *Applied Soil Ecology* 10:253–262.

Van der Putten, W. H., et al. 1993. Plant-specific soil-borne diseases contribute to succession in foredune vegetation. *Nature* 362:53–56.

van Riper III, C., et al. 1986. The epizootiology and ecological significance of malaria in Hawaiian land birds. *Ecological Monographs* 56:327–344.

Walsh, J. F., D. H. Molyneux, M. H. Birley. 1993. Deforestation: Effects on vector-borne disease. *Parasitology* 106:S55–S75.

Warner, R. E. 1968. The role of introduced diseases in the extinction of the endemic Hawaiian avifauna. *Condor* 70:101–120.

Williams, E. S., et al. 1988. Canine distemper in black-footed ferrets (*Mustela nigripes*) from Wyoming. *Journal of Wildlife Diseases* 24:385–398.

Williams, E. S., E. T. Thorne, S. L. Anderson, and J. D. Herriges Jr. 1993. Brucellosis in free-ranging bison (*Bison bison*) from Teton County, Wyoming. *Journal of Wildlife Diseases* 29(1):118–122.

Wilson, E. O. 1988. The current state of biological diversity. In E. O. Wilson (ed.), *Biodiversity*, pp. 3–19. Washington, D.C.: National Academy Press.

Wilson, E. O. 1992. *The Diversity of Life*. Cambridge, Mass.: Belknap Press.

Woodwell, G. M., and R. A. Houghton. 1990. The experimental impoverishment of natural communities effects of ionizing radiation on plant communities: 1961–1976. In G. M. Woodwell (ed.), *The Earth in Transition: Patterns and Processes of Biotic Impoverishment*, pp. 9–24. Conference, Woods Hole, Massachusetts, USA, October 1986. Cambridge, U.K.: Cambridge University Press.

Zaugg, J. L., S. K. Taylor, B. C. Anderson, D. L. Hunter, J. Ryder, and M. Divine. 1993. Hematologic, serologic values, histopathologic and fecal evaluations of bison from Yellowstone Park. *Journal of Wildlife Diseases* 29(3):453–457.

Zhu, Y. Y., et al. 2000. Genetic diversity and disease control in rice. *Nature* 406:718–722.

9

GLOBAL ENVIRONMENTAL CHANGE
Effects on Biodiversity

William H. Schlesinger, James S. Clark,
Jacqueline E. Mohan, and Chantal D. Reid

On December 19, 1999, the *New York Times* proclaimed, "1999 Continues Warming Trend Around Globe," thus confirming the reminiscence of many U.S. citizens of colder winters just a few decades ago. Supplementing the historical records from weather stations, surrogate measures of climate change confirm a rise in global temperature—showing a reduction in the extent and thickness of arctic sea ice (Johannessen et al. 1999; Rothrock et al. 1999), shorter ice-covered seasons on many northern lakes (Robertson et al. 1992), earlier arrival and activity of springtime plants and animals (Ahas 1999), and rising satellite-measured temperatures of Earth's upper atmosphere (Wentz and Schabel 2000). Changes in the distribution and abundance of Earth's biota also tell us that a warming trend is in progress. During the past few decades, the distributions of birds, butterflies, and trees have all shifted northward, consistent with a warming of Earth's climate. Scientists of the Intergovernmental Panel on Climate Change (IPCC) weigh the evidence that this change in climate is due to anthropogenic changes in the chemistry of our environment, namely the continuing rise of carbon dioxide (CO_2) and other trace gases in Earth's atmosphere (Santer et al. 1996; Tett et al. 1999).

As with past changes in climate, the current global warming may change the distribution of Earth's species, leading to entirely new assemblages of species in ecological communities. When climate change is slow, species can adapt or migrate to new habitats, changing their distribution or range to adjust to the new conditions. However, future changes in global climate, wrought by humans, may be so rapid as to overwhelm the natural ability of plants and animals to disperse to new habitats (Overpeck et al. 1991; Davis and Zabinski 1992). If species cannot change their distribution fast enough, will they go extinct?

In this chapter, we define global change more broadly than climate change alone. We focus on a variety of environmental changes that are global in extent (e.g., climate and rising CO_2) or global inasmuch as they affect large regions on several continents (e.g., rising concentrations of tropospheric ozone and nitrogen availability). Our objectives are (1) to review past changes in global climate for evidence of their effects on biodiversity, especially during the transition from glacial to interglacial periods; (2) to analyze what recent field experiments tell us about the effects of anticipated future environmental changes on biodiversity; and (3) to suggest research priorities in conservation biology that can help remove uncertainties about biotic responses to global change. We ask: Is the ongoing human-induced global environmental change a threat to biodiversity?

Community Structure:
Changes in Biodiversity with Stress and Subsidy

In a seminal paper entitled "Perturbation Theory and the Subsidy-Stress Gradient," Odum et al. (1979) noted that many of the changes seen when ecological communities are provided with supplemental resources (e.g., fertilizer) resemble changes seen when communities are exposed to toxic substances. Applications of fertilizer for more than one hundred years on the Parkgrass experimental plots at Rothamstead (figure 9.1) reduced the length of the dominance diversity curves (i.e., lowering richness) and increased their slope (i.e., decreasing equitability). (A dominance-diversity curve is a plot of the abundance of each species in a community of plants or animals as a function of its rank in abundance, from the most important species to the least.) Similar changes in dominance-diversity curves occur when communities are exposed to toxic effluents, such as from smelters (e.g., figure 9.2). In both cases, the changes in the dominance-diversity curves are the reverse of what is seen during the progressive recovery of communities following disturbance (e.g., Bazzaz 1975), leading Whittaker (1972) to call these trends "retrogressive."

FIGURE 9.1.   Plant species diversity in pasture in the Parkgrass experiment at Rothamstead, United Kingdom, following the continuous application of nitrogen fertilizer from 1856 to 1949. At each sampling, fewer species were present, and there was an increasing concentration of dominance in the most abundant species. From Kempton (1979).

FIGURE 9.2.   Cover of lichen species as a function of the decreasing rank of species in ground samples taken in Delaware Water Gap (unpolluted) and Lehigh Water Gap (polluted) environments in southern Pennsylvania. From Nash (1975).

At first glance, many of the global changes wrought by humans might appear to be beneficial in the context of Odum's stress-subsidy theory. Rising CO_2, warmer temperatures, greater precipitation, and increased nitrogen availability should stimulate the rate of plant growth in many regions. Moreover, rising plant production and a more clement climate might benefit many animal populations. However, nearly three decades ago, Rosenzweig (1971) postulated that enrichment would destabilize model ecosystems, lowering their diversity. Although subsequent experimental work has not always confirmed the theory (e.g., McCauley and Murdoch 1990), Rapport et al. (1985) found that a loss of species diversity was the most consistent indicator of ecosystem retrogression, caused by either stress or subsidy. Many ecosystem-level experiments simulating global change phenomena show losses of diversity that should warn us of the vulnerability of biodiversity to global environmental change. Prehistoric evidence of the retrogressive impacts of environmental change can be found in the species composition and dominance-diversity relationships of seafloor communities now preserved in the fossil record of mid-Cretaceous black shales (Sageman and Bina 1997). As human perturbation of the environment becomes global in extent, lower biotic diversity will result not only from the local disappearance of species, but also from global extinction.

The Last Great Global Warming: The End of the Pleistocene

The end of the last ice age, 10,000 years ago, provides an opportunity to examine how biodiversity responded to climate change at continental scales. However, in many ways, the transition to warmer postglacial conditions is not a perfect analog to the future greenhouse warming of our planet. The latitudinal distribution of incoming radiation (Kutzbach and Guetter 1986), the presence of continental ice sheets (Clark et al. 1999), an unusual pattern of ocean circulation (Broecker 1997), lower sea level (Fairbanks 1989), and a unique distribution of vegetation (Foley et al. 1994) produced a different distribution of climate change than what we expect from future greenhouse warming. Atmospheric CO_2 rose from ~200 ppm (parts per million by volume) to ~280 ppm between 15,000 and 9,000 years ago (Barnola et al. 1987), but that increase may not have had the same effects on biodiversity as the current trend, which will soon reach 600 ppm and above (Collatz et al. 1998). Because climate change depends on factors that control atmospheric circulation, we can expect the future greenhouse warming to differ from past global change in its rate and the seasonality of temperature and precipitation.

Despite the fact that the anticipated greenhouse warming differs from past climate changes, the end of the Pleistocene provides valuable perspective. Because it is recent, the fossil record is more complete for the last glacial period than it is for prior glacial cycles. The fossil record, ice-core evidence, and climate models agree that global mean temperature increased 4–7°C at the end of the Pleistocene (Kutzbach and Guetter 1986; Kutzbach et al. 1993; Wright et al. 1993). A temperature change of about the same magnitude is expected during the next century. Although the anticipated global warming will occur, on average, more rapidly than that of recent glacial cycles, some of the warming during deglaciation also happened quickly. For instance, a period of warming ~11,650 years ago occurred over a course of a few decades (Taylor et al. 1997), possibly similar to the rate of climate change anticipated during this century.

Extinctions

There is clear evidence for mass extinctions of large mammals at the end of the Pleistocene but limited evidence for the extinction of plant species. Paleontologists debate whether the animal extinctions resulted from changes in vegetation set in motion by climate change or from human hunting. Several overview volumes discuss this issue in detail (Martin and Klein 1984; MacPhee 1999).

The limited evidence for plant extinctions at the end of the Pleistocene may reflect, in part, limitations of the fossil record. An extensive record of fossil pollen provides evidence for distributions of many taxa (COHMAP 1988), but generally similar species cannot be distinguished, and the record is biased toward wind-dispersed species and those that occur in regions where lakes and mires preserve a sedimentary record. Packrat middens provide fossil records of plant occurrence in arid regions, but these records are often discontinuous in space and time. Thus, more extinctions may have occurred than are apparent from the fossil evidence. At least one extinction, that of the once extensive spruce *Picea critchfieldii,* has been documented from deposits of cones and needles in the southeastern United States (Jackson and Weng 1999). This species was apparently abundant across the region before ten thousand years ago and may have died out as temperatures began to rise. Unlike the case of faunal extinctions, humans probably played no direct role in its demise. Dispersal limitation is an obvious candidate for this species' failure to survive into the Holocene, suggesting that even widely distributed and abundant species can be at risk. The loss of this once extensive species has obvious implications for the losses of biodiversity that may accompany future global warming.

Evolutionary Change

Although the fossil record provides few examples of plant extinctions, molecular studies indicate losses of intraspecific genetic diversity since the Pleistocene. In many cases, the losses appear to be related to founder effects—the limited genetic diversity that arises when a new population is derived from only a few individuals (Ibrahim et al. 1996). Molecular data show that genetic diversity for many plant populations declines with increasing distance from refugia—areas where the populations survived during times of glacial expansion (Comes and Kadereit 1998). Migrational barriers appear to have slowed the northward spread of some genotypes and led to dominance by a subset of the haplotypes (groups of plants that are all descended from the same maternal ancestor) that may have been widely separated in different Pleistocene refugia. Modern haplotype distributions suggest that the east-west trending mountain ranges of western Europe, especially the Alps, were barriers for the spread of many genotypes (Hewitt 1999, 2000). A pattern of decreasing genetic diversity with distance from past refugia is consistent with the theoretical loss of genetic diversity that would be expected during long-distance founding events (LeCorre et al. 1997).

Not all evolutionary responses to past warming involved net losses of diversity. The mixing of previously segregated genotypes during migrations can also influence genetic diversity. For example, a large molecular data set from white oaks across Europe, which includes eight species and twenty-three haplotypes, suggests that migration pathways from southern European refugia crossed one another, thus producing mixtures of haplotypes (Dumolin-Lapeque et al. 1997). This pattern is especially evident in western Europe, where closely related species may have occupied as many as four different refugia, subsequently expanding into overlapping ranges in central and northern Europe. Taken together, the accumulating molecular evidence suggests that the broad-scale migrations that attend climate change are epochs of evolutionary change that may include both losses of diversity and additions of new genotypes.

Range Shifts

The fossil record shows that dramatic rearrangements of species have occurred during climate changes over areas of low relief, such as eastern North America (Davis 1981; Delcourt and Delcourt 1987) and western Europe (Huntley and Birks 1983). Ecologists have examined the Pleistocene record of these migrations to address the potential for biota to respond to current global warming (Pitelka et al. 1997; Clark 1998; Clark et al. 2001; Cain et al. 1998). "Isopoll maps," showing interpolated contours of equal pollen abundance in lake sediments, suggest rates of northward spread from 50 to 1,000 m per year for many trees since the end of the Pleistocene (Davis 1981; Huntley and Birks 1983; Ritchie and MacDonald 1986). Such rates were equaled or exceeded by some beetles (Coope 1987) and even flightless grasshoppers (Hewitt 1990). These rates are less than the rates of migration that would be needed to accommodate anticipated climate change—1,000 to

5,000 m per year during the next several centuries (Overpeck et al. 1991). The paleoestimates do not, however, rule out the potential for more rapid migration, because the distribution of past ecosystems may have been constrained by rates of climate change.

In many cases, ranges of populations both extended on their northern boundaries and retracted on their southern ones. However, migrations in response to global warming entail more than a northward shift in range limits; geography imposes barriers to dispersal and fragments opportunities for growth and survival. For example, the meadow grasshopper (*Chorthippus parallelus*) of Europe and Siberia appears to have occupied distinct glacial refugia in Spain, Italy, Greece, and Turkey. Today, the northern European populations of these grasshoppers are dominated by the Balkan genotype, perhaps because postglacial migration of populations in Spain and Italy was restricted by the Pyrenees and Alps (Hewitt 1999). Likewise, European beech (*Fagus sylvatica*) occurred in both Italy and the Carpathians, but the Alps appear to have blocked expansion of the Italian populations (Demesure et al. 1996); modern populations in northern Europe are derived only from Carpathian populations. Haplotype distributions for vertebrates suggest similar effects of dispersal barriers on postglacial spread (Hewitt 1999, 2000).

In many cases, species occupying similar regions today were allopatric (not occurring in the same region) during glacial periods, and many came to occupy their modern range by very different migrational pathways (Graham and Grimm 1990; Taberlet et al. 1998). Whereas European beech apparently failed to traverse the Alps from its Pleistocene refuge in Italy (Demesure et al. 1996), oaks expanded northward from Italian populations (Dumolin-Lapeque et al. 1997). In eastern North America, hemlock (*Tsuga*) and beech (*Fagus grandifolia*), which today are dominant species in northern hardwood forests, arrived at their modern distributions predominantly by way of the Appalachians and the Coastal Plain, respectively. The fact that the modern distribution of species appears to derive from different migration patterns in the past suggests that it may be difficult to ensure appropriate corridors for biotic migration in the future.

Rapid range expansions could have unexpected effects on biodiversity, especially if species escape from competitors, predators, pollinators, pathogens, and so forth. For example, during rapid postglacial warming, many insects reached Britain before most temperate tree species arrived. Coope (1987) suggests that the ground beetle (*Calosoma inquisitor*) must have preyed on caterpillar species that fed on birch trees that dominated British forests at the time, despite the fact that modern *Calosoma* prey almost exclusively on caterpillars that feed on oaks. The changes in assemblages indicated by the fossil record (e.g., Davis 1981; Graham and Grimm 1990) imply reorganization of trophic relationships and novel food webs.

The potential for the survival of species during climate change is higher in varied topography, such as in the western United States and southern Europe, than in regions of low relief (Van Devender and Spaulding 1979). Elevational shifts in the distribution of species in montane regions occurred throughout the globe as a result of gradients in temperature, moisture, and perhaps even CO_2 (Street-Perrott et al. 1997). For example, the highly dissected topography in northern Queensland provided the necessary habitat for rain forest species to persist in the face of regional aridity during the Pleistocene (Hopkins et al. 1990). Mesic valleys on the eastern edge of the Atherton tableland appear to have harbored rain forest, and expansion onto the tableland occurred as moisture levels increased in the Holocene (Walker and Chen 1987; Kershaw 1993).

Range expansion across areas where suitable habitats are scarce generally requires long-distance dispersal. In the case of forest trees, both fossil pollen and molecular data suggest that rare long-distance dispersal events have been important. The fossil record demonstrates dispersal across obstacles such as the North Sea (Reid 1899), the Baltic (Kullman 1996), and the Great Lakes (Webb 1987; Woods and Davis 1989). The dominance of an area by a single haplotype is consistent with models of long-distance dispersal, in which early arrivals rapidly come to dominate a population.

Summary

The fact that rapid spread may have occurred in the past is no insurance of future potential. The molecular evidence for mountain barriers to dispersal, and the extinction of at least one dominant forest species in eastern North America (Jackson and Weng 1999), indicate that the vegetation of the past did not uniformly track climate change. Moreover, the processes that make long-distance dispersal possible (involving winds, waters, and biotic vectors) are different today than they were in the past. Human activities could slow population spread by fragmenting the available habitat and by introducing alien species that affect the survival and spread of native species. Humans accelerate the spread of species through introductions and transport (Vitousek et al. 1996; Clark, Beckage et al. 2001). Some species benefit from accidental and intentional introductions by humans, whereas present or potential ranges of other species shrink due to competition, predation, and disease associated with exotics. As yet, the paleorecord of the response of biodiversity to global change has not provided us with general predictions that can be applied to current global warming. Given the critical importance of extreme dispersal events in mediating overall rates of population spread, future research should focus on factors responsible for long-distance dispersal.

One valuable insight gained from studies of paleoecology is an appreciation of the breadth of response to past climate change. While many species occupy new ranges sufficiently fast to track climate change, others do not. Lessons from the end of the Pleistocene include cautions that even widely distributed species may not survive rapid climate change.

Anticipated Global Climate Change

In this section we discuss projected changes in temperature and moisture, and the expected repercussions of these changes.

Temperature

The ranges of many plants and animals are correlated with temperature, particularly extremes of cold temperature that can kill most individuals in exceptional years. For instance, the northern distribution of creosotebush (*Larrea tridentata*) in the arid Southwest is correlated with minimum temperatures lower than −16°C, which cause freezing and cavitation of its xylem elements (Pockman and Sperry 1997). The deciduous trees of eastern North America extend to the −40°C minimum temperature isotherm, beyond which conifers of the boreal forest spread northward (Arris and Eagleson 1989). The northern edge of the winter range of the eastern phoebe (*Sayornis phoebe*) is correlated with the −4°C average minimum January temperature (Root 1988). Such correlations suggest that the ranges of species will shift northward with warming temperatures, consistent with recent studies of the historical distribution and activity of birds (Thomas and Lennon 1999; Dunn and Winkler 1999), butterflies (Parmesan et al. 1999; Roy and Sparks 2000), and white spruce (*Picea mariana*; Payette and Filion 1985). Satellite measures of "greenness" indicate a general northward expansion in the activity of vegetation with global warming (Myneni et al. 1997).

Similarly, the elevational distribution of plants and animals will change with changes in climate (Jobbagy and Jackson 2000). In the western United States, the distribution of chipmunks and ground squirrels on mountains is determined by the thermal tolerance of each species (Heller and Gates 1971; Hudson and Deavers 1973). With global warming, each species would be expected to occur at higher elevations, and some species may disappear from the highest portions of low mountains, as happened during the global warming that ended the last glacial period (Brown 1971). Similarly, the occurrence of lake trout in northern lakes is determined by cold waters—a habitat that may disappear when shallow lakes experience warmer temperatures (Schindler et al. 1990).

Rising temperatures of coastal waters are also a putative cause of a decline in the abundance of marine zooplankton (Roemmich and McGowan 1995), the poor recruitment of North Sea cod (*Gadus morhua*; O'Brien et al. 2000), and the widespread bleaching and loss of corals in many tropical seas (Glynn 1996; Brown 1997; Warner et al. 1999; Aronson et al. 2000). Predation by the sea star, *Pisaster,* in marine intertidal communities is lower during periods of warm ocean waters, reducing its role as a keystone predator (Sanford 1999). A resampling of marine intertidal communities at the Hopkins Marine Station (Monterey Bay) showed large changes in macroinvertebrate species during the past sixty years, apparently associated with warmer seawater temperatures (Sagarin et al. 1999). Most southern species increased in abundance, while northern species declined (Barry et al. 1995). At its extreme, global warming could lead to major changes in the circulation of ocean currents and flooding of coastal ecosystems and coral reefs by rising sea level.

Relatively few experiments have examined climatic warming at the ecosystem level. John Harte established one of the most realistic field studies at the Rocky Mountain Biological Laboratory in the mountains of central Colorado. Using overhead infrared heaters to apply 22 W/m^2 to the soil surface, Harte and Shaw (1995) were able to increase the total average downward radiation by 3 percent in replicate plots of alpine meadow, raising the surface temperature to values expected with a doubled-CO_2 atmosphere. Price and Waser (2000) report no changes in plant community composition over the first four years of the experiment, which they attribute to drier soils in the warmed plots. However, the field-warming experiment increased the length of the growing season, because spring snowmelt advanced by seven days in the warmed plots (Harte et al. 1995), where Price and Waser (1998) documented earlier flowering in several herbs. The importance of changes in the length of the growing season, rather than changes in temperature *during* the growing season, is also seen in studies that have manipulated the duration of the winter snowpack in alpine habitats (Bell and Bliss 1979; Galen and Stanton 1993, 1995).

Experimental manipulations of ecosystem temperature across a range of sites in arctic tundra show dramatic changes in the rate of plant development, also due to changes in the length of the growing season (Arft et al. 1999). Over a nine-year experiment in Alaska, elevated temperatures increased the growth of shrubs, at the expense of nonvascular plants, largely mosses (Chapin et al. 1995). Losses of species and steepening of dominance-diversity curves during the course of the experiment are indicative of biotic impoverishment (figure 9.3). Grime et al. (2000) report changes in plant community composition during a five-year soil warming experiment in British grasslands, with the largest changes seen in an early-successional field.

Whole-ecosystem manipulations of temperature are more difficult to establish in forest vegetation, but several researchers have attempted soil warming experiments (Peterjohn et al. 1993). On Cape Cod, a soil warming

FIGURE 9.3. Biomass of tundra plants plotted as a function of the decreasing rank of species importance in experimental plots in the tundra of Alaska. Experimental plots were warmed, whereas control plots received no manipulative treatment. From Chapin et al. (1995).

experiment produced a striking precocious bloom of Canada mayflower (*Maianthemum canadense*). At Harvard Forest, soil warming accelerated the development of herbaceous species, but it lowered the species richness in treated plots over a three-year period (Farnsworth et al. 1995). Other soil warming experiments resulted in changes in the composition of soil microbial communities (Harte et al. 1996; Zogg et al. 1997; Andrews et al. 2000), consistent with observations of greater soil CO_2 efflux and nitrogen mineralization in heated plots (Peterjohn et al. 1994; Rustad and Fernandez 1998; Lukewille and Wright 1997).

Moisture

Predicted changes in the global distribution and abundance of precipitation as a result of greenhouse warming of Earth's climate are much less certain than the predicted changes in temperature. Most global models suggest greater overall precipitation, owing to a general stimulation of the hydrologic cycle by higher temperatures (Mitchell et al. 1990, p. 138). However, several models suggest that during the transient period of global warming, there will be an increase in the extent and severity of midcontinental droughts (Rind et al. 1990).

Moisture availability certainly affects the distribution of plants and animals, and changes in the abundance and seasonal distribution of precipitation are likely to produce dramatic changes in species abundance in many areas. In 1988, severe drought reduced the density of birds by 61 percent in the northern Great Plains, but nearly complete recovery was seen in the next year, which had normal rainfall (George et al. 1992). A long-term increase in temperature or a decline in precipitation in the prairies of North America would surely be followed by a decline in the abundance of waterfowl (Poiani and Johnson 1991; Bethke and Nudds 1995; Sorenson et al. 1998), whereas increasing precipitation might expand their habitat.

In semiarid regions, the latitudinal and elevational distribution of vegetation may be more closely correlated with regional gradients in soil moisture than with temperature (Barton and Terri 1993; Hogg 1994; Tucker et al. 1994; Allen and Breshears 1998). During a warm, dry period of the mid-Holocene (~6,000 years ago), prairie and woodland vegetation in central North America expanded eastward, replacing forest (McAndrews et al. 1966; Clark, Grimm et al. 2001). Conversely, if future global warming is accompanied by greater rainfall, forest vegetation may spread to lower, drier elevations, as seen in the western Great Basin during the Pleistocene (DeLucia and Schlesinger 1990).

Examining the climate record for semiarid southern New Mexico, Neilson (1986) suggested that a shift in rainfall from summer to winter aided the expansion of arid-land shrubs, such as creosotebush and mesquite (*Prosopis glandulosa*), at the expense of warm-season grasses such as black grama (*Bouteloua eriopoda*). Brown et al. (1997) drew similar conclusions from long-term field studies in the Chihuahuan Desert in southeastern Arizona. Other factors, such as overgrazing, rising CO_2, and fire suppression may also be involved, but this change in community composition is associated with a loss of species diversity (figure 9.4).

Summary and Research Needs: Climate Changes

The predicted future changes in temperature and moisture associated with a greenhouse warming of Earth's atmosphere are likely to cause dramatic shifts in the distribution of plants and animals (Iverson and Prasad 1998). The climate changes will occur within a century. We know relatively little about the comparative dispersal ability of species, but to the extent that differentials exist, some species with poor dispersal ability may disappear, while the remaining species may be found in new assemblages for which there are no modern analogs (Iverson and Prasad 1998). To the extent that species have evolved in long-term associations with their neighbors, losses of biodiversity may cascade from low to higher trophic levels (Visser et al. 1998; Petchey et al. 1999). For instance, changes in the distribution and abundance

FIGURE 9.4. Plant species diversity in black grama grasslands, and in creosote-bush and mesquite shrublands in southern New Mexico. From L. F. Huenneke (unpublished).

of animals may depend more on changes in vegetation than on changes in the thermal conditions of the animals per se (Johnston and Schmitz 1997).

Conservation biology will benefit from more studies in which temperature and moisture are manipulated to study their effects on species diversity. These should be whole-ecosystem studies—not of warming of the soil alone. Long-term experiments are essential; in short-term experiments, warming may lead to a loss of native species at a rate greater than the rate of immigration of species that might flourish at warmer temperatures (table 9.1). Also, in nature, species are often more responsive to extreme events, whereas warming experiments typically impose only changes in mean annual temperature (for a welcome exception, see White et al. 2000). Future workers may wish to design field studies of species response to unusual, rather than altered mean, conditions. Experiments that impose warming should focus on high-latitude ecosystems, where climatic warming is projected to be greatest. Experiments that impose changes in moisture—by either additions or exclusions of rainfall—are most needed in semiarid ecosystems, where shifts in the moisture regime are likely to have critical effects on the persistence of species. These studies should also examine changes in the seasonal distribution and interannual variability of moisture, perhaps following the model used by Reynolds et al. (1999) for field experiments in the Chihuahuan Desert.

TABLE 9.1. Whole-Ecosystem Experiments Examining Changes in Plant
Biodiversity in Response to Anticipated Global Environmental Change

Experimental Perturbation Biome	Biodiversity Response	Reference
Ecosystem warming Alpine tundra	Decline No change	Harte and Shaw (1995) Price and Waser (2000)
Ecosystem warming Arctic tundra	Decline	Chapin et al. (1995)
Soil warming Temperate forest	Lower richness	Farnsworth et al. (1995)
High CO_2 Temperate grassland	No change in richness Higher equitability	Leadley et al. (1999)
High CO_2 Temperate grassland	Higher richness Higher equitability	Potvin and Vasseur (1997)
Nitrogen deposition Temperate grassland	Decline	Wedin and Tilman (1996)
Nitrogen deposition Temperate field	Decline	Bakelaar and Odum (1978)
Fertilization Temperate fields	Lower richness Lower equitability	Kempton (1979)

Rising Carbon Dioxide

During the last 420,000 years, the concentration of atmospheric carbon dioxide dipped below 200 ppm during each of the Pleistocene ice ages (Petit et al. 1999). At the end of the last continental glaciation, it increased to 270 ppm, where it remained with minor fluctuations until the beginning of the Industrial Revolution. The rapid increase to the current level of 360 ppm, due to the combustion of fossil fuels and deforestation of the tropics, is unprecedented in the geologic history of Earth (Schlesinger 1997). By the year 2050, the global CO_2 concentration is expected to reach 560 ppm (Schimel et al. 1995).

Carbon dioxide is the primary raw material for photosynthesis, and changes in its concentration have direct effects on the growth of plants, leading to changes in competitive abilities and potential impacts on biodiversity. For example, the decreased photosynthetic rate of C3 plants relative to that of C4 plants at low atmospheric CO_2 concentrations is believed to have triggered the expansion of C4 plants at the end of the Miocene, approximately seven million years ago (Cerling et al. 1993, 1997). When grown at low CO_2 concentrations, replicating conditions during the last continental glaciation (18,000 years ago), the C3 weed *Abutilon theophrasti* was unable to reproduce, while C4 plants flourished in the same conditions (Dippery et al. 1995). The rise in C3 plants during the Holocene has been attributed to the rise in atmospheric CO_2 at the end of the last glaciation, which increased their competitive ability (Cole and Monger 1994). In semiarid regions, a further decline in the C4 grasses during the past century may be due to a greater competitive ability of C3 shrubs and savannah trees under the rising concentrations of atmospheric CO_2 (Johnson et al. 1993; Polley et al. 1994).

In greenhouse experiments, elevated atmospheric CO_2 generally enhances plant productivity, with an average growth increase of 30–40 percent at 2X CO_2 and optimal growth conditions (Poorter 1993; Curtis and Wang 1998). In nature, where plant growth is often limited by nutrients, light, and/or soil moisture, the effect of CO_2 is smaller (Bazzaz 1990; DeLucia et al. 1999) and decreases with long-term exposure (Hattenschwiler et al. 1997; Idso 1999). Accelerated plant development with elevated CO_2 may explain the diminished growth response after long-term exposure (St. Omer and Horvarth 1983; Tissue, Thomas, and Strain 1997; Reid and Fiscus 1998; Ward et al. 2000). In their natural habitat, plants are subject to herbivory, competition, and mutualisms, so the effects of altered CO_2 on one plant are at least partly a function of its effect on neighboring plants, animals, and soil microbes.

Forest Ecosystems

In forests, a few studies show that early-successional, shade-intolerant "pioneer" species benefit from elevated CO_2 relative to shade-tolerant competitors, especially when grown under the low-light and/or low-nutrient conditions that typify most forest understories (Hattenschwiler and Korner 1996; Brown and Higginbotham 1986). Several physiological mechanisms explain how plants may require less light when grown under elevated CO_2 (Hirose et al. 1996; Pearcy and Bjorkman 1983). These experiments suggest that early-successional growth forms may retain their dominance and persist for longer periods under elevated CO_2 conditions. Since species diversity tends to increase during succession (Peet 1992), a delay in the successional process might be expected to decrease regional biodiversity, especially as mature forests become increasingly rare in human-dominated landscapes.

Pioneer (early-successional) species typically have higher innate relative growth rates than later-successional species (Bazzaz and Pickett 1980; Walters and Reich 1996, 2000). Growth enhancement by CO_2 fertilization is twice as great in fast-growing as in slower-growing species (Ackerly and Bazzaz 1995; Poorter 1998). Among slow-growing species, however, differences in relative growth rate are not correlated with the CO_2-growth response (Reid and Strain 1994). Thus, fast-growing, "weedy" plants may benefit most from future levels of atmospheric CO_2, leading to an increase in early-successional communities and a reduction in overall species diversity (Bazzaz 1996).

Many field studies, however, suggest the opposite conclusion. These studies show that shade-tolerant, later-successional tree species benefit from elevated CO_2 more than early-successional trees (Kubiske and Pregitzer 1996; Bazzaz et al. 1990; Williams et al. 1986), even when the seedlings are grown under low-light, low-nutrient, "forestlike" conditions (Bazzaz and Miao 1993; Hattenschwiler and Korner 2000). Various traits associated with shade tolerance may explain the tendency for these species to show a strong CO_2 growth response (Kerstiens 1998).

Field studies in the Duke Forest Free-Atmosphere Carbon Dioxide Enrichment (FACE) site, where elevated (560 ppm) and ambient (360 ppm) CO_2 conditions were maintained in a fifteen-year-old (in 1996) loblolly pine (*Pinus taeda*) forest, support the prediction that later-successional, shade-tolerant trees will be favored by future CO_2 levels (figure 9.5). In the second year of CO_2 treatment, seedlings of shade-tolerant understory trees grew 90 percent larger under elevated CO_2 compared to the ambient treatment. Seedlings of shade-intolerant and intermediately tolerant understory trees showed no significant CO_2 growth effects. These results suggest that during forest regrowth later-successional species will be favored and elevated CO_2 may increase species diversity.

In addition to different responses by species of differing shade tolerances and growth rates, different growth forms exhibit distinct growth responses to CO_2 enrichment. For example, vines, due to their lower carbon requirements for support tissues, typically exhibit a greater response to elevated CO_2 than trees (Bazzaz 1996; Sasek 1985). In addition, vines allocate more of their photosynthate to increase the photosynthetic area, causing a further enhancement in growth. In the Duke Forest FACE experiment, the understory vine poison ivy (*Toxicodendron radicans*) had a 71 percent increase in growth under elevated CO_2 when protected from mammalian herbivory. The growth enhancement of this vine far exceeds the 25 percent growth increase of the canopy trees (DeLucia et al. 1999).

Earlier work showed large growth enhancements (135 percent and 51 percent increases, respectively) when the exotic vines Japanese honeysuckle (*Lonicera japonica*) and kudzu (*Pueraria lobata*) were grown at elevated CO_2 under optimal greenhouse conditions (Sasek and Strain 1988, 1989, 1991). Both species are pernicious weeds in the southeastern United States. Sasek and Strain (1991) found only a 44 percent increase in the growth of the native, coral honeysuckle (*L. sempervirens*) at high CO_2—far less than that of the invasive, congenic competitor. Characteristics of successful exotic species (e.g., fast growth rate and low resource requirements) may also be associated

FIGURE 9.5. Relative growth rate of classes of tree seedlings in the Duke Forest Free-Air CO_2 Enrichment (FACE) experiment, in control plots and plots receiving a continuous 200-ppm supplement of CO_2. Seedlings of shade-tolerant species showed significant higher growth at elevated CO_2 (P < 0.02), while other species groups showed no effect. These growth rates were calculated from plants showing positive growth increments in uncaged plots, subject to mammalian herbivory. From J. E. Mohan (unpublished data).

with strong CO_2 growth responses. *L. japonica* has been linked to reduced tree recruitment and increased sapling mortality in eastern forests (Sasek 1985; Dillenburg et al. 1995) and is predicted to expand northward as a consequence of global warming (Sasek and Strain 1990). If vines become more competitive with elevated CO_2, future forest composition and species diversity will certainly be impacted.

In addition to the dramatic response of vines, tree species capable of harboring symbiotic nitrogen-fixing bacteria often show large increases in growth in response to elevated CO_2 (Arnone and Gordon 1990; Norby 1987; Thomas et al. 1991; Tissue, Megonigal, and Thomas 1997). Higher rates of plant photosynthesis at high CO_2 may allow N-fixing species to allocate additional carbohydrate to support their N-fixing symbionts and to avoid nutrient deficiencies that could otherwise result from accelerated growth.

Grassland Ecosystems

The effects of elevated atmospheric CO_2 on species diversity have also been studied in grassland ecosystems. Owensby et al. (1999) reported that C3 grasses declined and C3 forbs increased during an eight-year study of an ungrazed tallgrass prairie exposed to elevated CO_2. Although Leadley et al. (1999) found no CO_2 effect on the number of species in their four-year study of calcareous Swiss grassland, rare species became more abundant, so the overall diversity of the community increased slightly. In a three-year study of a Canadian pasture, Potvin and Vasseur (1997) observed higher species richness, increased equitability, and reduced extinction rates under elevated CO_2, but the resulting boost in species diversity favored pioneer species at the expense of later-successional species. Interestingly, in several studies, N-fixing species showed particularly large increases in abundance (Luscher et al. 1996; Leadley et al. 1999; Vasseur and Potvin 1998; Warwick et al. 1998) and changes in the population of N-fixing symbionts (Montealegre et al. 2000).

Higher Trophic Levels

Bezemer and Jones (1998) reviewed a large and growing literature on changes in the performance of insects feeding on plants grown under elevated CO_2. In nearly all cases, plant tissues grown at high CO_2 have lower nitrogen and higher carbohydrate concentrations, stimulating greater consumption by "leaf chewing" species (Docherty et al. 1996). Plants with the highest growth response to high CO_2 show the greatest increase in leaf carbon content, whereas slow-growing species often show increases in leaf defensive compounds as well as carbon (Lindroth, Kinney, and Platz 1993). The abundance of "phloem feeders" often increases when plants are grown at high CO_2 (Bezemer and Jones 1998), presumably due to higher levels of soluble carbohydrate available to support those insects. The changes in plant and insect populations are likely to mediate most changes in the abundance of vertebrates in response to rising CO_2, because there is no evidence of direct physiological effects of the predicted, elevated future CO_2 concentrations.

Soil Ecosystems

Rising concentrations of atmospheric CO_2 are unlikely to have a direct impact on soil organisms, which normally live in an environment of high CO_2. However, these organisms respond to increases in the input of labile carbon and plant residues to the soil (Sadowsky and Schortemeyer 1997; Jones et al. 1998; Degens et al. 2000), and the responses of decomposers may, in turn, affect plant growth (Naeem et al. 2000). Zak et al. (1996, 2000) reported higher microbial biomass but no apparent change in microbial community composition in soils used to grow poplar (*Populus grandidentata*) at elevated CO_2 (cf. Bruce et al. 2000). In contrast, using the same techniques, Ringelberg et al. (1997) found changes in the composition of microbial communities under white oak (*Quercus alba*). When grass species were grown at high CO_2, there was little change in soil microbial diversity (Hodge et al. 1998), although the number of N-fixing symbiotic bacteria increased in one study (Schortemeyer et al. 1996). Yeates et al. (1999) found a general decrease in nematode diversity and abundance, but an increase in the abundance of bacteria-feeding nematodes, in soils of a plant community exposed to high CO_2 near hydrothermal vents in New Zealand. On the other hand, Klironomos et al. (1996) found a large increase in the density of soil micro-fauna under desert shrubs grown at high CO_2 in fertile conditions. Zaller and Arnone (1999) found no effect of CO_2 enrichment on earthworm com-munities in calcareous Swiss pastures.

Mycorrhizae, the symbiotic fungi that enhance the nutrient uptake by higher plants, often increase when plants are grown under elevated CO_2 and low soil fertility (DeLucia et al. 1997; Walker et al. 1995; Klironomos et al. 1997). Klironomos et al. (1998) found that plant growth at elevated CO_2 affected mycorrhizal colonization, spore production, and hyphae growth, with different responses among four fungal varieties. Godbold et al. (1997) also found that elevated CO_2 affected mycorrhizal colonization, with the response depending on both the type of fungi and the species of tree that formed the symbiosis. Clearly, more work is needed to define the general response of mycorrhizal communities to the growth of their host plants at high CO_2.

Marine Ecosystems

The continuing rise in atmospheric CO_2 will lead to a drop in seawater pH, as apparently occurred during the early Miocene, 21 million years ago (Spivack et al. 1993). Because the formation of aragonite—the primary carbonate min-eral of corals—declines at lower seawater pH, decalcification may lead to the bleaching and degradation of coral reefs in many tropical areas. The forma-tion of biogenic aragonite, and thus coral reefs, is expected to decrease by 14–30 percent by the year 2050, as a direct result of rising atmospheric CO_2 (Kleypas et al. 1999; Leclercq et al. 2000). Thus, rising CO_2 may act in concert with rising temperature to increase the bleaching and degradation of coral reefs.

Marine phytoplankton are typically limited by the availability of nitrogen and phosphorus in seawater, so the growth of phytoplankton shows little response to changes in atmospheric CO_2 content. Nevertheless, some studies suggest both higher oceanic primary production as a result of increases in CO_2 and differential species' responses that might affect biodiversity (Hein and Sand-Jensen 1997; Raven 1994).

Summary and Research Needs: Rising CO_2

A limited number of field observations and experiments suggest that rising concentrations of atmospheric CO_2 will affect the composition and structure of ecological communities. Many, but not all, studies suggest that plant species with high innate growth rates will benefit from elevated CO_2, potentially lowering regional biodiversity. The competitive relations of species will be determined by the effect of CO_2 on plant photosynthesis, water relations, and nutrient uptake, and the translation of these effects to higher trophic levels. The potential for rapid evolution in response to elevated CO_2 is suggested by high intraspecific variation in plant response to CO_2 (Curtis et al. 1996; Ward and Strain 1997; Andalo et al. 1998; DeLucia et al. 1997; Wayne and Bazzez 1995, 1997) and by the results of selection experiments extending over several generations of short-lived plants (Tousignant and Potvin 1996; Ward et al. 2000). However, the evolutionary response of trees and other long-lived species is likely to lag behind the large rise in atmospheric CO_2 expected during this century.

Despite their high cost, Free-Air CO_2 Enrichment experiments provide the greatest insight into whole-ecosystem changes in ecosystem processes and species diversity that are likely to accompany the rise of CO_2 in Earth's atmosphere. The greatest insight will stem from FACE experiments that operate for a decade or more, and those in which the high CO_2 treatment is examined in conjunction with other factors, especially changes in nitrogen input. The response of biodiversity to CO_2 enrichment should be carefully examined in semiarid ecosystems, where changes in plant water-use efficiency in response to high CO_2 may have dramatic changes in the availability of soil water and the competitive ability of some species.

Nitrogen Availability and Acidification

The biota of most terrestrial and marine ecosystems has evolved in a state of chronic nitrogen deficiency, so changes in the availability of nitrogen would be expected to have dramatic effects on plant community structure. Through the production of synthetic nitrogen fertilizer, humans have roughly doubled the annual input of available nitrogen to the earth's surface, causing large increases in the concentration of nitrogen in surface and ground waters (Vitousek et al. 1997). Excessive nitrogen in runoff has caused losses of species in a variety of wetland and coastal habitats, the most well known being the loss of fish and shellfish in the "dead zone" in the Gulf of Mexico (Turner and Rabalais 1991; Rabalais et al. 1996). In coastal marine waters, human inputs of nitrogen have shifted nutrient limitation from nitrogen to silicon, resulting in shifts in the relative abundance of phytoplankton species (Turner and Rabalais 1994; Justic et al. 1995). Atmospheric transport of reactive nitrogen compounds (e.g., ammonia [NH_3] and nitric oxide [NO_x] carries human perturbation of the nitrogen cycle to all regions of the earth (Holland et al. 1997), as seen in ice-core records of nitrogen deposition in Antarctica (Mayewski and Legrand 1990) and Greenland (Mayewski et al. 1986). In the Netherlands, where deposition of nitrogen from the atmosphere is perhaps twenty-five times higher than one hundred years ago, formerly diverse pastures are now dominated by pure stands of bracken fern (*Brachypodium*), and *Calluna* heathlands have been replaced by depauperate grasslands of *Deschampsia* and *Molinia* (Heil and Diemont 1983). These community changes can be replicated by experimental additions of nitrogen fertilizer to field plots (Bobbink 1991; Willems and Peet 1993).

Nearly all field experiments show a decline in species diversity with nutrient enrichment (e.g., Hurd et al. 1971; Huenneke et al. 1990; Mun and Whitford 1989). Austin and Austin (1980) found lower species richness and equitability in experimental communities of grassland plants, resulting in changes in dominance-diversity curves that resemble those from the Parkland field plots (figure 9.1). Wedin and Tilman (1996) report over 50 percent reduction in plant diversity in field plots receiving nitrogen fertilizer over a twelve-year period. Bakelaar and Odum (1978) also show lower richness and equitability among the weed species of fertilized fields (figure 9.6). Although these changes in diversity were similar to those seen during ecosystem retrogression, Bakelaar and Odum note that the species that remained were those normally associated with later-successional stages. Ecologists have ample theory to explain these observations: the higher species diversity in natural communities persists when resources are low, so that each species is limited by a different resource (Levin 1970; Rosenzweig 1971; Tilman 1982). Any addition of a limiting resource to such a community will lead to the dominance of the species that can usurp that resource most effectively. Plant species with high inherent relative growth rates may benefit most from the synergistic effects of high CO_2 and soil nitrogen availability (Poorter 1998). Surprisingly, despite observations of higher plant production and lower plant diversity in fertilized, versus unfertilized, fields, Hurd et al. (1971) found an equal or higher diversity of herbivorous and carnivorous arthropods on the plants in fertilized fields.

Inasmuch as NO_x is removed from the atmosphere by reactions yielding nitric acid (HNO_3) in rainfall, human perturbations of the global nitrogen cycle may also be associated with a greater regional occurrence of "acid rain." Conversely, in some areas, anthropogenic emissions of ammonia (NH_3) may reduce the acidity of rainfall (Schlesinger 1997). The effects of acid rain on plants, fish, and amphibians are well known (Pough 1976; Schindler 1988).

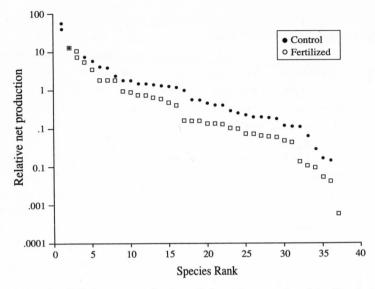

FIGURE 9.6. Relative net production of plants in control and fertilized plots in an abandoned field in Georgia, plotted as a function of the relative importance of species in the ecosystem. From Bakelaar and Odum (1978).

Air pollution and acid rain reduce plant species diversity and equitability in forest ecosystems (McClenahen 1978; Freedman and Hutchinson 1980). In many cases, the effect of acid rain on trees is mediated by changes in soil chemistry, rather than a direct effect on leaf tissues (Schlesinger 1997; Blake et al. 1999). Acid rain also causes indirect effects on higher trophic levels. For example, in the Netherlands, reproduction of the great tit (*Parus major*) is poor in forests subject to acid rain, owing to a decline in the abundance of snails, which are the main source of calcium for this bird's eggshells (Graveland et al. 1994).

Following the implementation of stringent air quality standards that reduced the emission of SO_2 in many developed countries, many aquatic ecosystems have recovered rapidly from previous, higher levels of acid deposition (Tipping et al. 1998; Stoddard et al. 1999). At the same time, nitric acid has become an increasingly important component of the anthropogenic acidity in rainfall, and high concentrations of NO_x are also associated with the formation of tropospheric ozone, which is a serious regional air pollutant.

Summary and Research Needs: Nitrogen Cycle

Humans will respond to the demand to feed ten billion of their own number, so the use of synthetic nitrogen fertilizer, and its inadvertent release to the environment, will undoubtedly increase during the next century. We believe that this human perturbation of the global nitrogen cycle will emerge as a serious threat to global biodiversity.

Field experiments that manipulate the availability of nitrogen are often easier to perform than those aimed at altering the temperature, moisture, or CO_2 regime. Experiments to examine changes in ecosystem processes (e.g., MaGill et al. 1997) could easily be maintained for long enough periods to also examine changes in biodiversity. Additional experiments should examine changes in biodiversity in response to the simultaneous exposure of ecosystems to high nitrogen and high CO_2. Special attention should be paid to the response of exotic species in nitrogen-rich habitats. It will be critical to establish large-scale experiments that evaluate the link between the nitrogen loading of fresh waters and subsequent changes in biodiversity and ecosystem function of the estuaries and coastal waters into which they drain. A recent analysis suggests that nutrient additions increase the production of phytoplankton in pelagic waters, without causing large changes at higher trophic levels (Micheli 1999).

Tropospheric Ozone

The photochemical production of ozone (O_3) from volatile organic carbon (VOC) compounds now dominates the ozone budget of the lower atmosphere—the troposphere (Levy et al. 1997). In urban areas, the reaction is catalyzed by high concentrations of nitrogen oxide gases derived from fossil fuel combustion. Low concentrations of NO_x formerly limited the production of ozone in most rural areas, but increasing atmospheric dispersal of NO_x from human activities (e.g., Cooper and Peterson 2000) and emissions from agricultural soils (Davidson et al. 1998) now cause regional ozone pollution in many of these areas as well (Chameides et al. 1994). Since the beginning of the Industrial Revolution, emissions of VOC and NO_x have doubled the O_3 background concentration (Fowler et al. 1998). Ozone concentrations in the lower atmosphere, averaging 10 nl l^{-1} from 1876 to 1910, are now up to 25–30 nl l^{-1} (Volz and Kley, 1988). Ozone concentrations are projected to increase 15 percent by the year 2035 (Thompson et al. 1990), approaching the critical 40 nl l^{-1} threshold level for vegetation (Fuhrer et al. 1997). High O_3 concentrations are now reported in remote areas of the South Pacific (Jiang and Yung 1996), in South America, and in Africa (Fishman et al. 1990). Emissions of NO_x from ships may more than double the background concentration of tropospheric O_3 over the North Atlantic and Pacific oceans (Lawrence and Crutzen 1999). Thus, the toxic effects of ozone are potentially a global problem, in both urban and rural areas. In forests, conifers appear more resistant to episodic O_3 exposures than broadleaf species—deciduous trees being the most sensitive (Reich 1987; Chappelka and

Samuelson 1998; Shärky et al. 1998). Black cherry (*Prunus serotina*) is among the most sensitive species in the United States (Rebbeck 1996), and in Europe, birch, aspen, and poplar are very sensitive to ozone damage (Shärky et al. 1998). Understory seedlings and saplings appear to incur less damage from O_3 than do canopy trees (Samuelson and Kelly 1997). In grasslands, fast-growing, invasive, and early-successional species are more sensitive to ozone than the slow-growing, late-successional species (Reiling and Davison 1992; Bungener, Nussbaum, et al. 1999).

A decrease in the cover of forbs is reported for grasslands exposed to O_3 on both acid (Ashmore and Ainsworth 1995) and calcareous (Ashmore et al. 1995) soils. In European grasslands, legumes are more sensitive to O_3 than other herbs and grasses (Fuhrer et al. 1994; Nebel and Fuhrer 1994; Warwick and Taylor 1995; Bungener, Balls, et al. 1999). A reduction in the legume clover with chronic O_3 exposure was associated with increases in grasses in the United States (Bennett and Runeckles 1977; Rebbeck et al. 1988). Davison and Barnes (1998) suggested that this sensitivity of legumes to O_3 could reduce N fixation and thus ecosystem nutrient availability.

Although these studies suggest species-specific responses to tropospheric O_3, few studies have addressed the effect of O_3 on species diversity per se. Westman (1979) reported that a decline in the biodiversity of coastal sage scrub was correlated with the concentration of airborne oxidants, which included O_3 (figure 9.7). Barbo et al. (1998) reported lower species diversity in an early-successional community exposed to elevated O_3 than in a similar community grown in charcoal-filtered air. Barnes et al. (1999) suggest that in many areas species that are sensitive to ozone have already been eliminated from natural communities. Reductions in biomass and acceleration of plant development and leaf senescence during chronic O_3 exposures (Pell et al. 1997; Reid and Fiscus 1998) are likely to decrease the reproductive potential of sensitive species. Ample examples of such reduced biomass and seed yield exist for crop species (e.g., reviews by Miller 1988; Heagle 1989). In natural communities, such changes are likely to alter species dominance and diversity.

Leaf conductance, which controls the uptake of CO_2 and O_3 by vegetation, is a main determinant of potential O_3 damage (Heath 1994). In studies examining a range of species, those most sensitive to O_3 had high stomatal conductances (Reich 1987; Nebel and Fuhrer 1994; Bungener, Balls, et al. 1999). Similarly, the greater O_3 sensitivity of canopy trees than of seedlings is associated with a greater leaf conductance in the trees (Samuelson and Kelly 1997). Chappelka and Samuelson (1998) suggest that stomatal conductance may be a good index for ozone tolerance in eastern tree species. The ranking of O_3 tolerance among conifers and broadleaf shade-tolerant and intolerant trees is

FIGURE 9.7. Lognormal distributions of species importance in polluted (B) and unpolluted (A) plots in the coastal sage scrub of Southern California. Reprinted with permission from Westman (1979). Copyright © 1979 American Association for the Advancement of Science.

correlated with differences in leaf conductance (Reich 1987). In addition, greater O_3 tolerance has been related to lower stomatal density (a determinant of leaf conductance) in comparisons among ash populations (Bennett et al. 1992), pine species (Evans and Miller 1972), and herbaceous plants (Evans et al. 1996).

Natural selection for O_3 tolerance has been observed in trees and herbs. A comparison of *Plantago major* populations from regions differing in O_3 concentration and climate suggests that O_3 is a major selection factor for O_3-resistant genotypes (Lyons et al. 1997). Whitfield et al. (1997) showed that genetic variation for O_3 resistance resulted in a resistant *Plantago* population, with lower leaf conductance, after four generations of selection. In two common garden experiments, Berrang et al. (1989) reported similar selection for O_3 resistance in poplar (*Populus tremuloides*) clones from several areas differing in O_3 concentration.

Little is known about O_3 effects on higher trophic levels. Studies of plant-animal interactions report higher rates of insect herbivory on O_3-injured than on noninjured leaves (Jones et al. 1994; Chappelka et al. 1997). Growth rates of spider mite populations are higher on clover (*Trifolium repens*) exposed to O_3 than on clover grown in charcoal-filtered air (Heagle et al. 1994). Feeding on O_3-injured leaves had no effect on gypsy moths (Lindroth, Reich et al. 1993) or aphids (Jones et al. 1994), but it decreased a beetle population (Jones et al. 1994). The increased herbivory on O_3-damaged tissue is likely to impact the long-term growth and survival of plants, and potentially their herbivores. Changes in plant diversity and community structure as a result of exposure to O_3 will likely affect food sources and protective cover for other trophic levels, including microbial communities in the soil (Islam et al. 2000).

Studies on the interactive effects of pollutant O_3 and CO_2 suggest that O_3 damage is ameliorated by elevated atmospheric CO_2. For example, photosynthesis is higher in crop plants exposed to a combination treatment of elevated CO_2 and O_3 than in plants exposed only to O_3 (McKee et al. 1995; Reid and Fiscus 1998). Similar results are reported for trees (e.g., beech [*Fagus sylvatica*], Grams et al. 1999; and Scots pine [*Pinus sylvestris*], Kellomäki and Wang 1997a, b). These studies also report a decrease in leaf conductance in a combined treatment of elevated CO_2 and O_3. Reduced flux of O_3 into the leaves because of reduced leaf conductance with elevated CO_2 partially explains the amelioration response (McKee et al. 1995; Fiscus et al. 1997).

Although the interactive effects of ozone and CO_2 on leaf physiology have been studied in trees and herbs, studies of their interactive effects on growth are limited to crop species (e.g., wheat [McKee et al. 1997; Mulholland et al. 1997, 1998], corn [Rudorff et al. 1996], and soybean [Fiscus et al. 1997; Miller et al. 1998]). Nevertheless, the similar physiological responses of crops and trees suggest that the growth of native species will also be ameliorated. A comparison of six perennial species, including trees and C3 and C4 grasses, suggests that amelioration may be widespread among perennials (Volin et al. 1998). However, the increase in growth with elevated CO_2 in O_3-fumigated air does not always translate into increased reproductive potential (Fiscus et al. 1997; Reid and Fiscus 1998). McKee et al. (1997) found no amelioration of wheat seed yield with elevated CO_2 when exposed to O_3-fumigated air, whereas Fiscus et al. (1997) found such an amelioration of seed yield in soybean. These authors further speculate that the enhancement in yield often reported with elevated CO_2 might be an artifact of lower damage from O_3 pollution. A review of the interactive effects of elevated CO_2 and O_3 on crop seed yield supports this hypothesis (Fiscus et al. 2000), but further investigations are needed.

Summary and Research Needs: Tropospheric Ozone

The effect of ozone on biodiversity resembles that of many other toxic effluents. The decrease in growth and seed set with O_3 exposure reduces plant biomass and species composition, which, in turn, adversely affects higher trophic levels. Differential sensitivity among species is likely to result in changes in community composition in regions of high exposure, potentially decreasing biodiversity. Unfortunately, no field experiments have examined changes in biotic diversity in natural systems with simultaneous exposure to both high CO_2 and O_3. The evidence to date suggests that interactive effects of O_3 and CO_2 will alter predictions based on responses to each gas separately. Field experiments, using the approach of Free-Air CO_2 Enrichment, will offer important insight into the changes in biodiversity expected in ecosystems that are exposed to O_3 for long periods, including acute short-term exposure to high concentrations. These O_3 studies should be designed to examine the interaction of O_3 with CO_2 and other environmental factors that are changing globally.

Increasing UV-B Radiation

While high concentrations of ozone in the troposphere have potentially toxic effects, high concentrations of ozone in the upper atmosphere—the stratosphere—protect the world's biota from potentially damaging effects of the sun's ultraviolet radiation (UV-B). Increasing UV-B radiation at the earth's surface has accompanied the appearance of the "ozone hole" over Antarctica (McKenzie et al. 1999) and losses of stratospheric ozone over temperate regions (Correll et al. 1992; Kerr and McElroy 1993). Increased UV-B radiation may affect the genome stability of plants (Ries et al. 2000).

UV-B radiation is absorbed by water, greatly attenuating its effects at depth. Nevertheless, greater UV-B radiation may already be reducing phytoplankton photosynthesis (Behrenfeld et al. 1993) and productivity in Antarctic waters (Smith et al. 1992; Neale et al. 1998). Various marine species are known to have differential sensitivity to UV-B radiation (Smith et al. 1992), but the effects of UV-B radiation on each species may be mediated by its seasonality and the vertical mixing of ocean waters (Hunter et al. 1981). McMinn et al. (1994) could find no evidence for unusual changes in the species composition of diatom communities in Antarctic waters during a recent twenty-year interval.

For land plants, Laakso and Huttunen (1998) note that pine seedlings are sensitive to short-term exposures to elevated UV-B radiation, but evidence for growth reductions in long-term field studies is limited to only one species, loblolly pine (Sullivan and Teramura 1992). Several studies have noted differential sensitivities of land plants to UV-B (Day et al. 1992), and one study found *enhanced* growth of the weed *Oenothera parviflora* in response to elevated UV-B exposure (Cybulski and Peterjohn 1999). Significant differences in the accumulation of UV-B protective flavonol compounds were noted among populations of clover grown in controlled conditions, with stress-adapted populations accumulating the largest amounts (Hoffman et al. 2000). Yakimchuk and Hoddinott (1994) noted that simultaneous exposure to elevated CO_2 and UV-B radiation reduced the seedling growth of three boreal forest species, but less so than exposure to high UV-B alone. Klironomos and

Allen (1995) note changes in the soil microbes associated with sugar maple (*Acer saccharum*) grown with supplemental UV-B.

Caldwell and Flint (1994) reviewed more than three hundred studies of the effects of UV-B radiation on terrestrial ecosystems. They note that studies showing differential species sensitivity have been conducted in controlled conditions (e.g., Gold and Caldwell 1983), but there are no long-term whole-ecosystem studies of the effects of UV-B on species diversity or ecosystem function. Subsequently, Day et al. (1999) reported increased growth of two vascular plants on the Antarctic Peninsula in response to experimental reductions in UV-B. Significantly, they found no interactive effects between UV-B levels and treatments imposing experimental warming. Recently, a one-year field experiment using plastic filters to reduce the high incidence of UV-B radiation on Tierra del Fuego, Argentina, resulted in only minor changes in plant growth in bogs and fens (Searles et al. 1999). Similarly, Norton et al. (1999) and Papadopoulos et al. (1999) report only small changes in artificial grassland communities exposed to ambient and elevated levels of UV-B. One study reports a greater abundance of freshwater algae in response to elevated UV-B, owing to its disproportional deleterious effects on the normal consumers of algal biomass (Bothwell et al. 1994).

Significant media attention has been devoted to studies suggesting that the worldwide decline in amphibian abundance may be related to the effects of increasing levels of UV-B radiation on amphibian embryos (Alford and Richards 1999; Blaustein et al. 1994). Other studies have produced conflicting results (Cummins et al. 1999), and the interaction of UV-B with other environmental factors and species-specific sensitivity is deserving of further study.

Conclusions

We concur with the recent biodiversity assessment that suggests that land use and climate change pose the most serious global threats to biodiversity (Sala et al. 2000). Beyond climate change, humans are causing a variety of changes in the chemistry and physical characteristics of Earth's environment. To the extent that species respond differently to those changes, some species may increase in abundance, while others may decline, leading to global changes in the composition and diversity of ecological communities. Many studies suggest that the responses of natural ecosystems to stress include losses of species and an increasing dominance of a few species that are competitive in the newly established environmental conditions. Studies of past environments confirm losses of species and genetic diversity during periods of environmental change; thus, we expect that global environmental change will lead to losses of biodiversity during this century. Policies to protect biodiversity from those changes must include international agreements that are successful in reducing the emission of greenhouse gases and pollutant levels of nitrogen and ozone to the environment. The policy-making process should be guided by large-scale, long-term field experiments that validate model projections of changes in biodiversity in response to global environmental change. Priority research areas are listed in box 9.1.

BOX 9.1. **Research Priorities for Ecosystem Studies Relevant to Conservation Biology**

1. Investigate global climate change, including long-term, whole-ecosystem effects of warming and altered moisture availability, particularly for the Arctic (warming) and desert (moisture).
   - Is the rate of species loss matched by the invasion of new species more suited to the altered environmental conditions?
   - Are alterations of the mean conditions (temperature and precipitation) more or less important to species persistence than alterations of extreme conditions?
2. Investigate rising carbon dioxide, including long-term, whole-ecosystem effects of high CO_2 on ecosystem function for all ecosystems, but particularly for localities with simultaneous exposure to high N deposition and O_3 and in desert ecosystems, where the effects of CO_2 and moisture availability interact.
   - Is there evidence for rapid selection of genotypes favored for growth at high CO_2?
   - How do simultaneous exposures to high temperature, nutrient additions, and O_3 affect ecosystem response to high CO_2?
3. Investigate the nitrogen cycle, including long-term, whole-ecosystem experiments with enhanced levels of nitrogen input, particularly for temperate grasslands, temperate forests, and estuarine and coastal ecosystems.
   - What is the effect of chronic levels of high N input on the biodiversity of terrestrial ecosystems?
   - Is the success of exotic species enhanced under high N inputs?
   - What is the effect of high N inputs on trophic level structure in estuarine systems?
4. Investigate tropospheric ozone, including long-term, whole-ecosystem studies of the response of biodiversity to high O_3, particularly in the temperate zone.
   - To what extent are the toxic effects of high O_3 on biodiversity mediated by simultaneous exposure to high CO_2?
   - What are the direct effects of high O_3 on higher trophic levels?
5. Investigate ultra-violet radiation, including the effects of high UV-B flux on biodiversity, particularly in high arctic terrestrial and marine ecosystems.
   - Are global increases in UV-B flux leading to a loss of amphibian biodiversity?
   - Are high-latitude increases in UV-B flux altering the structure and function of arctic ecosystems?

ACKNOWLEDGMENTS

We thank William Murdoch, Gordon Orians, and Lisa Dellwo Schlesinger for helpful reviews of the manuscript.

LITERATURE CITED

Ackerly, D.D., and F.A. Bazzaz. 1995. Plant growth and reproduction along CO_2 gradients: Non-linear responses and implications for community change. *Global Change Biology* 1: 199–207.

Ahas, R. 1999. Long-term phyto-, ornitho-, and ichthyophenological time-series analyses in Estonia. *International Journal of Biometerology* 42: 119–123.

Alford, R.A., and S.J. Richards. 1999. Global amphibian declines: A problem in applied ecology. *Annual Review of Ecology and Systematics* 30: 133–165.

Allen, C.D., and D. D. Breshears. 1998. Drought-induced shift of a forest-woodland ecotone: Rapid landscape response to climate variation. *Proceedings of the National Academy of Sciences USA* 95: 14839–14841.

Andalo, C., C. Raquin, N. Machon, B. Godelle, and M. Mousseau. 1998. Direct and maternal effects of elevated CO_2 on early root growth of germinating *Arabidopsis thaliana* seedlings. *Annals of Botany* 81: 405–411.

Andrews, J.A., R. Matamala, K.M. Westover, and W.H. Schlesinger. 2000. Temperature effects on the diversity of soil heterotrophs and the $\delta^{13}C$ of soil-respired CO_2. *Soil Biology and Biochemistry* 32: 699–706.

Arft, A.M., and the ITEX Group. 1999. Responses of tundra plants to experimental warming: Meta-analysis of the International Tundra Experiment. *Ecological Monographs* 69: 491–511.

Arnone, J.A., and J.C. Gordon. 1990. Effect of nodulation, nitrogen fixation and CO_2 enrichment on the physiology, growth and dry mass allocation of seedlings of *Alnus rubra* Bong. *New Phytologist* 116: 55–66.

Aronson, R.B., W.F. Precht, I.G. Macintyre, and T.J.T. Murdoch. 2000. Coral bleach-out in Belize. *Nature* 405: 36.

Arris, L.L., and P.S. Eagleson. 1989. Evidence of a physiological basis for the boreal-deciduous ecotone in North America. *Vegetatio* 82: 55–58.

Ashmore, M.R., and N. Ainsworth. 1995. The effects of cutting on the species composition of artificial grassland communities. *Functional Ecology* 9: 708–712.

Ashmore, M.R., R.II. Thwaites, N. Ainsworth, D.S. Cousins, S.A. Power, and A.J. Morton. 1995. Effects of ozone on calcareous grassland communities. *Water, Air, and Soil Pollution* 85: 1527–1532.

Austin, M.P., and B.O. Austin. 1980. Behaviour of experimental plant communities along a nutrient gradient. *Journal of Ecology* 68: 891–918.

Bakelaar, R.G., and E.P. Odum. 1978. Community and population level responses to fertilization in an old-field ecosystem. *Ecology* 59: 660–665.

Barbo, D.N., A.H. Chappelka, G.L. Somers, M.S. Miller-Goodman, and K. Stolte. 1998. Diversity of an early successional plant community as influenced by ozone. *New Phytologist* 138: 653–662.

Barnes, J., J. Bender, T. Lyons, and A. Borland. 1999. Natural and man-made selection for air pollution resistance. *Journal of Experimental Botany* 50: 1423–1435.

Barnola, J. M., D. Raynaud, Y. S. Korotevich, and C. Lorius. 1987. Vostok ice core provides 160,000 year record of atmospheric CO_2. *Nature* 329:408–414.

Barry, J.P., C.H. Baxter, R.D. Sagarin, and S.E. Gilman. 1995. Climate-related long-term faunal changes in a California rocky intertidal community. *Science* 267: 672–675.

Barton, A.M., and J.A. Terri. 1993. The ecology of elevational positions in plants: Drought resistance in five montane pine species in southeastern Arizona. *American Journal of Botany* 80: 15–25.

Bazzaz, F.A. 1975. Plant species diversity in old-field successional ecosystems in southern Illinois. *Ecology* 56: 485–488.

Bazzaz, F.A. 1990. The response of natural ecosystems to the rising global CO_2 levels. *Annual Review of Ecology and Systematics* 21: 167–196.

Bazzaz, F.A. 1996. *Plants in Changing Environments.* New York: Cambridge University Press.

Bazzaz, F.A., J.S. Coleman, and S.R. Morse. 1990. Growth responses of seven major co-occurring tree species of the northeastern United States to elevated CO_2. *Canadian Journal of Forest Research* 20: 1479–1489.

Bazzaz, F.A., and S.L. Miao. 1993. Successional status, seed size, and responses of tree seedlings to CO_2, light and nutrients. *Ecology* 74: 104–112.

Bazzaz, F.A., and S.T.A. Pickett. 1980. Physiological ecology of tropical succession: A comparative review. *Annual Review of Ecology and Systematics* 11: 287–310.

Behrenfeld, M.J., J.T. Hardy, H. Gucinski, A. Hanneman, H. Lee, and A. Wones. 1993. Effects of ultraviolet-B radiation on primary production along latitudinal transects in the south Pacific ocean. *Marine Environmental Research* 35: 349–363.

Bell, K.L., and L.C. Bliss. 1979. Autecology of *Kobresia bellardii*: Why winter snow accumulation limits local distribution. *Ecological Monographs* 49: 377–402.

Bennett, J.P., P. Rassat, P. Berrang, and D.F. Karnosky. 1992. Relationship between leaf anatomy and ozone sensitivity of *Fraxinus pennsylvanica* Marsh. and *Prunus serotina* Ehrh. *Environmental and Experimental Botany* 32: 33–41.

Bennett, J.P., and V.C. Runeckles. 1977. Effects of low levels of ozone on plant competition. *Journal of Applied Ecology* 14: 877–880.

Berrang, P., D.F. Karnosky, and J.P. Bennett. 1989. Natural selection for ozone tolerance in *Populus tremuloides*: Field verification. *Canadian Journal of Forest Research* 19: 519–522.

Bethke, R.W., and T.D. Nudds. 1995. Effects of climate change and land use on duck abundance in Canadian prairie-parklands. *Ecological Applications* 5: 588–601.

Bezemer, T.M., and T.H. Jones. 1998. Plant-insect herbivore interactions in elevated atmospheric CO_2: Quantitative analyses and guild effects. *Oikos* 82: 212–222.

Blake, L., K.W.T. Goulding, C.J.B. Mott, and A.E. Johnson. 1999. Changes in soil chemistry accompanying acidification over more than 100 years under woodland and grass at Rothamsted Experiment Station, UK. *European Journal of Soil Science* 50: 401–412.

Blaustein, A.R., P.D. Hoffman, D.G. Hokit, J.M. Kiesecker, S.D. Wallis, and J.B. Hayes. 1994. UV repair and resistance to solar UV-B in amphibian eggs: A link to population declines? *Proceedings of the National Academy of Sciences USA* 91: 1791–1795.

Bobbink, R. 1991. Effects of nutrient enrichment in Dutch chalk grassland. *Journal of Applied Ecology* 28: 28–41.

Bothwell, M.L., D.M.J. Sherbot, and D.M. Pollock. 1994. Ecosystem response to solar ultraviolet-B radiation: Influence of trophic-level interactions. *Science* 265: 97–100.

Broecker, W. S. 1997. Thermohaline circulation, the Achilles heel of our climate system: Will man-made CO_2 upset the current balance? *Science* 278:1582–1588.

Brown, B. 1997. Coral bleaching: Causes and consequences. *Coral Reefs* 16 (Suppl.): 129–138.

Brown, J.H. 1971. Mammals on mountaintops: Nonequilibrium insular biogeography. *American Naturalist* 105: 467–478.

Brown, J.H., T.J. Valone, and C.G. Curtin. 1997. Reorganization of an arid ecosystem in response to recent climate change. *Proceedings of the National Academy of Sciences USA* 94: 9729–9733.

Brown, K.R., and K.O. Higginbotham. 1986. Effects of carbon dioxide enrichment and nitrogen supply on growth of boreal tree seedlings. *Tree Physiology* 2: 223–232.

Bruce, K.D., T.H. Jones, T.M. Bezemer, L.J. Thompson, and D.A. Ritchie. 2000. The effect of elevated atmospheric carbon dioxide levels on soil bacterial communities. *Global Change Biology* 6: 427–434.

Bungener, P., G.R. Balls, S. Nussbaum, M. Geissmann, A. Grub, and J. Fuhrer. 1999. Leaf injury characteristics of grassland species exposed to ozone in relation to soil moisture condition and vapour pressure deficit. *New Phytologist* 142: 271–282.

Bungener, P., S. Nussbaum, A. Grub, and J. Fuhrer. 1999. Growth response of grassland species to ozone in relation to soil moisture and plant strategy. *New Phytologist* 142: 283–293.

Cain, M.L., H. Damman, and A. Muir. 1998. Holocene migration of woodland herbs. *Ecological Monographs* 68: 325–347.

Caldwell, M.M., and S.D. Flint. 1994. Stratospheric ozone reduction, solar UV-B radiation and terrestrial ecosystems. *Climatic Change* 28: 375–394.

Cerling, T.E., J.M. Harris, B.J. MacFadden, M.G. Leakey, J. Quade, V. Eisenmann, and J.R. Ehleringer. 1997. Global vegetation change through the Miocene/Pliocene boundary. *Nature* 389: 153–158.

Cerling, T., Y. Wang, and J. Quade. 1993. Expansion of C4 ecosystems as an indicator of global ecological change in the Late Miocene. *Nature* 361: 344–345.

Chameides, W.L., P.S. Kasibhatla, J. Yienger, and H. Levy. 1994. Growth of continental-scale metro-agro-plexes, regional ozone pollution, and world food production. *Science* 264: 74–77.

Chapin, F.S., G.R. Shaver, A.E. Giblin, K.J. Nadelhoffer, and J.A. Laundre. 1995. Responses of arctic tundra to experimental and observed changes in climate. *Ecology* 76: 694–711.

Chappelka, A., J. Renfro, G. Somers, and B. Nash. 1997. Evaluation of ozone injury on foliage of black cherry (*Prunus serotina*) and tall milkweed (*Asclepias exaltata*) in Great Smoky Mountains National Park. *Environmental Pollution* 95: 13–18.

Chappelka, A.H., and L.J. Samuelson. 1998. Ambient ozone effects on forest trees of the eastern United States: A review. *New Phytologist* 139: 91–108.

Clark, J. S. 1998. Why trees migrate so fast: Confronting theory with dispersal biology and the paleo record. *American Naturalist* 152:204–224.

Clark, J. S., C. Fastie, G. Hurtt, S. T. Jackson, C. Johnson, G. King, M. Lewis, J. Lynch, S. Pacala, I.C. Prentice, E. W. Schupp, T. Webb III, and P. Wyckoff. 1998. Reid's Paradox of rapid plant migration. *BioScience* 48:13–24.

Clark, J. S., M. Lewis, and L. Horvath. 2001. Invasion by extremes: Variation in dispersal and reproduction retards population spread. *American Naturalist* 157: 537–554.

Clark, J.S., E.C. Grimm, J. Lynch, and P.J. Mueller. 2001. Effects of climate change on the C4 grassland/woodland boundary in the northern Central Plains. *Ecology* 82: 620–636.

Clark, J.S., B. Beckage, J. HilleRisLambers, I. Ibanez, S. LaDeau, J. MacLachlan, J. Mohan, and M. Rocca. 2001. Dispersal and plant migration. In H. Mooney and J. Canadell (eds.), *Encyclopedia of Global Environmental Change,* Volume 8. London: John Wiley and Sons.

Clark, P. U., R. B. Alley, and D. Pollard. 1999. Northern Hemisphere ice-sheet influences on global climate change. *Science* 286:1104–1111.

COHMAP. 1988. Climatic changes of the last 18,000 years: Observations and model simulations. *Science* 241:1043–1052.

Cole, D.R., and H.C. Monger. 1994. Influence of atmospheric CO_2 on the decline of C4 plants during the last deglaciation. *Nature* 368: 533–536.

Collatz, G. J., J.A. Berry, and J.S. Clark. 1998. Effects of climate and atmospheric CO_2 partial pressure on the global distribution of C4 grasses: Present, past, and future. *Oecologia* 114:441–454.

Comes, H. P., and J. W. Kadereit. 1998. The effect of Quaternary climatic changes on plant distribution and evolution. *Trends in Plant Science Reviews* 3:432–438.

Coope, G.R. 1987. The response of late Quaternary insect communities to sudden climatic changes. In J.H.R. Gee and P.S. Giller (eds.). *Organization of Communities—Past and Present,* pp. 421–438. Oxford: Blackwell.

Cooper, S.L., and D.L. Peterson. 2000. Spatial distribution of tropospheric ozone in western Washington, USA. *Environmental Pollution* 107: 339–347.

Correll, D.L., C.O. Clark, B. Goldberg, V.R. Goodrich, D.R. Hayes, W.H. Klein, and W.D. Schecher. 1992. Spectral ultraviolet-B radiation fluxes at the earth's surface: Long-term variations at 39°N, 77°W. *Journal of Geophysical Research* 97: 7579–7591.

Cummins, C.P., P.D. Greenslade, and A.R. McLeod. 1999. A test of the effect of supplemental UV-B radiation on the common frog, *Rana temporaria* L., during embryonic development. *Global Change Biology* 5: 471–479.

Curtis, P.S., D.J. Klus, S. Kalisz, and S.J. Tonsor. 1996. Intraspecific variation in CO_2 response in *Raphanus raphanistrum* and *Plantago lanceolate*: Assessing the potential for evolutionary change with rising atmospheric CO_2. In C. Korner and F. Bazzaz (eds.), *Carbon Dioxide, Populations and Communities,* pp. 13–22. San Diego: Academic Press.

Curtis, P.S., and X. Wang. 1998. A meta-analysis of elevated CO_2 effects on woody plant mass, form, and physiology. *Oecologia* 113: 299–313.

Cybulski, W.J., and W.T. Peterjohn. 1999. Effects of ambient UV-B radiation on the above-ground biomass of seven temperate-zone plant species. *Plant Ecology* 145: 175–181.

Davidson, E.A., C.S. Potter, P. Schlesinger, and S.A. Klooster. 1998. Model estimates of regional nitric oxide emissions from soils of the southeastern United States. *Ecological Applications* 8: 748–759.

Davis, M. B. 1981. Quaternary history and the stability of forest communities. In D.C. West, H. H. Shugart, D.B. Botkin (eds.), *Forest Succession: Concepts and Application,* pp. 132–153. New York: Springer-Verlag.

Davis, M. B., and C. Zabinski. 1992. Changes in geographical range resulting from greenhouse warming: Effects on biodiversity in forests. In R. Peters and T. Lovejoy (eds.), *Global Warming and Biological Diversity,* pp. 297–308. New Haven, Conn.: Yale University Press.

Davison, A.W., and J.D. Barnes. 1998. Effects of ozone on wild plants. *New Phytologist* 139: 135–151.

Day, T.A., C.T. Ruhland, C.W. Grobe, and F. Xiong. 1999. Growth and reproduction of Antarctic vascular plants in response to warming and UV radiation reductions in the field. *Oecologia* 119: 24–35.

Day, T.A., T.C. Vogelmann, and E.H. DeLucia. 1992. Are some plant life forms more effective than others in screening out ultraviolet-B radiation? *Oecologia* 92: 513–519.

Degens, B.P., L.A. Schipper, G.P. Sparling, M. Vojvodic-Vukovic. 2000. Decreases in organic C reserves in soils can reduce the catabolic diversity of soil microbial communities. *Soil Biology and Biochemistry* 32: 189–196.

Delcourt, P.A., and H.R. Delcourt. 1987. *Long-Term Forest Dynamics of the Temperate Zone.* New York: Springer-Verlag.

DeLucia, E.H., R.M. Callaway, E.M. Thomas, and W.H. Schlesinger. 1997. Mechanisms of phosphorus acquisition for ponderosa pine seedlings under high CO_2 and temperature. *Annals of Botany* 79: 111–120.

DeLucia, E.H., J.G. Hamilton, S.L. Naidu, R.B. Thomas, J.A. Andrews, A. Finzi, M. Lavine, R. Matamala, J.E. Mohan, G.R. Hendrey, and W.H. Schlesinger. 1999. Net primary production of a forest ecosystem with experimental CO_2 enrichment. *Science* 284: 1177–1179.

DeLucia, E.H., and W.H. Schlesinger. 1990. Ecophysiology of Great Basin and Sierra Nevada vegetation on contrasting soils. In C.B. Osmond, L.F. Pitelka, and G.M. Hidy (eds.), *Plant Biology of the Basin and Range,* pp. 143–178. New York: Springer-Verlag.

Demesure, B., B. Comps, and R. J. Petit. 1996. Chloroplast DNA phylogeography of the common beech (*Fagus sylvatica* L.) in Europe. *Evolution* 50:2515–2520.

Dillenburg, L. R., A. H. Teramura, I. N. Forseth, and D. F. Whigham. 1995. Photosynthetic and biomass allocation responses of *Liquidambar styraciflua* (Hammamelidaceae) to vine competition. *American Journal of Botany* 82(4): 454–461.

Dippery, J.K., D.T. Tissue, R.B. Thomas, and B.R. Strain. 1995. Effects of low and elevated CO_2 on C3 and C4 annuals. I. Growth and biomass allocation. *Oecologia* 101: 13–20.

Docherty, M., D.K. Hurst, J.K. Holopainen, J.B. Whittaker, P.J. Lee, and A.D. Watt. 1996. Carbon dioxide-induced changes in beech foliage cause female beech weevil larvae to feed in a compensatory manner. *Global Change Biology* 2: 335–342.

Dumolin-Lapeque, S., B. Demesure, S. Fineschi, V. Le Corre, and R. J. Petit. 1997. Phylogeographic structure of white oaks throughout the European continent. *Genetics* 146:1475–1487.

Dunn, P.O., and D.W. Winkler. 1999. Climate change has affected the breeding date of tree swallows throughout North America. *Proceedings of the Royal Society of London* 266B: 2487–2490.

Evans, L.S., K. Albury, and N. Jennings. 1996. Relationships between anatomical characteristics and ozone sensitivity of leaves of several herbaceous dicotyledonous plant species at Great Smoky Mountains National Park. *Environmental and Experimental Botany* 36: 413–420.

Evans, L.S., and P.R. Miller. 1972. Comparative needle anatomy and relative ozone sensitivity of four pine species. *Canadian Journal of Botany* 50: 1067–1071.

Fairbanks, R. G. 1989. A 17,000–year glacio-eustatic sea level record: Influence of glacial melting rates on the Younger Dryas event and deep-ocean circulation. *Nature* 342:637–642.

Farnsworth, E.J., J. Nunez-Farfan, S.A. Careaga, and F.A. Bazzaz. 1995. Phenology and growth of three temperate forest life forms in response to soil warming. *Journal of Ecology* 83: 967–977.

Fiscus, E.L., J.E. Miller, F.L. Booker, A.S. Heagle, and C.D. Reid. 2001. The impact of ozone and other limitations on crop productivity response to CO_2 *Technology,* in press.

Fiscus, E.L., C.D. Reid, J.E. Miller, and A.S. Heagle. 1997. Elevated CO_2 reduces O_3 flux and O_3-induced yield losses in soybeans: Possible implications for elevated CO_2 studies. *Journal of Experimental Botany* 48: 307–313.

Fishman, J., C.E. Watson, J.C. Larsen, and J.A. Logan. 1990. Distribution of tropospheric ozone determined from satellite data. *Journal of Geophysical Research* 95: 3599–3617.

Foley, J. A., J. E. Kutzbach, M. T. Coe, and S. Levis. 1994. Feedbacks between climate and boreal forests during the Holocene epoch. *Nature* 371:52–54.

Fowler, D., C. Flechard, U. Skiba, M. Coyle, and J.N. Cape. 1998. The atmospheric budget of oxidized nitrogen and its role in ozone formation and deposition. *New Phytologist* 139: 11–23.

Freedman, B., and T.C. Hutchinson. 1980. Long-term effects of smelter pollution at Sudbury, Ontario, on forest community composition. *Canadian Journal of Botany* 58: 2123–2140.

Fuhrer, J., H. Shariat-Madari, R. Perler, W. Tschannen, and A. Grub. 1994. Effects of ozone on managed pasture. II. Yield, species composition, canopy structure, and forage quality. *Environmental Pollution* 86: 307–314.

Fuhrer, J., L. Skarby, and M.R. Ashmore. 1997. Critical levels for ozone effects on vegetation in Europe. *Environmental Pollution* 97: 91–106.

Galen, C., and M.L. Stanton. 1993. Short-term responses of alpine buttercups to experimental manipulations of growing season length. *Ecology* 74: 1052–1058.

Galen, C., and M.L. Stanton. 1995. Response of snowbed plant species to changes in growing-season length. *Ecology* 76: 1546–1557.

George, T.L., A.C. Fowler, R.L. Knight, and L.C. McEwen. 1992. Impacts of a severe drought on grassland birds in western North Dakota. *Ecological Applications* 2: 275–284.

Glynn, P.W. 1996. Coral reef bleaching: Facts, hypotheses, and implications. *Global Change Biology* 2: 495–509.

Godbold, D.L., G.M. Berntson, and F.A. Bazzaz. 1997. Growth and mycorrhizal colonization of three North American tree species under elevated atmospheric CO_2. *New Phytologist* 137: 433–440.

Gold, W.G., and M.M. Caldwell. 1983. The effects of ultraviolet-B radiation on plant competition in terrestrial ecosystems. *Physiologia Plantarum* 58: 435–444.

Graham, R. W., and E. C. Grimm. 1990. Effects of global climate change on the patterns of terrestrial biological communities. *Trends in Ecology and Evolution* 5:289–292.

Grams, T.E.E., S. Anegg, K.-H. Haberle, C. Langebartels, and R. Matyssek. 1999. Interactions of chronic ozone exposure to elevated CO_2 and O_3 levels in the photosynthetic light and dark reactions of European beech (*Fagus sylvatica*). *New Phytologist* 144: 95–107.

Graveland, J., R. van der Wal, J.H. van Balen, and A.J. van Noordwijk. 1994. Poor reproduction in forest passerines from decline of snail abundance on acidified soils. *Nature* 368: 446–448.

Grime, J.P., V.K. Brown, K. Thompson, G.J. Masters, S.H. Hillier, I.P. Clarke, A.P. Askew, D. Corker, and J.P. Kielty. 2000. The response of two contrasting limestone grasslands to simulated climate change. *Science* 289: 762–765.

Harte, J., A. Rawa, and V. Price. 1996. Effects of manipulated soil microclimate on mesofaunal biomass and diversity. *Soil Biology and Biochemistry* 28: 313–322.

Harte, J., and R. Shaw. 1995. Shifting dominance within a montane vegetation community: Results of a climate-warming experiment. *Science* 267: 876–880.

Harte, J., M.S. Torn, F.-R. Chang, B. Feifarek, A.P. Kinzig, R. Shaw, and K. Shen. 1995. Global warming and soil microclimate: Results from a meadow-warming experiment. *Ecological Applications* 5: 132–150.

Hattenschwiler, S., and C. Korner. 1996. Effects of elevated CO_2 and increased nitrogen deposition on photosynthesis and growth of understory plants in spruce model ecosystems. *Oecologia* 106: 172–180.

Hattenschwiler, S., and C. Korner. 2000. Tree seedling responses to *in situ* CO_2 enrichment differ among species and depend on understorey light availability. *Global Change Biology* 6: 213–226.

Hattenschwiler, S., F. Miglietta, A. Raschi, and C. Korner. 1997. Thirty years of *in situ* tree growth under elevated CO_2: A model for future forest responses? *Global Change Biology* 3: 463–471.

Heagle, A.S. 1989. Ozone and crop yield. *Annual Review of Phytopathology* 27: 397–423.

Heagle, A.S., R.L. Brandenburg, J.C. Burns, and J.E. Miller. 1994. Ozone and carbon dioxide effects on spider mites in white clover and peanut. *Journal of Environmental Quality* 23: 1168–1176.

Heath, R.L. 1994. Alterations of plant metabolism by ozone exposure. In R.G. Alscher and A.R. Wellburn (eds.), *Plant Responses to the Gaseous Environment,* pp. 121–145. London: Chapman and Hall.

Heil, G.W., and W.H. Diemont. 1983. Raised nutrient levels change heathland into grassland. *Vegetatio* 53: 113–120.

Hein, M., and K. Sand-Jensen. 1997. CO_2 increases oceanic primary production. *Nature* 388: 526–527.

Heller, H.C., and D.M. Gates. 1971. Altitudinal zonation of chipmunks (*Eutamias*): Energy budgets. *Ecology* 52: 424–433.

Hewitt, G. M. 1990. Divergence and speciation as viewed from an insect hybrid zone. *Canadian Journal of Zoology* 68: 1701–1715.

Hewitt, G. M. 1999. Post-glacial re-colonization of European biota. *Biological Journal of the Linnean Society* 68: 87–112.

Hewitt, G. 2000. The genetic legacy of the Quaternary ice ages. *Nature* 405: 907–913.

Hirose, T., D.D. Ackerly, M.B. Traw, and F.A. Bazzaz. 1996. Effects of CO_2 elevation on canopy development in the stands of two co-occurring annuals. *Oecologia* 108: 215–223.

Hodge, A., E. Paterson, S.J. Grayston, C.D. Campbell, B.G. Ord, and K. Killham. 1998. Characterization and microbial utilisation of exudate material from the rhizosphere of *Lolium perenne* grown under CO_2 enrichment. *Soil Biology and Biochemistry* 30: 1033–1043.

Hoffman, R.W., E.E. Swinney, S.J. Bloor, K. R. Markam, K.G. Ryan, B.D. Campbell, B.R. Jordan, and D.W. Fountain. 2000. Respones of nine *Trifolium repens* L. populations to ultraviolet-B radiation: Differential flavonol glycoside accumulation and biomass production. *Annals of Botany* 86: 527–537.

Hogg, E.H. 1994. Climate and the southern limit of the western Canadian boreal forest. *Canadian Journal of Forest Research* 24: 1835–1845.

Holland, E.A., B.H. Braswell, J.-F. Lamarque, A. Townsend, J. Sulzman, J.-F. Muller, F. Dentener, G. Brasseur, H. Levy, J.E. Penner, and G.-J. Roelofs. 1997. Variations in the predicted spatial distribution of atmospheric nitrogen deposition and their impact on carbon uptake by terrestrial ecosystems. *Journal of Geophysical Research* 102: 15849–15866.

Hopkins, M. S., A. W. Graham, R. Hewett, J. Ash, and J. Head. 1990. Evidence of late Pleistocene fires and eucalypt forest from a North-Queensland humid tropical rainforest site. *Australian Journal of Ecology* 15:345–347.

Hudson, J.W., and D.B. Deavers. 1973. Metabolism, pulmocutaneous water loss and respiration of eight species of ground squirrels from different environments. *Comparative Biochemistry and Physiology* 45(1A): 69–100.

Huenneke, L.F., S.P. Hamburg, R. Koide, H.A. Mooney, and P.M. Vitousek. 1990. Effects of soil resources on plant invasion and community structure in California serpentine grassland. *Ecology* 71: 478–491.

Hunter, J.R., S.E. Kaupp, and J.H. Taylor. 1981. Effects of solar and artificial ultra-violet-B radiation on larval northern anchovy, *Engraulis mordax*. *Photochemistry and Photobiology* 34: 477–486.

Huntley, B., and H.J.B. Birks. 1983. *An Atlas of Past and Present Pollen Maps for Europe 0–13,000 Years Ago*. Cambridge, Mass.: Cambridge University Press.

Hurd, L.E., M.V. Mellinger, L.L. Wolf, and S.J. McNaughton. 1971. Stability and diversity at three trophic levels in terrestrial successional ecosystems. *Science* 173: 1134–1136.

Hurrell, J.W., and K.E. Trenbert. 1997. Spurious trends in satellite MSU temperatures from merging different satellite records. *Nature* 386: 164–167.

Ibrahim, K. M., R. A. Nichols, and G. M. Hewitt. 1996. Spatial patterns of genetic variation generated by different forms of dispersal during range expansion. *Heredity* 77:282–291.

Idso, S.B. 1999. The long-term response of trees to atmospheric CO_2 enrichment. *Global Change Biology* 5: 493–495.

Islam, K.R., C.L. Mulchi, and A.A. Ali. 2000. Interactions of tropospheric CO_2 and O_3 enrichments and moisture variations on microbial biomass and respiration in soil. *Global Change Biology* 6: 255–265.

Iverson, L.R., and A.M. Prasad. 1998. Predicting abundance of 80 tree species following climate change in the eastern United States. *Ecological Monographs* 68: 465–485.

Jackson, S. T., and C. Weng. 1999. Late Quaternary extinction of a tree species in eastern North America. *Proceedings of the National Academy of Sciences* 96: 13847–13852.

Jiang, Y., and Y.L. Yung. 1996. Concentrations of tropospheric ozone from 1979 to 1992 over the tropical Pacific South America from TOMS data. *Science* 272: 714–716.

Jobbagy, E.G., and R.B. Jackson. 2000. Global controls of forest line elevation in the northern and southern hemispheres. *Global Ecology and Biogeography* 9: 253–268.

Johannessen, O.M., E.V. Shalina, and M.W. Miles. 1999. Satellite evidence for an arctic sea ice cover in transformaton. *Science* 286: 1937–1939.

Johnson, H.B., H.W. Polley, and H.S. Mayeux. 1993. Increasing CO_2 and plant-plant interactions: Effects of natural vegetation. *Vegetatio* 104/105: 157–170.

Johnston, K.M., and O.J. Schmitz. 1997. Wildlife and climate change: Assessing the sensitivity of selected species to simulated doubling of atmospheric CO_2. *Global Change Biology* 3: 531–544

Jones, C.G., J.S. Coleman, and S. Findlay. 1994. Effects of ozone on interactions between plants, consumers and decomposers. In R.G. Alscher and A.R. Wellburn (eds.), *Plant Responses to the Gaseous Environment,* pp. 339–364. London: Chapman and Hall.

Jones, T.H., L.J. Thompson, J.H. Lawton, T.M. Bezemer, R.D. Bardgett, T.M. Blackburn, K.D. Bruce, P.F. Cannon, G.S. Hall, S.E. Hartley, G. Howson, C.G. Jones, C. Kampichler, E. Kandeler, and D.A. Ritchie. 1998. Impacts of rising atmospheric carbon dioxide on model terrestrial ecosystems. *Science* 280: 441–443.

Justic, D., N.N. Rabalais, R.E. Turner, and Q. Dortch. 1995. Changes in nutrient structure of river-dominated coastal waters: Stoichiometric nutrient balance and its consequences. *Estuarine, Coastal and Shelf Science* 40: 339–356.

Kellomäki, S., and K.-Y. Wang. 1997a. Effects of elevated O_3 and CO_2 on chlorophyll fluorescence and gas exchange in Scots pine during the third growing season. *Environmental Pollution* 97: 17–27.

Kellomäki, S., and K.-Y. Wang. 1997b. Effects of elevated O_3 and CO_2 concentrations on photosynthesis and stomatal conductance in Scots pine. *Plant, Cell and Environment* 29: 995–1006.

Kempton, R.A. 1979. The structure of species diversity and measurement of diversity. *Biometrics* 35: 307–321.

Kerr, J.B., and C.T. McElroy. 1993. Evidence for large upward trends of ultraviolet B radiation linked to ozone depletion. *Science* 262: 1032–1034.

Kershaw, A. P. 1993. The development of rainforest-savanna boundaries in tropical Australia. In P.A. Farley, J. Proctor, and J.A. Ratter (eds.), *Nature and Dynamics of Forest Savanna Boundaries,* pp. 255–271. London: Chapman and Hall.

Kerstiens, G. 1998. Shade-tolerance as a predictor of responses to elevated CO_2 in trees. *Physiologia Plantarum* 102:472–480.

Kleypas, J.A., R.W. Buddemeir, D. Archer, J.-P. Gattuso, C. Langdon, and B.N. Opdyke. 1999. Geochemical consequences of increased atmospheric carbon dioxide on coral reefs. *Science* 284: 118–120.

Klironomos, J.N., and M.F. Allen. 1995. UV-B-mediated changes on below-ground communities associated with the roots of *Acer saccharum. Functional Ecology* 9:923–930.

Klironomos, J.N., M.C. Rillig, and M.F. Allen. 1996. Below-ground microbial and microfaunal responses to *Artemisia tridentata* grown under elevated atmospheric CO_2. *Functional Ecology* 10: 527–534.

Klironomos, J.N., M.G. Rillig, M.F. Allen, D.R. Zak, M. Kubiskes, and K.S. Pregitzer. 1997. Soil fungal-arthropod responses to *Populus tremuloides* grown under enriched atmospheric CO_2 under field conditions. *Global Change Biology* 3: 473–478.

Klironomos, J.N., M. Ursic, M. Rilig, and M.F. Allen. 1998. Interspecific differences in the response of arbuscular mycorrhizal fungi to *Artemisia tridentata* grown under elevated atmospheric CO_2. *New Phytologist* 138: 599–605.

Kubiske, M.E., and K.S. Pregitzer. 1996. Effects of elevated CO_2 and light availability on the photosynthetic light response of trees of contrasting shade tolerance. *Tree Physiology* 16: 351–358.

Kullman, L. 1996. Norway spruce present in the Scandes Mountains, Sweden at 8000 yr BP: New light on Holocene tree spread. *Global Ecology and Biogeography Letters* 5: 94–101.

Kutzbach, J. E., and P. J. Guetter. 1986. The influence of changing orbital parameters and surface boundary conditions on climate simulations for the past 18000 years. *Journal of Atmospheric Sciences* 43: 1726–1759.

Kutzbach, J. E., P. J. Guetter, P. J. Behling, and R. Selin. 1993. Simulated climatic changes: Results of the COHMAP climate-model experiments. In H.E. Wright Jr., J. E. Kutzbach, T. Webb III, W.F. Ruddiman, F.A. Street-Perrott, and P.J. Bartlein (eds.), *Global Climates since the Last Glacial Maximum,* pp. 24–93. Minneapolis: University of Minnesota Press.

Laakso, K., and S. Huttunen. 1998. Effects of ultraviolet-B radiation (UV-B) on conifers: A review. *Environmental Pollution* 99: 319–328.

Lawrence, M.G., and P.J. Crutzen. 1999. Influence of NO_x emissions from ships on tropospheric photochemistry and climate. *Nature* 402: 167–170.

Leadley, P.W., P.A. Niklaus, R. Stocker, and C. Korner. 1999. A field study of the effects of elevated CO_2 on plant biomass and community structure in a calcareous grassland. *Oecologia* 118: 39–49.

Leclercq, N., J.-P. Gattuso, and J. Jaubert. 2000. CO_2 partial pressure controls the calcification rate of a coral community. *Global Change Biology* 6: 329–334.

LeCorre, A., N. Machon, R. J. Petit, and A. Kremer. 1997. Colonization with long-distance seed dispersal and genetic structure of maternally inherited genes in forest trees: A simulation study. *Genetic Research* 69: 117–125.

Levin, S.A. 1970. Community equilibria and stability, and an extension of the competitive exclusion principle. *American Naturalist* 104: 413–423.

Levy, H., P.S. Kasibhatla, W.J. Moxim, A.A. Klonecki, A.I. Hirsch, S.J. Oltmans, and W.L. Chameides. 1997. The global impact of human activity on tropospheric ozone. *Geophysical Research Letters* 14: 791–794.

Lindroth, R.L., K.K. Kinney, and C.L. Platz. 1993. Responses of deciduous trees to elevated atmospheric CO_2: Productivity, phytochemistry, and insect performance. *Ecology* 74: 763–777.

Lindroth, R.L., P.B. Reich, M.G. Tjoelker, J.C. Volin, and J. Oleksyn. 1993. Light environment alters response to ozone stress in seedlings of *Acer saccharum* Marsh. and hybrid *Populus* L. 3. Consequences for performance of gypsy moth. *New Phytologist* 124: 647–651.

Lukeville, A., and R.F. Wright. 1997. Experimentally increased soil temperature causes release of nitrogen at a boreal forest catchment in southern Norway. *Global Change Biology* 3: 13–22.

Luscher, A., T. Hebeisen, S. Zanetti, U.A. Hartwig, H. Blum, G.R. Hendrey, and J. Nosberger. 1996. Differences between legumes and nonlegumes of permanent grassland in their responses to Free-Air Carbon Dioxide Enrichment: Its effect on competition in a multispecies mixture. In C. Korner and F.A. Bazzaz (eds.), *Carbon Dioxide, Populations, and Communities,* pp. 287–300. San Diego: Academic Press.

Lyons, T.M., J.D. Barnes, and A.W. Davison. 1997. Relationships between ozone resistance and climate in European populations of *Plantago major. New Phytologist* 136: 503–510.

MacPhee, R.D.E. 1999. *Extinctions in Near Time.* New York: Kluwer Academic.

MaGill, A.H., J.D. Aber, J.J. Hendricks, R.D. Bowden, J.M. Melillo, and P.A. Steudler. 1997. Biogeochemical response of forest ecosystems to simulated nitrogen deposition. *Ecological Applications* 7: 402–415.

Martin, P. S., and R. G. Klein. 1984. *Quaternary Extinctions.* Tucson: University of Arizona Press.

Mayewski, P.A., and M.R. Legrand. 1990. Recent increase in nitrate concentration of Antarctic snow. *Nature* 346: 258–260.

Mayewski, P.A., W.B. Lyons, M.J. Spencer, M. Twickler, W. Dansgaard, B. Koci, C.I. Davidson, and R.E. Honrath. 1986. Sulfate and nitrate concentrations from a south Greenland ice core. *Science* 232: 975–977.

McAndrews, J.H. 1966. Postglacial history of prairie, savanna, and forest in northwestern Minnesota. *Memoirs of the Torrey Botanical Club* 22: 1–72.

McCauley, E., and W.W. Murdoch. 1990. Predator-prey dynamics in rich and poor environments. *Nature* 343: 455–457.

McClenahen, J.R. 1978. Community changes in a deciduous forest exposed to air pollution. *Canadian Journal of Forest Research* 8: 432–438.

McKee, I.F., J.F. Bullimore, and S.P. Long. 1997. Will elevated CO_2 concentrations protect the yield of wheat from O_3 damage? *Plant, Cell and Environment* 20: 77–84.

McKee, I.F., P.K. Farage, and S.P. Long. 1995. The interactive effects of elevated CO_2 and O_3 concentration on photosynthesis in spring wheat. *Photosynthesis Research* 45: 111–119.

McKenzie, R., B. Connor, and G. Bodeker. 1999. Increased summertime UV radiation in New Zealand in response to ozone loss. *Science* 285: 1709–1711.

McMinn, A., H. Heijnis, and D. Hodgson. 1994. Minimal effects of UVB radiation on Antarctic diatoms over the past 20 years. *Nature* 370: 547–549.

Micheli, F. 1999. Eutrophication, fisheries, and consumer-resource dynamics in marine pelagic ecosystems. *Science* 285: 1396–1398.

Miller, J.E. 1988. Effects of photosynthesis, carbon allocation, and plant growth associated with air pollutant stress. In W.W. Heck, O.C. Taylor, and D.T. Tingey (eds.), *Assessment of Crop Loss from Air Pollutants,* pp. 287–314. London: Elsevier Science.

Miller, J.E., A.S. Heagle, and W.A. Pursley. 1998. Influence of ozone stress on soybean response to carbon dioxide enrichment. II. Biomass and Development. *Crop Science* 38: 122–128.

Mitchell, J.F.B., S. Manabe, V. Meleshko, and T. Ookioka. 1990. Equilibrium climate change—and its implications for the future. In J.T. Houghton, G.J. Jenkins, and J.J. Ephramus (eds.), *Climate Change: The IPCC Scientific Assessment,* pp. 131–164. Cambridge: Cambridge University Press.

Montealegre, C.M., C. van Kessel, J.M. Blumenthal, H-G. Hur, U.A. Hartwig, and M.J. Sadowsky. 2000. Elevated atmospheric CO_2 alters microbial population structure in a pasture ecosystem. *Global Change Biology* 6: 475–482.

Mulholland, B.J., J. Craigon, C.R. Black, J.J. Colls, J. Atherton, and G. Landon. 1997. Effects of elevated carbon dioxide and ozone on the growth and yield of spring wheat (*Triticum aestivum* L.). *Journal of Experimental Botany* 48: 113–122.

Mulholland, B.J., J. Craigon, C.R. Black, J.J. Colls, J. Atherton, and G. Landon. 1998. Growth, light interception and yield responses of spring wheat (*Triticum aestivum* L.) grown under elevated atmospheric CO_2 and O_3 in open-top chambers. *Global Change Biology* 4: 121–130.

Mun, H.T., and W.G. Whitford. 1989. Effects of nitrogen amendment on annual plants in the Chihuahuan Desert. *Plant and Soil* 120: 225–231.

Myneni, R.B., C.D. Keeling, C.J. Tucker, G. Asrar, and R.R. Nemani. 1997. Increased plant growth in the northern high latitudes from 1981 to 1991. *Nature* 386: 698–702.

Naeem, S., D.R. Hahn, and G. Schuurman. 2000. Producer-decomposer co-dependency influences biodiversity effects. *Nature* 403: 762–764.

Nash, T.H. 1975. Influence of effluents from a zinc factory on lichens. *Ecological Monographs* 45: 183–198.

Neale, P.J., R.F. Davis, and J.J. Cullen. 1998. Interactive effects of ozone depletion and vertical mixing on photosynthesis of Antarctic phytoplankton. *Nature* 392: 585–589.

Nebel, B., and J. Fuhrer. 1994. Inter- and intraspecific differences in ozone sensitivity in semi-natural plant communities. *Angewandte Botanik* 68: 116–121.

Neilson, R.P. 1986. High-resolution climatic analysis and Southwest biogeography. *Science* 232: 27–34.

Norby, R.J. 1987. Nodulating and nitrogenase activity in nitrogen-fixing woody plants stimulated by CO_2 enrichment of the atmosphere. *Physiologia Plantarum* 71: 77–82.

Norton, L.R., A.R. McLeod, P.D. Greenslade, L.S. Firbank, and A.R. Watkinson. 1999. Elevated UV-B radiation effects on experimental grassland communities. *Global Change Biology* 5: 601–608.

O'Brien, C.M., C.J. Fox, B. Planque, and J. Casey. 2000. Climate variability and North Sea cod. *Nature* 404: 142.

Odum, E.P., J.T. Finn, and E.H. Franz. 1979. Perturbation theory and the subsidy-stress gradient. *BioScience* 29: 349–352.

Overpeck, J.T., P.J. Bartlein, and T. Webb. 1991. Potential magnitude of future vegetation change in eastern North America: Comparisons with the past. *Science* 254: 692–695.

Owensby, C.E., J.M. Ham, A.K. Knapp, and L.A. Auen. 1999. Biomass production and species composition change in a tallgrass prairie ecosystem after long-term exposure to elevated atmospheric CO_2. *Global Change Biology* 5: 497–506.

Papadopoulos, Y.A., R.J. Gordon, K. McRae, R.S. Bush, G. Belanger, E.A. Butler, S.A.E. Fillmore, and M.M. Morrison. 1999. Current and elevated levels of UV-B radiation have few impacts on yields of perennial forage crops. *Global Change Biology* 5: 847–856.

Parmesan, C., N. Ryrholm, C. Stefanescu, J.K. Hill, C.D. Thomas, H. Descimon, B. Huntley, L. Kalla, J. Kulberg, T. Tammaru, W.J. Tennent, J.A. Thomas, and M. Warren. 1999. Poleward shifts in geographic ranges of butterfly species associated with regional warming. *Nature* 399: 576–583.

Payette, S., and L. Filion. 1985. White spruce expansion at the tree line and recent climatic change. *Canadian Journal of Forest Research* 15: 241–251.

Pearcy, R.W., and O. Bjorkman. 1983. Physiological effects. In E.R. Lemon (ed.), *CO₂ and Plants,* pp. 65–105. Boulder, Colo.: Westview Press.

Peet, R. K. 1992. Community structure and ecosystem function. In D.C. Glenn-Lewin, R.K. Peet, and T.T. Veblen (eds.), *Plant Succession: Theory and Predictions,* pp. 103–151. London: Chapman and Hall.

Pell, E.J., C.D. Schlagnhaufer, and R.N. Arteca. 1997. Ozone-induced oxidative stress: Mechanisms of action and reaction. *Physiologia Plantarum* 100: 264–273.

Petchey, O.L., P.T. McPhearson, T.M. Casey, and P.J. Morin. 1999. Environmental warming alters food-web structure and ecosystem function. *Nature* 402: 69–72.

Peterjohn, W.T., J.M. Melillo, F.B. Bowles, and P.A. Steudler. 1993. Soil warming and trace gas fluxes: Experimental design and preliminary flux results. *Oecologia* 93: 18–24.

Peterjohn, W.T., J.M. Melillo, P.A. Steudler, K.M. Newkirk, F.P. Bowles, and J.D. Aber. 1994. Responses of trace gas fluxes and N availability to experimentally elevated soil temperatures. *Ecological Applications* 4: 617–625.

Petit, J.R., J. Jouzel, D. Raynaud, N.I. Barkov, J.-M. Barnola, I. Basile, M. Bender, J. Chappellaz, M. Davis, G. Delaygue, M. Delmotte, V.M. Kotyakov, M. Legrand, V.Y. Kipenkov, C. Lorius, L. Pepin, C. Ritz, E. Saltzman, and M. Stievenard. 1999. Climate and atmospheric history of the past 430,000 years from the Vostok ice core, Antarctica. *Nature* 399: 429–438.

Pitelka, L. F., J. Ash, S. Berry, R.H.W. Bradshaw, L. Brubaker, J.S. Clark, M.B. Davis, J.M. Dyer, R.H. Gardner, H. Gitay, G. Hope, R. Hengeveld, B. Huntley, G.A. King, S. Lavorel, R.N. Mack, G.P. Malanson, M. McGlone, I.R. Noble, I.C. Prentice, M. Rejmanek, A. Saunders, A.M. Solomon, S. Sugita, and M.T. Sykes. 1997. Plant migration and climate change. *American Scientist* 85:464–473.

Pockman, W.T., and J.S. Sperry. 1997. Freezing-induced xylem cavitation and the northern limit of *Larrea tridentata*. *Oecologia* 109: 19–27.

Poiani, K.A., and W.C. Johnson. 1991. Global warming and prairie wetlands. *BioScience* 41: 611–618.

Polley, H.W., H.B. Johnson, and H.S. Mayeux. 1994. Increasing CO_2: Comparative responses of C4 grass *Schizachrium* and grassland invader *Prosopis*. *Ecology* 75: 976–989.

Poorter, H. 1993. Interspecific variation in the growth responses of plants to an elevated ambient CO_2 concentration. *Vegetatio* 104/105: 77–97.

Poorter, H. 1998. Do slow-growing species and nutrient-stressed plants respond relatively strongly to elevated CO_2? *Global Change Biology* 4: 693–697.

Potvin, C., and L. Vasseur. 1997. Long-term CO_2 enrichment of a pasture community: Species richness, dominance and succession. *Ecology* 78: 666–677.

Pough, F.H. 1976. Acid precipitation and embryonic mortality of spotted salamanders, *Ambystoma maculatum*. *Science* 192: 68–70.

Price, M.V., and N.M. Waser. 1998. Effects of experimental warming on plant reproductive phenology in a subalpine meadow. *Ecology* 79: 1261–1271.

Price, M.V., and N.M. Waser. 2000. Responses of subalpine meadow vegetation to four years of experimental warming. *Ecological Applications* 10: 811–823.

Rabalais, N.N., R.E. Turner, D. Justic, Q. Dortch, W.J. Wiseman, and B.K. Sen Gupta. 1996. Nutrient changes in the Mississippi River and system responses on the adjacent continental shelf. *Estuaries* 19: 386–407.

Rapport, D.J., H.A. Regier, and T.C. Hutchinson. 1985. Ecosystem behavior under stress. *American Naturalist* 125: 617–640.

Raven, J.A. 1994. Carbon fixation and carbon availability in marine phytoplankton. *Photosynthesis Research* 39: 259–273.

Rebbeck, J. 1996. Chronic ozone effects on three northeastern hardwood species: Growth and biomass. *Canadian Journal of Forest Research* 26: 1788–1798.

Rebbeck, J., U. Blum, and A.S. Heagle. 1988. Effects of ozone on the regrowth and energy reserves of a ladino clover-tall fescue pasture. *Journal of Applied Ecology* 25: 659–681.

Reich, P.B. 1987. Quantifying response to ozone: A unifying theory. *Tree Physiology* 3: 573–577.

Reid, C. 1899. *The Origin of the British Flora.* London: Dulau.

Reid, C.D., and E.L. Fiscus. 1998. Effects of elevated [CO_2] and/or ozone on limitations to CO_2 assimilation in soybean (*Glycine max*). *Journal of Experimental Botany* 49: 885–895.

Reid, C.D., and B.R. Strain. 1994. Effects of CO_2 enrichment on whole-plant carbon budget of seedlings of *Fagus grandifolia* and *Acer saccharum* in low irradiance. *Oecologia* 98: 31–39.

Reiling, K., and A.W. Davison. 1992. The response of native, herbaceous species to ozone: Growth and fluorescence. *New Phytologist* 120: 29–37.

Reynolds, J.F., R.A. Virginia, P.R. Kemp, A.G. DeSoyza, and D.C. Tremmel. 1999. Impact of drought on desert shrubs: Effects of seasonality and degree of resource island development. *Ecological Monographs* 69: 69–106.

Ries, G., W. Heller, H. Puchta, H. Sandermann, H.K. Seidlitz, and B. Hohn. 2000. Elevated UV-B reduces genome stability in plants. *Nature* 406: 98–101.

Rind, D., R. Goldberg, J. Hansen, C. Rosenzweig, and R. Ruedy. 1990. Potential evapotranspiration and the likelihood of future drought. *Journal of Geophysical Research* 95: 9983–10004.

Ringelberg, D.B., J.O. Stair, J. Ahmeida, R.J. Norby, E.G. O'Neill, and D.C. White. 1997. Consequences of rising atmospheric CO_2 levels for the belowground microbiota associated with white oak. *Journal of Environmental Quality* 26: 495–503.

Ritchie, J. C., and G. M. MacDonald. 1986. The patterns of post-glacial spread of white spruce. *Journal of Biogeography* 13: 527–540.

Robertson, D.M., R.A. Ragotzkie, and J.J. Magnuson. 1992. Lake ice records used to detect historical and future climatic changes. *Climatic Change* 21: 407–427.

Roemmich, D., and J. McGowan. 1995. Climatic warming and the decline of zooplankton in the California current. *Science* 267: 1324–1326.

Root, T. 1988. Energy constraints on avian distributions and abundances. *Ecology* 69: 330–339.

Rosenzweig, M.L. 1971. Paradox of enrichment: Destabilization of exploitation ecosystems in ecological time. *Science* 171: 385–387.

Rothrock, D.A., Y. Yu., and G.A. Maykut. 1999. Thinning of the Arctic sea-ice cover. *Geophysical Research Letters* 26: 3469–3472.

Roy, D.B., and T.H. Sparks. 2000. Phenoogy of British butterflies and climate change. *Global Change Biology* 6: 407–416.

Rudorff, B.F.T., C.L. Mulchi, E.H. Lee, R. Rowland, and R. Rausch. 1996. Effects of enhanced O_3 and CO_2 enrichment on plant characteristics in wheat and corn. *Environmental Pollution* 94: 53–60.

Rustad, L.E., and I.J. Fernandez. 1998. Experimental soil warming effects on CO_2 and CH_4 flux from a low elevation spruce-fir forest soil in Maine, USA. *Global Change Biology* 4: 597–605.

Sadowsky, M.J., and M. Schortemeyer. 1997. Soil microbial responses to increased concentrations of atmospheric CO_2. *Global Change Biology* 3: 217–224.

Sagarin, R.D., J.P. Barray, S.E. Gilman, and C.H. Baxter. 1999. Climate-related change in an intertidal community over short and long time scales. *Ecological Monographs* 69: 465–490.

Sageman, B.B., and C.R. Bina. 1997. Diversity and species abundance patterns in late Cenomanian black shale biofacies, western interior, U.S. *Palaios* 12: 449–466.

Sala, O.E., F.S. Chapin, J.J. Armesto, E. Berlow, J. Bloomfield, R. Dirzo, E. Huber-Sanwald, L.F. Huenneke, R.B. Jackson, A. Kinzig, R. Leemans, D.M. Lodge, H.A. Mooney, M. Oesterheld. N.L. Poff, M.T. Sykes, B.H. Walker, M. Walker, and D.H. Wall. 2000. Global biodiversity scenarios for the year 2100. *Science* 287: 1770–1774.

Samuelson, L.J., and J.M. Kelly. 1997. Ozone uptake in *Prunus serotina, Acer rubrum* and *Quercus rubra* forest trees of different sizes. *New Phytologist* 136: 255–264.

Sanford, E. 1999. Regulation of keystone predation by small changes in ocean temperature. *Science* 283: 2095–2097.

Santer, B.D., K.E. Taylor, T.M.I. Wigley, T.C. Johns, P.D. Jones, D.J. Karoly, J.F.B. Mitchell, A.H. Oort, J.E. Penner, V. Ramaswamy, M.D. Schwartzkopf, R.J. Stouffer, and S. Tett. 1996. A search for human influences on the thermal structure of the atmosphere. *Nature* 3982: 39–46.

Sasek, T.W. 1985. Implications of atmospheric carbon dioxide enrichment for the physiological ecology and distribution of two introduced woody vines, *Pueraria lobata* Ohwi (kudzu) and *Lonicera japonica* Thumb. (Japanese honeysuckle). Ph.D. dissertation, Duke University, Durham, N.C.

Sasek, T.W., and B.R. Strain. 1988. Effects of carbon dioxide enrichment on the growth and morphology of kudzu (*Pueraria lobata*). *Weed Science* 36: 28–36.

Sasek, T.W., and B.R. Strain. 1989. Effects of carbon dioxide enrichment on the expansion and size of kudzu (*Pueraria lobata*) leaves. *Weed Science* 37: 23–28.

Sasek, T.W., and B.R. Strain. 1990. Implications of atmospheric CO_2 enrichment and climate change for the geographical distribution of two introduced vines in the USA. *Climatic Change* 16: 31–51.

Sasek, T.W., and B.R. Strain. 1991. Effects of CO_2 enrichment on the growth and morphology of a native and introduced honeysuckle vine. *American Journal of Botany* 78: 69–75.

Schimel, D., I.G. Enting, M. Heimann, T.M.L. Wigley, D. Raynaud, D. Alves, and U. Siegenthaler. 1995. CO_2 and the carbon cycle. In J.T. Houghton, L.G. Meira Filho, J. Bruce et al. (eds.), *Climate Change 1994*, pp. 35–71. Cambridge: Cambridge University Press.

Schindler, D.W. 1988. Effects of acid rain on freshwater ecosystems. *Science* 239: 149–157.

Schindler, D.W., K.G. Beaty, E.J. Fee, D.R. Cruikshank, E.R. DeBruyn, D.L. Findlay, G.A. Linsey, J.A. Shearer, M.P. Stainton, and M.A. Turner. 1990. Effects of climatic warming on lakes of the central boreal forest. *Science* 250: 967–970.

Schlesinger, W.H. 1997. *Biogeochemistry: An Analysis of Global Change.* 2nd ed. San Diego: Academic Press.

Schortemeyer, M., U.A. Hartwig, G.R. Hendrey, and M.J. Sadowsky. 1996. Microbial community changes in the rhizospheres of white clover and perennial ryegrass exposed to Free Air Carbon Dioxide Enrichment (FACE). *Soil Biology and Biochemistry* 28: 1717–1724.

Searles, P.S., S.D. Flint, S.B. Diaz, M.C. Rousseaux, C.L. Ballare, and M.M. Caldwell. 1999. Solar ultraviolet-B radiation influence on *Sphagnum* bog and *Carex* fen ecosystems: First field season findings in Terra del Fuego, Argentina. *Global Change Biology* 5: 225–234.

Shärky, L., H. Ro-Poulsen, F.A.M. Wellburn, and L.J. Sheppard. 1998. Impacts of ozone on forests: A European perspective. *New Phytologist* 139: 109–122.

Smith, R.C., B.B. Prezelin, K.S. Baker, R.R. Bidigare, N.P. Boucher, T. Coley, D. Karentz, S. MacIntyre, H.A. Matlick, D. Menzies, M. Ondrusek, Z. Wan, and K.J. Waters. 1992. Ozone depletion: Ultraviolet radiation and phytoplankton biology in Antarctic waters. *Science* 255: 952–959.

Sorensen, L.G., R. Goldberg, T.L. Root, M.G. Anderson. 1998. Potential effects of global warming on waterfowl populations breeding in the northern Great Plains. *Climatic Change* 40: 343–369.

Spivack, A.J., C.-F. You, and H.J. Smith. 1993. Foraminiferal boron isotope ratios as a proxy for surface ocean pH over the past 21 Myr. *Nature* 363: 149–151.

Stoddard, J.L., D.S. Jeffries, A. Lukewille, T.A. Clair, P.J. Dillon, C.T. Driscoll, M. Forsius, M. Johannessen, J.S. Kahl, J.H. Kellogg, A. Kemp, J. Mannio, D.T. Monteith, P.S. Murdoch, S. Patrick, A. Rebsdorf, B.L. Skjelkvale, M.P. Stainton, T. Traaen, H. van Dam, K.E. Webster, J. Wieting, and A. Wilander. 1999. Regional trends in aquatic recovery from acidification in North America and Europe. *Nature* 401: 575–578.

St. Omer, L., and S.M. Horvath. 1983. Elevated CO_2 atmospheres and whole plant senescence. *Ecology* 64: 1311–1314.

Street-Perrott, F.A., Y. Huang, R.A. Perrott, G. Eglinton, P. Barker, L.B. Khelifa, D.D. Harkness, and D.O. Olago. 1997. Impact of lower atmospheric carbon dioxide on tropical mountain ecosystems. *Science* 278:1422–1426.

Sullivan, J.H., and A. H. Teramura. 1992. The effects of ultraviolet-B radiation on loblolly pine. 2. Growth of field-grown seedlings. *Trees* 6: 115–120.

Taberlet, P., L. Fumagalli, A.G. Wust-Saucy, and J.-F Cossons. 1998. Comparative phylogeography and post-glacial colonization routes in Europe. *Molecular Ecology* 7:453–464.

Taylor, K.C., P.A. Mayewski, R.B. Alley, E.J. Brook, A.J. Gow, P.M. Grootes, D.A. Meese, E. S. Saltzman, J. P. Severinghaus, M. S. Twickler, J. W. C. White, S. Whitlow, and G. A. Zielinski. 1997. The Holocene-Younger Dryas transition recorded at Summit, Greenland. *Science* 278:825–827.

Tett, S.F.B., P.A. Stott, M.R. Allen, W.J. Ingram, and J.F.B. Mitchell. 1999. Causes of twentieth-century temperature change near the earth's surface. *Nature* 399: 569–572.

Thomas, C.D., and J.J. Lennon. 1999. Birds extend their ranges northwards. *Nature* 399:213.

Thomas, R.B., D.D. Richter, H. Ye, P.R. Heine, and B.R. Strain. 1991. Nitrogen dynamics and growth of seedlings of an N-fixing tree (*Gliricidia sepium* [Jadq.] Walp.) exposed to elevated atmospheric carbon dioxide. *Oecologia* 88: 415–421.

Thompson, A.M., M.A. Huntley, and R.W. Stewart. 1990. Perturbations to tropospheric oxidants, 1985–2035. 1. Calculations of ozone and OH in chemically coherent regions. *Journal of Geophysical Research* 95: 9829–9844.

Tilman, D. 1982. *Resource Competition and Community Structure*. Princeton, N.J.: Princeton University Press.

Tipping, E., T.R. Carrick, M.A. Hurley, J.B. James, A.J. Lawlor, S. Lofts, E. Rigg, D.W. Sutcliffe, and C. Woof. 1998. Reversal of acidification in upland waters of the English Lake District. *Environmental Pollution* 103: 143–151.

Tissue, D.T., J.P. Megonigal, and R.B. Thomas. 1997. Nitrogenase activity and N_2 fixation are stimulated by elevated CO_2 in a tropical N_2 fixing tree. *Oecologia* 109: 28–33.

Tissue, D.T., R.B. Thomas, and B.R. Strain. 1997. Atmospheric CO_2 enrichment increases growth and photosynthesis of *Pinus taeda*: A 4-year experiment in the field. *Plant Cell and Environment* 20: 1123–1134.

Tousignant, D., and C. Potvin. 1996. Selective responses to global change: Experimental results on *Brassica juncea* (L.) Czern. In C. Korner and F.A. Bazzez (eds.), *Carbon Dioxide, Populations, and Communities*, pp. 23–30. San Diego: Academic Press.

Tucker, C.J., W.W. Newcomb, and H.E. Dregne. 1994. AVHRR data sets for determination of desert spatial extent. *International Journal of Remote Sensing* 15: 3547–3565.

Turner, R.E., and N.N. Rabalais. 1991. Changes in Mississippi River water quality this century. *BioScience* 41: 140–147.

Turner, R.E., and N.N. Rabalais. 1994. Coastal eutrophication near the Mississippi River delta. *Nature* 368: 619–621.

Van Devender, T.R., and W.G. Spaulding. 1979. Development of vegetation and climate in the southwestern United States. *Science* 204: 701–710.

Vasseur, L., and C. Potvin. 1998. Natural pasture community response to enriched carbon dioxide atmosphere. *Plant Ecology* 135: 31–41.

Visser, M.T., A.J. van Noordwij, J.M. Tinbergen, and C.M. Lessells. 1998. Warmer springs lead to mistimed reproduction in great tits (*Parus major*). *Proceedings of the Royal Society of London* 265: 1867–1870.

Vitousek, P.M., J.D. Aber, R.W. Howarth, G.E. Likens, P.A. Matson, D.W. Schindler, W.H. Schlesinger, and D.G. Tilman. 1997. Human alteration of the global nitrogen cycle: Sources and consequences. *Ecological Applications* 7: 737–750.

Vitousek, P. M., C. M. D'Antonio, L. L. Loope, and R. Westbrooks. 1996. Biological invasions as global environmental change. *American Scientist* 84:468–478.

Volin, J.C., P.B. Reich, and T.J. Givnish. 1998. Elevated carbon dioxide ameliorates the effects of ozone on photosynthesis and growth; species respond similarly regardless of photosynthetic pathway or plant functional group. *New Phytologist* 138: 315–325.

Volz, A., and D. Kley. 1988. Evaluation of the Montsouris series of ozone measurements made in the nineteenth century. *Nature* 332: 240–242.

Walker, D., and Y. Chen. 1987. Palynological light on tropical rainforest dynamics. *Quaternary Science Reviews* 6: 77–92.

Walker, R.F., D.R. Geisinger, D.W. Johnson, and J.T. Ball. 1995. Enriched atmospheric CO_2 and soil P effects on growth and ectomycorrhizal colonization of juvenile ponderosa pine. *Forest Ecology and Management* 78: 207–215.

Walters, M.B., and P.B. Reich. 1996. Are shade tolerance, survival and growth linked? Low light and nitrogen effects on hardwood seedlings. *Ecology* 77: 841–853.

Walters, M.B., and P.B. Reich. 2000. Trade-offs in low-light CO_2 exchange: A component of variation in shade tolerance among cold temperate tree seedlings. *Functional Ecology* 14: 155–165.

Ward, J.K. and B.R. Strain. 1997. Effects of low and elevated CO_2 partial pressures on growth and reproduction of *Arabidopsis thaliana* from different elevations. *Plant Cell and Environment* 20: 254–260.

Ward, J.K., B.R. Strain, J. Antonovics, and R.B. Thomas. 2000. Is atmospheric CO_2 a selective agent on model C3 annuals? *Oecologia* 123: 330–341.

Warner, M.E., W.K. Fitt, and G.W. Schmidt. 1999. Damage to photosystem II in symbiotic dinoflagellates: A determinant of coral bleaching. *Proceedings of the National Academy of Sciences USA* 96: 8007–8012.

Warwick, K.R., and G. Taylor. 1995. Contrasting effects of tropospheric ozone on five native herbs which coexist in calcareous grassland. *Global Change Biology* 1: 143–151.

Warwick, K.R., G. Taylor, and H. Blum. 1998. Biomass and compositional changes occur in chalk grassland turves exposed to elevated CO_2 for two seasons in FACE. *Global Change Biology* 4: 375–385.

Wayne, P. M., and F. A. Bazzaz. 1995. Seedling density modifies the growth responses of yellow birch maternal families to elevated carbon dioxide. *Global Change Biology* 1: 315–324.

Wayne, P. M., and F. A. Bazzaz. 1997. Light acquisition and growth by competing individuals in CO_2 enriched atmospheres: Consequences for size structure in regenerating birch stands. *Journal of Ecology* 85: 29–42.

Webb, S. L. 1987. Beech range extension and vegetation history: Pollen stratigraphy of two Wisconsin lakes. *Ecology* 68:1993–2005.

Wedin, D.A., and D. Tilman. 1996. Influence of nitrogen loading and species composition on the carbon balance of grasslands. *Science* 274: 1720–1723.

Wentz, F.J., and M. Schabel. 2000. Precise climate monitoring using complementary satellite data sets. *Nature* 403:414–416.

Westman, W.E. 1979. Oxidant effects on California coastal sage scrub. *Science* 205: 1001–1003.

White, T.A., B.D. Campbell, P.D. Kemp, and C.L. Hunt. 2000. Sensitivity of three grassland communities to simulated extreme temperature and rainfall events. *Global Change Biology* 6: 671–684.

Whitfield, C.P., A.W. Davison, and T.W. Ashenden. 1997. Artificial selection and heritability of ozone resistance in two populations of *Plantago major. New Phytologist* 137: 645–655.

Whittaker, R.H. 1972. Evolution and measurement of species diversity. *Taxon* 21: 213–251.

Willems, J.H., and R.K. Peet. 1993. Changes in chalk-grassland structure and species richness resulting from selective nutrient additions. *Journal of Vegetation Science* 4: 203–212.

Williams, W.E., K. Garbutt, F.A. Bazzaz, and P.M. Vitousek. 1986. The response of plants to elevated CO_2. IV. Two deciduous forest tree communities. *Oecologia* 69: 454–459.

Woods, K. D., and M. B. Davis. 1989. Paleoecology of range limits: Beech in the upper peninsula of Michigan. *Ecology* 70:681–696.

Wright Jr., H. E., J. E. Kutzbach, T. Webb III, W. F. Ruddiman, F. A. Street-Perrott, and P. J. Bartlein. 1993. *Global Climates since the Last Glacial Maximum.* Minneapolis: University of Minnesota Press.

Yakimchuk, R., and J. Hoddinott. 1994. The influence of ultraviolet-B light and carbon dioxide enrichment on the growth and physiology of three conifer species. *Canadian Journal of Forest Research* 24: 1–8.

Yeates, G.W., P. Newton, and D.J. Ross. 1999. Responses of soil nematode fauna to naturally elevated CO_2 levels influenced by soil pattern. *Nematology* 1: 285–293.

Zak, D.R., D.B. Ringelberg, K.S. Pregitzer, D.L. Randlett, D.C. White, and P.S. Curtis. 1996. Soil microbial communities beneath *Populus grandidentata* grown under elevated atmospheric CO_2. *Ecological Applications* 6: 257–262.

Zak, D.R., K.S. Pregitzer, P.S. Curtis, and W.E. Holmes. 2000. Atmospheric CO_2 and the composition and function of soil microbial communities. *Ecological Applications* 10: 47–59.

Zaller, J.G., and J.A. Arnone. 1999. Earthworm responses to plant species loss and elevated CO_2 in calcareous grassland. *Plant and Soil* 208: 1–8.

Zogg, G.P., D.R. Zak, D.B. Ringelberg, N.W. MacDonald, K.S. Pregitzer, and D.C. White. 1997. Compositional and functional shifts in microbial communities due to soil warming. *Soil Science Society of America Journal* 61: 475–481.

White, T.A., B.D. Campbell, P.D. Kemp, and C.L. Hunt. 2000. Sensitivity of three grassland communities to simulated extreme temperature and rainfall events. *Global Change Biology* 6:671–684.

Whitfield, C.P., A.W. Davison, and T.W. Ashenden. 1997. Artificial selection and heritability of ozone resistance in two populations of *Plantago major*. *New Phytologist* 147:645–658.

Whittaker, R.H. 1972. Evolution and measurement of species diversity. *Taxon* 21:213–251.

Williams, J.H., and R.K. Peet. 1990. Changes in chalk-grassland structure and species richness resulting from selective nutrient additions. *Journal of Vegetation Science* 4:205–212.

Williams, W.E., K. Garbutt, F.A. Bazzaz, and P.M. Vitousek. 1986. The response of plants to elevated CO$_2$. IV. Two deciduous forest tree communities. *Oecologia* 69:454–459.

Woods, K.D., and M.B. Davis. 1989. Paleoecology of range limits: Beech in the upper peninsula of Michigan. *Ecology* 70:681–696.

Wright, H.E., J.E. Kutzbach, T. Webb III, W.F. Ruddiman, F.A. Street-Perrott, and P.J. Bartlein. 1993. *Global Climates since the Last Glacial Maximum*. Minneapolis: University of Minnesota Press.

Yakimchuk, R., and J. Hoddinott. 1994. The influence of ultraviolet-B light and carbon dioxide enrichment on the growth and physiology of three conifer species. *Canadian Journal of Forest Research* 24:1–8.

Yeates, G.W., R. Newton, and D.J. Ross. 1999. Responses of soil nematode fauna to naturally elevated CO$_2$ levels influenced by soil pattern. *Nematology* 1:285–293.

Zak, D.R., D.B. Ringelberg, K.S. Pregitzer, D.L. Randlett, D.C. White, and P.S. Curtis. 1996. Soil microbial communities beneath *Populus grandidentata* grown under elevated atmospheric CO$_2$. *Ecological Applications* 6:257–262.

Zak, D.R., K.S. Pregitzer, P.S. Curtis, and W.E. Holmes. 2000. Atmospheric CO$_2$ and the composition and function of soil microbial communities. *Ecological Applications* 10:47–59.

Zaller, J.G., and J.A. Arnone. 1999. Earthworm responses to plant species' loss and elevated CO$_2$ in calcareous grassland. *Plant and Soil* 208:1–8.

Zogg, G.P., D.R. Zak, D.B. Ringelberg, N.W. MacDonald, K.S. Pregitzer, and D.C. White. 1997. Compositional and functional shifts in microbial communities due to soil warming. *Soil Science Society of America Journal* 61:475–481.

10

MAKING SMART CONSERVATION DECISIONS

H. P. Possingham, S. J. Andelman,
B. R. Noon, S. Trombulak, and H. R. Pulliam

In little more than a decade, conservation biology has generated a large and growing body of theory aimed at predicting and managing the impacts of anthropogenic activities on populations, species, and ecosystems (e.g., Meffe and Carroll 1997; Ferson and Burgman 2000). However, to succeed in its mission, conservation biology must do more than generate theories and principles. Conservation biologists must use their theories to deliver effective, science-based decision tools for practical use by managers and policy makers. Furthermore, to demonstrate the efficacy of conservation biology theory and applications to skeptical managers, conservation biologists must also establish feasible yet reliable monitoring programs to detect key trends and effects and determine the effectiveness of decisions (Kareiva et al. 1999; Simberloff 1999). Transforming pure science—whether theoretical or empirical—into information that actually be used by managers and decision makers to address the sorts of conservation problems outlined throughout this volume remains a major challenge in conservation biology. To meet this challenge, the next generation of conservation biologists will need to embrace more economics, more management science, more decision theory, and more operations research.

This chapter has two broad objectives. First, we outline the key elements for how to make smart conservation decisions and measure their benefits—an outline that is applicable to the sorts of conservation problems discussed in this book. With this framework in mind, we then highlight priorities and directions for future research.

We begin by discussing some specific problems generic to species and ecosystem management that have resisted rational, science-based solutions. These examples illustrate some of the gaps in traditional approaches to conservation biology research and provide a platform to illustrate the benefits of decision-theory approaches. Next we give examples of the application of decision theory to solving conservation problems. While decision-making methods can account for uncertainty and errors, monitoring the results of conservation actions is essential for improved decision making in the future.

Ultimately, our objective is to encourage the development of active adaptive management programs on real conservation problems (Parma et al. 1998). Within an adaptive management framework, once an action has been initiated, a monitoring program must be implemented that evaluates and measures the performance of that action. The performance evaluation then informs the decision-making process by reducing uncertainty and ignorance. In this paper we use the term *monitoring* to refer to a repeated assessment of some environmental attribute for the purpose of detecting change within a defined area over time (Thompson et al. 1998). Specifically, the type of monitoring we describe is "performance evaluation." That is monitoring explicitly designed to assist decision making and management. We conclude with an overview of research opportunities for monitoring and decision making, emphasizing the challenge of developing robust rules of thumb for making decisions.

Where Conservation Biology Has Failed to Provide Useful Advice

Where in space should habitat be protected from destruction, and where should it be restored? This question is fundamental to conservation, and conservation biology should provide the answer. A range of theories generally addresses this question, beginning with "island biogeography theory" (MacArthur and Wilson 1967) and continuing more recently with "metapopulation theory" (Hanski 1999) and "source-sink theory" (Pulliam 1988). Let us see what answers these theories provide to this question.

Island biogeography theory generated a series of rules, most notably:

- big reserves are better than small reserves; and
- connected (or close) reserves are better than unconnected reserves.

Metapopulation theory tells us that there are several ways to increase the likelihood that a metapopulation will persist:

- decrease local extinction rates;
- increase between-patch colonization rates;
- increase the number of suitable patches (or, conversely, minimize patch loss); and

- increase the number of occupied patches.

Source-sink theory emphasizes the importance of identifying habitat where population growth rates are consistently positive, hence the rule: Protect source populations and ignore sink populations.

We also have the empirically derived rule: Reserves with a low edge-to-area ratio are better than reserves with a high edge-to-area ratio.

All these rules have been useful in providing general conservation principles, but their use is more limited for dealing with specific conservation and management problems. To provide useful practical advice, conservation biologists must have a clear understanding of the questions that managers want answered. For example, consider the question of where it is best to restore habitat. A land manager interested in habitat restoration will typically have a finite set of resources and financial constraints. She may envisage revegetating 10 percent of an existing site. The question for conservation biologists is: Which 10 percent should be restored to maximize biodiversity benefits?

Let us assume that two patches of native vegetation exist on the site, one larger than the other. The specific question then is: To which of the following options should revegetation efforts be directed?

1. Make the bigger patch bigger.
2. Make the smaller patch bigger.
3. Connect the two patches with a broad corridor.
4. Make one new large patch.
5. Make many new small patches evenly scattered over the property.

Island biogeography theory and metapopulation theory provide few insights to help us choose among these options, except, perhaps, ruling out option 5—although without further information about the distribution of species and communities across the property, we might not want to eliminate that option. The reason these conservation biology theories have limited utility for such practical decisions is that the general principles have not been couched within a decision-making framework. The question of where to focus restoration efforts raises the immediate question of what biodiversity the land manager is trying to maintain or recover. What is the objective? A simple question that is rarely answered.

Even when an explicit objective is specified, none of our theories provide explicit predictions as to how that objective is most likely to be met. If the biodiversity objective is to minimize the likelihood of a species becoming extinct, we might test alternative habitat reconstruction scenarios in a population viability model (something that is rarely done given the lack of data and the demands on the time of managers). If the biodiversity objective is to restore a representative sample of habitats, then information on soil types and topography would be vital. Given the vast conservation literature on habitat fragmentation generated over the past decade, it is an embarrassment that so little of this research (empirical or theoretical) provides answers to the most basic practical questions concerning optimal habitat reconstruction for biodiversity conservation. The theories generate a list of generally useful things we might do, but they offer little insight into how best to choose among the alternative options (Possingham 1997).

The example provided is deliberately simple. Most conservation problems are more complex, and a land manager may envisage spending her conservation dollars in other nature conservation activities, such as control of introduced predators and weeds in the existing habitat patches. The question then becomes: What fraction of our conservation effort should be directed to each of several management activities: revegetation, predator control, and weed control? What conservation biology theory helps us juggle these alternatives?

A second common question in conservation biology relates to the problem of reintroductions and captive breeding. A large amount of science is focused on reintroduction and captive breeding (Ralls and Ballou 1986), but few protocols exist for answering basic questions like:

- When, in the decline of a threatened species, should we initiate a captive breeding program?
- How many individuals should be brought in to found a captive breeding program?
- Which individuals should constitute the initial captive population (i.e., what combination of ages and sexes)?
- When should individuals be released into the wild, and how many?

Again, the existing theory provides general rules, such as: reintroduction of a large number of individuals will typically be better than a smaller reintroduction. Empirical reintroduction research yields essential data and experience but provides only a limited framework for extending experience with a particular species or site to new species or uncertain circumstances. Maguire et al. (1987), Maguire and Servheen (1992), and Lubow (1996) address some of these reintroduction and captive breeding questions for specific circumstances using decision-theory tools, but they are lonely beacons.

Decision Theory and How It Has Been Used in Conservation

Decision theory is a framework within which people responsible for management attempt to achieve explicitly stated objectives while acknowledging the levels of uncertainty involved with the decision process (Clemen 1996). Other professionals such as engineers (Kulkarni et al. 1993; Ravirala and Grivas 1995; Teng and Tzeng 1996) and financial advisers (Beenhakker 1975) use this sort of theory to make decisions. In some cases decision theory relies on complex mathematical tools that go under the generic name of "mathematical programming methods." However decision theory can be much broader and often includes qualitative tools such as "ecological risk assessment" or "multicriteria decision analysis." The common thread within the theory is a disciplined protocol for problem solving that includes the following seven steps (Shea et al. 1998):

1. Specify the management objectives, or at least list the indicators of policy performance (e.g., minimize the risk that numbats become extinct in the next fifty years). Where there are multiple objectives, utility theory (Maguire and Servheen 1982; Guikema and Milke 1999) is employed to deal with the problem of how to maximize multiple objectives that are measured in different currencies (e.g. money, biodiversity loss, and risk to human health).

2. List the management options and express them as control variables (e.g., release x animals in year t, expend $y\$$ on baiting predators, etc.).

3. Specify the variables that describe the state of the system (e.g., population size, predator abundance).

4. Develop a conceptual model of the dynamics of the system being managed, and if possible develop equations to describe the dynamics of the state variables (e.g., equations for numbat and predator population codynamics). This step will often involve collaboration between a field biologist and an ecological modeler.

5. Specify constraints that bound the decision variables and state variables (e.g., a recovery program budget, the social acceptability of predator reduction strategies, etc.).

6. Specify the range of uncertainty for all the parameters.

7. Find solutions to the problem. Once the problem is defined by these components the manager may often need a decision-making protocol and/or mathematical programming package to find the best, or at least some good, solutions.

To illustrate how decision-theory thinking might be integrated with traditional conservation biology approaches, we now present three examples.

Example 1: Optimal Fire Management for a Large Conservation Park

Managing disturbance processes in conservation areas is a contentious issue throughout the world. In many large and seminatural terrestrial systems fire is the main force that disturbs vegetation, and it is often managed with fuel reduction burning, attempts to reduce ignition frequency, and prescribed burning to increase fire frequency. The pro- and anti-fire advocates often take highly polarized positions based on preconceived ideas, conflicting goals, and nonpredictive mental models of an area's natural history. The fear of making the wrong decision invariably means we make the decision to do nothing.

Here we summarize Richards et al.'s (1999) use of decision theory to determine an optimal fire management strategy for a large conservation reserve. The description of the problem fits into the seven-point framework above:

1. *Statement of objective(s)*. Richards et al. set the broad objective of maintaining a balance of the different successional states. Specifically, they set the target that at least 20 percent of the park should be in each of the three successional states of the vegetation—early, mid, and late. (This is a surrogate objective for the more fundamental objective of minimizing species loss. However, there are so many species, and limited information on each, that the ecosystem-level objective is expedient and more likely to be implemented and monitored.)

2. *List of management options*. Each year the park manager has three options: do nothing, attempt to stop all wild fires, and prescribe burns in some of the park.

3. *State variables*. The state of the system is the percentage of the park in each of the three successional states—early, mid, and late.

4. *State dynamics*. Using basic probability theory, Richards et al. constructed transition probabilities that predicted the chance of the park moving from any one state to any other state in a single year, given that the manager had chosen a particular management option for that year. The two forces that drove the dynamics were fire and vegetation succession, and the final state dynamics model is a Markov chain model.

5. *Constraints*. Richards et al. did not constrain the ability of the manager to enact any of the three management options, although they did assume that the fire suppression option would not be completely effective.

6. *Uncertainty*. Richards et al. varied the definition of early-, mid-, and late-successional habitat, and the relationship between fire frequency and habitat state, to do a limited test of the robustness of their conclusions. For more detailed fire management, a whole host of uncertainties regarding ability to control fire, risk to life and property, and even variability in the political climate, could be used to enhance the robustness of the results.

7. *Solution methods*. Richards et al. used stochastic dynamic programming, a common mathematical programming method (Intriligator 1971), to find the optimal solution.

The main outcome was a state-dependent fire management diagram that tells a manager what to do given a particular state of the park (see figure 10.1, from Possingham and Shea 1999). Where the resources to explore options for a particular park are limited, figure 10.1 *could* be thought of as a rough rule of thumb for fire management in a large seminatural park, although its robustness would require considerable further testing (see point 6 above).

Unexpectedly, Richards et al. found that a manager often needs to do something even when the park is in a "good" state, and sometimes a manager should do nothing when the park is in a "bad" state. Exactly when a manager should take no action is not easy to predict without a model. The stochastic dynamic programming method is able to accurately integrate decisions in a stochastic world over a longer time frame than human intuition can. There is no fixed optimal policy—it depends on the state of the park.

One question raised by the work was the social and economic cost of adopting different policies. This was not dealt with in detail, yet it may be important in some cases. Utility theory would need to be applied to generate relationships between different "values"—conservation value, economic

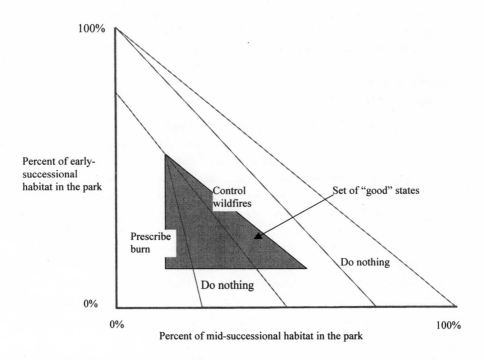

FIGURE 10.1.   A state-dependent decision diagram for fire management in a large conservation reserve. The result is based on an objective of trying to keep at least 20 percent of the park in each of the three successional states—early, mid, and late. "Do nothing," means that fires are neither started nor extinguished. From Possingham and Shea (1999).

cost, and social perception of fire. This issue is discussed in the context of another example below.

Example 2: Conservation Program Planning

Guikema and Milke (1999) consider the problem of allocating funds between different conservation projects. They use decision analysis to show how an agency would go about fund allocation. This use of decision theory is at quite a different scale from the example above and relies heavily on resource allocation theory developed in other industries. The work is briefly described below.

Envisage an agency that needs to allocate funds between a variety of projects. For example, it may be a federal agency that must decide how to allocate funds among one hundred projects on different threatened species, or it may be a nongovernmental organization that must decide how much money to invest in acquisition of new protected areas vs. how much to invest in management and restoration of existing reserves. Guikema and Milke (1999) summarize the conservation program planning procedure using a single diagram (figure 10.2).

Again we frame the problem in our seven-step outline:

1. *Statement of objective(s).* A typical agency-based management objective will represent a balance between recreation and biodiversity objectives, such as: number of visitor days, level of visitor satisfaction, species protection, and ecosystem protection. Much of Guikema and Milke (1999) is devoted to defining a utility function for each objective and combining them into one objective function. They show how some probability is required to modify the value of different objectives where the chance they will occur is uncertain. For species protection this might require some population viability analysis (Beissinger and Westphal 1998). For an objective such as number of visitor days this may require some expert judgment and statistical modeling.

2. *List of management options.* The decision is how to allocate the total budget among the projects. The control variables will be the amount of money allocated to each project.

3 and 4. *State variables and their dynamics.* The relationship between the funds allocated to each project and the achievement of each objective needs to be modeled. Often this requires submodels that may include expert opinion. These submodels will incorporate definitions of state dynamics, state variables, and the impact of funding allocation on the state variables and dynamics.

5. *Constraints.* The primary constraint is the total budget. The sum of the allocated funds must be less than, or equal to, the total budget. The user may wish to allocate varying amounts of funds to each project or just allow a project to occur, or not, with one fixed budget.

FIGURE 10.2. Components of the proposed conservation program planning provedure. From Guikema and Milke (1999).

6. *Uncertainty*. Systematically changing uncertain parameters would involve modifying the weights given to different objectives, or the parameters determined by expert opinion, etc.

7. *Solution methods*. In this case, the problem is linear (the objective function is a linear combination of the control variables, as are the constraints), so many standard mathematical programming packages can solve this problem quickly (even with a spreadsheet).

Guikema and Milke (1999) focused most of their discussion on step 1, framing the objective using multicomponent decision analysis. This task in government agencies is extremely complex and should include a wide range of social and human management issues that we have glossed over. While the scale and scope of this problem is quite different from that of example 1, we can apply the same seven-step framework.

Example 3: Reserve System Design

There is now a large literature on optimal reserve system design (Margules and Pressey 2000). Some of the problem formulation approaches and mathematics involved are reviewed in Possingham et al. (2000). Most ecologists are aware of the problem that is outlined in the seven-step decision-theory framework below; the social and economic aspects of reserve system design are in most need of further research:

1. *Statement of objective(s)*. The objective is to adequately represent a suite of biodiversity features (usually species or habitat types) in a reserve system for the minimum cost. The minimum cost could be expressed in terms of the area, land cost, opportunity cost for other uses, management cost, and/or boundary length of the entire reserve system eventually chosen.

2. *List of management options.* The control options are to acquire, or not, each parcel of land or sea (called a planning unit) for the reserve system. There is a control variable for each planning unit. Current interest in off-reserve conservation and marine zoning would mean that the list of options for any parcel of land or sea should be broader and include nonpurchase options (like covenanting) or partial protection (through zoning).

3. *State variables.* This is usually a static problem where the key parameters are whether or not a biodiversity feature occurs in a planning unit.

4. *State dynamics.* There are normally no dynamics in reserve design problems, although this is an area of active research.

5. *Constraints.* The primary constraints are the biodiversity constraints. Usually the problem is formulated so that all biodiversity feature targets (e.g., conserve 15 percent of habitat A, or 500 individuals of species B) must be met. Planning units are constrained to one of a few states: already in the reserve system, available for conservation designation, or unavailable. Where we are interested in multi-use zoning of a region (e.g., the Great Barrier Reef Marine Park), a planning unit could be allocated to one of many zones.

6. *Uncertainty.* What happens if some of the GIS data layers have errors, or species lists are incomplete and inaccurate? Recent work (Andelman and Meir in press; Meir and Andelman in press) has been carried out on the impacts of uncertain, incomplete, or unreliable data.

7. *Solution methods.* These sorts of problems often become so large that classical mathematical programming methods fail. Simple heuristic algorithms or less classical optimization methods like simulated annealing have been used (Possingham et al. 2000).

Where reserve system design is implemented, it quickly becomes apparent that society is not a single entity with a single value system. Whereas a conservationist may value a particular site because it contains habitat for an endangered species, a timber company may value that site because of the potential revenue that might be generated from harvesting trees, while a horse-riding group may value that site for its recreational values. Perhaps one of the greatest benefits of decision theory is that it provides a protocol to help distinguish among, and integrate, various goals held by multiple sectors of the public.

For example, for the Great Barrier Reef Marine Park Possingham and colleagues are working on designing a reserve system to conserve representative samples of about eighty marine habitats and reefs in a fully protected zone. If conserving biodiversity is all that matters, this problem is hard, but not as hard as a problem that takes into account other values. Real-world marine reserve design has to include the interests of others: commercial fishers, recreational fishers, the tourist industry, etc. Another complication for reserve system design is multiple zoning, where any planning unit needs to be assigned to one of several zones, each of which has implications for conservation, scientific, social, and economic use. Multiple zoning problems raise many questions. From the biodiversity perspective, would two reefs closed for fishing but not tourism compensate for one reserved reef opened up for fishing? A question like this is hard to answer for a single species, let alone for biodiversity in its entirety. We are a long way from developing fully integrated and adaptive reserve system planning tools, yet most countries need them now.

The reader will have noticed that the seven steps to decision making fall short of taking action. Two further essential steps now need to be considered, enacting the consequences of the decision making and then monitoring their consequences. Monitoring and consequent reevaluation and reformulation of our underlying data and models should lead us to the cycle of active adaptive management.

Monitoring and Performance Evaluation

Once a decision has been made, a monitoring program needs to be implemented, with the aim of evaluating and refining the solution or management action chosen. Ideally, the monitoring targets biodiversity assets and processes of greatest concern and uncertainty. Although monitoring activities are widespread in conservation biology and applied ecology, data from existing monitoring programs are underused for a number of reasons:

- many monitoring programs are poorly designed (e.g., the questions they are intended to answer are not well defined, and/or the methods selected are inappropriate to the objectives), resulting in suboptimal data;
- the spatial and temporal scales of monitoring activities are frequently inappropriate to the scales at which inferences from those monitoring activities are made;
- statistical and technological tools for analyzing monitoring data remain relatively primitive;
- little is known about the population and natural history needs of the organisms of concern; and
- there is a perception, particularly among agencies, that data collection, rather than analysis and communication of trends and patterns in those data, is the goal of monitoring.

Detecting ecologically meaningful trends and distinguishing the effects of human activities from those of natural disturbances—particularly at large scales—are notoriously difficult (Dayton et al. 1998; Bradshaw et al. in review). Recent controversies over the magnitude and causes of declines of neotropical migratory songbirds (e.g., James et al. 1996; Thomas 1996) and amphibians (Blaustein et al. 1994; Alford and Richards 1999 and references therein) illustrate some of these difficulties. Furthermore, historically, monitoring has not been viewed as a legitimate research activity by many academic institutions, and maintaining adequate funding for monitoring over the requisite temporal and spatial scales has been difficult. Nevertheless, there is no way around the fact that successful management of biodiversity requires substantial monitoring.

Just as the common thread within decision theory involves a disciplined protocol consisting of seven steps, designing an effective monitoring program also involves a series of seven systematic steps (Mulder et al. 1999):

1. Specify goals.
2. Identify stressors.
3. Develop conceptual model.
4. Select indicators.
5. Establish sampling design.
6. Define methods of analysis.
7. Ensure link to decision making.

Below we briefly discuss each of these stages, highlighting key steps as well as areas where more research is needed.

Specify Goals

As discussed in the context of decision making, many conservation projects go wrong at this step. The same is true of monitoring projects. We cannot emphasize enough the importance of providing an explicit statement of the questions the monitoring program is intended to answer. Without such a statement, it is impossible to evaluate the effectiveness of the monitoring program.

Identify Stressors and Develop Conceptual Model

A conceptual model provides an overview of scientific understanding of how the ecosystem or population works and the state variables that describe the system. This knowledge is used to determine what measures of system performance are likely to be useful. The conceptual model should also include the effects of natural and human-induced stressors on the system. Measurements and inferences to biological systems are affected by the scale of observation, thus the temporal and spatial scales at which processes operate and populations and communities respond must be estimated and clearly identified in the conceptual model (Noon et al. 1999).

Select Indicators and Establish Sampling Design

No monitoring design can feasibly encompass all ecological processes and species (Bradshaw et al. in review). Therefore, the design of monitoring programs requires careful consideration of candidate species and processes for measurement. Selection of variables should be made in the light of the overall goals and underlying conceptual model of the ecosystem of interest (NRC 2000). Although biodiversity has many dimensions, since we are limited in the number of things we can measure, and there is pressure for quick answers, it is popular to rely on monitoring surrogates such as indicators, keystones, and umbrella species (Landres et al. 1988; Simberloff 1998; Andelman and Fagan 2000).

Despite their popularity, particular umbrella and indicator species have often been chosen as management and monitoring surrogates without adequate testing of their performance (Simberloff 1998; Caro and O'Doherty 1999). This, in turn has led to confusion about when a species is being monitored for its own sake, and when a species can appropriately function as an accurate indicator of other organisms or environmental attributes. Where the performance of surrogates has been evaluated, results suggest that the utility of existing approaches to selecting surrogates is limited and almost certainly scale dependent (Andelman and Fagan 2000; Williams et al. in press). Another analysis, focusing on the northern spotted owl, suggests that that species may act as an adequate habitat surrogate for many other species (Noon and Bingham in press).

Despite concerns, the fact remains that it is impossible to assess the status and trend of all species of interest and that some surrogate-based approach is necessary. In addition, some land management agencies, such as the U.S. Forest Service, are legally required to use "management indicator species" to assess the impacts of any proposed management action on the system as a whole. This reality lead a recent committee of scientists to propose the use of focal species, defined as those species whose status and trend allow insights into the integrity of the larger ecological system to which they belong. However, the committee noted that currently there is no definitive algorithm for unambiguously identifying focal species. This remains a priority area for accelerated ecological research.

Define Methods of Analysis and
Ensure Link from Monitoring to Decision Making

For monitoring programs to be useful they must be efficient, informative, and reliable (Thomas 1996). For example, given an objective of detecting a

population decline and a fixed level of effort or resources, how many sites
should be monitored? How many samples are required to detect a trend in
population size when there is uncertainty in observations and significant
environmental variation? How likely is it that a particular sampling pr.ce-
dure will detect a rare species?

Most frequently used statistical techniques do not answer these questions
because they are rooted in the tradition of null-hypothesis testing. Typical
management programs hinge on the implicit assumption that if no problem
is observed, none exists. The burden of proof rests with monitoring programs. If a monitoring program fails to detect a trend or a rare species, that
impact is assumed to be absent. In such circumstances, the reliability of the
monitoring system becomes critically important. This reliability depends on
statistical power, the ability of a method to detect real outcomes, often against
a background of natural environmental variation, measurement error, and
ignorance of biological processes.

Given the importance of reliable monitoring, the persistent failure of conservation biologists and ecologists to explicitly incorporate reliability considerations into the design of monitoring programs is notable (Fairweather
1991; Mapstone 1995). The task of stipulating an appropriate level of statistical power and an acceptable effect size is not simply a statistical decision. It
entails judgments about the biological importance of an effect (Mapstone
1995). These judgments may be influenced, in part, by the social, political,
and economic implications of the ensuing management or policy decisions.
Such considerations are critically important because analysis indicates that
without reliability calculations, experimenters often are overly optimistic
about the reliability and representativeness of their samples. The risk of false
optimism may be compounded by experimental designs that fail to account
for the independence of replicates (Hulbert 1984).

These observations highlight a general failing in the interpretation of scientific data. The implications of this failing are especially important in conservation biology, where the consequences of wrong decisions may be
irreversible (e.g., destruction of critical habitat or extinction of species). Thus,
conservation biologists should seek to apply methods that minimize incorrect
decisions. Such methods are readily accessible (Fairweather 1991; Thomas
1996) yet seldom applied.

One potential explanation for this lack of application suggests that different sorts of research may be needed if we are to improve the effectiveness of
monitoring programs and the linkages between monitoring, decision making, conservation management, and policy. Research on human judgment
under uncertainty suggests that humans have a widespread tendency to perceive patterns where none exist (Redelmeier and Tversky 1996). Efforts to

link monitoring programs to conservation decision making may need to explicitly consider such fundamental human psychological constraints.

Recently, there has been a chorus of calls for establishment of global monitoring of biodiversity trends. While well intentioned, such programs have the potential to waste time and use limited resources. Continued use and advocacy of unproven methods, regardless of scale, serve only to undermine the credibility of conservation biology as a legitimate scientific discipline, in academic, policy, and other public fora. More seriously, given the magnitude of current rates of human exploitation of biodiversity, a consensus is needed as to the priority questions monitoring should aim to answer. Once those questions have been identified, the development of credible, extensive, and selectively intensive monitoring is essential, to provide warnings and adequate opportunities to avoid biotic catastrophe (Beddington 1995).

Future Directions for Research

The examples provided here illustrate a few specific ways in which decision theory and monitoring can apply to conservation on the ground (Box 10.1). The application of decision theory to identifying solutions and evaluating alternatives is limited only by the imagination of conservation practitioners. Virtually every practical, real-world problem in nature conservation would benefit from application of a decision theory and adaptive management approach. Several aspects of decision theory as applied to conservation problems would benefit from further research to strengthen its potential as a practical tool for resource managers and conservation biologists. These include:

1. *Develop ways to assess social values and use them to define composite utilities to allow for the complex integration of ecology, social sciences, and economics in specific situations.* This has rarely been achieved, and most of the examples described here have dealt with simple biodiversity objectives, or multiple biodiversity objectives, while blithely ignoring the realities of real-world decision making. Multidisciplinary research in this area would be fruitful, yet the tradition of specialization within both the natural and the social sciences may mitigate against tackling these hard problems.

2. *Develop models that can tractably evaluate multiple biological and social objectives that are measured in different currencies (e.g., representation, probabilities of extinction, frequency and magnitude of disturbance events) and that may also interact with one another (e.g., cascading effects through food webs, differential influence of habitat fragmentation on population densities of species).* With such models, decision theory can be used to assess optimal strategies for achieving multiple objectives that collectively are considered to be integrated indices of, for example, ecological health or biological integrity. In this way, conservation

strategies can begin to move away from their tendency toward single-species management.

3. *Develop models that allow for assessment of the trade-offs involved for a wider, but more realistic, range of management options than what has been accomplished to date.* For example, with respect to the implementation of an ecological reserve system, such a set of realistic options might include acquiring a new reserve for the system, enlarging an existing reserve, or restoring an existing reserve to improve its quality. Moreover, the value of each of these management options may change depending on the timing of implementation. These same approaches could then be used to evaluate a wider range of management options for an existing reserve, independent of other reserves in the system, such as revegetation, species reintroduction, eradication of exotics, road closure, and restoration of disturbance regimes.

4. *Explicitly test a specific decision-theory model, in terms of whether the hypothetical optimal solution generally performs better than random, status quo, or no management.* This would be impossible for some objectives, like minimizing extinction probabilities. However, it would be possible for testing management activities that are repeated frequently in space and time, like weed management.

5. *Develop the tools that can tractably evaluate the relative sensitivity of the solutions to uncertainty in the true state of the system and the dynamics of the state variables.* This is important for giving resource managers insight about how robust predicted solutions are and where priorities should be placed for collecting additional data.

6. *Develop user interfaces that will allow resource managers to change models and quickly generate new solutions.* Those who use the models to generate solutions will rarely be the ones to develop the models in the first place, yet they need to be able to alter models to input new goals, new management options, new values for state variables, new descriptions of the dynamics among state variables, and new constraints. This happens not only because managers want to evaluate alternative scenarios and the robustness of solutions, but also because the political and economic realities of management never remain constant.

7. *Develop and test rules for the selection of surrogate and indicator species.* Since we can feasibly measure only a handful of environmental attributes, we urgently need a set of robust rules to help determine which attributes will be most informative.

8. *Develop and test rules of thumb for conservation practice and implementation of active adaptive management.* Conservation practitioners must make many decisions, and their resources for modeling, experimental design, and evaluation are limited. While it is important for conservation biology researchers to pro-

vide guidance on best practice in decision making and monitoring, we must also deliver robust rules of thumb that can be generically applied and improve the consequences of conservation management in the short term.

ACKNOWLEDGMENTS

We are very grateful for the ideas of fellow workshop participants, the editors, and others who have commented on the paper: Dee Boersma, Drew Tyre, and Bill Murdoch.

LITERATURE CITED

Alford, R.A., and S. Richards. 1999. Global amphibian declines: A problem in applied ecology. *Annual Review of Ecology and Systematics* 30: 133–165.
Andelman, S.J., and W.F. Fagan. 2000. Umbrellas and flagships: Efficient conservation surrogates or expensive mistakes? *Proceedings of the National Academy of Sciences* 97:5954–5959.
Andelman, S.J., and E. Meir. In press. Breadth is better than depth: Biodiversity requirements for adequate reserve networks. *Conservation Biology.*

BOX 10.1. **Research Priorities for Conservation Biology in Decision Making and Monitoring**

Summary priority: Develop and implement integrated programs of adaptive decision making and monitoring (active adaptive management):

1. Develop and test rules for the selection of surrogate indicator species.
*2. Determine whether decisions based on meeting a single biodiversity objective meet the needs of other biodiversity objectives.
*3. Develop and test rules of thumb for conservation practice.
4. Create usable composite objectives that incorporate social and economic values.
5. Improve the way uncertainty and risk are incorporated into decision making.
6. Determine what kinds of decision-making methods work.
*7. Isolate gaps in data that impede performance evaluation for making decisions.
*8. Determine factors impeding collaboration between managers and researchers.
9. Transform environmental policy (e.g., National Environment Policy Act) to trigger monitoring that enhances decision making.

Note: Asterisks indicate the most urgent priorities.

Beddington, J. 1995. Fisheries—The primary requirements. *Nature* 374:213–214.

Beenhakker, H.L. 1975. *Capital Investment Planning for Management and Engineering.* Rotterdam: Rotterdam University Press.

Beissinger, S. R., and M. I. Westphal. 1998. On the use of demographic models of population viability in endangered species management. *Journal of Wildlife Management* 62:821–841.

Blaustein, A., D. Wake, and W. Sousa. 1994. Amphibian declines: Judging stability, persistence and susceptibility of populations to local and global extinction. *Conservation Biology* 8: 60–71.

Bradshaw, G.A., M.-J. Fortin, and H.C. Biggs. In review. Large-scale monitoring of terrestrial and aquatic systems: Matching what science can do with what society wants. *Conservation Biology.*

Caro, T.M., and G. O'Doherty. 1999. On the use of surrogate species in conservation biology. *Conservation Biology* 13:805–814.

Clemen, R.T. 1996. *Making Hard Decisions: An Introduction to Decision Analysis.* Pacific Grove, Calif.: Duxbury Press.

Dayton, P. K., M. J. Tegner, P. B. Edwards, and K. L. Riser. 1998. Sliding baselines, ghosts, and reduced expectations in kelp forest communities. *Ecological Applications* 8:309–322.

Fairweather, P. G. 1991. Statistical design and power requirements for environmental monitoring. *Australian Journal of Marine and Freshwater Research* 42:555–567.

Ferson, S., and M.A. Burgman. 2000. *Quantitative Methods for Conservation Biology.* New York: Springer-Verlag.

Guikema, S., and M. Milke. 1999. Quantitative decision tools for conservation programme planning: Practice, theory and potential. *Environmental Conservation* 26:179–189.

Hanski, I. 1999. *Metapopulation Ecology.* Oxford: Oxford University Press.

Hulbert, S.H. 1984. Pseudoreplication and the design of ecological field experiments. *Ecological Monographs* 54:187–211.

Intriligator, M. D. 1971. *Mathematical Optimisation and Economic Theory.* Englewood Cliffs, N.J.: Prentice-Hall.

James, F.C., C.E. McCullough, and D.A. Wiedenfield. 1996. New approaches to the analysis of population trends in land birds. *Ecology* 77:13–27.

Kareiva, P.K., S.J. Andelman, D. Doak, B. Eldred, M. Groom, J. Hoekstra, L. Hood, F. James, J. Lamoreux, G. LeBuhn, C. McCulloch, J. Regetz, L. Savage, M. Ruckelshaus, D. Skelly, H. Wilbur, K. Zamudio, and the NCEAS HCP Working Group. 1999. *Using Science in Habitat Conservation Plans.* Santa Barbara, Calif.: National Center for Geological Analysis and Synthesis; Washington, D.C.: American Institute of Biological Sciences.

Kulkarni, R.B., R.L. Burns, J. Wright, B. Apper, T.O. Baily, and S.T. Noaek. 1993. Decision analysis of alternative highway alignments. *Journal of Transportation Engineering* 119:317–332.

Landres, P., J. Verner, and J. Thomas. 1988. *Conservation Biology* 2:316–328.

Lubow, B. C. 1996. Optimal translocation strategies for enhancing stochastic metapopulation viability. *Ecological Applications* 6:1268–1280.

MacArthur, R.H., and E.O. Wilson. 1967. *The Theory of Island Biogeography.* Princeton, N.J.: Princeton University Press.

Maguire, L. A., U. S. Seal, and P. F. Brussard. 1987. Managing a critically endangered species: The Sumatran rhino as a case study. In *Viable Populations for Conservation,* ed. M. E. Soulé, pp. 141–158. Cambridge, U.K.: Cambridge University Press.

Maguire, L.A., and C. Servheen. 1992. Integrating biological and social concerns in endangered species management: Augmentation of grizzly bear populations. *Conservation Biology* 6:426–434.

Mapstone, B. D. 1995. Scalable decision rules for environmental impact studies—Effect size, type-I and type-II errors. *Ecological Applications* 5:401–410.

Margules, C. R., and Pressey, R.L. 2000. Systematic conservation planning. *Nature* 405:243–253.

Meffe, G.K., and C.R. Carroll. 1997. *Principles of Conservation Biology,* 2nd ed. Sunderland, Mass.: Sinauer Associates.

Meir, E., and S. J. Andelman. In press. Biodiversity data requirements for adequate reserve system design. *Ecological Applications.*

Mulder, B.S., B.R. Noon, C.J. Palmer (and others), tech. coords. 1999. *The Strategy and Design of the Effectiveness Monitoring Program for the Northwest Forest Plan.* General Technical Report, PNW-GTR. Washington, D.C.: USDA, Forest Service.

Noon, B.R., and B.B. Bingham. In press. The northern spotted owl as an umbrella species for late seral forests of the Pacific Northwest. In *Single-Species Approaches to Ecosystem Conservation,* eds. P. Foster-Tukey, P. Kelly, and R. Medellin. Washington, D.C.: Island Press.

Noon, B.R., T.A. Spies, and M.G. Raphael. 1999. Conceptual basis for designing an effectiveness monitoring program. In Mulder, B.S., B.R. Noon, C.J. Palmer (and others), tech. coords., *The Strategy and Design of the Effectiveness Monitoring Program for the Northwest Forest Plan,* pp. 17–39. General Technical Report, PNW-GTR. Washington, D.C.: USDA, Forest Service.

NRC (National Research Council). 2000. *Ecological Indicators for the Nation.* Washington, D.C.: National Academy Press.

Parma, A., P. Amarasekare, M. Mangel, J. Moore, W. W. Murdoch, E. Noonburg, M. A. Pascual, H. P. Possingham, K. Shea, W. Wilcox, and D. Yu. 1998. What can adaptive management do for our fish, food, forests and biodiversity? *Integrative Biology* 1:16–26.

Possingham, H. P. 1997. State-dependent decision analysis for conservation biology. In *The Ecological Basis of Conservation: Heterogeneity, Ecosystems and Biodiversity,* eds. S.T.A. Pickett, R.S. Ostfeld, M. Shachak, and G.E. Likens, pp. 298–304. New York: Chapman and Hall.

Possingham, H., and K. Shea. 1999. The business of biodiversity. *Australian Zoologist* 31:3–10.

Possingham, H. P., I. R. Ball, and S. Andelman. 2000. Mathematical methods for reserve system design. In *Quantitative Methods for Conservation Biology,* eds. S. Ferson and M. Burgman, pp. 291–306. New York: Springer-Verlag.

Pulliam, H.R. 1988. Source, sinks and population regulation. *American Naturalist* 132:652–661.

Ralls, K., and J. Ballou. 1986. Captive breeding programs for populations with a small number of founders. *Trends in Ecology and Evolution* 1:19–22.

Ravirala, V., and D.A. Grivas. 1995. Goal-programming methodology for integrating pavement and bridge programs. *Journal of Transportation Engineering* 121:345–351.

Redelmeier D.A., and A. Tversky. 1996. On the belief that arthritis pain is related to the weather. *Proceedings of the National Academy of Science USA* 93:2895–2896.

Richards, S.A., H.P. Possingham, and J. Tizard. 1999. Optimal fire management for maintaining community diversity. *Ecological Applications* 9:880–892.

Shea, K., P. Amarasekare, M. Mangel,, J. Moore, W. W. Murdoch, E. Noonburg, A. Parma, M. A. Pascual, H. P. Possingham, W. Wilcox, and D. Yu. 1998. Management of populations in conservation, harvesting and control. *Trends in Ecology and Evolution* 13:371–374.

Simberloff, D. 1998. Flagships, umbrellas and keystones: Is single species management passé in the landscape era? *Biological Conservation* 83:247–257.

Simberloff, D. 1999. The role of science in the preservation of forest biodiversity. *Forest Ecology and Management* 115:101–111.

Teng, J., and G. Tzeng. 1996. A multi-objective programming approach for selecting non-independent transportation investment alternatives. *Transportation Research* 30B:291–307.

Thomas, L. 1996. Monitoring long-term population change: Why are there so many analysis methods? *Ecology* 77:49–58.

Thompson, W.L., G.C. White, and C. Gowan. 1998. *Monitoring Vertebrate Populations*. New York: Academic Press.

Williams, P.H., N.D. Burgess, and C. Rahbek. In press. Flagship species, ecological complementarity, and conserving the diversity of mammals and birds in sub-Saharan Africa. *Animal Conservation*.

I I

ECOLOGICAL RESTORATION
A Key to Conservation Biology's Future

James A. MacMahon and Karen D. Holl

> *The success or failure of most restoration projects depends heavily on such ecological processes as demographic stochasticity, environmental stochasticity, intra- and interspecific interactions, habitat scale, and habitat destruction. Hence, restoration ecology should draw heavily from and contribute to these ecological principles. The question becomes one of identifying those areas of ecology that offer promise and deserve scrutiny within restoration ecology.*
> —J. S. Brown

Research in restoration ecology has grown exponentially in the past twelve years (Young 2000; Soulé and Kohm 1989). This growth is the result of increasing recognition that restoration is a necessary component of conservation, as much of the world is disturbed, and intact natural systems are disappearing. In contrast, abandoned and degraded lands are increasing and may become some of the only lands available for conservation. Moreover, as the opening quote from Brown identifies, besides producing a desired change in a system, attempts at restoration may also reveal insights into ecosystem functioning, and, therefore, may help in preventing or minimizing damage in the first place.

That said, we must reiterate at the outset a few oft-stated cautions regarding the potential for restoration. Although restoration may further conservation efforts, it is a second choice to preserving more intact habitats and should not serve as an alternative for preservation. Because restoration is difficult and often unsuccessful in restoring native ecosystem structure and functioning, restoration ecologists should stress the unparalleled value of preservation. Restoration will necessarily compete for funding with other conservation efforts, and it is important to carefully consider the best expenditure of funds to further conservation on a regional scale.

As Ehrenfeld (2000) notes, restoration efforts are motivated by a broad spectrum of reasons, ranging from permit compliance to reducing the threat of extinction of a single species to restoring ecosystem disturbance patterns across entire watersheds. To achieve restoration for conservation ends requires meeting lofty goals, such as restoring connectivity with the surrounding landscape, self-maintaining disturbance regimes such as flooding (Cairns and Heckman 1996), and a variety of hydroperiods that can influence community composition, such as amphibian species diversity (Snodgrass et al. 2000). Selecting the most appropriate goals for restoration can be highly controversial, and the task is particularly difficult because of the common lack of undisturbed reference systems or pre-impact data (Palmer et al. 1997). This book being focused on research priorities in conservation biology, our highest-priority research questions are those that must be answered so that restoration efforts can maximize the conservation of biodiversity.

We focus the discussion on the community and ecosystem levels of integration, despite the focus of much of conservation biology on the restoration of populations of particular species (Bowles and Whelan 1994; Montalvo et al. 1997; cf. Simberloff et al. 1999). We feel restoration efforts must necessarily aim to restore entire ecosystems, as it is difficult to restore species populations without restoring the abiotic environment conducive to a species persistence and reproduction and interactions with many other species; conversely, biotic communities strongly influence the abiotic environment. While there are some differences in applying restoration to terrestrial as opposed to aquatic systems (Mitsch et al. 1998), we fuse both in our discussion and highlight a few of the differences where appropriate. Because of the interdependence of terrestrial and aquatic ecosystems, their restoration should be integrated (NRC 1992); it is impossible to restore rivers, lakes, wetlands, or marine systems if material and energy fluxes from adjacent terrestrial systems are not restored to natural levels.

Multiple Faces of Restoration

To provide common ground and clarify goals, we briefly define several types of activity that fall under the general umbrella of restoration. Strictly speaking, restoration means bringing an ecological system back to its original or former state (Bradshaw 1997). In the context of conservation biology, it is distinct from such actions as preservation, conservation, and stewardship. The common thread between restoration and these other activities is that they all involve efforts devised to overcome alterations caused by natural or anthropogenic forces.

The term *restoration* implies that the system's original state was good or perfect (Francis et al. 1979). In contrast, *rehabilitation* is an attempt to restore elements of a structure or functioning of an ecosystem without necessarily attempting complete restoration to a prior condition. *Reclamation* involves the rehabilitation of severely degraded sites. Generally, the result of this is not a return to a prior condition, but achievement of a condition where some stabilization of the degradation process occurs and inroads are made toward restoring certain species and ecosystem processes. *Re-creation* is an attempt at wholesale reconstruction of an ecosystem on sites so severely disturbed that there is little left to restore; the product may not represent the historical system. This is often a difficult activity, but it can yield important insights into ecological processes as one develops model ecosystems.

Finally, simply allowing a disturbed system to follow its natural trajectory is referred to as *ecological recovery* (succession). In many cases, ecological recovery is undesirable, particularly in terrestrial systems, because of the time involved and the uncertainty of the result; but it may be the de facto restoration strategy when resources are limited. Some aquatic systems, however, are sufficiently resilient that if stresses are removed, they recover fairly quickly (Cairns and Heckman 1996).

All of these processes of restoration typically require intensive management, and their success should be defined by an unambiguous statement of the goal of the process (MacMahon 1997; Michener 1997; White and Walker 1997). In this discussion we generally refer to restoration, rehabilitation, and reclamation under the general term *restoration*. Restoration per se will not be our sole focus, simply because it is extremely difficult, time consuming, and, therefore, expensive if done correctly (Jackson et al. 1995). Moreover, there is much more gray area among these three terms than most restoration ecologists acknowledge. Although it is important to define how restoration is used in specific case studies, lengthy semantic debates are not productive (Hobbs and Norton 1996; Ehrenfeld 2000). In this chapter, we do not strongly focus on *de novo* habitat creation, as extensive evidence suggests that substituting created habitat for existing habitat seldom furthers conservation efforts, although it provides insights into what extreme management actions we may be forced to employ in the future. We do discuss one form of habitat creation, designer ecosystems, but it is recommended only under special, extreme circumstances to conserve certain species.

Research Priorities in Restoration Ecology

Rather than outlining research priorities specific to certain ecosystem types, we suggest general areas of urgency, using examples from a range of ecosystems. We refer readers to two other recent efforts to prioritize research areas in restoration ecology. First, a workshop was held at the National Center of Ecological Analysis and Synthesis in Santa Barbara entitled "Research Priorities in Restoration Ecology" (Allen et al. 1997), the ideas from which are summarized in a special feature of *Restoration Ecology* (vol. 5, issue 4, 1997). In addition, the book *Restoration of Aquatic Ecosystems* (NRC 1992) outlines a number of research priorities for aquatic ecosystem restoration. We draw heavily on both these sources, although we have necessarily condensed their ideas and couched them in a conservation biology context.

We organize our research questions into five somewhat overlapping or related themes: the recovery process, using succession as a template for thinking about restoration; introduction of species; scale; evaluation of success; and policy. The latter topic requires collaborative research between natural and social scientists to successfully implement restoration. Those research questions that workshop participants agreed were of highest priority for restoration are listed in box 11.1. Below we provide a brief discussion of the underlying rationale for those questions.

Recovery Processes

Much research in restoration ecology has been focused on documenting observed patterns. Restoring ecosystems requires an understanding of the processes that lead to their development. Studies that have investigated the assembly rules governing the order of species entry into disturbed ecosystems suggest that there may be multiple potential endpoints for ecosystems (Hobbs and Norton 1996; Lockwood 1997). Understanding the factors that lead to different successional trajectories is essential to designing specific plans to restore areas.

Since succession is generally viewed as the natural recovery of altered ecological systems, logic dictates that we turn to succession, as first described by Clements (1916), to organize our initial thoughts about restoration (MacMahon 1998) and ask how our knowledge of succession might apply to the attainment of conservation goals. Succession begins with some disturbing event that is characterized by its areal extent, intensity, type, timing, and other attributes (White and Pickett 1985).

The most important aspect of the disturbance process is what is left afterward (residuals or legacies; cf. Franklin and MacMahon 2000), because those remaining components and organisms are the starting point for recovery. Following a disturbance, a variety of organisms and nonliving materials may move ("migrate") or be moved to the site, depending on the migrants' distance from the disturbed site and their vagility. A low-vagility species is not likely to colonize the site, but one with high vagility could arrive in a matter of minutes to hours, if it was sufficiently nearby. Recently disturbed sites become a mix of residuals and migrants. Clements aptly pointed out that these organisms will not necessarily be part of the subsequent communities and ecosystems as they develop. Environmental conditions in the post-disturbance habitat edit out those migrants not suited for the site at that time, as well as those survivors whose world is so altered that their tolerance capacity is exceeded.

Understanding migration and establishment is an important facet of restoration, especially since species in restoration projects are usually rare species (Maina and Howe 2000), and dispersal from nearby areas can augment the recovery process. During restoration, we can act as agents of migration, and we can also create conditions to permit establishment. Thus, these two processes are among the most amenable to management. One difference between terrestrial and aquatic systems may be the rate of reintroduction of many species. Disturbed streams, because of the power and continuity of currents, may carry large objects, allowing recolonization to occur faster than on land, where air movement generally carries smaller loads and is less continuous. Organisms that live on land may require active migration more often than those in water that can be passively moved; however, lakes and ponds with few inlets may require migration processes more like those in terrestrial systems than those in flowing water.

Eventually, migrants and residuals establish a variety of interactions that act as a further filter in determining the species composition of the evolving community or ecosystem. For example, a predator or parasite that has the capacity to remove one of the migrants or residuals can change the community composition (Louda 1994). On the other hand, a species may be successful and capable of establishment only if it is associated with another species. Numerous mutualisms, such as pollination, seed dispersal, protection from herbivory, and mycorrhizal interactions, can have significant impact on an evolving community (Handel 1997). For example, the association of mycorrhizal fungi with plants can confer an advantage on both that allows them to remain a part of the developing community when individuals that do not develop a mutualistic relationship disappear (Allen 1991).

Once established, migrants and residuals alter the site, in the "reaction" process of Clements (1916). Such an altered site may develop characteristics that no longer favor some of the successful residuals or early migrants, and a new mixture of migrants may be favored. In many instances, some residuals are unable to withstand the early post-disturbance environment but persist in the form of long-lived propagules that experience a suitable habitat after the process of reaction and are thus able to establish as a component of the developing system (Franklin et al. 1985). If the constantly changing environment and the consequent substitution of species come to a point of relative stability, the system appears, at least to our superficial perception, to remain unchanged. Obviously, we no longer believe that most systems show true stability; however, the sense of stability is occasioned by the persistence of long-lived woody perennials. A forest often does not change in gross appearance for decades or more, despite the fact that, other than a dozen tree species, all of its species populations turn over rapidly, changing the composition and the nature and magnitude of a variety of dynamic processes appreciably.

To alter the successional trajectory of a system to conserve certain species, it may be necessary to alter or direct one or more of those successional processes. Often this is done by speeding up succession, by preventing it from moving beyond a desired stage, or by introducing a species that might not colonize naturally. For example, lack of seed dispersal by animals is a major factor impeding the recovery of tropical forests in disturbed areas. Therefore, restoration efforts in such places have focused on planting trees and shrubs to attract animals, thus enhancing seed dispersal and accelerating succession (Holl and Kappelle 1999). Clearly, detailed knowledge of succession is necessary to restore communities and ecosystems to support conservation goals.

Assembly Rules

Since restoration is the reassembly of communities through human intervention, the search for rules in nature is of considerable importance to restoration ecologists. A recent perspective on ecological assembly rules (Weiher and Keddy 1999) shows the divergence of opinion as to whether rules actually exist. These range from the opinion that there has been a failure to demonstrate simple, general assembly rules to date, to the opinion that assembly rules, based on the assembly of guilds rather than species, are indeed robust and demonstrable in nature, confirming Diamond's (1975) suggestions. A search of the literature suggests that assembly rules probably do exist, but that there is sufficient variation among systems that they do not give exact predictions. Nonetheless, many attributes of developing systems predispose them to being invaded by a new species, including exotics, or to extinction of a resident species, thus causing compositional turnover. If we can quantify these attributes or determine likelihood of invasion and/or extinction events, we may be able to dramatically increase the efficiency of the restoration process. This is especially important to conservation goals if assembly rules are general, deterministic, and mechanistic (Belyea and Lancaster 1999), and such rules may even include attributes of the past history of a site (Kettle et al. 2001; Anderson et al. 2000).

In attempting to describe some aspects of ecological assembly, Keddy (1999) listed three principles that exhibit our understanding of the interplay of environmental factors and ecological patterns. First, any particular community or ecosystem is produced by multiple environmental factors acting simultaneously. Second, there are quantitative relationships between environmental factors and the properties of communities. Third, multiple factors that produce a community or ecosystem will change through time. If one examines these three, it is clear why we often have moving targets in the process of restoration and difficulty in its use as a valuable tool for conservation, either because of the time frame in which we operate, or because of the changing composition of species that are involved. One way to deal with some of these problems is to simplify communities through the use of guilds or functional groups. If this approach is applied with discretion and with the recognition that guilds are not necessarily natural entities, such an approach can lead to useful results during restoration (Hawkins and MacMahon 1989).

Structure and Function

Conservation biology has traditionally focused on conserving native species rather than preserving ecosystem processes. Likewise, the endpoint to the restoration process is often defined by the community composition rather than the establishment of normal processes. Although there is not always one-to-one correspondence between species composition and ecosystem processes (Zedler and Calloway 1999), it is clear that they are intricately related. Feedback loops between structure and function may help to maintain restored areas in certain states or may make restoration of degraded sites difficult. For example, Boyer and Zedler (1998) found that low nitrogen levels due to improper sediment texture in salt marshes in Southern California resulted in low cordgrass growth. Short cordgrass resulted in low populations of a native predatory cocinellid beetle that requires tall cordgrass during high tides. This resulted in an outbreak of a scale insect on the cordgrass that was normally controlled by the beetle. The failure to restore an appropriate sediment type led to an altered trajectory of recovery.

These feedback loops between structure and function are poorly understood, and research to elucidate these relationships is critically needed (Ehrenfeld and Toth 1997). In particular, little research has addressed the ability of soil and aquatic microbial communities to recover both composition and functional processes such as decomposition (discussed further in chapter 6). There has been a great deal of debate recently about the related question of whether some species are redundant from a functional standpoint (Tilman et al. 1997; Hooper and Vitousek 1997). Recent studies (Nijs and Impens 2000) suggest that increasing the number of functionally analogous species enhances ecosystem reliability (probability of functioning), while increasing the number of functions these species support has a negative influence. These questions are particularly important in the context of deciding whether to try to restore all species (Palmer et al. 1997) or merely a functionally important subset of the total community.

Understanding the relationship between structure and function is important not only for conserving species but also for to gaining widespread public support for restoration. While conservation biologists generally support conservation of species for their inherent worth, if restored ecosystems can also provide processes critical to humans, such as flood control or nutrient uptake, it will be easier to obtain funding for such restoration.

Community Architecture

Many of our conservation activities, as well as restoration activities, emphasize the management of individual species by directly operating on their population characteristics or a variety of their species-specific requirements. It is not possible to find the funds to look at every species in a particular community or ecosystem. Rather than trying to manipulate individual species, we suggest that manipulating the architecture of the system may produce both the target ecosystem processes and patterns of ecosystem components in some cases. This means both the horizontal distribution of entities, often referred to as dispersion, and the vertical distribution of entities, frequently referred to as architecture or form. There are many examples in which architecture has been used to manage either species or communities. In native tallgrass prairie, the persistence of three-dimensional vegetation structure does not change over time, yet the species composition does. The stability of consumers relates to structure in these systems, not to species composition of the vegetation (Collins 2000). Simply thinning plants in agricultural or forestry contexts (changing the density and dispersion pattern) changes the composition of the fauna and flora of a particular area. While thinning directly affects processes such as competition, it also alters the structure of the system dramatically, and that structural change alone may be the direct cause of the organisms' responses. If we want more wood ducks, we generally put out more wood duck breeding boxes, under the assumption that this cavity nesting species is limited by the number of tree holes available and not by the food in an area. In aquatic systems the structure provided by downed woody debris is critical to invertebrates, fish, and the control of sediment movement (Harmon et al. 1986).

The idea of managing ecosystem components rather than species is hotly debated (Goldstein 1999; Walker 1999; Risser 1999). There should be little question about whether architecture is important; rather, we need to know where and under what circumstances it is critical.

Our conservation efforts need to take into account the importance of system architecture. Under many scenarios, successful restoration of a system may depend as much on structure as it does on species composition. We need to know the extent of such relationships.

Interaction with Other Human-Induced Changes

Potential effects of climate change, elevated CO_2, and increased N availability on conservation goals are addressed in chapter 9. It is important to note, however, that all these factors are likely to affect the trajectory of succession and disproportionately affect disturbed ecosystems. For example, recently germinated seedlings are particularly susceptible to changes in precipitation (Woodward 1992; Larcher 1995), and increased N availability is likely to favor non-native species (Maron and Connors 1996). Likewise, interactions between species, such as nurse-plant effects (Ashton et al. 1997) and mycorrhizal associations (Egerton-Warburton and Allen 2000), may be altered by changes in climate and increased N availability. As we discuss later, most restoration work has focused on small scales and has not considered the effects of other human-induced change such as climate change, elevated CO_2, and increased nitrogen availability. Conversely, much of the research on the latter questions has focused on later-successional ecosystems. More research is needed that considers the interactions between other human-induced environmental changes and ecosystem recovery from disturbance. Such knowledge is important to the development of conservation plans. For example, if we propose to restore a certain species in its former range, but the conditions for establishment have been altered due to a change in climate, the plan will have little probability of success.

Introduction of Species

The introduction or reintroduction of species to disturbed sites raises a number of research questions that must be addressed to ensure that restoration results in net positive rather than negative effects on conservation.

Reintroduction of Species

Reintroduction of species to disturbed sites has a number of important consequences for the success of restoration and for maintenance of genetic integrity, thus having significant conservation implications (discussed in detail in Montalvo et al. 1997). First, a great deal of research has demonstrated that local adaptation to site conditions results in higher fitness (Bradshaw 1984; Huenneke 1991), suggesting that local adaptation is important to restoration success. But the spatial scale of local adaptation can range from an order of meters for metal-tolerant plants (Antonovics and Bradshaw 1970) to many kilometers for plants adapted to a climatic gradient (Knapp and Rice 1998), and very little is known about this phenomenon. Second, introduction of nonlocal seed can have important consequences for the genetic integrity of surrounding populations (Montalvo et al. 1997). Managers are often forced to make decisions regarding the question of how local is local, which has important economic consequences given that seed collection is time consuming and expensive (Lippit et al. 1994). Clearly, detailed studies of genetics will not be possible for all species, which means that research on the degree of local adaptation must be compiled to try to predict rules of thumb for scales of genetic variability based on life history and dispersal characteristics of species. For example, for *Elymus glaucus,* a primarily self-pollinating grass species with highly localized genetic differences, seed collection may need to be more spatially restricted than for conifer species that are wind pollinated and less genetically distinct (Knapp and Rice 1996). Questions of whether localized adaptation increases restoration success can be tested in restoration experiments (Montalvo et al. 1997).

Invasive Species

Invasive exotic species have an enormous effect on the establishment and survival of native species in disturbed areas, and their control is often the focus of restoration efforts (Berger 1993). Predicting invasions of exotic species and their potential to spread and have unwanted impacts is of particular importance to restoration given that disturbed areas are particularly susceptible to invasion (D'Antonio et al. 1999). Susceptibility to invasion may relate inversely to diversity (Levine 2000), but contrary to previous assertions, exotics invade areas of high native plant diversity (Stohlgren et al. 1999). This question is also important given that exotic species are sometimes introduced at early stages in restoration to ameliorate stressful conditions for later-successional species (Kuusipalo et al. 1995; Chapman and Chapman 1996). Therefore, guidelines are needed for when the use of exotics is appropriate depending on their potential impacts on other species. These research questions are addressed in more detail in chapter 4.

Designer Communities

More land is being altered at an ever increasing rate in the world than at any time in the past. This suggests that to "conserve" anything will require restoration at some level. The most damaged sites are often ignored for restoration projects because of the difficulty of restoring them to any semblance of a natural community. Even though regulations in many countries demand that restoration be attempted, efforts are often unsuccessful because the goals set are unreasonable for the particular site. An alternative might be to design ecosystems that contain groups of species that do not generally exist together in nature but support some conservation goal. These "designer" ecosystems would be constructed of groups of species that are selected so that each has a specified purpose in a conservation plan. Many who review this idea presume that scientists are attempting to play God and dismiss it out of hand.

It seems reasonable, in severely disturbed sites, to ask the question, Can this site become more useful for the purposes of conservation if we create a unique assemblage of organisms? Such combinations may not persist on their own and may require constant management (Bridgewater 1988). We are painfully aware of the effect of introducing alien species. In the case of designer ecosystems, the optimal situation would be to combine species that are native in a general area, but usually do not occur together, to meet some goal. These need not even be different species; they could be different varieties or genotypes of the same species (Lesica and Allendorf 1999). In the extreme, however, if we knew enough about the biology of a particular group of species, we might put together a completely novel system whose components were drawn from around the world but whose characteristics had been studied so that the species could not cause problems elsewhere. Obviously, this is a difficult and risky venture that must be well studied and controlled from the outset. We would have to negate the environmental and economic costs of invaders (Pimentel et al. 2000) by predicting their invasiveness (Reichard and Hamilton 1997; Pysek 1998; Ruesink et al. 1995).

We often have a static view of communities as having a particular species composition. In fact, fossil data for both plants and animals suggest that many species do not associate now as they did in the past, and that community composition is highly changeable over time. The idea of forming communities of species that do not normally associate for some purpose has been around for a long time, and many successes have been obtained, even though the goal has not always been one of self-consciously developing a designer ecosystem. An example of considerable current interest to conservation biologists is the development of the traditional shade type of coffee plantations, in which layers of vegetation are fostered, as opposed to the open uni-layer of sun-grown coffee. The layers of traditional plantations contain native and non-native species, yet the structure mimics that of natural forests, and many animals and native plants form new combinations that did not typically occur (Greenberg et al. 1997). When the structure is altered to a form that is more open, the industrial non-shade coffee plantation, species diversity is lost (Perfecto et al. 1996; Roberts et al. 2000).

Scale

One of the most important questions in restoration ecology is how to prioritize restoration at larger scales. Funding for restoration as a conservation
tool will always be limited, which means that managers will need to make
decisions about where to restore (Clewell and Rieger 1997). Most restoration
efforts have focused on small spatial and temporal scales (NRC 1996, 1997;
Lockwood and Pimm 1999), the selection of sites often largely dictated by
political and economic concerns rather than by ecological principles. The
success of such restoration efforts in conserving species and ecosystems will
depend on their position in the landscape relative to the condition, use, and
community composition of land in the surrounding area (NRC 1992; Hobbs
et al. 1993; Hansson et al. 1995). Restoring a natural trophic structure may
require reintroducing large vertebrates, the movement of which is dictated
by landscape patterns (Simberloff et al. 1999). Moreover, restoration of many
disturbance-dependent ecosystems, such as ponderosa pine forests (Moore et
al. 1999) and riparian systems (Sparks et al. 1998), requires managing disturbances at large scales. Also, ecological processes occurring at larger scales
(e.g., dispersal) may affect local conservation and restoration efforts (Baillie
et al. 2000).

Landscape Mosaic

The spatial arrangement of landscape elements influences ecosystem structure and functioning, and thus influences the attainment of conservation
goals. A number of basic ideas from conservation biology and ecosystem
ecology should help to inform restoration efforts directed to conservation
goals. Davies et al. (chapter 5) list four ways in which the matrix surrounding a habitat fragment can affect population dynamics within the fragment;
the surrounding habitat can: (1) alter dispersal and colonization rates,
(2) provide habitat for focal species, (3) provide habitat for invasive species,
and (4) affect biophysical processes. Restored areas can also have these interactions with the surrounding landscape matrix, although only recently have
such ideas been applied in a restoration context. For example, Hobbs (1993)
discusses options for allocating revegetation efforts. Revegetation buffers
around existing patches may serve to ameliorate the microclimate in the
existing patch, whereas restoring a corridor may provide connectivity
between existing patches. Prioritizing areas to restore for specific animal
species requires knowledge of their movement patterns. Based on measurements of movement patterns of the endangered Fender's blue butterfly
(*Icaricia icarioides fenderi*), Schultz (1998) recommended restoring a series of
stepping stone patches rather than a continuous corridor of habitat. Restoring wetlands requires restoring hydrological connectivity, as well as proxim-

ity to source populations of plants and animals. Although it has rarely been done, restoration provides an excellent opportunity to test ideas regarding the effect of the surrounding landscape experimentally (Hobbs 1993).

Patchiness and Disturbance

One of the greatest challenges of restoration is restoring the spatial and temporal variability of ecosystems. Many ecosystems are adapted to periodic disturbances that happen at a range of scales; for instance, grassland may be disturbed by rodent burrowing on an annual basis on the order of square centimeters and by fire on the scale of tens of years and many square kilometers. Some species even require disturbance in order to persist (Pavlovic 1994). The question arises as to whether there is a minimum size or time scale, for specific ecosystems, below which restoration efforts are futile (Ehrenfeld and Toth 1997). The massive Kissimmee River restoration project in Florida is aimed at 70 km of river channel and 11,000 ha of floodplain wetlands (Koebel 1995) and restoring the natural meandering of the river. Is restoring river processes at a smaller scale possible? Pink shrimp often use different areas of estuaries during different parts of the year, which requires restoring areas with a range of water depths to provide the temporal variability of habitats necessary (NRC 1994). Pickett and Thompson (1978) suggest that there is a "minimum dynamic area" that must be maintained to allow for a natural disturbance regime to provide sufficient source populations for recolonization, but such areas are rarely known for different species or ecosystem types (see also Baillie et al. 2000).

Quantitative Tools

Landscape-level processes take place at spatial and temporal scales that are not amenable to traditional methodologies of experimental design and inferential statistics. Studying restoration at large scales will require integration of information from smaller-scale experiments and ongoing restoration with modeling efforts. It will require the use of a range of quantitative techniques such as time series analysis, multivariate statistics, geostatistical approaches, simulation modeling of dispersal behavior, materials flux, and other spatial processes, and meta-analysis for unreplicated or nonexperimental data (Arnqvist and Wooster 1995; Michener 1997; Chapman 1999). More research is needed on the ability of different combinations of experimental and quantitative tools to predict system behavior at large scales. For example, Schultz and Crone (1998) used parameters estimated from small-scale burn plots to predict the outcome of different fire management strategies to restore habitat for the endangered Fender's blue butterfly. They also used sensitivity analysis to determine which parameters had the most influence on the model in order to prioritize future data collection. In a special feature of *Restoration Ecology* (vol. 5, issue 4S, 1997), authors present various mapping techniques for prioritizing riparian restoration efforts that consider geomorphology (topography, channel properties, and sediment properties), hydrology, and existing vegetation. A similar approach has been used successfully for terrestrial landscapes (Palik et al. 2000). The accuracy of these approaches for predicting the outcome of restoration should be tested in actual restoration projects to determine their utility.

Evaluating Effectiveness of Restoration

Conservation planning and monitoring are discussed elsewhere in this volume (chapter 10), but monitoring is such a critical component of restoration that we feel it is important to highlight a couple of research questions. Setting clear goals for restoration is absolutely essential to select criteria on which to base progress and determine success (Hobbs and Norton 1996). For restoration to be successful as a conservation tool, progress toward clearly specified goals needs to be monitored and the results incorporated to modify subsequent restoration actions. Although there has been some research into developing criteria for monitoring restoration success (PERL 1990; Sutter 1996), much more work is needed in this area. Monitoring will need to be done at temporal and spatial scales sufficiently large to incorporate natural variability (discussed above; NRC 1992; White and Walker 1997).

It is important to evaluate the success of a restoration project, although this is not always done. Lockwood and Pimm (1999) examined eighty-seven restoration projects that met a number of criteria they considered to be important with regard to meeting the stated goals of the individual project, e.g., sustaining habitat, restoring productivity, and controlling erosion. In their study, using their personal interpretation of the goals of the projects, seventeen (20 percent) of the projects satisfied all the prescribed goals. An additional seventeen projects were generally completed unsuccessfully, leaving 60 percent of the projects judged to be partially successful. Lockwood and Pimm found that 97 of 184 (52 percent) specific goals were achieved. Sixty-one percent of the 80 functional goals and 104 structural goals were considered achieved. Although there are limitations to this review, it seems clear that the success rate is good enough that we should incorporate restoration into our conservation planning. We must emphasize, however, that not all of the goals listed by the practitioners in these examples are goals that would be appropriate for conservation biology, and some of conservation biology's goals might be more difficult to achieve.

An important aspect of the Lockwood and Pimm study is their finding that 48 percent of restoration projects stopped monitoring before achieving stated goals, suggesting that most monitoring programs are conducted for too short a time. Given the common funding constraints on monitoring (Clewell and Rieger 1997), it is essential to carefully select variables or indicators that allow evaluation of the stated goals for a particular project.

Policy

Cairns and Heckman (1996) note that restoration ecology "is a bridge between the social and natural sciences." Although this volume focuses on biological research questions for conservation, it is impossible to separate biology and policy questions in restoration ecology. Prioritizing and researching these questions will necessarily require interdisciplinary research between natural scientists (biologists, chemists, hydrologists) and social scientists (political scientists, economists, sociologists). We list a few important topics.

Paying for Restoration

Although our ecological knowledge base for restoration is generally weak, the primary factor limiting restoration efforts is typically funding (Holl and Howarth 2000). Implementing research on restoration will require collaboration with economists and political scientists to develop creative ways for implementing restoration. One such example is the Partners for Fish and Wildlife Program of the U.S. Fish and Wildlife Service, which provides matching funds for landowners who voluntarily undertake restoration projects on their land (USFWS 1999). Likewise, in Costa Rica landowners are given tax breaks for reforesting logged land with native tree species (Butterfield and Fisher 1994). For each conservation project, it is important to determine what the most appropriate sources of funding should be.

Legislation

Legislation is an important control on restoration and conservation efforts, but often it is not designed to accommodate the dynamic nature of ecosystems. The difference in ecological and political time scales presents a major obstacle to legislating and financing restoration (NRC 1992). Usually, the success of restoration is judged after a relatively short time compared to that of natural ecosystem recovery. For example, success of mine reclamation in eastern hardwood forests is commonly judged after five years (Holl and Cairns 1994), a much shorter time than is necessary for recovery from natural disturbance. Moreover, the aggressive, non-native, herbaceous species commonly planted to achieve five-year cover requirements have been shown to inhibit the long-term development of the vegetational communities (Brenner et al. 1984; Burger and Torbert 1990; Holl, unpublished data). More work is

BOX 11.1. **Research Priorities in Restoration**

Questions are organized into categories paralleling the discussion.

Investigate Recovery Processes
1. What are the rate-limiting steps to recovery in different biomes?
2. To what degree does the species pool of early-establishing species affect the trajectory of recovery during restoration?
3. Does the sequence in which species are introduced to human-disturbed sites affect the successional trajectory or the speed of recovery?
4. To what degree does establishing appropriate abiotic conditions result in the establishment of the complement of native species? For example, since it would be impossible to reintroduce all soil microorganisms, does restoration of appropriate soil texture, nutrients, and temperature result in the reestablishment of the majority of species?
5. To what degree do we need to restore ecosystem processes such as nutrient cycling and water fluxes to restore native species and vice versa?
6. To what degree does the physical architecture of a community's components determine its species composition?
*7. How is the trajectory of succession during restoration affected by other human-induced changes, such as climate change, elevated CO_2, and increased N availability? For example, will altered precipitation patterns favor the establishment of seedlings of non-native species in restored areas, thereby reducing the potential for recovery of native species?

Analyze and Synthesize the Results of Species Introductions
1. What are the effects of introduction of genotypes on restoration success and surrounding populations?
2. Are there rules of thumb for maintaining local adaptations?
3. What is the influence of numbers of individuals and genetic variation of the founding population on the success of reestablishing native species?

needed on how to develop mechanisms that recognize and respond to the inherent uncertainty involved in restoration, such as some form of "restoration insurance" if restoration efforts initially appear to be unsuccessful (Holl and Howarth 2000).

Regional Planning

Given the limited funding available and the ecological consequences of piecemeal restoration, work is needed on how to better coordinate restora-

4. Can we predict the invasion probability of species from easily defined traits?
5. Can we create "designer ecosystems" composed of species that normally do not co-occur that allow us to address specific conservation goals yet do not, themselves, impose threats (e.g., the escape of non-native species)?

Consider Spatial scale

*1. How does the surrounding land-use matrix affect movement of organisms and materials into and population dynamics in restored areas and vice versa? For example, could restored areas serve as sink populations? Or, can restoration of habitat surrounding protected areas help to minimize edge effects within the protected area?
2. At what scale do we need to manage for natural temporal and spatial variability in restoration? Can we develop rules of thumb for minimum sizes of areas to restore for certain types of disturbance regimes or species?
3. What combination of tools (e.g., small-scale experiments, large-scale experiments, models) are useful in predicting success of restoring species at larger scales?

Implement and Analyze Monitoring

1. Which parameters should be selected for monitoring?
*2. To what degree do early indications of restoration success suggest long-term establishment of native species?

Consider Policy Questions

1. What funding tools are most appropriate to ensure long-term success of restoration?
2. How can legislation be designed to better accommodate the dynamic nature of ecological systems?
3. What strategies could be used to coordinate policies and legislation to aim for regional restoration plans?

Note: Those priorities marked with an asterisk are those we consider to be most important.

tion at a regional scale. For example, the Coastal Wetlands Planning, Protection and Restoration Act calls for a planning group composed of local, state, and federal government agencies; private groups; scientists; and the public (Good 1993). Each year the planning group prioritizes project proposals based on their ecological importance and cost effectiveness; the proposed projects are reviewed at public hearings and ultimately approved by the state legislature and by Congress. Although the number of regional planning efforts is increasing, numerous private and public agencies involved in restoration are still not coordinating efforts. Such coordination is particularly important in marine ecosystems, where reducing stresses to allow for ecosystem recovery necessarily means crossing numerous jurisdictional boundaries (NRC 1997).

Closing Thoughts

In 1987 Bradshaw wrote that "the successful restoration of a disturbed ecosystem is the acid test of our understanding of that system." We have made progress in the field of restoration in designing experiments to better understand controls over ecosystem structure and function (Young 2000). That understanding leads to more effective conservation techniques and strategies. Yet restoration ecology has not realized its full potential to further our knowledge of ecological processes and dynamics and consequently to contribute to conservation initiatives. As mentioned throughout this chapter, restoration ecology provides the potential to do large-scale replicated experiments, but such experiments are surprisingly few. Much basic ecological theory, such as metapopulation dynamics and the ecology of exotic invaders (chapter 4; Simberloff et al. 1999), is not widely applied in restoration, despite its effective use in conservation.

Our experience and that of others (Allen et al. 1997) suggest that one of the main issues leading to the gap between theory and practice is the separation between people who are doing research and those who are implementing restoration projects. Many government agencies and nonprofit organizations fund implementation of restoration projects with a minimal research component, even though these agencies need the research to inform their management actions. Likewise, government science-funding agencies often support research on basic principles that underlie restoration, or that could be applied to restoration, but are unlikely to provide funding to work at the scale at which many restoration projects are implemented. We suggest that partnerships between research universities and public and private land management agencies be encouraged or in some cases required. This would give researchers incentive to work on "real-world" questions and management agencies incentive to design their restoration efforts in a manner to test specific hypotheses. For example, the U.S. National Park Service, which has traditionally been in the business of promoting tourism, is increasingly working to facilitate research in national parks (NPS 2000; M. Soukup, personal communication). Building such partnerships is critical to answering many of the questions outlined above and to ensuring the long-term success of conservation efforts globally.

We hope it is clear that developments in conservation biology and restoration ecology are interdependent in this rapidly changing world. Each field of study requires and gains from intellectual movement in the other. Together they provide data and approaches that are directed to some of humankind's most challenging problems in both natural and managed systems (Daily 1995). Human needs and the ecological goals of restoration have a positive, reciprocal relationship (Geist and Galatowitsch 1999).

Finally, from our discussion, it is clear that much research is required for us to develop restoration strategies that support conservation efforts. Interestingly, such studies also add to the corpus of knowledge about the characteristics and rules that govern natural systems and should inform our efforts to conserve biodiversity in more intact ecosystems.

ACKNOWLEDGMENTS

We appreciate helpful comments on earlier drafts from E. Allen, C. D'Antonio, and L. Meyerson, as well as valuable input from other workshop participants. Linda Finchum agonized over many drafts of the manuscript.

LITERATURE CITED

Allen, E.B., W.W. Covington, and D.A. Falk. 1997. Developing the conceptual basis for restoration ecology. *Restoration Ecology* 5:275–276.

Allen, M.F. 1991. *The Ecology of Mycorrhizae.* Cambridge: Cambridge University Press.

Anderson, R.C., J.E. Schwegman, and M.R. Anderson. 2000. Micro-scale restoration: A 25-year history of a southern Illinois barrens. *Restoration Ecology* 8:296–306.

Antonovics, J., and A.D. Bradshaw. 1970. Evolution in closely adjacent plant populations. VIII. Clinal patterns at a mine boundary. *Heredity* 25:349–362.

Arnqvist, G., and D. Wooster. 1995. Meta-analysis: Synthesizing research findings in ecology and evolution. *Trends in Ecology and Evolution* 10:236–240.

Ashton, P.M.S., S. Gamage, I.A.U.N. Gunatilleke, and C.V.S. Gunatilleke. 1997. Restoration of a Sri Lankan rainforest: Using Caribbean pine *Pinus caribaea* as a nurse for establishing late-successional tree species. *Journal of Applied Ecology* 34:915–925.

Baillie, S.R., W.J. Sutherland, S.N. Freeman, R.D. Gregory, and E. Paradis. 2000. Consequences of large-scale processes for the conservation of bird populations. *Journal of Applied Ecology* 37 (Suppl. 1):88–102.

Belyea, L.R., and J. Lancaster. 1999. Assembly rules within a contingent ecology. *Oikos* 86:402–416.

Berger, J.J. 1993. Ecological restoration and nonindigenous plant species: A review. *Restoration Ecology* 1:74–82.

Bowles, M.L., and C.J. Whelan (eds.). 1994. *Restoration of Endangered Species.* Cambridge: Cambridge University Press.

Boyer, K.E., and J.B. Zedler. 1998. Effects of nitrogen additions on the vertical structure of a constructed cordgrass marsh. *Ecological Applications* 8:692–705.

Bradshaw, A.D. 1984. Ecological significance of genetic variation between populations. In R. Dirzo, and J. Sarukhan (eds.), *Perspectives on Plant Population Ecology,* pp. 213–228. Sunderland, Mass.: Sinauer Associates.

Bradshaw, A.D. 1987. Restoration: An acid test for ecology. In W.R.I. Jordan, M. Gilpin, and J.D. Aber (eds.), *Restoration Ecology,* pp. 23–29. Cambridge: Cambridge University Press.

Bradshaw, A.D. 1997. What do we mean by restoration? In K.M. Urbanska, N.R. Webb, and P.J. Edwards (eds.), *Restoration Ecology and Sustainable Development.* chapter 2, pp. 8–14. Cambridge: Cambridge University Press.

Brenner, F.J., M. Werner, and J. Pike. 1984. Ecosystem development and natural succession in surface coal mine reclamation. *Minerals and Environment* 6:10–22.

Bridgewater, P.B. 1988. Synthetic plant communities: Problems in definition and management. *Flora* 180:139–144.

Brown, J.S. 1994. Restoration ecology: Living with the prime directive. In M.L. Bowles and C.J. Whelan (eds.), *Restoration of Endangered Species,* pp. 355–380. Cambridge: Cambridge University Press.

Burger, J.A., and J.L. Torbert. 1990. Mined land reclamation for wood production in the Appalachian region. In J. Skousen, J. Sencindiver, and D. Samuel (eds.), *Proceedings of the 1990 Mining and Reclamation Conference and Exhibition,* vol. 1, pp. 159–163. Morgantown, W.V.: West Virginia University.

Butterfield, R.P., and R.F. Fisher. 1994. Untapped potential for native species reforestation. *Journal of Forestry* 92:37–40.

Cairns J. Jr., 1989. Restoring damaged ecosystems: Is predisturbance condition a viable option? *Environmental Professional* 11:152–159.

Cairns Jr., J., and J.R. Heckman. 1996. Restoration ecology: The state of an emerging field. In R.H.Socolow (ed.), *Annual Review of Energy and the Environment,* vol. 21, pp. 167–187.

Chapman, C.A., and L.J. Chapman. 1996. Exotic tree plantations and the regeneration of natural forests in Kibale National Park, Uganda. *Biological Conservation* 76:253–257.

Chapman, M.G. 1999. Improving sampling designs for measuring restoration in aquatic habitats. *Journal of Aquatic Ecosystem Stress and Recovery* 6:235–251.

Clements, F.E. 1916. *Plant Succession: An Analysis of the Development of Vegetation.* Carnegie Institution of Washington, Publication Number 242, 1–512.

Clewell, A., and J.P. Rieger. 1997. What practitioners need from restoration ecologists. *Restoration Ecology* 5:350–354.

Collins, S.L. 2000. Disturbance frequency and community stability in native tallgrass prairie. *American Naturalist* 155:311–325.

Daily, G.C. 1995. Restoring value to the world's degraded lands. *Science* 269:350–354.

D'Antonio, C.M., T.L. Dudley, and M. Mack. 1999. Disturbance and biological invasions: Direct effects and feedbacks. In L. Walker (ed.), *Ecosystems of Disturbed Ground,* pp. 413–452. New York: Elsevier.

Diamond, J.M. 1975. Assembly of species communities. In M.L. Cody and J.M. Diamond (eds.), *Ecology and Evolution of Communities,* pp. 342–444. Cambridge, Mass.: Harvard University Press.

Egerton-Warburton, L.M., and E.B. Allen. 2000. Shifts in arbuscular mycorrhizal communities along an anthropogenic nitrogen deposition gradient. *Ecological Applications* 10:484–496.

Ehrenfeld, J.G. 2000. Defining the limits of restoration: The need for realistic goals. *Restoration Ecology* 8:2–9.

Ehrenfeld, J.G., and L.A. Toth. 1997. Restoration ecology and the ecosystem perspective. *Restoration Ecology* 5:307–317.

Francis, G.R., J.J. Magnuson, H.A. Regier, and D.R. Talhelm. 1979. *Rehabilitating Great Lakes Ecosystems.* Ann Arbor, Mich.: Great Lakes Fishery Commission.

Franklin, J.F., and J.A. MacMahon. 2000. Perspectives. *Science* 288:1183–1185.

Franklin, J.F., J.A. MacMahon, F.J. Swanson, and J.R. Sedell. 1985. Ecosystem responses to the eruption of Mount St. Helens. *National Geographic Research* 1:198–216.

Geist, C., and S.M. Galatowitsch. 1999. Reciprocal model for meeting ecological and human needs in restoration projects. *Conservation Biology* 13:970–979.

Goldstein, P.Z. 1999. Functional ecosystems and biodiversity buzzwords. *Conservation Biology* 13:247–255.

Good, B. 1993. Louisiana's wetlands. *Restoration and Management Notes* 11:125–133.

Greenberg, R., P. Bichier, A.C. Angon, and R. Reitsma. 1997. Bird populations in shade and sun coffee plantations in central Guatemala. *Conservation Biology* 11:448–459.

Handel, S.N. 1997. The role of plant-animal mutualisms in the design and restoration of natural communities. In K.M. Urbanska, N.R. Webb, and P.J. Edwards (eds.), *Restoration Ecology and Sustainable Development,* chapter 7, pp. 111–132. Cambridge: Cambridge University Press.

Hansson, L., L. Fahrig, and G. Merriam. 1995. *Mosaic Landscapes and Ecological Processes.* New York: Chapman and Hall.

Harmon, M.E., J.F. Franklin, F.J. Swanson, P. Sollins, S.V. Gregory, J.D. Lattin, N.H. Anderson, S.P. Cline, N.G. Auman, J.R. Sedell, G.W. Lienkaemper, K. Cromack Jr., and K.W. Cummins. 1986. Ecology of coarse woody debris in temperate ecosystems. *Advances in Ecological Research* 15:133–302.

Hawkins, C.P., and J.A. MacMahon. 1989. Guilds: The multiple meanings of a concept. *Annual Review of Entomology* 34:423–451.

Hobbs, R.J. 1993. Can revegetation assist in the conservation of biodiversity in agricultural areas? *Pacific Conservation Biology* 1:29–38.

Hobbs, R.J., and D.A. Norton. 1996. Towards a conceptual framework for restoration ecology. *Restoration Ecology* 4:93–110.

Hobbs, R.J., D.A. Saunders, and G. Arnold. 1993. Integrated landscape ecology: A western Australian perspective. *Biological Conservation* 64:231–238.

Holl, K.D. 1996. Restoration ecology: Some new perspectives. In A. Noble and R. Noble (eds.), *Preservation of Natural Diversity in Transboundary Protected Areas: Research Needs/Management Options,* pp. 25–35. Washington, D.C.: National Research Council.

Holl, K.D., and J. Cairns Jr. 1994. Vegetational community development on reclaimed coal surface mines in Virginia. *Bulletin of the Torrey Botanical Club* 121:327–337.

Holl, K.D., and R.B. Howarth. 2000. Paying for restoration. *Restoration Ecology* 8:260–267.

Holl, K.D., and M. Kappelle. 1999. Tropical forest recovery and restoration. *Trends in Ecology and Evolution* 14:378–379.

Hooper, D.U., and P.M. Vitousek. 1997. The effects of plant composition and diversity on ecosystem processes. *Science* 277:1302–1305.

Huenneke, L.F. 1991. Ecological implications of genetic variation in plant populations. In D.A. Falk and K.E. Holsinger (eds.), *Genetics and Conservation of Rare Plants*, pp. 31–44. New York: Oxford University Press.

Jackson, L.L., N. Lopoukhine, and D. Hillyard. 1995. Ecological restoration: A definition and options. *Restoration Ecology* 3:71–75.

Keddy, P. 1999. Epilogue: From global exploration to community assembly. In E. Weiher and P. Keddy (eds.), *Ecological Assembly Rules: Perspectives, Advances, Retreats*, pp. 393–402. Cambridge: Cambridge University Press.

Kettle, W.D., P.M. Rich, K. Kindscher, G.L. Pittman, and P. Fu. 2000. Land-use history in ecosystem restoration: A 40-year study in the prairie-forest ecotone. *Restoration Ecology* 8:307–317.

Knapp, E.E., and K.J. Rice. 1996. Genetic structure and gene flow in *Elymus glaucus* (blue wildrye): Implications for native grassland restoration. *Restoration Ecology* 4:1–10.

Knapp, E.E., and K.J. Rice. 1998. Comparison of isozymes and quantitative traits for evaluating patterns of genetic variation of purple needlegrass (*Nassella pulchra*). *Conservation Biology* 12:1031–1041.

Koebel Jr., J.W. 1995. An historical perspective on the Kissimmee River Project. *Restoration Ecology* 3:149–159.

Kuusipalo, J., G. Adjers, Y. Jafarsidik, A. Otsamo, K. Tuomela, and R. Vuokko. 1995. Restoration of natural vegetation in degraded *Imperata cylindrica* grassland: Understorey development in forest plantations. *Journal of Vegetation Science* 6:205–210.

Larcher, W. 1995. *Physiological Plant Ecology*. New York: Springer-Verlag.

Lesica, P., and F.W. Allendorf. 1999. Ecological genetics and the restoration of plant communities: Mix or match? *Restoration Ecology* 7:42–50.

Levine, J.M. 2000. Species diversity and biological invasions: Relating local process to community pattern. *Science* 288:852–854.

Lippit, L., M.W. Fidelibus, and D.A. Bainbridge. 1994. Native seed collection, processing, and storage for revegetation projects in the western United States. *Restoration Ecology* 2:120–131.

Lockwood, J.L. 1997. An alternative to succession. *Restoration and Management Notes* 15:45–50.

Lockwood, J.L., and S.L. Pimm. 1999. When does restoration succeed? In E. Weiher and P. Keddy (eds.), *Ecological Assembly Rules: Perspectives, Advances, Retreats*, pp. 363–392. Cambridge: Cambridge University Press.

Louda, S.M. 1994. Experimental evidence for insect impact on populations of short-lived, perennial plants, and its application in restoration ecology. In M.L. Bowles and C.J. Whelan (eds.), *Restoration of Endangered Species,* pp. 118–138. Cambridge: Cambridge University Press.

MacMahon, J.A. 1997. Ecological restoration. In G.K. Meffe and C.R. Carroll (eds.), *Principles of Conservation Biology,* 2nd ed,. chapter 14, pp. 479–511. Sunderland, Mass.: Sinauer Associates.

MacMahon, J.A. 1998. Empirical and theoretical ecology as a basis for restoration: An ecological success story. In M. Pace and P. Groffman (eds.), *Successes, Limitations, and Frontiers in Ecosystem Science,* pp. 220–246. New York: Springer-Verlag.

Maina, G.G., and H.F. Howe. 2000. Inherent rarity in community restoration. *Conservation Biology* 14:1335–1340.

Maron, J.L., and P.G. Connors. 1996. A native nitrogen-fixing shrub facilitates weed invasion. *Oecologia* 105:302–312.

Michener, W.K. 1997. Quantitatively evaluating restoration experiments: Research design, statistical analysis, and data management considerations. *Restoration Ecology* 5:324–337.

Mitsch, W.J., X. Wu, R.W. Nairn, P.E. Weihe, N. Wang, R. Deal, and C.E. Boucher. 1998. Creating and restoring wetlands: A whole-ecosystem experiment in self-design. *BioScience* 48:1019–1030.

Montalvo, A.M., S.L. Williams, K.J. Rice, S.L. Buchmann, C. Cory, S.N. Handel, G.P. Nabhan, R. Primack, and R.H. Robichaux. 1997. Restoration biology: A population biology perspective. *Restoration Ecology* 5:277–290.

Moore, M.M., W.W. Covington, and P.Z. Fule. 1999. Reference conditions and ecological restoration: A southwestern ponderosa pine perspective. *Ecological Applications* 9:1266–1277.

Nijs, I., and I. Impens. 2000. Biological diversity and probability of local extinction of ecosystems. *Functional Ecology* 14:46–54.

NPS (National Park Service). 2000. *Cooperative ecosystem studies units.* http://www.cesu.org/cesu/introduction/brochure.html. Accessed April 5, 2000.

NRC (National Research Council). 1992. *Restoration of Aquatic Ecosystems.* Washington, D.C.: National Academy Press.

NRC (National Research Council). 1994. *Restoring and Protecting Marine Habitat.* Washington, D.C.: National Academy Press.

NRC (National Research Council). 1997. *Striking a Balance: Improving Stewardship of Marine Areas.* Washington, D.C.: National Academy Press.

Pacific Estuarine Research Laboratory (PERL). 1990. *A Manual for Assessing Restored and Natural Coastal Wetlands with Examples from Southern California.* California Sea Grant Report No. T-CSVCP-021, La Jolla, California.

Palik, B.J., P.C. Goebel, L.K. Kirkman, and L. West. 2000. Using landscape hierarchies to guide restoration of disturbed ecosystems. *Ecological Applications* 10:189–202.

Palmer, M.A., R.F. Ambrose, and N.L. Poff. 1997. Ecological theory and community restoration ecology. *Restoration Ecology* 5:291–300.

Parker, V.T., and S.T.A. Pickett. 1997. Restoration as an ecosystem process: Implications of the modern ecological paradigm. In K.M. Urbanska, N.R. Webb, and

P.J. Edwards (eds.), *Restoration Ecology and Sustainable Development,* chapter 3, pp. 17–32. Cambridge: Cambridge University Press.

Pavlovic, N.B. 1994. Disturbance-dependent persistence of rare plants: Anthropogenic impacts and restoration implications. In M.L. Bowles and C.J. Whelan (eds.), *Restoration of Endangered Species,* pp. 159–193. Cambridge: Cambridge University Press.

Perfecto, I., R.A. Rice, R. Greenberg, and M.E. Van der Voort. 1996. Shade coffee: A disappearing refuge for biodiversity. *BioScience* 46:598–608.

Pickett, S.T.A., and J.N. Thompson. 1978. Patch dynamics and the design of nature reserves. *Biological Conservation* 13:27–37.

Pimentel, D., L. Lach, R. Zuniga, and D. Morrison. 2000. Environmental and economic costs of nonindigenous species in the United States. *BioScience* 50:53–65.

Pysek, P. 1998. Is there a taxonomic pattern to plant invasions? *Oikos* 82:282–294.

Reichard, S.H., and C.W. Hamilton. 1997. Predicting invasions of woody plants introduced into North America. *Conservation Biology* 11:193–203.

Risser, P.G. 1999. Examining relationships between ecosystem function and biodiversity: Reply to Goldstein. *Conservation Biology* 13:438–439.

Roberts, D.L., R.J. Cooper, and L.J. Petit. 2000. Use of premontane moist forest and shade coffee agroecosystems by army ants in western Panama. *Conservation Biology* 14:192–199.

Ruesink, J.L., I.M. Parker, M.J. Groom, and P.M. Kareiva. 1995. Reducing the risks of nonindigenous species introductions. *BioScience* 45:465–477.

Schultz, C.B. 1998. Dispersal behavior and its implications for reserve design in a rare Oregon butterfly. *Conservation Biology* 12:284–292.

Schultz, C.B., and E.E. Crone. 1998. Burning prairie to restore butterfly habitat: A modeling approach to management tradeoffs for the Fender's blue. *Restoration Ecology* 6:244–252.

Simberloff, D.J., D. Doak, M. Groom, S. Trombulak, A. Dobson, S. Gatewood, M.E. Soulé, M. Gilpin, C. Martinez del Rio, and L. Mills. 1999. Regional and continental restoration. In M.E. Soulé and J. Terborgh (eds.), *Continental Conservation: Scientific Foundations of Regional Reserve Networks,* pp. 65–98. Washington, D.C.: Island Press.

Snodgrass, J.W., M.J. Komoroski, A.L. Bryan Jr., and J. Burger. 2000. Relationships among isolated wetland size, hydroperiod, and amphibian species richness: Implications for wetland regulations. *Conservation Biology* 14:414–419.

Soulé, M.E., and K.A. Kohm. 1989. *Research Priorities in Conservation Biology.* Washington, D.C.: Island Press.

Sparks, R.E., J.C. Nelson, and Y. Yin. 1998. Naturalization of the flood regime in regulated rivers. *BioScience* 48:706–720.

Stohlgren, T.J., D. Binkley, G.W. Chong, M.A. Kalkhan, L.D. Schell, K.A. Bull, Y. Otsuki, G. Newman, M. Bashkin, and Y. Son. 1999. Exotic plant species invade hot spots of native plant diversity. *Ecological Monographs* 69:25–46.

Sutter, R.D. 1996. Monitoring. In D.A. Falk, C.I. Millan, and M. Olwell (eds.), *Restoring Diversity: Strategies for Reintroduction of Endangered Plants,* pp. 235–264. Washington, D.C.: Island Press.

Tilman, D., J. Knops, D. Wedin, P. Reich, M. Ritchie, and E. Siemann. 1997. The influence of functional diversity and composition on ecosystem processes. *Science* 277:1300–1302.

USFWS (United States Fish and Wildlife Service). 1999. *Partners for Fish and Wildlife Program: Cumulative Accomplishments.* http://www.fws.gov/r9dhcpfw/ ACCOMPS/Accomps.htm. Accessed March 11, 1999.

Walker, B. 1999. The ecosystem approach to conservation: Reply to Goldstein. *Conservation Biology* 13:436–437.

Weiher, E., and P. Keddy (eds.). 1999. *Ecological Assembly Rules: Perspectives, Advances, Retreats.* Cambridge: Cambridge University Press.

White, P.S., and S.T.A. Pickett. 1985. Natural disturbance and patch dynamics: An introduction.In S.T.A. Pickett and P.S. White (eds.), *The Ecology of Natural Disturbance and Patch Dynamics,* pp. 3–13. New York: Academic Press.

White, P.S., and J.L. Walker. 1997. Approximating nature's variation: Selecting and using reference information in restoration ecology. *Restoration Ecology* 5:338–349.

Woodward, F.I. 1992. A review of the effects of climate on vegetation: Ranges, competition, and composition. In R.L. Peters and T.E. Lovejoy (eds.), *Global Warming and Biological Diversity,* pp. 105–123. New Haven, Conn.: Yale University Press.

Young, T.P. 2000. Restoration ecology and conservation biology. *Biological Conservation* 92:73–83.

Zedler, J.B., and J.C. Callaway. 1999. Tracking wetland restoration: Do mitigation sites follow desired trajectories? *Restoration Ecology* 7:69–73.

12

CONSERVATION BIOLOGY RESEARCH
Its Challenges and Contexts

Michael E. Soulé and Gordon H. Orians

Eighty percent of success is showing up.
—Woody Allen

What is conservation biology for? The conventional answer to that question has been that its purpose is to provide guidance to conservationists, so that they can do their work of protecting nature armed with the best available tools and knowledge. That answer reflects the belief that the main task of conservationists is to design, construct, and manage protected areas, a belief based on the correct perception that conservation of nature requires whole ecological systems, with their dynamic, changing, complex interactions over space and through time.

Reserves have two main functions in a conservation strategy. First, they can represent and house the biodiversity of a region. Second, they can shield the biodiversity they contain from processes that may threaten its persistence. However, reserves can fulfill these roles only if they contain a relatively complete, unbiased sample of a region's biodiversity and if they are large enough to protect their contained biodiversity from external threats.

Unfortunately, existing reserves contain a highly biased sample of Earth's ecosystems (Margules and Pressey 2000). For example, most of the areas that are protected in the United States have poor soils or low rainfall or are high-elevation terrain covered by rock and ice (Scott et al. 2000). In other words, the least valuable and productive areas have been favored for protection. Similarly, many marine protected zones have been established in areas with little commercial value, either because they were originally areas of low productivity or because they had already been fished out. Also, there are too few reserves and they are too small (see the preface). Important efforts are underway to correct for these deficiencies, but they are likely to achieve only partial success.

Therefore, although protected areas are the essential core of conservation strategies, by themselves they are not sufficient for the task of preserving nature. Some have argued that we also need to change the ways we use lands and waters that are not given protected status, the so-called matrix lands, which must play a more important role in conservation than they have heretofore. Unfortunately, the intensification of economic uses of nonprotected lands is steadily decreasing their value for nature protection and their use as buffer zones (Groom et al. 1999). Other components of the conservation toolbox, such as *ex situ* conservation facilities (zoos, aquariums, botanic gardens), have value, but their overall contributions are more likely to be as adjuncts and suppliers to protected areas and in furthering public education than in protecting nature per se. In part, this is because these institutions are severely limited in the number of species they can house or grow. Frozen collections of gametes, other cells, DNA, etc., have much value for comparative research, but they don't preserve ecological and evolutionary processes.

Unfortunately, the public and the decision makers they elect poorly understand how all of these components are related in an overall conservation strategy. Therefore, conservationists, and their partners (conservation biologists, funders, some philosophers and theologians, etc.), have the additional responsibility of educating decision makers so that their policies are better informed. This a relatively new arena for conservation biologists and one that demands new, bold initiatives.

New Directions in Conservation Biology: Crosscutting Themes

One of the changes that has occurred in conservation biology during the last two decades, and is reflected in the research priorities identified in many chapters of this book, is the growing appreciation of the complexity of interactions in nature and of the varied spatial and temporal scales and contexts in which they operate. Twenty years ago the tendency was to study relatively pristine systems in isolation from the human and global matrix in which they were imbedded. Today there is a developing awareness of the complex feedbacks at all levels of biological organization, between relatively remote parts of the earth, between developed lands and wildlands, and between human beings and nonhuman biotas.

Complex, multiscale, multidisciplinary, and global interactions have motivated many of the research questions and priorities that are prominent in this volume. For heuristic reasons, here we classify many of the issues under the heading of "crosscutting" themes because they straddle many subdisciplines. Some of these themes, which are now prominent in the thinking of conservation biologists, were hardly mentioned in the earlier volume on this subject (Soulé and Kohm 1989). Therefore, they represent a considerable advance in the field. The most important crosscutting themes include the following:

- positive feedbacks and multiple direct and indirect effects;
- the ecological and cultural context of issues (the "it depends" problem);
- lag effects;
- cumulative effects;
- the role of the "matrix" of inhabited and exploited lands that surround protected areas;
- introduced, invasive exotic species;
- the complexity of pollutants (including such agents as hormonally active substances);
- the role of ecological restoration, particularly at regional scales;
- the identification and use of focal species; and
- monitoring and evaluation.

Although these categories overlap, they nonetheless highlight many of the emerging themes in conservation biology. In the following sections, we elaborate on these themes.

Positive Feedbacks and Multiple Direct and Indirect Effects

The chapters of this book contain frequent references to feedbacks between processes that affect ecosystems and populations at all scales. At the local or population scale, for example, there are interactions between genetics, demography, and environmental variation that affect the viability of populations and that require attention (chapters 2 and 5).

At an intermediate (regional) scale, we are becoming more aware of cascading interactions between trophic levels in food webs and the need to clarify the relative frequencies of strong and weak ecological interactions among species (chapter 3). For example, there is increasing interest in top carnivores and other strongly interacting species and the consequences of their disappearance for the diversity and resilience of ecosystems (Terborgh et al. 1999).

Some conservation biologists are calling for regional and continental networks of wildlands that specifically depend on the stabilizing presence of large carnivores (Soulé and Noss 1998). They are declaring that the current criteria for recovery of endangered species fail to acknowledge the need to restore ecological interactions of "keystone" species throughout much of their former ranges (Soulé, et al. in preparation).

At continental and global scales, conservation biologists are increasingly concerned about the likely interactions between sea surface temperatures in the Pacific (El Niño and La Niña), the devastating fires in tropical rain forests, and the role that logging plays in exacerbating those fires. The possibility that deforestation will cross a threshold that could produce a biologically devastating and economically disastrous decline in precipitation in the Amazon Basin is considered a real one.

Other kinds of complex interactions between spatially separated marine and terrestrial ecosystems are becoming more apparent. The important role of salmon, for example, as conveyors of oceanic nutrients and calories to temperate forests is now gaining attention and causing concern as salmon disappear and nitrogen subsidies from marine to terrestrial ecosystems are reduced. This example is representative of the growing literature on nutrient and organic matter exchanges between ecosystems and on the global transport of pollutants, pathogens, and, possibly, hormonally active substances (chapter 8).

The Ecological and Cultural Context of Issues

Because natural systems (ecosystems) are heterogeneous, infinitely variable, nonlinear, dynamic, open, corruptible by invasion by exotic species, frequently chaotic, and strongly influenced by both nearby and remote processes and human activities, they cannot be managed successfully by a "cookbook" approach. Thus, simple rules of thumb have limited value (chapter 10). This means that conservationists who are designing and managing protected area networks have few useful generalizations available to guide them. And the practice of conservation—not to mention the development of the science that supports it—is likely to become even more complex in the future.

Therefore, the answer to most questions about nature protection is, unfortunately, "It depends." For example, it is evident that the nature and penetration of edge effects depend not only on local human culture but also on landscape characteristics (chapter 5). The "it depends" dilemma plagues ecological management and is one of the reasons for the frequent plea for adaptive management (chapter 10). We hasten to add, however, that this growing realization is healthy, because progress is unlikely if we continue to believe that complex, embedded systems are amenable to simplistic descriptions and formulaic prescriptions.

For example, conservation biologists are often asked about the ecological and architectural rules that should govern the design and placement of wildlife corridors: how long? how wide? should they follow ridge tops or watercourses? should they provide food and shelter? should they avoid roads and highways? Such obvious questions need to be evaluated and answered in two different ways. One concerns basic biology—that is, the specific functions of the linkages and the natural history of the target species. A useful literature to guide the design of corridors based on biological criteria is rapidly developing (Earn, Levin, and Rohani 2000). Similar issues arise when considering questions related to the design of protected areas.

The other perspective concerns the context—both biophysical and cultural—on which many decisions hinge. For example, even if it were established that individuals of the target species would actually use the corridors, it might not be advisable to establish them because a high potential for poaching, a cultural factor, might exist. If poaching is now (or is likely to become) a threat to the survival of animals using a corridor for dispersal or migration, then human behavior must be considered as a major element in determining the desirability of establishing corridors, their placement, and their dimensionality. Finally, it is necessary to consider how the context, including climate, human technologies, and land uses, might change in the future.

The research priorities identified in some of the previous chapters also reflect the recognition that research results that are fundamentally important for conservation may not be used unless the results are cast in ways that help managers make their on-the-ground decisions. Note in this regard that the Society for Conservation Biology is now publishing a new journal, *Conservation Biology in Practice,* that responds to this need.

Another problem for managers is that the optimal solutions suggested by basic research may not be feasible for a variety of reasons. Therefore, managers need help in using scientific information to make choices from among the set of suboptimal options available to them. In other words, conservation biologists need to frame their theories and empirical results in ways that deliver effective decision-making tools that can be used by managers (chapters 2 and 10). This is a major challenge in conservation biology.

Lag Effects

There is increasing recognition that many of the consequences of land conversion and environmental change do not materialize until well after the changes have happened. That is, there are often important lag effects. For example, small, isolated populations of long-lived species may not actually disappear for decades after their fate has been sealed (chapters 2, 3, and 5). And exotic species may not become invasive for several years or decades after being introduced (chapter 4). Moreover, the indirect, cascading effects of extinctions may not be detectable for long periods (chapter 3). In the oceans, the disappearance of large pelagic fish and great whales, resulting from over-fishing, including bycatch, can cause systems to switch between stable states, effects that may also have long lags (e.g., chapter 7). Decision-theory approaches can provide guidance where the consequences of management actions and inactions are expressed only after long delays. For example, a fire suppression policy has serious consequences, but they may not be expressed in time for human intuition to perceive, much less provide, useful solutions to them (chapter 10).

Cumulative Effects

Most environmental problems are not generated primarily by single large perturbations—the smoking gun scenario. Rather, they are the result of many, often unrelated, insults that combine to cause major changes—the "death from a thousand small wounds" syndrome. Acid precipitation is caused not by the discharges from one power plant but by the discharges from many of them. Extinction is caused not by the alteration of one small tract of land but by the progressive loss of habitat, small piece by small piece. Government agencies that manage public lands often ignore deleterious and additive or multiplicative impacts caused by different kinds of land-use practices, or spillover effects on adjacent lands. Instead, agencies tend to focus on the potential short-term effects of a new proposal or use of a partic-ular site (chapter 2). Different kinds of disturbances may also combine to yield deleterious effects that are more serious and qualitatively different from the effects caused by any one of them. In the seas, cumulative effects are often the consequence of overfishing combined with pollution, and exacer-bated by pathogens that may be more virulent in physiologically weakened hosts (chapters 7 and 8).

The Role of the "Matrix" of Inhabited and Exploited Lands That Surround Protected Areas

One of the consequences of the rapid growth in human numbers is a general ecological deterioration of the so-called matrix lands that support the human economy and that surround protected areas. These areas have the potential to support many species and are, therefore, relevant components of effective conservation strategies. Unfortunately, matrix lands are being converted into ecological sinks for most species (chapter 5). In the oceans, overfishing by both industrially advanced and impoverished developing nations is devastating huge areas on continental shelves, and coral reefs are being destroyed by the use of dynamite, cyanide, and bleach fishing (chapter 7).

Nonprotected areas may be managed so intensively for production of marketable commodities that little habitat remains for wildlife, and what habitat exists is increasingly harsh. Moreover, the new communication technologies that allow people to carry on their professional activities from rural areas such as Bozeman, Montana, for example, and the wealth that enables people to build vacation homes in such remote places are contributing to a new kind of sprawl (e.g., "ranchette pox" or "bungalow blight") that is more detrimental to wildlife than the farms and ranches it is replacing. These forces, in addition to ordinary uses of matrix lands, including agricultural intensification, are reducing the utility of buffers around protected areas.

Introduced, Invasive Exotic Species

Many observers believe that invasive exotic species are overtaking habitat loss and fragmentation as the major threat to biodiversity. Combating invasive species is one of the most challenging problems faced by marine, aquatic, and terrestrial conservationists (chapters 4, 6, and 7). We still understood too little about the dynamics and interactions of invaders and how they come to dominate ecosystems. And there is increasing concern about species used in mariculture, whether or not they have been genetically modified, that are cultivated in areas remote from their native ranges. For example, Atlantic salmon, which are farmed extensively on the Pacific Coast of North America, have escaped in large numbers, and reproductively mature individuals are now found in some of the region's rivers.

The Complexity of Pollutants

As noted above, pollutants can sometimes exacerbate other drivers (climate change, disease) in the destruction of biodiversity. We need to know more about how pollutants affect food webs (chapter 3) and ecosystem productivity. The oceans are the ultimate recipients of materials introduced into terrestrial and freshwater ecosystems. Coastal and continental-shelf marine ecosystems receive large quantities of materials derived from agricultural lands, among which nitrogen is especially important. In some places the resulting eutrophication is producing expanding dead zones of decomposing algae and plant material (chapters 6, 7, and 9). The threats to birds and other organisms from persistent pollutants are increasing in many parts of the world. And we know far too little about the ecological consequences of industrial chemicals, including the constituents of plastics (chapter 8).

The Role of Ecological Restoration, Particularly at Regional Scales

Restoration ecology is an increasingly important discipline, and attention is shifting from restoring relatively small, highly disturbed sites, where non-native plants were frequently used, to restoring natural assemblages and interactions of species over large areas (chapter 11). Increasingly, the focus in conservation will be restoration at regional scales (Ricketts et al. 1999; Simberloff et al. 1999), the reintroduction of native animals having large spatial requirements, the effects of large-scale disturbances on food webs (chapter 3), and jurisdictional issues between agencies with overlapping responsibilities (chapter 10).

The Identification and Use of Focal Species

There is increasing interest in how to identify and use organisms that can serve as surrogates for the integrity, diversity, and resilience of ecosystems (chapter 2). Although current methods are varied and in a state of flux, species judged to be useful include those that signal habitat quality, keystone species (chapter 3), umbrella species, flagship species (chapter 2), and sentinel species (for early detection of new diseases, chapter 8). Focal species are also increasingly used as key elements in monitoring protocols (chapter 10).

Monitoring and Evaluation

Long-term monitoring is known to be essential for detecting trends, identifying emerging problems, testing hypotheses, and evaluating the efficacy of management interventions, but proposals for funding long-term monitoring are often not competitive with requests for experiments designed to advance the frontiers of knowledge. Fortunately, long-term monitoring is likely to receive increasing support because of the capabilities of new remote sensing and spectral analysis technologies. Some of these technologies are likely to be quickly adopted for monitoring ecological changes (chapter 10). For a few remotely sensed measurements, twenty-five-year time series of high-quality data of nearly global coverage already exist (NRC 2000a). For measurements over large spatial scales, remote sensing offers, for the first time, an affordable means of sampling.

Research Priorities as Checklists for Conservationists

We stated above that the complexity of nature severely limits the types and scopes of useful generalizations that are likely to be generated by conservation biology research in the near future. So, while the unsatisfying answer "It depends" will persist, it behooves conservation biologists to point out, as well, that the relevant variables and conditions (that "it depends" on) are accessible; they are often obvious to competent ecologists and wildlife biologists. In fact, the research priorities found at the end of each of the core chapters in this book can be easily converted to checklists of factors that should be considered by designers and managers of protected areas when faced with conservation problems. Therefore, conservationists can use this book as a guide to the factors and phenomena that are likely to be germane to their work, even if the chapters do not offer simple answers to the questions being asked.

The Sociopolitical Context in Which Research Priorities Are Implemented

Scientists, such as the authors of this book, who think about research priorities for conservation biology, assume that good science is necessary, or at least helpful, for good conservation. And many scientists believe that they are the ones to decide on the ranking of research questions. Certainly, a consensus on research priorities can advance the field by guiding the funding decisions of private and public foundations looking for ways to leverage dollars into knowledge that will help save nature during the current human-caused sixth huge extinction event. In addition, research priorities aid both young, passionate conservation biologists and older ecologists who are in a state of despair because their discipline is rapidly losing its subject matter.

Nevertheless, it is neither the scientists nor the on-the-ground conservationists who decide what research will be funded. Actually, the relative priorities of research themes identified by scientists typically are only one, and a minor one at that, of the inputs to the processes by which research funds are allocated. In democratic societies, politicians are charged with making such decisions.

Most readers of this book believe that knowledge—and, therefore, research—is a good thing. It is well to bear in mind, however, that this attitude is not universal. People's decisions can be more strongly motivated by expediency and self-interest than by knowledge, especially when the facts do not support their beliefs and desires. Politicians are lobbied every day to ignore the best science available because the answers given by science do not always serve vested interests. Because politicians respond to pressures from multiple agendas that have different levels of financial backing, the discourse on national research priorities is often dominated by fear and short-term (getting reelected and the desire for control) and longer-term exigencies (security, and healthy, long lives). These pressures result in much more research money being allocated to studies on human diseases and weaponry than to solving environmental problems. This is one of the reasons most scientific disciplines have found it necessary to have offices in national capitals.

Most scientists, either by choice or by default, are poorly informed about political processes. They may grumble about the shortage of research funds;

they may even be indignant that funding agencies and foundations don't share their views. Sadly, though, few scientists engage in political processes directly. There are many reasons for this reluctance, some of them good ones. But the cost of scientists' withdrawal from politics, whether conscientious or selfish, ignorant or thoughtful, is that policies continue to be more strongly influenced by those who show up when and where decisions are being made. If conservation biologists hope to see their knowledge recognized and their agendas funded, they will have to join other scientists who have learned the truth of the Woody Allen adage that appears at the opening of this chapter.

It is simply not enough for scholars to deliberate and agree on research priorities. They must also organize themselves to lobby for their funding and to educate decision makers about the merits of making scientifically informed conservation and natural resource decisions and also about the costs of failing to do so. A consensus on research priorities is just one step, albeit an essential one, in a more complex sociopolitical process—one that is characterized by multiple agendas and multiple motivations, not all of which are honorable. These elements cannot be eliminated, but they can sometimes be neutralized by an energetic, persistent educational and lobbying campaign.

The point is that unless decision makers are educated (and lobbied) about the long-term benefits of nature-friendly policy options, and thereby develop an appreciation of the ethical, spiritual, psychological, physical, and economic values of nature, they will continue to support programs that are harmful for both society and nature.

Sustainable Development and Conservation

The dominant view in western societies today is that economic growth, provided that it is "sustainable," is essential for raising the quality of life for people. Furthermore, it is widely assumed that encouraging trade is the most powerful way of stimulating economic growth. For the last two decades, the notion of sustainable development has replaced, in international conservation and development circles, the older idea of protected areas—such as national parks—as the mainstay of biodiversity protection. One result is that governments and banks that control development in the tropics and elsewhere are now encouraging the "sustainable use" of protected areas, and the extraction of commodities from them for the global market, notwithstanding the rapid accumulation of evidence that exploitation and protection may, in many cases, be incompatible (Terborgh 1999).

For example, currently, there is a movement in the New World tropics to subsidize the colonization of national parks by rural and indigenous peoples. This exacerbates the problem that many national parks are "paper parks" that lack effective enforcement of laws and regulations essential for their protection. At the same time, tropical forests within a few kilometers of roads are silenced by the extirpation of primates, birds, and other edible, marketable animals (Redford 1992).

Thus, although free trade and economic growth have the potential to serve conservation agendas, they have equally great potential to undermine efforts to preserve Earth's biodiversity. Currently, the latter process is dominant. A frequent consequence of free trade is the conversion of important ecosystems (such as coastal estuaries and tropical forests) to aquaculture or livestock growing. The destruction and degradation of these habitats often are irreversible. In addition, economically driven decisions are tacitly made to maximize short-term profits, whether or not they are accompanied by unsustainable practices.

Sustainability is an essential vision for the future relations between humankind and its environment. It is yet to be shown, however, that sustainable development, as currently financed and practiced by institutions engaged in international lending and development assistance, is benefiting nature. Most surveys suggest the opposite (see Soulé 2001 for a review). It is time for conservation biologists to speak out on this subject, arguing constructively for the kinds of institutions and facilities that will be needed to minimize the grievous destruction being caused by population growth, rapidly proliferating technologies, and an increasingly globalized economy.

The Future

Even though human behavior poses the fundamental threat to biodiversity preservation, only people can act to reduce the threats to Earth's biodiversity. Therefore, we believe that there is a need for another workshop and book of this kind that focuses explicitly on the "human dimension" of the biodiversity-nature crisis. This will be a difficult workshop because the social and economic challenges (poverty, disease, injustice, corruption, collapsing infrastructure, fatalism, and so on) are formidable. Moreover, there is a tendency for discussions of the human dimension of biodiversity to be politicized by creeping anthropocentrism—to shift from discussions of what is best for nature to what is best for indigenous and poor people. In a world where human beings receive more than 99 percent of the development aid and charity, it is increasingly difficult to find resources to help the other species with whom we share the earth.

Fortunately, there are many opportunities that could be explored during such a workshop on research priorities in the human dimensions of conservation. For example, some parts of the world are becoming depopulated as global economic pressures, soil degradation, and water supply problems stimulate human emigration from those areas. Eastern and central Europe and parts of the U.S. plains and short-grass prairies, for example, will offer many options for conservation as people vacate rural areas. Moreover, some governments are creating positive incentives for land-owning communities and individuals to become better stewards of their land.

Opportunities also exist to build upon some promising initiatives that are beginning to explore more creative ways in which to harness market forces to assist conservation (Daily 1997; Daily et al. 2000). In May 2000, an Australian firm called Earth Sanctuaries, Ltd., was listed on the Australian Stock Exchange (see www.esl.com.au), becoming the world's first conservation company to go public. The company buys up land (90,000 hectares as of July 2000) and restores native vegetation and wildlife. It earns its income from tourism, consulting, and sales of wildlife products. The company lobbied for and won a change in Australian accounting law so as to include its rare native animals as assets. The Sydney Futures Exchange is positioning itself to be a global leader in the trading of ecosystem services, including carbon sequestration and nontraditional environmental products such as credits for clean water and biodiversity.

In Costa Rica, which has long been a leader in devising imaginative conservation policies, the government has been paying landowners for such ecosystem services as carbon sequestration, watershed protection, preservation of biodiversity, and scenic beauty. The payments are financed in part by a tax on fossil fuels and are achieving significant conservation and restoration of forests (Castro et al. 1998). Costa Rica has also sold carbon sequestration credits to several European countries (Gámez 1999).

Assigning value to nature's various services provides a way of organizing information (Baskin 1997), but it is only one of many tools that need to be employed in making decisions in today's complex political climate. However, even approximate estimates of value may suffice to identify those initiatives for which the benefits greatly outweigh the costs, such as preserving a natural watershed rather than building a water filtration plant (Chichilnisky and Heal 1998; NRC 2000b), and, hence, to provide powerful advice to municipalities. In addition, such initiatives are stimulating development of integrated ecological-social-economic approaches to managing environmental assets. The future potential of these approaches is substantial.

These and other promising governmental and nongovernmental initiatives are growing rapidly and are receiving increasing support from the private sector. There is no reason to believe that the initiatives that have been tried so far exhaust the possibilities. There are, to be sure, risks associated with putting price tags on nature, and extreme care must be taken to ensure that these initiatives really serve the purpose for which they are intended (Daily 1997). However, the alternative is to continue with what has been the implicit value heretofore placed on nature's services, namely that they have no value, that they are "free."

Encouraging signs also are provided by current trends in several areas of environmental research. Physical and biological scientists are joining together to craft the new discipline of earth system science (Jacobson et al. 2000), a field that employs powerful analytical techniques to analyze environmental processes at large spatial and temporal scales. Ecologists are also directing more attention to the study of patterns and processes at large spatial scales, a field known as macroecology (Brown 1995; Maurer 1999). These efforts are developing solutions to a major problem, namely that of the high standards for experimental design and adequate controls that legitimately dominate thinking in small-scale science being incompatible with the kinds of research necessary to understand large-scale patterns.

For example, there is no controlled way to manipulate the biodiversity of a large region even if it were morally defensible to do so. Much of the most important work done in conservation biology is based on large-scale studies for which controls are virtually impossible. Studies of top-down regulation by keystone species have to be carried out over many years and over large areas. Studies of the factors affecting the decline in salmon include variables such as fishing pressure for which there can be no replication. Studies of ecological restoration at a regional scale cannot be controlled adequately.

These problems are not unique to earth system science, ecology, or conservation biology. They apply equally, for example, to astronomy, but this has not prevented astronomers from learning a great deal about stars, solar systems, and galaxies. Astronomy owes its success to its strong theoretical foundation in physics, on the basis of which testable predictions are made about the patterns to be expected in complex galaxies. Similarly, ecology and conservation biology, fields that have a strong theoretical foundation in behavioral and population ecology, are now poised to make testable predictions about macroscopic patterns in nature. We earnestly hope that the concepts and ideas in this book will help us make rapid progress in that direction.

LITERATURE CITED

Baskin, Y. 1997. *The Work of Nature*. Washington, D.C.: Island Press.
Brown, J. H. 1995. *Macroecology*. Chicago: University of Chicago Press.
Castro, R., F. Tattenbach, L. Gamez, and N. Olson. 1998. The Costa Rican Experience with Market Instruments to Mitigate Climate Change and Conserve Biodiversity. Fundecor and MINAE, San Jose, Costa Rica.
Chichilnisky, G., and G. Heal. 1998. Economic returns from the biosphere. *Nature* 391:629–630.

Daily, G. C. (ed.). 1997. *Nature's Services: Societal Dependence on Natural Ecosystems.* Washington, D.C.: Island Press.

Daily, G. C., T. Söderqvist, S. Anlyar, K. Arrow, P. Dasgupta, P. R. Ehrlich, C. Folke, A. Jansson, B.-O. Jansson, N. Kautsky, S. Levin, J. Lubchenco, K.-G. Mäler, D. Simpson, D. Starrett, D. Tilman, and Brian Walker. 2000. The value of nature and the nature of value. *Science* 289:395–396.

Earn, D. J. D., S. A. Levin, and P. Rohani. 2000. Coherence and conservation. *Science* 290:1360–1364.

Gámez, R. 1999. *De Biodiversidad, Gentes y Utopias.* Santo Domingo, Costa Rica: INBio.

Groom, M. J., D. B. Jensen, R. L. Knight, S. Gatewood, Lisa Mills, D. Boyd-Heger, L. S. Mills, and M. E. Soulé. 1999. Buffer zones: Recognizing the benefits and dangers of compatible stewardship. In M. E. Soulé and J. Terborgh (eds.), *Continental Conservation: Scientific Foundations for Regional Conservation Networks,* pp. 171–198. Washington, D.C.: Island Press.

Jacobson, M. C., R. J. Charlson, H. Rodhe, and G. H. Orians (eds.). 2000. *Earth System Science: From Biogeochemical Cycles to Global Change.* New York: Academic Press.

Margules, C. R., and R. L. Pressey. 2000. Systematic conservation planning. *Nature* 405:243–253.

Maurer, B. A. 1999. *Untangling Ecological Complexity: The Macroscopic Perspective.* Chicago: University of Chicago Press.

NRC (National Research Council). 2000a. *Ecological Indicators for the Nation.* Washington, D.C.: National Academy Press.

NRC (National Research Council). 2000b. *Watershed Management for Potable Water Supply.* Washington, D.C.: National Academy Press.

Redford, K. H. 1992. The empty forest. *BioScience* 42:414–422.

Ricketts, T. H., E. Dinerstein, D. M. Olson, and C. J. Loucks. 1999. *Terrestrial Ecoregions of North America: A Conservation Assessment.* Washington, D.C.: Island Press.

Scott, J. M., R. J. F. Abbitt, and C. R. Groves. 2000. What are we protecting? *Conservation Biology in Practice* 2(1):18–19.

Simberloff, D., D. Doak, M. Groom, S. Trombulak, A. Dobson, S. Gatewood, M. E. Soulé, M. Gilpin, C. Martinez Del Rio, and L. Mills. 1999. Regional and continental restoration. In M. E. Soulé and J. Terborgh (eds.), *Continental Conservation: Scientific Foundations for Regional Conservation Networks,* pp. 65–98. Washington, D.C.: Island Press.

Soulé, M. E. 2001. Does sustainable development help nature? *Wild Earth* (Winter): 56–64.

Soulé, M. E., J. Estes, J. Berger, and C. Martinez del Rio. In preparation. Ecologically effective population size and distribution.

Soulé, M. E., and K. A. Kohm. 1989. *Research Priorities for Conservation Biology.* Washington, D.C.: Island Press.

Soulé, M. E., and R. K Noss. 1998. Rewilding and biodiversity as complementary tools for continental conservation. *Wild Earth* (Fall):18–28.

Soulé, M. E., and J. Terborgh. (eds.). 1999. *Continental Conservation: Scientific Foundations of Regional Reserve Networks.* Washington, D.C.: Island Press.

Terborgh, J. 1999. *Requiem for Nature.* Washington, D.C.: Island Press.

Terborgh, J., J. A. Estes, P. C. Paquet, K. Ralls, D. Boyd-Heger, B. Miller, and R. K. Noss. 1999. Role of top carnivores in regulating terrestrial ecosystems. In M. E. Soulé and J. Terborgh, (eds.), *Continental Conservation: Scientific Foundations of Regional Reserve Networks,* pp. 39–64. Washington, D.C.: Island Press.

ABOUT THE CONTRIBUTORS

Michael E. Soulé is professor emeritus in environmental studies, University of California, Santa Cruz, and science director of the Wildlands Project. His Ph.D. in biology is from Stanford University and his research interests include island biogeography, community ecology, and conservation genetics. He was founder of the Society for Conservation Biology and the Wildlands Project. He has written and edited nine books on biology, conservation biology, and the social context of conservation. He is a fellow of the American Association for the Advancement of Science, has received a Guggenheim Fellowship, is the sixth recipient of the Archie Carr Medal, and was named by *Audubon* magazine in 1998 as one of the one hundred Champions of Conservation of the 20th century.

Gordon H. Orians, an ecologist, is professor emeritus of zoology at the University of Washington in Seattle. He received his B.S. in zoology from the University of Wisconsin, Madison, and his Ph.D. in zoology from the University of California, Berkeley, in 1960. His research has focused on the behavioral ecology of vertebrates (foraging, habitat selection, mate selection, and relationships between social systems and the environment), the structure of ecological communities, and the science-policy interface. He was director for eleven years of the Institute for Environmental Studies at the University of Washington, and serves on the board of directors of the World Wildlife Fund–U.S. He is an elected member of the National Academy of Sciences and the American Academy of Arts and Sciences.

A. Alonso Aguirre, D.V.M., M.S., Ph.D., is director for conservation medicine at Wildlife Trust and affiliated professor at Tufts and Columbia universities. Dr. Aguirre is former health director for Protected Species Investigation for the National Marine Fisheries Service in Hawaii. His professional interests include wildlife epidemiology, conservation medicine, and international training of conservation biologists and wildlife veterinarians.

Sandy Andelman is deputy director of the National Center for Ecological Analysis and Synthesis and adjunct associate professor in the Department of Ecology, Evolution and Marine Biology, University of California, Santa Barbara. She served as secretary of the Society for Conservation Biology and currently is a member of the editorial board for *Conservation Biology*.

Jonathan Baillie is a Ph.D. student based at the Institute of Zoology, Zoological Society of London, and at Imperial College, London. His research focuses on the extinction filter role humans have played in the process of colonizing oceanic islands, as well as the ecological and evolutionary traits that make some species more vulnerable than others to human threats.

Steven R. Beissinger is an associate professor of conservation biology in the Department of Environmental Science, Policy, and Management at the University of California, Berkeley, where he teaches courses in conservation and population biology. Beissinger studies the behavior, demography, and population viability of threatened and exploited birds. He instructs training courses in small-population biology for the U.S. Fish and Wildlife Service.

P. Dee Boersma is professor of zoology at the University of Washington and the past president of the Society for Conservation Biology. She studies Magellanic penguins at Punta Tombo, Argentina, and other seabirds to understand environmental variation and perturbations. She is a fellow of the American Ornithology Union and AAAS, a member of the advisory board of the Puget Sound Environmental Learning Center, and a member of the Peregrine Fund Board.

James S. Clark is professor of biology in the Division of Earth and Ocean Sciences at Duke University, where his research focuses on how global change affects forests and grasslands. Clark received a B.S. in entomology from the North Carolina State University (1979), an M.S. in forestry and wildlife from the University of Massachusetts (1984), and a Ph.D. in ecology from the University of Minnesota (1988).

Alan Covich is professor of fishery and wildlife biology at Colorado State University. He is past president of the American Institute of Biological Sciences and the North American Benthological Society and is an AAAS fellow. He studies food-web dynamics of temperate and tropical streams and has investigated the benthic-community response to drought and hurricane disturbances in Puerto Rico for the last twelve years as part of the Long Term Ecological Research Program.

Carla M. D'Antonio is an associate professor in the Department of Integrative Biology at UC Berkeley. She completed her dissertation research at UC Santa Barbara and has worked on the dynamics and impacts of plant invaders in California and Hawaii for fifteen years as both an academician and a professional biologist. She has served on several advisory boards related to the research and management of invasive species.

Kendi F. Davis is a postdoctoral fellow in sustainable ecosystems at the Centre for Arid Zone Research in Alice Springs, Australia. She spends the southern summer studying the ecology of Antarctic islands. She has a B.Sc. (1993) and a Ph.D. (1999)

from the Australian National University. Her current research interests are the management of fragmented landscapes, beetle ecology, and the ecological dynamics of Antarctic islands.

Julie S. Denslow is research ecologist and leader of the Invasive Species Unit with the Institute of Pacific Islands Forestry, USDA Forest Service, Hawaii. She is past president of the Organization for Tropical Studies and the Association of Tropical Biology. She has worked for many years on the dynamics of neotropical forests in Costa Rica and Panama and more recently on invasive plants in tropical and warm temperate forests.

Andy Dobson teaches conservation biology and the ecology of infectious diseases at Princeton University and undertakes research in Scotland, East Africa, and Yellowstone in conservation biology and the ecology of infectious diseases. His current research focuses on the role of pathogens in natural ecological systems and on how the geographical distribution of biodiversity determines both threats and conservation solutions.

Claude Gascon is the vice president for Field Support Programs at Conservation International. Previously, he was the project director and Scientific Coordinator for the Biological Dynamics of Forest Fragments Project in Brazil. He studies Amazonian biodiversity, especially amphibians. His publications encompass conservation and forest management in the Amazon and wildlife management.

Mark A. Hixon is professor of marine ecology and conservation in the Department of Zoology at Oregon State University. His research focuses on the processes driving and regulating population dynamics of reef fishes, as well as the natural mechanisms that maintain biodiversity on reefs. His current conservation interest is marine reserves, especially on coral reefs and along the West Coast of the United States.

Karen Holl is an assistant professor in the Environmental Studies Department at the University of California, Santa Cruz. She studies ecosystem recovery from both natural and human-caused disturbances and uses what she learns to restore ecosystems. She has done research in a range of ecosystems, including eastern hardwood forest, tropical rain forest, California chaparral, grassland, and riparian systems.

Malcolm "Mac" Hunter is the Libra Professor of conservation biology in the Department of Wildlife Ecology at the University of Maine. He is president of the Society for Conservation Biology, a Pew Conservation Fellow, and the author of five books.

Georgina Mace is the director of science at the Institute of Zoology, Zoological Society of London. She has worked extensively on the development of systems for assessing the degree of threat faced by wild species, especially in relation to the IUCN Red List. She now studies the evolutionary aspects of species extinction, especially the biological traits that make species more vulnerable to extinction pressures.

Jim MacMahon is a professor of biology and vice president for university advancement at Utah State University. He is a past president of the Ecological Society of

America, a fellow of AAAS, and a member of several National Research Council committees as well as the Board on Environmental Studies and Toxicology. He studies plant-animal interactions, disturbances in deserts and subalpine areas, and restoration of anthropogenic disturbances such as strip mines.

Chris Margules is officer in charge of the CSIRO Tropical Forest Research Centre and leader of the Sustainable Ecosystems Tropical Forest Ecology Program at Atherton in Queensland, Australia. He is also deputy CEO of the Rainforest Cooperative Research Centre, based in Cairns, Queensland. His current research interests are systematic conservation planning and reintegrating fragmented landscapes.

Laura A. Meyerson is an adjunct assistant professor in ecology and evolutionary biology and environmental studies at Brown University. She studies the ecosystem-level effects of biological invasions and the role of foundation species in ecological restoration. She is a Switzer Environmental Fellow and the author of a review paper for the Ecological Society of America on research priorities for introduced species in the United States.

Fiorenza Micheli is assistant professor of biological sciences at Hopkins Marine Station, Stanford University. She is a marine community ecologist, with a focus on the ecology and conservation of coastal marine ecosystems. Her research interests include species interactions and food-web dynamics in aquatic ecosystems, the effects of multiple disturbances on populations and communities, the conservation and restoration of marine habitats and species, and the design and evaluation of marine protected areas.

Jaqueline E. Mohan earned her baccalaureate degree at the University of Chicago, and her master's of environmental management (MEM) at Duke University, where she is currently in the pursuit of her doctorate in the Department of Biology. Her research focus is the role of biodiversity in terrestrial ecosystems.

Barry Noon is a professor in the Department of Fishery and Wildlife Biology and the Graduate Degree Program in Ecology at Colorado State University in Fort Collins. He studies conservation planning for threatened and endangered species, science-based management of public lands to conserve biological diversity, population dynamics and viability analyses for at-risk species, vertebrate demography and life history, and methods to study natural populations of vertebrate wildlife, ecological monitoring, and biometrics.

Elliott A. Norse is president of Marine Conservation Biology Institute and is a marine and forest conservation biologist. He defined the concept of biological diversity in 1980 and now focuses on creating a comprehensive national network of fully protected marine reserves to protect marine biodiversity.

Richard S. Ostfeld is associate scientist and animal ecologist at the Institute of Ecosystem Studies in Millbrook, New York. His research focuses on linkages between community ecology and the dynamics of infectious diseases. Current projects include the impacts of reduced species diversity on transmission of Lyme disease and other vector-borne diseases, and how forest fragmentation influences communities of hosts for ticks.

Gary Allan Polis was professor and chair of the Department of Environmental Sciences and Policy of the University of California, Davis. He made major contributions to the theory and empirical study of food webs, particularly on the roles of spatial subsidies, cannibalism, and intraguild predation in food-web dynamics. He did extensive work on the ecology of island and desert communities in California, Mexico, and Namibia, and he was a world authority on scorpion and spider ecology. Gary died in the spring of 2000 during a tragic boating accident in the Gulf of California.

Mary Poss, D.V.M., Ph.D., is a member of the faculty of the Division of Biological Sciences and the Wildlife Biology Program at the University of Montana. She is a veterinary pathologist and has degrees in biochemistry and experimental pathology. Her areas of research include virus evolution, ecology, and pathogenesis.

Hugh Possingham is professor of zoology and mathematics and director of the Ecology Centre at the University of Queensland, Australia. He is a population modeler and ecological theorist with a particular interest in applications of decision theory to conservation biology. Among a variety of public service and advocacy roles, he is the chair of the national Biological Diversity Advisory Committee in Australia.

H. Ronald Pulliam is Regent's Professor of Ecology at the University of Georgia. His major research contributions have been in the areas of biological diversity, community ecology, behavioral ecology, source-sink theory, and population dynamics in heterogeneous landscapes. He has also served as president of the Ecological Society of America, science advisor to the United States Secretary of Interior, and director of the National Biological Service.

Kent H. Redford is director of biodiversity analysis and synthesis for the Wildlife Conservation Society based in New York. Prior to that he directed the Conservation Science and Stewardship Program of The Nature Conservancy's Latin American and Carribbean Division and taught at the University of Florida. His interests are in park-based conservation, subsistence use of wildlife, and resource use by indigenous peoples.

Chantal D. Reid is a research scientist in the Department of Biology at Duke University, where she is studying direct effects of elevated CO_2 and tropospheric ozone on plant growth and productivity. She received a B.Sc. in biology from McGill University (1980), an M.S. in biology from San Diego State University (1985), and a Ph.D. in botany from Duke University (1990).

William H. Schlesinger is James B. Duke Professor of Biology at Duke University, where he holds a joint appointment in the Division of Earth and Ocean Sciences and the Nicholas School of the Environment. He studies global biogeochemical cycles, focusing on the response of soils in global change. He received an A.B. from Dartmouth College in 1972 and a Ph.D. from Cornell University in 1976.

Paul Snelgrove is an associate chair in Fisheries Conservation at Memorial University of Newfoundland in Canada. He studies the early life histories of marine organisms, particularly the invertebrates that live in sediments and the factors that regulate their diversity.

Gary Tabor, a wildlife veterinarian and an ecologist, is head of the Wilburforce Foundation's Yellowstone to Yukon Program, which promotes science and conservation to maintain ecological connectivity in the Rocky Mountain region of the United States and Canada. He served as founding director of the Center for Conservation Medicine, an ecological health collaborative. His international conservation background includes seven years in East Africa and a year each in Latin America and Asia.

Steve Trombulak is a professor of biology and environmental studies at Middlebury College in Vermont. His research includes topics in ecological reserve design, vertebrate ecology, and conservation education. He is also interested in the role of conservation biology in conservation advocacy.

Diana Wall is professor and director of the Natural Resource Ecology Laboratory, an international ecosystem research center at Colorado State University, and past-president of the Ecological Society of America, the American Institute of Biological Sciences, and the Society of Nematologists. She studies the consequences of human activities on soil sustainability in agricultural, grassland, and Antarctic ecosystems.

INDEX

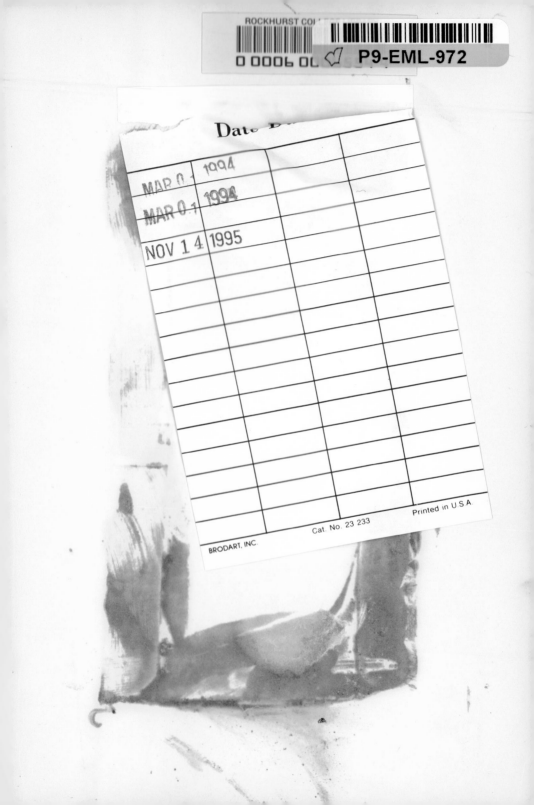

Date Due

MAR 0 1 1994			
MAR 0 1 1994			
NOV 1 4 1995			

BRODART, INC. Cat. No. 23 233 Printed in U.S.A.

THEORIES OF KNOWLEDGE

THEORIES
OF KNOWLEDGE

REGINALD F. O'NEILL, s.j.
Weston College, Weston, Massachusetts

ENGLEWOOD CLIFFS, N. J.
PRENTICE-HALL, INC. 1960

Imprimi Potest
  James E. Coleran, s.j.
  Provincial, New England Province

Nihil Obstat
  Russell J. Collins
  Diocesan Censor

Imprimatur
  † Richard Cardinal Cushing
  Archbishop of Boston

August 24, 1959

Printed in the United States of America
91317

PREFACE

This book presents an introductory study of human knowledge which is positive rather than polemical in its approach, adequate both in the number of topics discussed and in the individual analyses, and indicative of the richness of the content of a philosophical Realism.

All too often a realistic theory of knowledge has been developed by the negative technique of setting up "adversaries" to be refuted; this procedure has amounted to saying that since this or that theory is wrong, its opposite is right and should be accepted. Here that polemical approach has been avoided by the simple device of dividing the book into two distinct but interrelated parts.

Part One presents an integrated and carefully developed analysis of a realistic explanation of the human capacity for truth and of the various elements and sources of knowledge.

Part Two studies, analyzes, and evaluates other theories which have been proposed. Obviously, in any such study a comprehensive consideration of each theory is impossible; rather, the main themes have been selected and analyzed and an effort made to understand what each theory sought to accomplish, what contribution it made, and where it was inadequate.

Hence, it can be seen that Part Two is not meant to be a history of philosophy. In fact, the presentation of topics parallels as much as possible the problems considered in Part One, instead of following a strictly chronological order.

This division of the book allows for great flexibility in handling the subject matter. Some may want to study the chapters in the order they appear. Others may prefer to treat simultaneously the correlated chapters of both parts. Still others may use either portions or all of Part Two as supplementary reading.

The lists of suggested readings and the selected bibliography at the back of the book should prove helpful as a starting point for anyone interested in further study.

If the work appears a bit difficult, it should be remembered that the subject matter itself is not easy, and a book which oversimplifies and dilutes the necessary analyses may conceivably do more harm than good: it can leave the student convinced of the inadequacy of a realistic theory of knowledge, rather than lead him to the realization that Realism is in no wise synonymous with the naive or the simple.

Finally, I wish to acknowledge my indebtedness, particularly in the composition of the first part of this book, to the writings of Reverend Francis Morandini, S.J., Gregorian University. I also want to express my appreciation and gratitude to my fellow professors and to my students at Weston College, and in a special way to Reverend John D. Donoghue, S.J. of Fairfield University for the inspiration and assistance I received from them.

—REGINALD F. O'NEILL, S.J.

TABLE
OF CONTENTS

Contents

Part Two • Other Theories of Knowledge

INTRODUCTION

Presupposing that the student has some generic notion of what philosophy in general aims at accomplishing with respect to being, God, man, and the world, we can immediately assert that the fundamental subject matter to be considered in this book is human knowledge and its relation to things known.

In traditional terminology, then, we can say that:

The *material object* is human knowledge.

The *formal object* is the truth or reliability of human knowledge.

METHOD OF APPROACH

Open-Mindedness. Although the aim of this guide is to lead the student to an explicitation of much that he now takes for granted, and more specifically, to an admission of the validity of that philosophical position which can be called "Realism," at the beginning the only demand made is that he approach the subject with honesty, openness, and a readiness to admit what the facts of the case themselves demand. Therefore, he will not explicitly opt for Realism or Idealism, for Subjectivism or Objectivism, for Relativism or Absolutism; nor will he prejudge the issue by choosing to begin with a deliberate and positive doubt either about his personal capacity for truth or about the reliability of his prephilosophic intellectual experiences.

An honest, open-minded approach will in no wise prejudge the issue, nor render it impossible to respond to what presents itself as factual; nor will it render the finding of an explicit answer logically inconsistent. What is basically needed is fairness, not naiveté, nor arbitrary assumptions.

No Material Measurement. The *special difficulty* encountered in this study is that, unlike any other matter, what is to be considered here cannot be placed before one for measurement and dissection. In most of the empirical sciences, the matter to be examined can be spread out right before the investigator, and he proceeds to apply the appropriate techniques. Obviously in dealing with knowledge as such, this is not possible.

The Fact of Judging. Hence, all that is asked at the beginning of this study is that the student be willing to start with the only "matter" available for him to work on—*acts of knowledge.* Each one is well aware that he makes acts which he admits to be knowing acts, and these are called *judgments.*

The only procedure open to us, then, is to accept this fact (that we make judgments), and to analyze any concrete act of judgment for its implications with respect to the validity of human knowledge.

Such procedure must, then, be personal, reflective, descriptive, and evaluative. Its results should be a systematic, coordinated, and verified theory of knowledge.

Hence:

Our *problem* concerns human knowledge in respect to its truth value.

Our *aim* is a systematic theory of knowledge.

Our *procedure* will be to analyze a concrete act of knowledge or concrete acts of knowledge to see what such acts reveal about the mind's ability to know reality truthfully.

Other theories of knowledge will be discussed later on (in Part Two), and their strong and weak points indicated. These latter studies should bring about a better understanding of the basic positions taken by "Realism" as well as indicate the complexity of this problem of human knowledge and the special care and open-mindedness so necessary if one is to arrive at a truly balanced theory.

Two things must be kept in mind from the start:

(1) *Identity and Diversity.* It is *human* knowledge which we are trying to explain; hence all the factors which actually figure in human knowledge must be finally admitted. The concrete data of one's own experience is present to him, and a valid theory must take all these data into account. To reject some aspect or other, or to explain it away rather than to explain it, and that because of

some preconceived assumption, cannot result in the balance needed in this matter. Many schools, as we shall see in Part Two, oversimplify by the simple process of neglecting to consider or by denying some of the data relevant to the human experience of knowing.

For example, a distinction must be made between *pointing out* the fact that human knowledge seems to demand both an identity and a diversity between knower and known, and *explaining* how both characteristics can be simultaneously present or how they can be reconciled. Whatever the explanation, the fact is clear that when I say that *I know something*, the experience includes a *subject* and an *object* * known, and these are experienced as *diverse* and as two different terms of a relational act. The "I" is clearly distinguished from the "something" known; were they totally identical there would be no ground for the experience. The experience is not that of an act knowing itself, but of a knower who, by an act of knowledge, knows an "other-than-himself."

However, at the same time it must be admitted that were the object totally *other*, and were it to remain totally "outside" the act of knowledge and the knower, there could not be any knowledge present at all. Somehow what is known must enter and become one with the knower, which is to say that *identity* of knower and known is also necessarily involved in the act of knowing.

Various theories of knowledge have sought in various ways to reconcile these apparently clashing characteristics. (1) Some have denied the diversity and insisted on the total homogeneity of knower and known. (2) Others have denied identity, and so considered the two terms as irreducibly different from each other. (3) Still others have distinguished and found identity in one order of being and diversity in another order. These three views are mentioned here merely by way of illustration; as our study proceeds they will receive more careful consideration.

(2) *Relation to Metaphysics.* Despite some serious efforts to isolate Epistemology completely from Metaphysics, it will gradually become clear that the analysis of the knowing act cannot be completely divorced from an attitude toward "being" itself. Hence, while emphasis is to be placed on knowledge itself, the

---

* Throughout this work the term "object" is used as synonymous with "thing." The same word is often used technically in philosophy to signify something slightly, but importantly, different.

relation of this study to a general Metaphysics will inevitably become manifest.

One illustration of this relationship which might be offered has to do with the nature of man. Thus if I hold that the object known is totally intramental and that the only thing which I immediately know is an idea, my view of man's nature will be quite different from that held if I maintain that "outside" things are immediately known. In the first case the experienced "self" would be said to be a mind, somehow (maybe) linked to a body, and "body" and "soul" might be considered two separate and quite distinct beings. In the second case, the self would be said to be only one being in which mind and body not only work somehow as a unity, but also are not really two in any ordinary sense; they must constitute a special sort of composite unity. So a theory of knowledge can imply the position that man is just "mind" with an illusion of a body; or that he is just body having no "superior" principle; or that he is a unique "unity" of both mind and matter, the total being only one being; or finally that he is really two distinct and separate beings somehow contingently united.

Part One

A REALISTIC THEORY
OF KNOWLEDGE

Carefully analyzing what the fact of making judgments means, I find that I know some definite underlying truths (Chapter One), and that I simultaneously know my capacity for truth (Chapter Two). Further reflection on this procedure shows that such positive analyses of concrete acts of judgment provide a successful way to approach the question of my capacity for truth (Chapter Three).

• Suggested Readings—Chapters One
Through Three

Aquinas, St. Thomas, *Basic Writings of Saint Thomas Aquinas*, ed. Anton C. Pegis, vol. I, pp. 840–41. New York: Random House, Inc., 1945.

———, *On Truth*, 3 vols., translated by R. Mulligan, vol. I, q. 1, aa. 9–11. Chicago: Henry Regnery Company, 1952–1954.

Aristotle, *The Basic Works of Aristotle,* ed. Richard McKeon, pp. 737–43. New York: Random House, Inc., 1941.

Bahm, Archie J., *Philosophy: An Introduction,* ch. 2. New York: John Wiley & Sons, Inc., 1953.

Maritain, Jacques, *The Degrees of Knowledge,* translation under the direction of G. Phelan, ch. 2. New York: Charles Scribner's Sons, 1959.

———, *The Range of Reason,* ch. 1, pp. 3–18. New York: Charles Scribner's Sons, 1952.

de Raeymaker, Louis, *The Philosophy of Being,* translated by E. Ziegelmeyer, pt. I, ch. 2, pp. 11–23. St. Louis: B. Herder Book Co., 1954.

Régis, L. M., O. P., *Epistemology,* translated by I. C. Byrne, pp. 175–308. New York: The Macmillan Company, 1959.

Wild, John (ed.), *Return to Reason,* ch. 6, pp. 152–66. Chicago: Henry Regnery Company, 1953.

Wilhelmsen, Frederick D., *Man's Knowledge of Reality,* chs. 1–7, pp. 3–71. Englewood Cliffs, N. J.: Prentice-Hall, Inc., 1956.

2

CHAPTER ONE

A GENERAL ANALYSIS

OF JUDGMENT

I • THE SITUATION

From Acts to Potency. We have already stated that in this matter it will be necessary to proceed by analyzing concrete acts of knowledge to see what they reveal in respect to the capacity for truth, and what we hope to show early and incontrovertibly is that I do actually know certain truths. Once this is recognized, the admission of my capacity for truth becomes an easy step to make. Just as I can say that a certain man has the ability to paint, if I see him actually painting, or say that John has the ability to play the piano, if I see and hear him actually playing the piano, so, too, if I do actually have veridical knowledge, then I must admit the ability or capacity or potency to know truth. In this matter, one incontrovertible instance is actually sufficient.

Precise Content Not Important. Again, if I wish to proceed to a valid theory of knowledge, I must start with what is accessible to me; hence, I must take as my datum some act of knowing proper to myself; or in other words, the only thing I need to "presuppose" is one or other of my own acts of judging. It may be as common as the judgment that "this house is red," or "the book on the table is huge," or "that apple is tasty." What we are after cannot, at this point, be concerned precisely with the particular content of the judgment selected. Moreover, it is with an act of judgment that

3

we must begin, since, as will be seen in Chapter Nine, the problem of truth or falsity does not arise in regard to the act of simple apprehension.

II · THE ANALYSIS

(a) *Seek Meaning of "To Judge."* I have made a judgment, what is generally admitted to be a concrete act of human knowledge.

Now I must ask myself what that really *means*—not just in the sense that I have said that the house is red or the apple tasty, and so forth, but in the sense of finding out if any such act of judging really *means* something much more profound.

(b) *I Say Something about Something.* Every time that I intellectually judge as above, the absolute minimum that is implied in the very act of my judging is that *I am saying something about something.* Hence, even if what I explicitly say in the judgment happens to be "false," still the very making of the judgment implies that minimum. If that were not true, the function of judging would be totally inexplicable or even pernicious.

In other words, the natural and spontaneous inclination of myself to judge involves more than the mere assertion that certain characteristics are found in some object or other, for it indicates also that I recognize and by my very act admit that I can say something about something.

(c) *Three Truths Grasped.* Reflecting upon and analyzing this minimum more carefully, I come to see that there are three elements of the act of knowing which can be further elucidated:

(1) *Judgment is about something.* Whether what is judged is an act of intellect or of will, of internal or of external sensation, in any case it is a *something other* than my present act which is in question. Even the admission of error indicates this, for by admitting the possibility of error I admit the need of conforming to *what is.* Hence the minimum involved here is that a judgment involves a *being,* existing in some manner or other, as that about which my judgment is concerned. In that sense the *truth of being* is inevitably grasped in every judgment.

(2) *Something determinate is said.* I make any explicit predica-

tion only because more radically I recognize and admit that whatever is is what it is, exists in a determinate way, and cannot simultaneously not be as it is. Whether the particular object judged actually is as it is explicitly judged to be or not, the only explanation of my making any assertion is the natural, implicit, but formal grasp of the underlying truth that a thing cannot simultaneously be and not be, or be and not be what it is. Radically, the spontaneous grasp of this truth is why I judge, why I distinguish between "truth" and "falsehood," and why a sceptical attitude ever attracted philosophers. And this basic truth, grasped formally but implicitly in every judgment is known as the *principle of contradiction*.

(3) *I am saying something determinate.* In this act of judging (as well as in other acts of feeling, knowing, and willing) I grasp the act or acts as modifications of myself as subject. These are *my* acts, somehow proceeding from, united in, and existing by virtue of *me*. Hence, I also implicitly yet formally know the truth of *my own existence*.

(d) *My Truth-Capacity Grasped.* In any concrete act of judgment, these are not the only underlying truths which are implicitly and formally known by me, but even these suffice for establishing the point at issue. The point is whether or not I can know truth—and that in its most general sense. The fact gathered from an analysis of any concrete act of judgment, regardless of the truth or falsity of the explicit judgment, is that the very fact of my judging is radically explained only by the recognition of three fundamental implied truths: being, my own existence, and the principle of contradiction.

Granted the knowledge of these three, which are implicitly but actually known in every concrete judgment, and without which no act of judging would ever be attempted, I must also grant that I implicitly but actually also know my capacity to know truth. In other words, since I *do* actually know some truths, I recognize that I *can* know truth. Thus I implicitly know my truth capacity in so far as I grasp it actually operating.

This does not mean, however, that I have as yet grasped the metaphysical nature of the elements involved.

III • SUMMARY OF THE ANALYSIS

(a) I am spontaneously conscious of certain acts, such as willing, knowing, feeling, and so forth, but especially I am aware of judging, that is, of saying something determinate about something. This implicitly involves knowledge of the truth of being, of my own existence, and of the principle of contradiction. Therefore, I know these truths implicitly.

(b) These are known in *every* act of judging, hence they are known prior to all proper demonstration; that is, they are *immediately* known.

(c) Since all particular truths rely on them, they are also called *fundamental* truths.

IV • THE STATEMENT IN BRIEF

I naturally know certain immediate and fundamental truths.

V • DEFINITIONS

I: a realization of myself as an experiencing subject. The same can also be said of others in so far as they possess the same faculties and abilities as I.

Know: taken in the ordinary and generally admitted sense of that activity whereby some object is made present to a subject. Knowledge is distinguished into (a) *sensitive,* whereby I attain things colored, resistant, odorous, and so forth, and (b) *intellectual,* whereby I judge and reason.

Naturally: spontaneously apprehended, as distinguished from *methodically,* which is arrived at by planned study.

Truth: taken in the general sense of a conformity between mind and object known. It is opposed to *falsity,* which means a positive difformity between mind and object.

Immediate: known without the sort of medium had in reasoning.

Fundamental: necessarily required for the knowledge of other truths.

Certain: some. I select only a few principal truths, which are:

(a) *Being,* or that which is. In knowledge it is *not nothing* but something to which my cognition is conformed.

(b) *Principle of Contradiction:* a thing cannot simultaneously be and not be.

(c) *My existence:* the reality of myself as the knowing, willing, feeling subject.

Explicit: principally and terminally known.

Implicit: involved in and known along with what is principally known.

VI • SUPPLEMENTARY NOTES

(a) If what is said above is true, it would seem that no one could ever have *doubted* these immediate truths; and yet they have actually been doubted.

REPLY: The merely *verbal* expression of doubt in these matters does not involve a real intellectual doubt. In fact, doubt itself as a state of mind is inexplicable except against the background of an implicit knowledge of these basic truths. Furthermore, it is usually in the more subtle applications to difficult problems, for example, motion and change, that issue is really taken. Finally, even a fully sceptical position is taken only because of a recognition of these truths, as will be seen later.

(b) Since scientific knowledge requires demonstration, it would seem that in order to get a truly systematic doctrine of human knowledge one should *demonstrate* the basic truths.

REPLY: No such demonstration of the basic truths is either required or possible. Demonstration means the process of arriving at true and certain conclusions from previously known true and certain premises; and it must be admitted that anything demonstrable should be demonstrated. However, the burden of our analysis has been to show that some basic truths are really known in *every* judgment, and so there cannot be any prior premises from which they can be deduced. All that need or can be done, then, in this matter is to point out explicitly and distinctly how what is self-evident manifests its own truth and is known as such by the human knower.

Involved in the second objection lies a confusion between a *postulate* and a *principle.*

A *Postulate* is a proposition whose truth one is asked to concede, although it is neither self-evident nor demonstrated.

A *Principle* is a proposition which is self-evidently true.

A *Conclusion* is a proposition which is accepted because it has been demonstrated.

CHAPTER TWO

THE KNOWLEDGE
OF MY TRUTH-CAPACITY

I • THE SITUATION

In Chapter One the analysis indicated that I know some concrete truths, and hence have a capacity to know truth. Further consideration must now be given to the precise way in which I grasp or come to know this capacity. It has been argued that I must first know my capacity for truth before I can accept any particular truth; and that, on the other hand, I cannot know my capacity for truth except by way of the veridical acts themselves. Hence it is concluded that we are involved in an impossible situation. In fact, it must be conceded that were that true, no solution would be possible. Hence the present analysis must show what the priorities are in this matter, how I actually know this capacity for truth, and how the above dilemma is to be avoided.

II • THE ANALYSIS

(a) *Priority of Existence.* I can first recall the example of the painter in order to illustrate a point in priority. It must be admitted that if the man does not have the ability to paint, he will not actually paint. Hence, ontologically speaking, the capacity must exist prior to the act. Hence, too, in the matter of knowing truth, I must first have the capacity before I can actually know any truth.

(b) *My Knowledge of That Existence.* However, my considera-

8

tion is concerned with the question of how I come *to know* the existence of my capacity for truth. There are three possibilities: (1) I can first know the capacity, then the act; (2) I can first know the act, then the capacity; (3) I can come to know both of them simultaneously. To find out which of these is actually the explanation, I shall have to consider what it means to accept any judgment as certainly true.

(c) *Awareness of the Object.* Acceptance of a judgment as certain involves on my part awareness that the object known is as I judge it to be. Until I am conscious of this, there is hesitation and the fear of error, but once this awareness is present, I reject all hesitation and fear, and make the judgment. Thus the sure and firm acceptance of a truth requires my awareness that the object is as I judge it to be.

(d) *Awareness of My Own Act.* Furthermore, this awareness or consciousness means that I recognize the noetic or cognitive nature of my act of judging, since it carries with it the realization that my judgment itself is conformed to what is.

(e) *Awareness of Myself.* This means that I also know the subject of the act, since I know that the act of judging is mine.

SUMMARY: The acceptance of a judgment as certainly true cannot be made unless it is connected with the awareness that the object is known as it is, that the act is a cognoscitive one, and that the one judging is the subject of the act. These three elements together make up what is known as *complete reflection.*

(f) *Simultaneous Complete Reflection.* Having admitted, therefore, that complete reflection is an element necessarily found in the admission of any judgment as true and certain, we must next determine whether this reflection takes place prior to or concomitant with the very act of judgment. It will help our analysis to recall here that in the previous chapter we saw that there are some truths which are *immediately* known.

Now, if these are *immediately* known, it must be admitted that everything necessary for their acceptance is also *immediately* known or grasped. Since complete reflection is a necessary element (cf. *Analysis* [c], [d], [e]), it must be present in the very exercise of the act, that is, it must be present simultaneously in the act whereby the truth itself is grasped. Hence in admitting that we know some immediate truths, we also admit that we are, in one and the same act, simultaneously aware of the three elements involved in complete reflection. By simply considering the fact that the same elements must also be found in any and every judgment which is accepted as true and certain, the extension of this fact can be made to all other judgments.

(g) *Negative Confirmation.* The same analysis can be confirmed in a negative way by considering what the situation would be if such reflection were not present implicitly in the act itself, but were to be present only in a subsequent judgment. That subsequent reflex judgment would (in the supposition) validate the previous judgment, but it itself would be in need of validation by a subsequent judgment, and so on indefinitely.

(h) *Simultaneous Knowledge of My Truth-Capacity.* All that has thus far been said can now be applied to the initial question concerning the way in which I actually know that my mind is capable of truth. This capacity of my mind is known by me only by my knowing that my mental activity is conformed to reality; and this conformity is known by me precisely through complete reflection. Hence it is because I have complete reflection that I know my capacity for truth. In other words, in every act of judgment, I am concomitantly implicitly aware of my ability to possess truth.

(i) *Objective Norm for Truth.* From what has been seen thus far, it can be stated now (leaving until later a more detailed consideration) that there must be *some objective norm* of truth. Since there is truth, there must be a norm for distinguishing it from falsity; and since truth requires my being aware of the object as it is, that norm must be some quality of the object manifesting itself.

III · SUMMARY OF THE ANALYSIS

(a) Acceptance of a judgment as true involves awareness that (1) the object is as I judge; (2) my act of knowing is conformed to the object; (3) I am the subject of the act. But these three elements constitute complete reflection. Hence acceptance of a judgment as true involves complete reflection. Cf. *Analysis (c), (d), (e).*

(b) There are some immediately known truths. As *known truths*, these require complete reflection. As *immediately known* truths, these require that reflection be also immediately present in the very act itself. Hence in some judgments complete reflection is immediately or simultaneously present. Cf. *Analysis (f)*. Moreover, if complete reflection demands a reflex judgment, the process would have to go on to infinity, and there could be no true judgment. Cf. *Analysis (g)*.

(c) My intellectual capacity for truth is known, if I know that my intellectual activity is conformed to reality. But this conformity is known through complete reflection. Hence my intellectual capacity for truth is known through complete reflection. Cf. *Analysis (h)*.

IV • THE STATEMENT IN BRIEF

In the very act of judging I am aware of my intellectual capacity for truth.

V • DEFINITIONS

Judging: that operation of the mind which affirms or denies the identity of some subject with some predicate. In every judgment certain truths are known explicitly and expressly (*in actu signato*), whereas others are known only implicitly because they are involved in the very exercise of the act of judging (*in actu exercito*). Complete reflection is one of the elements included in the very exercise of the act.

Reflex Judgment: one in which the affirmation or negation bears on a previous judgment, as when I say: "My judgment that 'the house is red' was false." All other judgments are called *direct* judgments.

I Am Aware: consciousness of the self knowing an object, called *complete reflection,* is found in every judgment.

Reflection: the return of the knower upon himself.

Complete Reflection: a simultaneous awareness touching the whole process of knowing, that is, the object, the act, and the subject knowing. This chapter indicates that such reflection is *natural,* since it is possessed by reason of the very nature of my cognoscitive faculty.

Truth: the conformity of intellect with thing.

Capacity For Truth: the aptitude of my mind to be conformed to reality. It can be viewed: (a) *potentially:* as conformable to reality; (b) *reduced to act:* as conformed to reality.

This chapter states that I implicitly know my capacity for truth as here and now *reduced to act.*

VI • SUPPLEMENTARY NOTES

(a) To proceed in this matter, the capacity of the mind for truth must be *presupposed*, and therefore the process is illegitimate.

REPLY: It is true that the capacity must first exist, and in that sense it is ontologically presupposed; but in the order of knowledge, that is, logically, it is simultaneously grasped.

(b) Since the ability of the mind to know truth is what is in question, it should not be asserted in the very data used to settle the question.

REPLY: When this capacity is said to be *in question,* it must be remembered that here there is no *problem* in the *strict* sense of the word. Rather the burden of the analysis has been to show that the reply is *naturally* known to all. Hence it must be said that this capacity is implicitly asserted in the data examined (being naturally known) and the function of the analysis is to render it explicitly and distinctly known.

CHAPTER THREE

THE QUESTION
OF METHOD

I • THE SITUATION

Without having gone explicitly into a study of the metaphysical nature of the mind, of the act of cognition, or of the object known, we have already analyzed concrete acts of what is commonly and spontaneously recognized as knowledge, and we have seen that we implicitly but formally know certain truths and simultaneously know also our actuated capacity for truth. Since some have felt that the question considered should never have been raised by philosophers, and since others have proposed other methods of approach (cf. Part Two, Chapter Sixteen), it will be helpful to reconsider from those points of view what we have accomplished.

II • THE ANALYSIS

(a) *One Formulation of the "Critical Question."* Historically speaking, the critical question can be summed up somewhat as follows: "Granted that the immediate objects of my knowledge are my own ideas, is it possible or how is it possible for me to know any 'outside' or transcendent reality?" In the face of that or similar formulations Scholastic philosophers have usually insisted that the very question rests on a false supposition, and hence the critical question never should have been asked and should not now be raised.

Our previous and subsequent analyses show in effect that any such formulation of the critical question is unacceptable because it prejudges the issue and is not consistent with the data of the noetic experience.

(b) *Our Formulation of the Question.* In the present study the question which we have faced actually has had no such suppositions implied in its formulation, even though it hopes to meet and answer the points raised by modern philosophers on the question of human knowledge. What has been asked is *whether the human intellect has the capacity for truth.* This is a very simple, honest question, which implies no false suppositions and merely asks for whatever answer is offered by the facts. At the same time, it does not require the retraction of any previously known truths, nor does it in any way imply the negation of any naturally known truths.

(c) *The Need for Explicitation.* Furthermore, although we have seen that the capacity for truth is naturally known in every judgment, the question asked really needed to be raised for purposes of *explicitation.* Although our knowledge-acts are very personal to each of us, we can easily fail to recognize their implications. The difficulty is all the more pronounced because "things" easily catch our attention and occupy our interest, whereas introspection and the analysis of our internal acts is not an easy task. Hence it is important that we reflect seriously and explicitly on our acts of knowledge to find out what they mean and what they necessarily involve.

(d) *A Fundamental Question.* Again, it is the work of the philosopher to seek the fundamental explanations of things, and so in the question of human knowledge it is his work to seek the foundations of knowledge, the basic truths upon which it rests.

(e) *The Proper Method.* Since Descartes' famous *Discourse on Method,* the question of just how a theory of knowledge should be approached has been one of primary importance. Descartes' method will be considered later. For the present we can say that any reflex method, if it is to offer any possibility of success, must (1) not prejudge the issue, and (2) be fundamentally natural.

This means that deliberate choice of a method should be such that it will let the facts of the situation speak for themselves, and will not arbitrarily set up requirements which will predetermine the answer to the question in the method chosen; that is, it must not preclude the logical possibility of a successful solution.

Secondly, since the point at issue is to determine the veridical nature of human judgment, the correct method must be somehow naturally known; that is, it should be in keeping with the nature of the intellect. Thus that method which is to be reflexively and explicitly followed in this matter must reflect the way in which the mind naturally knows its own capacity for truth.

(f) *Analysis of Judgments.* The only positive approach or method to be followed explicitly in this work will have to be the method of analysis, since demonstration is here impossible. And the only "matter" we have to analyze is a concrete act of judgment, since what is in question is human knowledge. Positively this means that, having decided to focus our attention precisely on our experienced act of *knowing*, we will not approach this with a denial of the *possibility* of truth, with a denial that things can be known *as they are*, with a denial that the act of knowing terminates in an object somehow *independent* of the act itself, or with any other preconceived determination to exclude part of the complex factors which are found in the concrete knowing situation. Rather, with open-minded honesty we should be prepared to face the experience of human knowing, with all the complexity and all the limitations that may involve, and be ready to *recognize* (not create) and to express just what is found.

To describe our knowledge *as it would be* were it divine or "angelic," to lay down conditions for knowledge which are not fulfilled in our concrete experiences, to seek to force the facts and to fit them within some *a priori* metaphysical theory, is not to explain *human knowledge* as it is, but rather to explain it away or to describe what it would be were it other than it is. This, as will be shown, has been the shortcoming of some other theories of knowledge.

(g) *Results of This Method.* This method of analyzing concrete acts of knowledge will necessarily result (as it has) in our recognizing the naturally and immediately known truths which lie at the roots of the very meaning of our making any act of judgment. That is to say, the analysis of any act of judgment will ultimately lead to answering the problem in question here by revealing the immediate and naturally known truths involved in it. If the judgment is a mediate one, it will have to be resolved into some immediate cognition; and if it is an "artificial" one, it must be founded on "natural" cognition. Hence the solution can be found only in the discovery of some immediate and natural knowledge.

(h) *A Negative Approach.* This "critical" question can also be said to be answered by another method—negative and indirect (whereas our method has been positive and direct). This would consist in analyzing the position which denies the mind's capacity for truth, and hence denies our certain grasp of any truth. Here the analysis consists in pointing out the contradiction involved between the negative position explicitly taken and the implicit affirmation of the very fundamental truths which are denied. However, while this method appears different, it is basically the same as the one we have proposed above. It amounts to the statement that even if an act of judgment were to go so far explicitly as to deny the very possibility of truth, that judgment would nonetheless, because it states something determinate about something, implicitly affirm the same fundamental truths we have already explicitated. In other words, any act of judgment, one affirming being, one offering mere conjecture or probability, or even one explicitly denying the very possibility of truth must have some determinate meaning and hence implies what it may seek explicitly to reject.

III • SUMMARY OF THE ANALYSIS

(a) The question of the intellect's capacity for truth is one which should be raised by a philosopher, since it
   (1) is a sincere question,
   (2) implies no false suppositions or arbitrary retractions of truths, and
   (3) seeks the fundamentals of human knowledge. Cf. *Analysis (a), (b), (c), (d).*

(b) The proper method must be one which open-mindedly analyzes a concrete act of judgment.
   (1) This is the only "matter" available for analysis.
   (2) If it is mediate, it must be resolved into the immediate.
   (3) If it is artificial, it must rest on the natural.
Hence the proper method will reveal the immediate and

naturally known cognition which underlies all judgment. Cf.
Analysis (e), (f), (g).

(c) An analysis can also be made of the negation of my capacity
for truth. This will reveal:

(1) an implicit affirmation of fundamental truths, and hence

(2) the contradictory of what is explicitly stated.

Hence this indirect or negative method can also be used to
answer the "critical" question. Cf. *Analysis (h).*

IV • THE STATEMENT IN BRIEF

*The critical question is a philosophical one, and the solution
to it is found only by an analysis of concrete acts of judgment.*

V • DEFINITIONS

Question: an interrogation as to whether or not a certain predicate
is to be affirmed of a certain subject.

In the Strict Sense: when the answer is actually unknown.

In the Wide Sense: when the answer is actually implied, but explicitation is needed.

Of a True or False Supposition: insofar as it excludes or includes
error in its formulation.

Philosophical or Scientific: insofar as it seeks ultimate or proximate
causes.

Critical Question: an interrogation asking whether the human intellect has the capacity for truth. (This is the formulation of the
question which we have actually faced and analyzed in general. Our analysis implies the rejection of the validity of what
has frequently been proposed as the Critical Question. Cf.
Analysis [a] for the latter formulation.)

Analysis: a method which distinctly explicitates what is only implicitly and confusedly known. (This is to be distinguished
from the *process* of strict demonstration which uses an extrinsic
middle term and results in new knowledge.)

Direct or Indirect Analysis: insofar as it proceeds positively by
analyzing the data, or negatively by analyzing some false
position taken.

VI • SUPPLEMENTARY NOTE

The main objection to our analysis would insist that our position is not impartial.

REPLY: True impartiality in a question as fundamental as the one we have considered would require us to study and evaluate various solutions offered, but it cannot require us to adopt an initial attitude which would render any solution logically impossible. Hence the wholesale rejection of all our previously held judgments or the entertaining of universal doubt does not mean impartiality. It represents an *a priori* position taken rather than an unprejudiced examination.

Reflecting on what it means to be capable of possessing truth, I find that to have consistent meaning truth must be said to be in some sense "absolute" or objective (Chapter Four), and to have real being as its object (Chapter Five).

• *Suggested Reading—Chapters Four and Five*

Gilson, Etienne, *Being and Some Philosophers,* ch. 6, pp. 190–215. Toronto: Pontifical Institute of Medieval Studies, 1949.

———, *The Unity of Philosophical Experience.* New York: Charles Scribner's Sons, 1937.

von Hildebrand, Dietrich, *The New Tower of Babel,* ch. 2, pp. 57–102. New York: P. J. Kenedy & Sons, 1953.

Plato, *Plato's Theory of Knowledge,* translated by F. M. Cornford, pp. 29–102. London: Routledge & Kegan Paul, Ltd., 1951.

———, *The Dialogues of Plato,* translated by B. Jowett, vol. II, pp. 143–217, especially 153–183. New York: Random House, Inc., 1937.

Wild, John, *Introduction to Realistic Philosophy,* ch. 18, sect. 4, pp. 407–412. New York: Harper and Brothers, 1948.

Wilhelmsen, Frederick D., *Man's Knowledge of Reality,* chs. 3–4, pp. 17–41. Englewood Cliffs, N. J.: Prentice-Hall, Inc., 1956.

CHAPTER FOUR

TRUTH KNOWN

AND THE KNOWER

I • THE SITUATION

Although all who refuse to despair over man's ability to know truth are in fundamental agreement that truth involves some sort of conformity between the mind and reality, not all are agreed on precisely what that means. Some hold that truth itself is entirely relative; others that it has an absolute character. Some find the object known to be totally intramental; others see it has having some independence of mind. Some again admit the extramentality of the object, but insist that it is not immediately grasped; others deny this. These views must be discussed in order, so that the facts of the situation may be recognized. As our first step, we shall now consider the nature of relativity only as it applies to knowledge.

II • THE ANALYSIS

(a) *Variability of My Grasp of Truth.* That an element of relativity (that is, of variability according to the knower) is characteristic of my knowledge is a fact which hardly needs any discussion. It is easily admitted that age, education, environment, and intellectual acumen all enter into my greater or lesser penetration and grasp of various truths. Thus I may know that aspirin will relieve my headache; later on, after becoming a physician or a biologist, I may know the same truth, but now I know more what "aspirin" is, what it does, how it helps, and so forth. Relative to my education in this matter, there is now a much greater understanding of the situation.

Put another way, I can see that things are knowable in so many different ways that no one act of mine can exhaust the total intelligibility of any one of them. Hence, one man's knowledge can be more exhaustive or inclusive than another's, and at one period I may know less about an object than at another time.

(b) *Invariability As an Aspect of Truth.* The question now arises whether everything has been said on this subject or whether some note of nonrelativity is to be admitted. To answer this we must reconsider what we have already done, but from a slightly different point of view.

We have seen that any judgment necessarily involves predicating something determinate about something. That is to say, in the very act of judging there is intended a submission to *being;* there is a statement that *what is* actually *is.* Maybe *what* I say is not *all* there is, maybe what is said is a very minimal part of what there actually is; however, the very statement that *this is* or *this is such and such* indicates that my capacity for truth is synonymous with my capacity to know and to affirm *what is,* not with just what I *want* to be or what I try to make exist. It is rather a submission to *what is.* "What is" may be grasped under one or another formality, with more or less penetration, and hence with more or less comprehensiveness—but in any case the *object known* in this or that way is precisely *what is known* and affirmed.

In other words, human knowledge is not knowledge at all unless it means the affirmation of *what is* and the negation of *what is not;* it means a confrontation with being, not a partial or total creation of what is known. What is affirmed in a true judgment is precisely what is found in reality, whether it is affirmed by me or by another, at one period of my life or at another.

We can take as an example the statement, "John is here." This judgment can be made by his mother who loves him profoundly and for whom the mere mention of John can call up memories of tender and loving association. It can be made by someone who just met him for the first time and scarcely knows him. It can be made by someone who despises John. In each case, however, *what is affirmed* is true in reality: John is here. However much or however little John may mean to me, his presence remains a "stubborn fact," unyielding in the face of varying ways in which it is grasped.

(c) **What *Is Known and* the Way in Which *It Is Known.*** It is in this sense that truth must be said to be *absolute* rather than merely *relative.* This does not mean that the object known cannot change, nor that my intellectual grasp of the object is total or exhaustive; but it does mean that *what is affirmed* in a true judgment *actually is* in reality, or *what is denied* actually *is not* in reality. In that sense *what I affirm* is not dependent on my knowledge nor on my affirmation, nor does it vary with my subjective conditions. Only the *way in which* I know things is subjectively mutable and dependent on my conditions.

(d) *Application to Some Cases.* Reflecting upon what has already been seen, we said that the principle of contradiction and my own existence are truths which I formally know in every judgment. If that statement has any significance it must mean that whether I think of it or not, hence independent of my judgment, "being is"; and "being is what it is"; and "I exist." These are *objectively true* and my knowing them does not change what is known. Hence, it is not that these are true because I say so, but rather I say so because they are objectively and factually true.

Therefore when we said in Chapter One that I have a capacity for truth, that truth must be characterized as *absolute truth* in the sense just explained.

Truth, then, has some objective stability and independence outside of the knower, even when what is known is some momentary and transitory event. My way of knowing this or that truth is mine alone, but what I know can also be known by another, so that we both know the same thing: for example, that John is a murderer, that it is raining, or that being is not non-being.

(e) *The Word "Absolute."* The word "absolute" in this context has caused much trouble, so it should be used with caution and accuracy. It means freed (*soluta*) from (*ab*) or independent of my subjective conditions. It means that truth does not vary with my subjective conditions, that it is not a creation of my mind, that it is something which I admit and to which I submit.

(f) *Negative Explanation.* Were this not the case (thus putting the explanation in a negative way), then all truth would be totally relative with respect to what is known. From this it would follow that I could never know anything as it is, I would never have conformity with being. What is known would depend on each knower, and each one's knowledge would be so individualized and personal that no communication of it would be possible. I would have my truth, you would have yours, and he his; but no one of us would know anything as it is in itself.

(g) *Communication.* By truth, then, even in our first chapter, we meant an objective conformity of mind to what is, which makes possible communication between men and the development of common bodies of scientific knowledge or truth. Finally, it must be added that what has been seen in this chapter in no way conflicts with the statement that truth is a "relation" of conformity (cf. Chapter Nine) between the mind and the object known. For truth is a property of the judgment whereby what is enunciated is related to or "measured" by that which is known. This, however, in no way involves the relativity of *what is known,* which is discussed in the present chapter.

III • SUMMARY OF THE ANALYSIS

(a) The way in which I know is dependent on and relative to my personal situation. Cf. *Analysis (a).*

(b) Truth, however, with respect to what is known, necessarily involves an admission of and a submission to what is, that is, to being.

(1) Thus the truth of the Principle of Contradiction or of my own existence is affirmed by me *because it is objectively so;* my affirmation does not make it true.

(2) Were this not so, anything at all could be made to be true by my mere affirmation of it.

(3) In this sense truth is said to be objective or absolute. Cf. *Analysis (b), (c), (d), (e), (f).*

(c) Hence communication of truth is possible, and men can work cooperatively to build up bodies of scientific knowledge. Cf. *Analysis (g).*

IV • THE STATEMENT IN BRIEF

Truth, with respect to what is known, must be said to be absolute.

V • DEFINITIONS

Truth: a conformity of the mind with an object.

What Is Known: that which is affirmed or denied of reality. This must *be* in order that my affirmation may be justified. This is what many may simultaneously know.

The Way I Know: how I grasp that which is known. In human cognition, this is admittedly limited, abstractive, and dependent on internal and external conditions.

Absolute: not dependent on subjective conditions; conformity with the object as it is in itself.

Relative Truth: dependent on subjective conditions even with respect to what is known, so that *what is* is never known as it is.

VI • SUPPLEMENTARY NOTES

(a) If the intellect were ordered to absolute truth, there would not be so many disagreements and errors in the works of philosophers.

REPLY: This objection is based upon the false notion that to be capable of absolute truth and to be incapable of error are synonymous. The human intellect can and does fail often enough to reach truth, especially in questions which are difficult. However, it must be noted that the above objection would not even be proposed were it not based on some absolute truths; for example:

(1) The clear knowledge of the fact of error;

(2) The clear knowledge of some fundamental truths; and

(3) The recognition of the mind's ability to proceed logically.

Moreover, those who oppose our explanation propose their own as the objectively true one and therefore admit what we mean by absolute truth.

From the fact of error and disagreement the proper conclusions to be drawn are: (1) although truth is absolute, it is imperfectly grasped in human cognition; (2) although scientific knowledge is possible, it is not easy to acquire; (3) hence we should be careful to proceed logically, being especially careful about initial positions taken.

(b) Reality itself is in constant flux and hence cannot be known in any scientific or stable way.

REPLY: Although it is true that the things we meet are mutable, this does not mean that there is nothing in reality but pure "becoming." Nor does it mean that there is no stability in nature, since each thing is what it is at any moment, naturally operates regularly according to its being, and retains its own essential nature in accidental changes.

CHAPTER FIVE

TRUTH

AND THE KNOWN

I • THE SITUATION

Thus far the analysis has shown that I actually do know some truths, that I therefore have a capacity for truth, and that truth in this context has some stability and objectivity so that it is not subject to the whim or caprice or variability of the knower. Hence truth, so far as the content is concerned, means the intellectual submission to and acceptance of what is, or the denial of what is not. This means briefly the *objectivity of the content* of a true judgment—*what is said* is *what is.* This leads to the need for explicitating one more element in human knowledge. At least theoretically it can be seen that with respect to the object known, to *what is,* a double possibility can be conceived. The object may be said either to exist with a purely mental existence or to exist extramentally and independently of the mind. If it exists merely intramentally, then all that exist are minds and their modifications, all extramental existence being a projection of the mind and hence an illusion; if it exists extramentally, then the apparently normal experience of man is valid, but difficulties of explanation do arise.

II • THE ANALYSIS

(a) *Apparently Conflicting Notes.* That man naturally and spontaneously experiences objects known as being other than and independent of himself is a fact admitted by all. In this sense it is admitted that man has a native tendency to admit some doctrine of Realism. Should this realistic tendency be corrected or merely explained? It would have to be corrected, if the object known were purely intramental; on the other hand, a realistic explanation would not be easy because it would have to show just how an *independent* object can simultaneously be outside the mind in existence and inside the mind as known.

In other words, this question really has to do with one problem mentioned in the introduction. Both identity and diversity of knower and known are primary *data* in the noetic experience of man. I experience the object as *other* or diverse, and feel that I know an other-than-myself; at the same time I feel that I do know it, hence that I possess it in a very special way, that it is *present to me,* and that it somehow "enters" my mind and becomes one with it in being known. How can these be reconciled?

(b) *The Denial of Diversity.* One possible answer is to deny the validity of the experienced *diversity* aspect of knowledge. This position would mean that since knowledge is an act of the mind and an intellectual assimilation of an object, the object must be entirely proportioned to mind, that is, it must itself be mental. In this way, it is easy to explain the experience of identity between knower and known; there is an obvious proportion. As to the nature of the object, it is the same sort of being as mind itself. It is purely intramental and mind-dependent.

Hence any "realistic" tendency in man is to be corrected, because what is known is mind-existent, produced by the very act of knowing. This must be the case, since what has only mental existence must be the product of the mind.

Such an explanation, however, cannot be reconciled either with the data of human experience or with what we have already seen about truth. We shall proceed to show therefore: (1) what our previous analyses require in this matter; (2) what the data themselves demand; and (3) how this relates in a general way to the existence even of sensible reality.

(c) *Transcendence. From previous analyses* it has become clear that some transcendence (independence) is necessarily to be admitted with respect to the object of our knowledge. (This transcendence is a characteristic of the object known. A transcendent object is simply one which is not mind-produced or mind-depend-

ent, but which exists whether or not I think of it, and which is not
itself changed by becoming known.) This means that at least in
certain areas a basic Realism is implied in what has already been
seen. Thus, for example, we have seen that the following must be
admitted: (1) being is not non-being; (2) I exist; (3) truth is not
totally relative. These and similar judgments have been admitted
to be true, and their denial cannot be allowed. That is to say,
precisely because they affirm *what* the actual situation *is*, we must
admit them. Hence they are not totally mind-dependent, and to
that extent what they affirm is affirmed because of what is. My
affirmation does not *make them true,* but my affirmation is true because it affirms what is. Thus, as was said above, in at least this
minimal sense Realism alone can give consistent meaning to what
we have already seen. Were this not so, were everything purely
internal to the mind and a product of the mind, it would make no
difference what I affirmed, because it would become true by my
affirmation. This, however, is not my experience. Hence, I realize
that I must submit to and admit *what* is, precisely because *it is.*

Thus, without going immediately into the consideration of the
nature, grades, or kinds of being, we can say that human judgment,
as an act of knowledge, and the very existence of human truth, at
least in the instances we have seen, involves transcendence. It is
not the same to affirm Scepticism as it is to deny it, or to affirm
Idealism as to deny it. We are not free to judge in these matters
just as we please, but we must admit the situation as it is and
affirm what is. The statement that being is the object of our intellect becomes clear in the light of this consideration.

(d) *The Proper Question To Ask. Reflection on the data themselves:* That our analysis has been proceeding correctly can also
be seen by reflecting further on the situation, and by examining the
precise question which should be asked in this matter.

The function of the philosopher is to seek the fundamental explanation of reality, not as he would like it to be nor as it might
be, but as it is actually given. So, in evaluating human knowledge,
he should first seek to describe it as it is and then to explain the
facts of the concrete situation. That we do experience being as
"other," that we seek to conform to the other as it is, is not only
the normal everyday experience of man, but is also the explanation

of the experiments and studies of every scientific endeavor. That complex, concrete experience is what is to be explained. Explanation cannot mean the suppression or elimination of evidence, because once that is done, the experienced situation is no longer being explained.

Hence, the precise question which should have been and should now be raised is, *how* can human knowledge be knowledge of an independent being? *How* can the elements of identity and diversity be reconciled? Both are data. They are equally elements in the situation, so to deny identity is really to deny the experience, or to deny diversity is to deny, not to explain, the experience.

(e) *Real and Intentional Orders.* Since, in the concrete acts of knowledge, the reconciliation is actually simultaneously present, the function of a theory of knowledge is to provide a coherent explanation of the facts. Now if both elements under discussion are said to be in the same order, the assumption of a philosophical attitude called reductionism [1] is inevitable; and this cannot be admitted. Hence, we are forced to recognize and admit the fact that part of the complexity in human knowledge is found in the need for distinguishing clearly between being in the "real" order and being in the order of cognition or the "intentional" order. What is known has both real existence and intentional existence. Only by the recognition of this dual existence of one and the same "nature" or "thing" can both the identity-diversity and the dynamic-static or active-passive characteristics of cognition be preserved in a theory of knowledge. Thus, knower and known are *diverse* in the real order, whereas there is *identity* only in the intentional order. Passivity or receptivity characterizes knowledge in the sense that the mind must receive from what is and conforms to it; the dynamism or activity of the intellect is seen in that the mind gives the object

---

[1] *Reductionism:* This term is used in reference to those schools which admit only one order or level of being (for example, merely material or merely spiritual; merely mental or merely corporeal), and which then seek to explain *all* data in the light of the one "order" accepted. In the matter at hand this attitude is found in the attempt to accept only one part of the data of the knowledge situation (the identity of knower and known) and to reject the other (diversity), thus equating knowledge with total immanence of the object. Thereby the physical order is "reduced" to the intentional order, and the experienced facts are not explained, but evaded. This is an oversimplification.

a new "intentional" existence by making what was merely *knowable* actually *known*. This is far from a creation of the object known, since it just renders *what is* now a *what is known*.

(f) **Intentional Representation.** The explanation thus far still leaves a region of obscurity. Physical or real diversity means that the object really remains "other" and does not in its physical entity enter the mind. How then can there be intentional identity? There is one and only one explanation possible, and since cognition is a fact, it must be accepted. It is that the object enters the mind by way of an intentional likeness or representation. Only in this way can the data of the situation be explained.

Even this explanation, however, cannot be accepted without qualifications, since it would seem that in the last analysis the thing actually known would be the representation of reality and not reality itself, and this would fail to explain knowledge as it is experienced. Hence the meaning and function of the "likeness" or "representation" must be clarified.

A Unique Sort of Representation. First: in this matter a representation must not be naively equated with an ordinary photograph or copy. Such a view is too imaginative and crude to fit the facts in the case of knowledge. Ordinarily a "copy" is itself a something which I can handle and see and which I know in itself, only secondarily and by means of it do I know something about that which is copied. The copy functions as an *instrumental sign*. In knowledge, however, two things must be kept in mind: (1) there is conformity to *what is,* not to a copy of what is; (2) things are intentionally present by representations, not in their physical independent reality. Hence, the representation here is unique and quite unlike any other type of representation. Since knowledge is true and reliable, it does represent what is; since, however, reality is itself known, the representation functions only as that *whereby* or *in which* the object is known, rather than as the object known. Hence, it is said to be a *formal sign.*

A Unique Sort of Means. Second: The representation can also be called a *means* necessary for intentional union with an object. Here again the notion of "means" needs the same qualifications. Elsewhere "means" are things which are themselves taken and used as they are in order to obtain some result or other. In knowledge, the sole function of the representation is to be that *whereby* I grasp the object known, not to be grasped in itself. In other words, representation or means or whatever word is used to describe the function of this element of knowledge must be seen to have a very special meaning, not applicable to any other sphere. Nor is this surprising. Since knowledge is a unique, *sui generis* sort of experience, it is an oversimplification to expect to be able to transfer to it notions taken from other spheres of reference, unless qualifications are added.

(g) *Existence of Sensible Things. Concerning sensation in general:* A more detailed treatment of sensible qualities will be postponed to the next chapter, but we can easily apply what has just been said to the question of the existence of sensible things and of our knowledge of the existence of sensible reality.

This existence of sensible things is a *given* in my experience, hence the question whether or not they exist should never arise. One can ask about the *nature* of material things (in Cosmology), or about *how I know them;* but that they do exist is as evident to me as knowledge itself, the truth of my own existence, the principle of contradiction, or the knowledge that truth is possible of attainment.

In fact, a serious doubt about the existence of sensible reality strikes at the very roots of knowledge, and amounts to a rejection of the ability of the human mind to know truth. The mind, as we have seen, can know truth, that is, can be conformed to *what is.*

Moreover, reflection shows that the mind in all its activities is intimately linked with sensation. What we know directly and easily are sense things, whose images we possess; when we think of "non-sensible" things we always conceive them in relation to or comparison with sensible things. As a matter of fact, even my own acts and existence are known in connection with my sense acts.

In addition, we judge by natural inclination and with the greatest conviction that sensible things do exist. Were the intellect unreliable in this matter, it would quite simply be unreliable in its supposed ability to know what is. It would of its very nature be erroneous, and this cannot be reconciled with what was seen in the early chapters.

In a word, the existence of sensible things is an evidence, a source or principle, of knowledge. It is no more dubious and no more capable of being demonstrated than is the ability of the mind itself to know truth. Hence we said that the question is not whether sensible reality exists, but *how we know it.*

(h) *How I Know Sensible Existents.* To that last question basically the same answer must be given as was given in the general consideration above. Existent sensible things obviously do not physically enter the mind (a thought about the color green is not a green thought, nor is the idea of weight a heavy idea). Hence there is an intentional representation of existent sensible things in the mind.

This representation is not itself the object known, nor is it a means *which* is *first* grasped and which then leads to an "other." Rather it is a means only in the sense that it is that *whereby* the sensible object is immediately known. Hence immediacy in this matter takes on a new meaning. It does not exclude complexity, and it does allow for a unique sort of representation or means.

(i) **What I Know** *Is the Existent Sensible Object. How I know it* is by means of a unique sort of representation.

This provides a coherent explanation of the easily experienced but difficult-to-explain fact of my knowledge of things other than the knower. Moreover, it allows for the immediacy of my knowledge of things, since the representation itself is not made into an object known, as happens in the "copy" theory of knowledge. Thus the human *way of knowing* is complex and unique, but through and in that human way, I immediately know reality itself.

III · SUMMARY OF THE ANALYSIS

(a) Human knowledge seems to require both identity and diversity between knower and known. Can these apparently conflicting characteristics be explained? Cf. *Analysis (a).*

(b) The denial of diversity results in a coherent explanation but cannot be reconciled either with our experience or with what has been seen in previous analyses. Cf. *Analysis (b).*

(c) *From previous analyses* we have seen that some judgments are *true*, that they must be affirmed because of what the situation is, and that my affirmation does not *make* them true. To this minimal degree, some transcendence of the object is implied in what we have already seen. Cf. *Analysis (c).*

(d) Turning to the *data themselves*, we find that both identity and diversity are given or experienced. The proper question, then, is *how* they can be reconciled, or *how* an independent object can become known. Cf. *Analysis (d)*.

(e) This necessitates a distinction between the "real" and the "intentional" orders of being. Diversity of knower and known is true of the "real" order; identity is found in the "intentional" order. In this way, also, the dynamic and static aspects of knowledge can be reconciled. Cf. *Analysis (e)*.

(f) Intentional identity is made possible by the existence in the mind of some representation or likeness of the object. Otherwise there would be physical identity. This likeness, moreover, is something unique and peculiar to human knowledge. It is not a "copy" in any ordinary sense, and cannot be called the object known. It is rather a representation *whereby* or *in which* the transcendent object is immediately known. With the same qualifications, it can be called a *means* whereby or in which the object is immediately known. The immediacy of knowledge, then, does not exclude complexity in our way of knowing things. Cf. *Analysis (f)*.

(g) Considering more specifically the existence of sensible things, we can say that:

(1) Their existence is indubitable and indemonstrable.

(2) Serious doubt about this would undermine the mind's ability to know truth.

(3) All my knowledge is intimately linked to the knowledge of sensible existents.

(4) I naturally and spontaneously judge that they exist, and error here would affect the very nature of the mind.

(5) In human knowledge, the existence of sensible reality is an evidence, a principle or source of knowledge. Cf. *Analysis (g)*.

(h) But the question "how do I know such sensible existents?" can rightly be asked. The reply is that although they remain physically diverse, I grasp them through some representation or likeness which, as was seen above, must be not the object known but the means or sign *whereby* or *in which* I immediately know sensible existents. This means that human knowledge is at once complex and immediate. Cf. *Analysis (h), (i)*.

IV • THE STATEMENT IN BRIEF

The transcendence of the object of knowledge is immediately known. The way in which I know, however, is complex and involves a unique sort of means or representation.

V • DEFINITIONS

Transcendence: a characteristic of the object known, meaning its *independence* with respect to the cognitive activity whereby it is known. (Thus a transcendent object is one whose reality is not due to cognition, but which precedes and regulates knowledge. This term is used elsewhere in philosophy with a much less restricted meaning and in general it involves the character of surpassing a given level or limit.)

Immediately Known: indemonstrable because self-evident.

Intentional Order: the order of knowledge or the logical order, seen precisely as *tending into* or towards reality.

Real Order: the physical order of actual existents.

Instrumental Sign (or a means *which* is used): a thing which is itself first grasped and then leads to the knoweldge of something else, as for example, smoke with respect to fire.

Formal Sign (or a means *whereby* or *in which*): a sign whose whole function is relative, so that it itself is not *first* grasped, but rather in or through it the thing represented is immediately grasped.

Complex: not simple, something which has several elements or parts.

VI • SUPPLEMENTARY NOTES

The main objections to the analysis presented in this chapter rest on misunderstandings as to (a) the nature of the representation or likeness which has been proposed; (b) the internal presence of the object known; and (c) the active-passive aspects of knowledge. Hence for purposes of clarity the following points can be repeated:

(a) The likeness is not the object known, hence knowledge is not mediate. Its whole function is to make the object intentionally present. Nor should this explanation be confused with the "copy" theory, according to which knowledge really is only the representation of a representation. That no parallel to this sort of formal sign or means can be found elsewhere is nothing strange, since human knowledge is not found elsewhere. And we must be careful to avoid the sort of reductionism which seeks to explain one sort of experience, such as cognitional "possession," in terms of a different level of experience, such as material "possession."

(b) In our theory, the object known is at once intentionally immanent and physically transcendent, so that the act of knowledge is entitatively an internal perfection of the knower, and is, at the same time, objective with respect to what is represented. So there is no physical "going out" to the object on the part of the subject, nor any physical "entering in" on the part of the object. Such views are only the result of a too simplified or imaginative approach.

(c) Finally, the analysis given preserves both the active and passive aspects of knowledge. Knowledge is passive only in the sense that it must acknowledge what is and be regulated by and receptive to being; it is active in the sense that it makes the object actually known by virtue of the likeness or representation. This it does, not by an act of creation, but by actively conforming itself to what is.

Thus far our analyses have considered our basic capacity for knowing truth, and we have gradually come to an explicit admission of the fact that a careful reflective analysis of human knowledge as it is concretely experienced must lead to the explicit recognition of my human capacity for truth in the sense that truth is absolute or objective, that its object is transcendent, and that my manner of knowing the object is at once complex and immediate.

We shall now turn to a more careful consideration of the various elements or sources of human knowledge. Apart from the fact that each of the elements to be considered has provided abundant problems for discussion in the history of thought, each of them should be considered carefully in order that we may complete a well-balanced theory of knowledge, and face the specific questions which can arise in these various areas. Thus in the following chapters, we shall turn our attention to: sensation (Chapter Six), concepts (Chapters Seven and Eight), judgments (Chapter Nine), and reasoning (Chapters Ten through Twelve). Our knowledge seems to rest on sense experience, to proceed through the formation of new ideas, and to be expressed in judgments. We seem also to be able to acquire new truths by various sorts of ratiocination. Hence, all these elements must be considered for their relation to truth.

• *Suggested Readings Chapters Six*
Through Twelve

Aquinas, St. Thomas, *Basic Writings of Saint Thomas Aquinas,* ed. Anton C. Pegis, vol. I, q. 16, pp. 168 ff. New York: Random House, Inc., 1945.

Hoenen, Peter, S.J., *Reality and Judgment According to St. Thomas Aquinas.* Chicago: Henry Regnery Company, 1952.

Joyce, George H., S.J., *Principles of Logic,* chs. 14–15, 20, pp. 215–251, 320–336. London: Longmans, Green & Co., Ltd., 1926.

Regis, L. M., O.P., *Epistemology,* translated by I. C. Byrne, pp. 311–64. New York: The Macmillan Company, 1959.

Wild, John, *Introduction to Realistic Philosophy,* pp. 413–39, 441–68. New York: Harper and Brothers, 1948.

Wilhelmsen, Frederick D., *Man's Knowledge of Reality,* chs. 8–13, pp. 75–164. Englewood Cliffs, N. J.: Prentice-Hall, Inc., 1956.

CHAPTER SIX

SENSATION AND TRUTH

I • THE SITUATION

(a) *Sensation Attacked.* Although it would seem true that all our knowledge somehow arises from sensation, the reliability of the senses has been variously and vigorously attacked during the whole course of the history of philosophy. The attack has been particularly acute in the modern era, due to the influence of Rationalism and philosophical Empiricism, and especially to misunderstandings arising from the findings of the empirical sciences, particularly from those of experimental psychology. Since this question is so vast in its implications, our limited purpose of discovering the basic elements of a coherent theory of knowledge must be kept in mind. Hence we cannot be expected to go into details which pertain more to Psychology, Cosmology, or the empirical sciences.

(b) *Senses Are Very Limited in Range.* Many difficulties would be avoided and many objections easily answered, if the following fact were remembered: the senses are material, very limited, and highly specialized organs of perception. They are built to function in a very limited range and in respect to selected stimuli. To expect them to be reliable beyond their extremely limited horizon is to ask the unreasonable, and to impugn them for their limitations is merely a refusal to accept them as they are. To want the eye to see sound or the ear to hear color is patently absurd, but to expect the eye to see subvisual particles and to complain because it does not is also to fail to recognize the natural limitations of each sense in respect to what it can perceive.

(c) *Proper Conditions Needed.* Thus we can say in general that since the senses are *material organs*, they must be in a healthy condition and rightly disposed in order to function reliably. Any disease or defect will naturally interfere with their normal functioning. The very recognition of abnormalities in this matter is an indication that the intellect is capable of knowing what is and what is not normal.

Moreover, since the object perceived remains other than the particular sense involved, its impulse is usually transmitted through *some medium* or other. An improper medium or a defect in the medium will naturally interfere with proper perception.

Finally, *the object* perceived must be proportioned to the sense. If it is too distant or otherwise disproportioned to the sense in question, it will not be rightly sensed.

Hence, certain conditions must normally be fulfilled on the part of the organ, the medium, and the object; and it is the work of an intelligent subject to recognize these.

(d) *Divisions.* Traditionally, what we perceive has been divided into two classes called *Proper Sensibles* and *Common Sensibles.*

A *Proper Sensible* is that which is naturally perceived by only one sense. Thus the object of sight is color; of hearing, sound; of smell, odor; of taste, savor; of touch, heat or cold or resistance, and so forth.

A *Common Sensible* is that which is naturally perceived by several senses, such as movement, rest, number, figure, and magnitude.

John Locke popularized a different terminology so that what had been known as a proper sensible became known as a *Secondary Quality*, whereas the common sensible was called a *Primary Quality*. (The slight changes which he made in his lists of qualities need not detain us here.)

(e) *Objectivity of Qualities.* In the previous chapter we saw that the existence of sensible things is an immediate datum of our experience, and hence it need not and cannot be demonstrated. It is a source of knowledge, not a conclusion. In the present chapter we want to consider more closely the sensible qualities themselves, in order to assess their objectivity and to learn how reliable the senses may be said to be in perceiving them. For purposes of orderliness and clarity we shall turn first to the common sensibles.

II • THE ANALYSIS

Common Sensibles

(a) *On Extension.* The explicit consideration here can be limited to the quality of extension alone, since the other primary qualities are intimately linked up with and dependent on extension. Moreover, the present analysis can very easily be applied to each of the others in turn.

We can describe extension as that property of sensible things by virtue of which they are shaped one way or another, or are divisible into parts.

(b) *Intelligible Notion.* Reflecting now upon my experience of extension and my judgments concerning it in the light of what has already been seen about my intellectual capacity for truth, I recognize certain facts. First, I do have an intelligible notion of extension. Second, I predicate that notion regularly and evidently of things.

That I have an intelligible notion of extension is clear from the fact that I can describe what I mean by it, and can do so in a way which distinguishes it from every other quality and every other thing. Moreover, my notion is such that it is not limited to any one extension or extended thing, but can be predicated indifferently of many very different things, such as apple, tree, brick, house, man. In this sense it is truly an *intelligible* notion in that it is not limited to this or that particular extended thing, but can be and is predicated of any one or all of several things.

(c) *Predicated of Reality.* Moreover, it is an incontrovertible fact that I regularly and naturally judge material reality to be extended. The actions of ordinary living, such as walking and riding are constant confirmations of my acceptance of the fact of extension as something given and something constantly present to me. This is something of which I am ever aware and on which the conduct of my life and action rests without hesitation.

(d) *Reliability of the Predication.* Reflecting upon these facts, and viewing them in the light of our first chapters, it must be admitted that we are here presented with an element whose validity or falsity is intimately connected with the intellect's capacity for truth. The intellect naturally and spontaneously grasps and affirms the fact of extension. Of its very nature it applies this intelligible notion to reality, and has no grounds or even possibility of detecting error in this, its natural inclination. Hence this sort of predication is as natural to the mind as anything can be, and were it to be deceived in such evident and ever recurrent predications, the error would be an inherent and native characteristic of the mind. This, however, cannot be admitted, since it would not be consistent with the natural capacity of the mind for truth. To admit such error to be of the nature of the intellect would be to deny that it is an intellect at all, because insofar as it is an intellect, it is capable of knowing being, whereas insofar as it would be said to err of its very nature in respect to extension, it would of its very nature be erroneous, hence noncognoscitive of being.

Hence, careful reflection merely explicitates in this matter what is naturally recognized: namely, that extension is an immediate datum of our experience, and that the intellect cannot of its nature be erroneous in the judgments whereby it predicates extension of reality.

Proper Sensibles

(e) *Difficulties Proposed.* Turning to a consideration of the proper sensibles (or secondary qualities), it must first be admitted that many difficulties concerning their objectivity can be and have been proposed. Thus, color blindness, the feeling of pain in an amputated limb, the variety of sense reactions to electrical stimuli, and similar experiences have awakened doubts and problems about the reliability of sensation. Later we shall briefly consider some of these objections, but for the present it should be borne in mind that a difficulty is vastly different from a well-founded doubt and that even when a solution cannot be apodictically proposed, it can be sufficient to show that no other proposed solution meets the difficulty in question. In this present question our analysis will proceed in two steps to show: (1) that the secondary quality somehow really exists in the object which is sensed; and (2) that the quality should normally be said to exist formally in reality.

(f) *Causal vs. Formal Existence.* At least theoretically, various possibilities are open regarding the existence of proper sensibles. They can be said to be purely and totally subjective; they can be said to exist in reality only in the sense that a something or other is there which causes the subjective impressions; or they can be said to exist formally in reality as they are perceived. However, since their *purely mental* existence can in no way be reconciled with the analysis in Chapter Five, it remains to determine whether they exist only causally or formally in reality.

Mere causal extramental existence would mean that there exists in the object sensed some power or other, not sensed in itself, but somehow capable of causing in the sense organ that which is perceived as sound, color, taste, and so forth. This would mean that such qualities are not actually qualities of the object sensed, but subjective reactions to and interpretations of the stimulus received.

Formal extramental existence would mean that this wall is really white, this sound really loud or harsh, or this water really warm. The function of the senses would be not to interpret an unknown impulse from the object, but to perceive what is characteristic of the object and admit what objectively is in reality.

(g) *Passivity of the Senses.* First of all, it can be stated that sensible cognition appears to be clearly aroused by and dependent on a sense object. If I open my eyes, I see something there and I can will not to see it as seriously and vigorously as possible, but without success. This and countless similar experiences indicate that my senses are passive or receptive in relation to their objects.

In other words, my senses are *activated* by some activity of the object which influences and determines them. Thus, the healthy eye is capable of seeing or the ear is capable of hearing, but to actuate that ability there is required some action or some causal influence from the object in respect to the color, sound, or other sense quality which is perceived. Hence, the color or sound must somehow or other exist in the object itself.

(h) *Common Sensible Grasped Through Proper Sensible.* This explanation, however, does not yet determine clearly enough just how the quality is said to be in the object, whether formally or merely causally. It has already been seen that the common sensible is truly objective, and the same should be said of the proper sensible, since I do actually perceive both. In fact, I *see* extension precisely insofar as it is colored, and the loss of the ability to see a colored thing involves the loss of the ability to see an extended thing, because it literally means the loss of the ability to see. The senses, in other words, witness to and reveal what is; and their witnessing reveals not only the common but also the proper sensible. Hence both are objectively or formally in reality. This does not mean that my knowledge of proper sensibles presupposes my knowledge of common sensibles or that I establish the objectivity of the proper by reliance on the common. Rather, what has been said of the objectivity of the common sensibles can also be said, even *a fortiori,* of the proper sensibles, since in my grasp of sensibles the latter have a priority. (Cf. Wild, *Introduction to Realistic Philosophy,* Chapter Eighteen.)

(i) *A Natural Conviction.* Again, reflection reveals our natural conviction that secondary qualities are real. In this context it is difficult (even impossible) to see what the mere causal existence of secondary qualities can mean or how it can actually differ from their being there formally. If by "causally" is meant "eminently," then *what is* in reality *is not sensed,* and sensation ceases to testify to *what is,* since in that case it would give only the "appearance" of what is, leaving the object itself impenetrable. If by "causally" is meant any more than that, it coincides with saying that the qualities sensed exist "formally" in reality, and thus the objectivity of sensation is preserved.

(j) *Predication of an Intelligible Notion.* Our analysis in this matter could actually take the form of the reflection used under (b), (c), and (d). Thus, we do have an intelligible notion of each proper sensible, since we can recognize each and distinguish it from all else. These intelligible notions we do naturally and regularly predicate of sensible things (as is experienced by all and admitted even by those who might deny our position because of certain obscurities and difficulties in this question). Were the intellect wrong in this matter, its very nature as a faculty cognoscitive of what is would have to be seriously doubted or denied.

III • SUMMARY OF THE ANALYSIS

(a) I have an intelligible notion of extension, and this I constantly predicate as a self-evident property of reality. Cf. *Analysis (b), (c).*

(b) Since the intellect is naturally trustworthy in this sort of predication, its judgment as to the validity of extension is trustworthy. Cf. *Analysis (d).*

(c) Proper sensibles must be said to exist either causally or formally in the object sensed. This is clear from an analysis of the passivity of the senses in respect to what is sensed about the object. Cf. *Analysis (e), (f), (g)*.

(d) With the same clarity the senses actually witness to and reveal the formal existence in reality of both the common and the proper sensibles. In fact, perception of the common sensibles presupposes perception of the proper sensibles. Hence, the proper sensibles exist formally in reality. Cf. *Analysis (h)*.

(e) What can mere "causal" existence of the qualities mean? If it means "eminently," the senses give us only phenomena and leave things unknowable as they are. If they are still said to be sensed as they are, then "causally" in this context really means the same as "formally." Cf. *Analysis (i)*.

(f) Here also it is true that we have an intelligible notion of the proper sensible, and that we do naturally and regularly predicate it of things. Error in such a case would touch the very nature of the intellect as a capacity for truth. Cf. *Analysis (j)*.

IV · THE STATEMENT IN BRIEF

Both common and proper sensibles exist formally in reality.

V · DEFINITIONS

Sensible: the object of a sense organ.

Proper Sensible (Secondary Quality): that which is naturally perceived by one sense only, such as color, sound, heat, resistance, and so forth.

Common Sensible (Primary Quality): that which is naturally perceived by two or more senses, such as extension, magnitude, rest, motion, and so forth.

Sense: an organic faculty the object of which is material being.

Extension: that property of sensible things whereby they have shape and are divisible into parts.

Exist Formally in Reality: that which is sensed does objectively characterize reality.

Exist Causally in Reality: that which is sensed is not objectively

in reality, but there exists some power or other in the object
capable of producing the sensation.

VI • SUPPLEMENTARY NOTES

(a) The senses do frequently err, and yet the analyses just given
do seem to exclude the possibility of error.

REPLY: Error is strictly found only in the intellect judging, and
hence is not found in the senses. However, in the sense that the
senses do frequently provide the occasion for intellectual error, it
can be admitted that they frequently err. This can readily be
conceded. However, in view of the intellect's capacity for truth
and its natural reliance on the senses, it cannot be admitted that
the senses are of their very nature erroneous. Rather, error is due
to various other factors, such as inadvertence or precipitance of
the intellect in judging, to a disease or defect in the senses, to improper media, or to some disproportion in the object. As was
said, the senses are highly specialized and limited faculties, and
should not be expected to function beyond their capacity. Man is
intellectually capable of recognizing and respecting these limits,
but he often in practice goes beyond the evidence and thus errs.
This does not mean that the senses are not basically cognoscitive
and reliable, but does mean that caution should be observed. (On
this matter cf. St. Thomas, *On Truth*, q. 1, a. 11.)

(b) The microscopically discoverable discontinuity of sensible
matter would seem to contradict the objectivity of real extension.

REPLY: There is no contradiction here for two reasons: (1) two
quite different sorts of analyses are confused in the objection offered, since we are asking what is to be said about *sensibly perceived* extension, whereas the objection has to do with the infrasen-
sible; and (2) even in the microscopic view there is still extension,
hence there is confirmation rather than denial of our position.
Moreover it must be remembered that the study of the *nature* of
extension will be treated in Cosmology.

(c) One of the factors contributing to the assertion that proper
sensibles are really subjective reactions only, is the fact that much
progress has been scientifically made by treating qualities merely
as quantified. That is to say, color, sound, and so forth, can be

measured with great accuracy according to wave lengths and thus easily classified. This clearly involves a quantitative aspect in respect to these qualities. It is then an easy step to the assertion that the so-called quality is identical with the quantity or measurable frequency recorded.

REPLY: As a matter of fact, it can be admitted that for all practical purposes in such empirical studies this method suffices and has been fruitful of results. And insofar as the scientist makes only scientific statements, he is quite justified. However, to extend his findings to determine the ontological categories of being is not the work of the empirical sciences but of that science which analyzes the grades of being, that is, of philosophy. From that perspective quality and quantity are irreducibly different categories. I do see an extended color, but the color is not the extension itself, nor is the extension the color. In fact, so far as knowledge is concerned, it is by virtue of the color that *I* see the extension, although ontologically it is by virtue of the extension that the color can *be* in the object. Hence if proper perspectives are maintained, there is no real conflict on this point.

(d) The proper sensible cannot be formally in the object, since, for example, pain or pleasure is not in the fire which either burns or gives warmth.

REPLY: A whole barrage of similar difficulties can be put forward concerning color, sound, and the other sense stimuli. Such objections would stress the point that these sensations are merely reactions of the organism to its environment and hence the qualities themselves are formally only in the organism.

In general it can be said that these involve a confusion between the sensation of a quality and the quality sensed. To say that *sensation* is formally in the sense is a truism, and if the quality for example, color, is made identical with the sensation of color, then the quality is purely subjective. However, this cannot be justified since sensation is the organic reaction to and reception of what is given. It is true that nothing can be sensed unless it is sensed, still it is not justifiable to say that what is sensed is the product of the act whereby it is sensed. On the level of sensation this parallels the famous "egocentric predicament" discussed and dismissed by R. B. Perry, when he so clearly pointed out that although it is true that the *ego* is involved in all knowledge in the sense that nothing can be known without a knower, still from that it does not follow that what is known is the product of its being known. So, the actual sensation is formally in the organism which senses, but the actual quality sensed is formally in the object sensed.

Finally, it can be admitted that the "otherness" of what is sensed is more evident in the "higher" senses such as sight and hearing than it is in the "lower" or more material senses such as touch or taste. In the latter the sense itself is more entitatively and physically affected than in the former; as for example, in feeling heat, the organism itself not only feels the warmth but itself becomes warm. However, despite this greater "materiality," the same basic principle of objectivity must be admitted, although its application naturally requires qualifications according to the sense in question.

(e) Many individual facts would seem to show that what is sensed is not in the object. Thus pain is felt in an amputated limb; the same stimulus applied to the eye and ear produces the sensation of light in the one and of sound in the other; a sensation of cold is felt when a cold spot is touched even with a warm object; some people are "red-green blind," others, who are totally color-blind, equate every color with a definite neutral gray.

REPLY: In general these difficulties can be resolved to some extent by the following considerations:

(1) the stimulus may be so complex that it does actually contain the quality which affects several senses; or

(2) the stimulus by somehow exciting the nerves merely occasions the recall of previous images; or

(3) abnormal structure of a sense naturally means defective reception of the objective quality.

In all these instances the intellect has its part to play by recognizing the defects in the situation, which might arise either from the object, or medium or sense, and by judging accordingly.

CHAPTER SEVEN

THE VALIDITY OF CONCEPTS

I • THE SITUATION

(a) *True Judgments and Their Elements.* Having briefly analyzed sense experience, and having recognized its natural and basic reliability as a source of knowledge about material things, we can now consider the intellectual way in which we come to know reality. In this matter it must be repeated that the act of which I am conscious and which I recognize as cognoscitive is the act of judgment. That I have judgments is an immediate datum of my experience; that I have some true judgments has already been seen in the first chapters. Moreover, since any judgment of predication is recognized as a complex act made up of a subject, a predicate and the assertion of a relation between them, we can already admit in a general sort of way that the elements of the judgment are themselves reliable and objective.

(b) *Special Problem.* However, there is a very special problem which arises in connection with some of our ideas, and it must be carefully analyzed. On the one hand every sensible thing we meet seems to be totally individual, singular, changing, and contingent; on the other hand we seem to have ideas which are universal, stable, and necessary. Hence, the questions naturally arise: Are there really any universal ideas? If there are, is there anything in reality which corresponds to them? This is the famous problem of the *universals* on which John of Salisbury even in the twelfth century said that "more time has been consumed in discussion than the Caesars spent in conquering the world and more money than Croesus ever had." Porphyry would hardly touch the subject, since he considered it too complicated for an introductory work. Modern philosophy has handled the question in various ways, and solutions have varied from the denial of all universality to the denial of any real contingency.

(c) It must be admitted that we can now propose a coherent, satisfactory and valid solution to the vexing problem of the "one and the many" in its logical implications for a theory of knowledge, precisely because of the efforts at analysis, both successful and unsuccessful, on the part of our intellectual progenitors.

(d) *Limitation of the Question.* In approaching this question, two things are to be kept in mind: (1) We are speaking of judgments of predication, for example, this tree is an oak, John is a man, the house is white, this water is hot, and so on. (2) Although we recognize various types of ideas, as, for example, *singular* (this house), or *abstract* (whiteness, longevity, ability, or coldness), we are concerned precisely and solely with those which are called *universal*. These are concrete and univocal. Examples are man, house, dog, blue, green, tree, table, chair, tall, fat, gold, clean, and so forth—hence anything which can be used directly as a predicate in a judgment of predication.

II • THE ANALYSIS

(a) *Universality of Words, Ideas, Things.* Reflecting on the question of the existence of the "universals," it can be seen *a priori* that many solutions are at least theoretically possible. Thus, since the elements involved in knowledge and in its communication are the *words* used, the *things* about which one judges, and the *ideas* related in a judgment, it is theoretically possible to decide that there are:

(1) universal words, singular ideas, and singular things; or
(2) universal words, universal ideas, and singular things; or
(3) universal words, universal ideas, and universal things; or finally,
(4) universal words, universal ideas, and things somehow universal and yet singular.

(b) *Oral Predication and Ideas.* To determine whether there are universal ideas and whether these correspond to reality, it is necessary to start from something recognized and admitted by all.

That minimum is the fact that we do use one and the same *word* in *oral predication* to refer to many individuals. Thus orally I use the term "man" to refer to John, Peter, Joseph, or the word "chair" to refer to several objects.

This fact can now be related to the purpose and the function of human speech, which is to communicate to others what is in our minds. Words, therefore, manifest ideas, and various sorts or types of words must represent varying sorts of ideas. Thus singular words reveal singular ideas, and collective words reveal collective ideas, such as *flock* or *family*. Now we do use words such as *man, tree, dog, table,* and so forth, which cannot be characterized either as singular, since they are applied univocally to many, nor as collective, since they are applied distributively to individuals but not to groups or collectivities. Hence, if these are, as they must be, revelatory of what is in the mind, there must be in the mind ideas which are neither singular nor collective, but of another kind. These we call universal ideas, since they represent and can be predicated of many individuals singly and univocally.

Since the purpose of language is to manifest ideas in the mind, and since there are words which are universal, there are universal ideas in the mind.

(c) *Objectivity of Universal Concepts.* Moreover, in any true judgment the subject and predicate must represent what is in reality, in the sense that *what is asserted* is true of the object known. Since the mind is capable of truth, it follows that its concepts in true judgments are objective; and since what we call universal concepts are referred to and predicated of reality, it follows that universal concepts, when truly predicated, are objective. Without this objectivity, scientific knowledge would consist purely of fabrications quite divorced from reality.

(d) *The Way in Which the Universal Is Conceived.* Further analysis, however, of judgments of predication demands a new qualification. If you consider the judgments John is a man, Peter is a man, Joseph is a man, it is clear that what is predicated must be identical with the subject in each case. Hence *what* I conceive in grasping the concept "man" is found concretely identified with John, Peter, and Joseph.

But in order that *what* is predicated of John may also be predicated of Peter and Joseph, it is necessary that the notion of "man"

be known in a manner which makes such predication possible.
How is it possible? It is possible only if the notion "man" is
grasped in a way which does not restrict it to John alone or to
Peter alone, that is, it must not include that which singularizes and
concretizes it as it is in reality. All that makes man to be *this*
man must be excluded from the notion. It is in this sense that we
must say that the notion is grasped without the notes or characteristics which individuate it.

Thus *what* is conceived (man) exists in reality and the concept
can be predicated of reality; *the way in which* the notion is conceived is peculiar to our human way of knowing, for it entails
the exclusion of individuating notes, and this is not true of reality.
One and the same nature exists individuated in reality, and nonindividuated in the mind. If this were not true, predication would
not be possible—and we know that predication is possible because
it is a fact.

(e) *Need for Distinction.* All this, however, gives rise to a further question. If *what* is conceived and predicated of a thing, for
example, John is white, is objectively true of the thing, then both
"John" and "white" must be identified or one in reality. If, on the
other hand, I must conceive "white" in such a *way* that it can also
be predicated of "Joseph" and of "book" and of "paper," it would
seem that "John" and "white" must also be somehow separated or
nonidentified. So a distinction or separation must somehow be
introduced between "man" and whatever singularizes it or makes
it to be "this" man.

Now, understanding *distinction* to mean the absence of identity
between two or more things, and a *real distinction* to be one
which exists in reality, for example, between chair and table, and
a *rational distinction* to be one which is introduced by the mind
where there is none in reality, we can say immediately that between *man* and *this man* there is and can be no real distinction.
True predication demands real identity between the subject and
predicate. Thus, to say that "John is a man" is to say that all that
makes up the comprehension of man is found concretely and
singularly in and identified with John. "Man" is predicated of him
as a whole, not as a part.

(f) *Foundation for Rational Distinction.* Thus, any distinction
between "man" and all that concretely makes man "this man" must

be said to be a rational distinction, one introduced by the mind between "things" really identical.

Is this a purely arbitrary function of the human mind, or is there some foundation in things themselves which justifies this rational distinction? That it is not purely arbitrary or whimsical is already admitted by the recognition of the objectivity and validity of judgments of predication. More concretely, the precise reason for saying that the mind legitimately makes this distinction is to be found in the fact that things are objectively and really similar. Thus, there is a real similarity between John and James in that both are men, or tall, or white; between the wall and the book in that both are green in color; between an elephant and a mouse in that both are animal, or gray, or four-footed, and so on.

Since two or more things can be recognized as objectively similar in many aspects, and yet one is not the other, it must also be admitted that they are dissimilar in other aspects, and in fact that they are dissimilar in everything which makes each to be an individualized singular being. Thus, things have *common* characteristics, and also *proper* or individual characteristics. For example, every man has blood and bones and flesh, and these are common aspects in which Joseph and Peter equally share, yet each of them differs from the other in that each one's body is limited and determined and proper and thus individually different.

Going one step beyond this, one can say that the foundation for the rational distinction between the common nature and the individuating principles in question is to be found in the fact that each material thing is a *composite* being. That is to say, beings *are* similar and they *are* also dissimilar, and that whereby they are similar must be really distinct from that whereby they are dissimilar. Otherwise there could be no objective differences between them, and they would not be really two or more individuals, but one only individual. Hence, there must be some real composition in a material thing, the components being: (1) that reason or source or principle which accounts for similarity and (2) that principle which accounts for their dissimilarity.

Later, in Metaphysics, it will be seen that this fundamental composition is one of matter and form or of act and potency, but for the present it is enough to see that the individual is not really simple but a composite being, and that *some real composition* in each individual is the basis and the objective justification for the rational distinction made between the "nature" and the "individuating principles" in the conception of the universal.

III • SUMMARY OF THE ANALYSIS

(a) By admitting that oral predication is meaningful only in the sense that it is manifestative of knowledge, one admits also that to words of irreducibly different types there correspond ideas of irreducibly different types. Thus, to singular words there correspond singular ideas, and to collective words, collective ideas. Since words such as *man, tree,* and *white* are neither singular nor collective, there must correspond another sort of idea, called universal, since it is one idea which is predicable of many individuals distributively. Cf. *Analysis (b).*

(b) *That which* is grasped in a universal concept is true of reality, since this follows from the mind's ability to know and judge reality and to build up scientific knowledge. Cf. *Analysis (c).*

(c) *The way in which* the nature or notion is grasped is one which excludes all that individuates the nature. Otherwise it would be limited to one being only, for example, to John alone, and could not be predicated of others. Hence, *the way* or *manner* of conceiving things is peculiar to our way of knowing and is neither predicated nor predicable of what is in reality. Cf. *Analysis (d).*

(d) True predication requires that the nature conceived (the universal) be really identified or one with the individuating principles. Cf. *Analysis (e).*

(e) Hence, the distinction between a nature and all that individuates or singularizes it is a rational distinction. This rational distinction is not purely arbitrary or fictional, but has a foundation in singular things. The foundation is the fact that things are objectively similar in many aspects, and hence are *composite beings,* having both a principle whereby they are similar and a principle whereby they are dissimilar. Cf. *Analysis (f).*

IV • THE STATEMENT IN BRIEF

There are universal concepts which are (a) objective in respect to what *is conceived, (b) not objective in respect to* the way in which *they are conceived. What is conceived is really identical with and rationally distinct from the individuating principles.*

V • DEFINITIONS

Universal: capable of being related to many, for example, the cause of several things, the blueprint of a chair in respect to many chairs made according to it, a word used to designate several objects, and so forth.

Universal in Being: capable of being in many.

Universal in Predication: capable of *being predicated* of many.

Objective: corresponding to reality, or existing in things.

What is Conceived: the aspect, note, or nature which is affirmed or denied.

The Way in Which: how the mind grasps or apprehends, according to the nature of the mind.

Distinction: the absence of identity between several things.

Real Distinction: one which is had in reality, as for example, John is really distinct from Joseph, or a table from a chair, or my foot from my arm.

Rational Distinction: one which is due to the operation of the mind. This can be done either with or without a foundation in reality for the distinction made.

Individuating Principles: those determinations whereby the individual is actually individuated. These are more essential than the mere accidents of size, shape, place, time, and so forth, and refer to the very material being itself in all that makes it radically "this."

CHAPTER EIGHT

THE VALIDITY OF
CONCEPTS *(Continued)*

I • THE SITUATION

(a) From the previous chapter it can be seen that in respect to universality and singularity we can say that we do use universal *words*, we do have universal *concepts*, and that *things* are formally or actually singular, but that their objective similarities provide a real foundation for universal concepts. In that sense they can be said to be *fundamentally universal*.

(b) Granted, therefore, that we have valid universal concepts, the question naturally arises as to *how these are acquired*. Since the nature apprehended is formally singular in reality and is rendered formally universal only by the mind (under the direction, it is true, of reality, and guided by the recognized structure of beings), what is the process which makes this possible while preserving the objectivity of knowledge?

II • THE ANALYSIS

(a) *Explanation Must Respect Fact of Predication.* To arrive at a balanced and coherent answer to this question, it will be necessary to recall just what we have to work on, and to follow each step carefully, keeping in mind that we are speaking of *intellectual* knowledge, and so being careful not to let the imagination hamper our investigation by demanding a "picture" of something which is beyond the sense level.

What we have to work on and what we are fully conscious of possessing is a judgment of predication, such as John is a man, Peter is a man, the table is green, the book is green, the water is cold, the house is cold, and so forth.

I predicate one and the same concept or predicate of several different subjects. Hence, however I acquired the predicate, it had to be in a way which makes such predication possible.

(b) *Two Ways of Apprehending.* Yet, reflecting upon the meaning and the implications in the judgments John is a man, Peter is a man, I find that "man" as used here has almost conflicting characteristics. On the one hand, in order to be predicated of many, it must first be seen precisely *as common to* and predicable of many. That is to say, the notion "man" must be apprehended in a way that includes the note of universality as such. In this way I grasp the *species* man, which as such cannot exist in reality, since only individual men exist. This, then, is a mental category which classifies beings but which I cannot predicate; I cannot say that "John is man (the species)."

On the other hand, since I do predicate man of John, I must also have apprehended this nature in another way; namely, in a way which does not include either universality as such nor singularity as such. Were it to include universality, it could not be predicated of individuals (one cannot say that John is the universal man). Were it to include singularity, it could not be predicated of more than that one singular being (one does not say that John is man in a sense applicable to no one else).

Hence, it must be admitted that in apprehending natures or things under one or another aspect, I apprehend that aspect or note in two quite different ways. In one apprehension I conceive the nature or note merely according to what it means (its comprehension), omitting all reference either to singularity or plurality, and hence, this gives me a grasp of the nature which is not formally universal, and which can therefore be predicated of things. All that is conceived is the nature *absolutely* considered, that is, not limited to any one individual nor yet formally universal. This is commonly known as the *direct universal: direct* because it is directly acquired from things and directly predicable of things, and *universal* only in the sense that it can be given the form of universality.

In the other apprehension I conceive the nature or note not only in its comprehension, but also according to its extension, that is, precisely as common to many. Without this conception there could be no predication of *that which* is grasped in the direct universal. To predicate *what is grasped* of many, it must first be seen that it is predicable of the many, and hence must receive the form of universality. This is known as the *reflex universal: reflex* because it presupposes a prior act of the subject in apprehending the nature absolutely, and *universal* because it is actually one and many simultaneously. This is the *universal in being* and it can exist only in the mind.

(c) Briefly, what I am explaining is the fact of predication. I am fully conscious of judging, that is, of affirming or denying the identity of a subject and a predicate. To do this, I must first apprehend them. To affirm the predicate of the subject, I must apprehend it as predicable, and this means: (1) grasping *what is to be predicated,* which is the nature or intelligible aspect—the direct universal, and (2) grasping it as *predicable of many*—the reflex universal. True, I am not conscious of these as such, but without them a judgment cannot be explained.

(d) The next question which arises is: *by what process does the mind form or apprehend these universals?* To answer this question is merely to explicitate what has already been seen.

(e) *Acquisition of Direct Universal.* Thus, since everything which exists is singular and individuated, and since the direct universal is some nature or intelligible aspect conceived according to its comprehension only, and thus totally free of individuating principles, the intellectual process of apprehending it must be one which separates that nature from the individuating principles. Abstraction in general means the separation of one thing from another. Hence, the direct universal is formed by some sort of mental abstraction or separation.

Moreover, this abstraction, or separation, is most unusual in the sense that nothing at all is positively apprehended except the nature according to its comprehension. That is to say, the individuating principles are not apprehended at all; even the fact that they are omitted is not grasped. If it were, the nature or note could not be predicated of the subject, because predication would mean, for example, that "John is a man without individuating principles," and this is not true. But "John" is and has all that "man" signifies, and that is all that is grasped through the sort of abstraction we are considering.

This sort of intentional abstraction is known as *total precision: total* because what is grasped is true of and predicable of the whole object; and also because the individuating principles are *totally* unknown, and *precision* because it involves an intentional separation in the order of simple apprehension of two things which in reality are identical.

Briefly, the direct universal is formed by the process of *total precision*. This, then, is the way in which human knowledge of sense reality originates, this is how our intellect spontaneously, naturally, and preconsciously proceeds in the acquisition of knowledge; and although we know *in this way*, we are totally unaware of it as we actually proceed in the act of knowing.

(f) **Acquisition of Reflex Universal.** So far as the reflex or formal universal is concerned, we see that it refers to one and the same nature now apprehended in its relation to the many individuals in which it is or can be. Hence, all that is added is the form of universality, and that is added by the mind because of the objective similarity existing between things.

So the reflex universal exists only in the mind and only when it is known. It is formed merely by the mind conceiving the direct universal as related to many, by simply comparing the one nature with the individuals in which it is or can be found.

This process is called *simple comparison: comparison* because one nature is compared or related to many individuals and *simple* because it is merely in the order of apprehension, not in the composite act of judgment. This latter must be true, since judgment itself already presupposes the reflex universal. No affirmation or negation is required, but merely the intellectual apprehension of the nature as capable of being found in many numerically distinct individuals. In other words, predicating the nature of many presupposes the apprehension of that nature as *predicable* of many or as related to many, and this is apprehended by simple comparison.

(g) **Existence in Two Orders.** From what has been seen, we can say that the nature or note which is conceived exists individuated and singular in reality, and the same nature exists in the intentional order in a nonindividuated or universal way. So it can be considered either absolutely, apart from singularity or universality, or as singular, in reality, or as universal, in the mind.

(h) **Relation of Both to Reality.** Finally we can add that while both universals are possible and valid because things are really similar to one another, yet the direct universal is proximately founded on reality, since it is immediately abstracted from reality and can be predicated of it; whereas the reflex universal is only remotely founded on reality, in that it is directly concerned with the direct universal.

III • SUMMARY OF THE ANALYSIS

(a) In seeking to determine *how* the universal concept is acquired, I must realize that ultimately I am explaining how one nature can be predicated of several individuals, and hence the way of acquisition must be such as to allow for the possibility of such predication. Cf. *Analysis (a)*.

(b) What is apprehended, a nature, note, or intelligible aspect, is grasped in two ways: (1) *absolutely*, that is, without reference either to singularity or to universality, hence according to comprehension only—this nature does exist in reality and hence can be predicated of reality; (2) *relatively* to the individuals in which the nature is or can be found, hence including the note of universality. This is the "one common to many," which exists formally only in the mind. It can itself be predicated of nothing, but because of it, I can predicate the direct universal of things. Cf. *Analysis (b), (c)*.

(c) The direct universal is grasped by a process called total precision. This means that what is grasped is the nature or intelligible aspect according to its comprehension only; I not only do not grasp the individuating principles, but I am not even aware that they are being omitted.

This must be so, because *what* is apprehended is predicated as a whole of an entire being, and the *omission* of individuating principles cannot be predicated. Hence this *omission* is the *natural way* in which the mind spontaneously apprehends, but it is not part of *what* is apprehended. Cf. *Analysis (e)*.

(d) The reflex universal is the same nature grasped in its relation to many. As such it is a mental product, since no nature in reality is simultaneously one and many. To form it, only the form of universality or "predicability-of-many" need be added. This is added by *simple comparison*, that is, by the simple apprehension of the nature as capable of being in many. The judgment of predication itself presupposes the reflex universal, hence judgment cannot be required for its formation. Cf. *Analysis (f)*.

(e) The direct universal is predicated of reality and is directly founded on reality; the reflex universal is remotely founded on reality through the direct universal. Cf. *Analysis (h)*.

IV • THE STATEMENT IN BRIEF

There is a twofold universal: namely, direct and reflex. The direct universal is formed by total precision and is proximately founded on reality. The reflex universal is formed by simple comparison and is remotely founded on reality.

V • DEFINITIONS

Direct Universal: a nature or intelligible aspect, not having individuating notes, considered only in its comprehension, and hence, although not actually singular or universal, capable of being universalized.

Reflex Universal: a nature or intelligible aspect not having individuating notes, considered according to its comprehension and extension, and hence actually universal.

Total Precision: the conception of one thing in such a way that something really one or identical with it is not conceived. This is a sort of abstraction.

Abstraction: is the separation of one thing from another. If this separation takes place or exists in reality, it is *physical abstraction;* if it is done by the mind it is *intentional abstraction.*

Intentional Abstraction: can be either *negative*, in a negative judgment: John is not a lawyer, or *precisive*, in the order of apprehension.

Precision can be either total or partial:

Partial: if some part is considered, the remaining parts being known but not attended to, for example, abstract ideas such as whiteness, tallness, sobriety, and so forth.

Total: if some whole is considered under one or another aspect without conceiving the individuating notes.

Comparison: is the consideration of one thing in its relation to another. Comparison can be either *simple,* in the order of simple apprehension, or *complex,* in the order of judgment.

VI • SUPPLEMENTARY NOTES

Without going into the many particular difficulties which can be urged against this analysis, it may be profitable to express some of the basic distinctions which lie at the roots of most of the confusion in this matter.

(a) *What is grasped* in the universal is objective and therefore predicated. *The way in which it is grasped* is proper to the intellect apprehending, is not objective, and therefore is not predicable.

(b) What is grasped in the universal is *not a part,* but rather the whole object under one or another aspect; that is, a subject and its form is grasped. Hence it is predicated as a whole of a whole. Thus it is true of the whole being "John" that he is "man," or "tall," or "bald-headed," or "fat," or "lame."

(c) The individuating principles are not positively included in the concept, else it would be a singular and predicable of just one being. Nor are they positively excluded, else it would not be predicable of that particular individual. So they may be said to be included *in a common way.* Thus the idea of "man" includes flesh, blood, and bones in general or in common, since there cannot be a man without them; but it is applicable to each and every man because it neither positively includes nor positively excludes any one of all the possible individual men. In this sense the universal is said to prescind from individual matter but not from common matter.

(d) The direct universal is neither singular nor universal in its reference, but is in potency to either.

(e) The reflex universal is either a genus, species, specific difference, property, or logical accident.

CHAPTER NINE

TRUTH

AND THE JUDGMENT[1]

I • THE SITUATION

Since the first chapter we have been occupied with judgments and with truth as being found in them. At this point we want to determine more precisely the relation that exists between them. Where is truth formally or essentially to be found, only in judgments, in other operations of the mind, or even outside the mind? Actually, by analyzing and answering this question we shall also learn what the nature of truth is.

II • THE ANALYSIS

(a) *Formal Truth.* In order to understand the nature of truth and where it is formally to be found, we can first consider in general those things to which "truth" is applied, and then study them for their interrelationships. We shall thus see that truth is applied essentially and formally to one of them, whereas it is applied to the others in a secondary or derivative sense.

In various ways we can say that *words* are true, that *things* are true, and that *knowledge* is true. From experience it would seem that this exhausts the possibilities for the appellation of truth.

(1) *Words and truth:* From one point of view we say that

---

[1] For a more detailed study of judgment and truth read, F. Wilhelmsen, *Man's Knowledge of Reality*, chs. 10–13, pp. 101–164.

words are *true* in the sense that they are the correct or right ones to use for a given thing. In this sense this is a question of the *true meaning* had in the mind rather than one of mistaking, for example, chrome for silver.

From the moral point of view speech is said to be true when it really manifests what is in the mind, that is, there is no deliberate lie.

Clearly in both of these cases there is a relation of the words to the truth of the intellect, and even a dependence on intellectual truth.

(2) *Things and truth:* Omitting consideration of essential ontological truth which will be discussed in another tract and which involves some conformity with the divine mind, we can say that *things* are called true in two ways. Thus the artist or artificer can say that his creation or artifact is *true* to his conception or idea; and anything at all can be said to be true in the extrinsic sense that it can be known and is thus conformable to the human mind.

Again in both of these cases, it is in respect to intellectual truth that these things are said to be true.

(3) *The mind and truth:* The intellect is said to be true or to have truth in so far as it is conformed to what is. Hence, although words and things are called true by their relation to intellectual truth, the intellect is said to be true by its relation to *being itself*, to what is.

Reflecting on these points we can easily see where essential or formal truth is to be found. Things and words are called true in a dependent and participative sense, deriving from the mind's truth; whereas intellectual truth is the proper perfection of the mind, toward which the mind is directed of its very nature. It does not presuppose the *truth* but the *being* of other things, and it actually accounts for their being called true.

Hence, formal truth, truth in its primary and essential meaning, is one with logical or intellectual truth.

(b) *Formal Truth Found Only in Judgment.* The next point to be clarified has to do with the precise mental act in which truth is found. The choice lies between the acts of simple apprehension and of judgment.

As a perfection which perfects the mind according to its intellectual nature, truth can be said to be formally or actually or essentially present only when it is experienced intellectually, that is, only

when it is *known to be possessed.* Apart from its *being known* it
cannot be said actually to perfect the intellect the nature of which
is to know.

Moreover, it is only in the *intentional* (knowledge) *order* that
the conformity or adequation of mind and reality can be attained;
hence it is only in so far as it is known that the *conformity which is
truth* can exist.

Thus the *existence* of truth is one with its *being known,* and it
cannot be said to be known in a simple apprehension where nothing is affirmed or denied, but only in an act of judgment where
the intellect not only is proportioned to reality but is aware of this
conformity and affirms it by saying that the thing is as it is judged
to be.

Briefly, truth, as a perfection of the intellect, exists only as known,
and is formally known only in the act of judging; hence truth formally exists only in the judgment.

(c) *Truth and Simple Apprehension.* Simple apprehension can
be said to be true in the same sense in which it is said to be knowledge. It is preparatory and imperfect and ordained to judgment;
so it has truth inchoatively or in its initial stages. It can be said
to be *true cognition* in the sense that it does validly represent in
the intentional order some aspect or note of what is; but since
there is no complete reflection had and hence this conformity is not
known, it cannot be said to be the *cognition of truth.*

(d) *Truth as Relational.* Finally, since formal truth is a relation or a conformity between mind and object known, a more
refined question can be raised concerning the precise elements
which are to be called the terms of the relation.

In the widest sense the two terms are *knower* and *known,* that
is, when I have some truth, *I* am conformed to *what is.* But since
I am not conformed to the object by my arm or my foot, greater
precision can be had.

First, it is by virtue of *my intellect* that I am related to reality
known. Moreover, it is my intellect precisely as having representations of reality and as *judging* about reality which is related.
Hence the *act of judgment* is one term of the relation. More
precisely still, it is not the act of judging as an ontological perfection of the mind which is related, for this is immaterial or spiritual.
Moreover, if this act in its physical entity were conformed to what
is, the knowledge of silver or of black would be a silver thought

or a black thought, and so on. Hence in the most precise formulation, one term of the relation which is truth is the act of judging *in respect to that which is said in the judgment.*

On the other hand, the other term of the relation cannot be the internal representation of the object, as we saw in chapter five; hence it is the object as it exists independently of the mind.

Briefly, truth involves a relation or conformity of two things, and these in general are the knower and the known. More accurately the one term is the knower in respect to *that which the intellect says* by judging, and the other term is the thing known just as it exists in reality.

III · SUMMARY OF THE ANALYSIS

(a) Words, things, and intellect can be said to be *true*. However, truth as predicated of words and things derives from, participates in, and depends on intellectual truth. Hence, truth is primarily or formally predicated of the intellect. Intellectual or logical truth refers not to the *truth* of things, but to their *being*. Thus essential or formal truth is intellectual rather than moral or ontological. Cf. *Analysis (a)*.

(b) Formal truth exists only in the judgment. This is true because as an intellectual perfection it must exist in the mind according to the cognoscitive nature of the mind, that is, precisely *as known*. Although the simple apprehension is ontologically conformed to what is, this conformity is not yet known. Hence formal truth is not yet possessed. Cf. *Analysis (b), (c)*.

(c) The precise terms of the relation of conformity which is truth itself are: (1) *the thing* as it exists in reality and (2) *that which the intellect says* in its judgment. One term is the thing as it is in reality; otherwise only a representation of the thing would be known. The other term is what the intellect *says* in judging, because that which is said is precisely proper to the intellect and is also true of reality. Cf. *Analysis (d)*.

IV · THE STATEMENT IN BRIEF

Intellectual truth is formal truth, and this is found only in the act of judgment. Truth is a relation of conformity, and the terms of that relation are: (a) the thing as it exists extramentally, and (b) that which the intellect affirms or denies of the thing.

V • DEFINITIONS

Moral Truth: the conformity of speech to the mind.

Ontological Truth: (omitting essential ontological truth) is the conformity of an artifact or creation to the mind of the artificer. It can also mean the conformability of anything to the human intellect.

Logical Truth: the conformity of the mind to reality.

Formal Truth: that form or perfection in virtue of which everything said to be true is true.

Judgment: the act whereby the intellect affirms or denies one thing of another.

Simple Apprehension: the act whereby the intellect forms an idea or conception.

Relation: the reference of one thing to another.

Terms of a Relation: two objects or things referred to one another.

VI • SUPPLEMENTARY NOTE

Having seen that formal truth is found in the judgment, we can now turn to a brief consideration of the contrary of truth, namely, error or falsity. We shall examine the meaning and the sources of error.

(a) *Meaning of Error.* When we speak of error it must be kept in mind that we do not mean merely that the human mind often fails to know something which exists. It cannot be denied that our knowledge is partial and fragmentary; it proceeds by knowing an object now under this and now under that aspect, and thus it gradually acquires a fuller and more complex understanding of the object. This means that at any given moment there are many areas of ignorance and many unknown aspects in respect to any given reality. This, however, does not constitute falsity or error, but merely ignorance or a lack of knowledge.

Error actually means *the positive disconformity of judgment to reality.* By saying that it is a *positive disconformity* we mean that something is asserted about reality which is not so in reality, and thus the mind in judging falsely lacks the perfection of conformity which it should have. This *lack of a perfection which should be present* is called a *privation*. In respect to the intellect judging, this privation consists in saying that something is what it is not or that it is not what it is; in other words it attributes a characteristic to an object which it does not have, or denies one which the object does have.

Moreover, just as truth is found formally only in the judgment, so, too, and for the same reasons, is falsity. Only in judgment is there conscious or known affirmation of the conformity between the mind knowing and the object known, so only in judgment can there be positive disconformity between mind and object.

Neither in sensation as such nor in simple apprehension as such can there be formal error, for these operations merely grasp what they grasp, and do not and cannot make any reference of their content to reality. When the color-blind person sees various colors merely as different shades of gray, and when he on any of those occasions forms the concept of gray, he does actually have that sensation, he does actually form that concept. These are not false, for they simply are what they are. Falsity enters when he asserts, "That house is gray and that fence is gray," whereas the house is really white and the fence green. In this case he asserts of an object a characteristic which does not belong to it.

(b) *Sources of Error.* Now if the intellect not only is capable of true judgments but is made to know truth, and if it knows only what is presented to it, how can there be error? If all we can know is what we encounter in our experience, and if we express in judgments what we do encounter, how can error be explained? For an answer we can consider some of the factors which influence our judgments.

(1) *The Senses.* Although the senses must never be likened to hostile beings which deceive us, it can be admitted that often enough they do occasion error. Thus a diseased or malfunctioning sense makes it impossible to reflect back to the material being as it is. The intellect can recognize this defect and compensate for it, as is seen, for example, in the case of those color-blind who know their condition. In this matter also the distance of the object or its obscurity, the imperfection of the medium between object and sense can provide occasions for error. But in any such case the formal error is not due to the sense but to the intellect judging without being sufficiently determined by the object as it is.

(2) *Other Factors.* Only brief reflection is needed to see that we err for various other reasons, such as lack of proper *attention*, owing to boredom, day-dreaming, and the like; *prejudices*, whether political, religious, personal, or scientific, which can make us accept appearances for reality; *carelessness* owing to an unwillingness to work hard and apply oneself until the object is really known, or until a reasoning process is completed; confusion of mind owing to anger, hatred, love, or other strong *emotions*. These and similar influences can be recognized in ordinary daily life as contributing to the subject's forming judgments which are afterwards frequently seen as erroneous.

(3) *The Will.* Although all this is true, the question still remains as to why or how the intellect *as an intellect*, that is, as a faculty for knowing what is, can assent to what is not. Basically the answer to this must be sought in the *will*, the object of which is the good. When the object presents itself with full self-evidence, the intellect is compelled to assent. However, this is not often the case, because of the complexity of beings and the limitations of the knowing subject. To wait for full evidence at all times would be the ideal in knowledge; however, the exigencies of living and the need for immediate action make it practically impossible. Here the influence of the will enters. To judge, to give firm assents is good for the intellect and for man. There is a peace and satisfaction experienced with a definitive stand and the assertion that "this is so." Thus swayed by one or several of the influences mentioned earlier, joined with the intellect's drive to know and to judge, the will determines the intellect to act. Under the circumstances the motives seem to be sufficient, so the decision should not be any longer postponed, and the intellect is determined to place the act of judging which perfects it.

This means that, faced with insufficient evidence, yet urged on by the pressure of emotion, impatience, or practical need, we freely decide that on this or that occasion it is good for us to take the risk involved in judging and so we actually assert that "this is so." What we say *may* be true, but often enough, as experience shows, it is false. When the latter is the case, we have said that something is what it is not, or that something is not what it is. We have acquiesced not in a partial truth, but in falsity.

CHAPTER TEN

REASONING
IN GENERAL

I • THE SITUATION

Reasoning or ratiocination has been considered so characteristic of man that he has traditionally been defined as a rational animal, and in this ability both the power and the weakness of the human intellect is seen. It indicates man's power in the sense that from certain known truths he claims to derive new truths; it indicates his weakness since the process of reasoning is frequently a slow and laborious one and is always a complex way of arriving gradually at truth, whereas an immediate, intuitional, and comprehensive grasp of reality would point to a more powerful and more penetrating sort of intellect.

The fact that man does ratiocinate is a clear datum of experience; so the aim of these considerations is to learn by reflective analysis just how reasoning proceeds and to show that the methods of reasoning are basically valid, fruitful, and hence valuable processes of the mind. That both deduction and induction have at times been carelessly used to acquire unjustified conclusions, or that in many questions no certain solution has been reached or is likely to be reached, do not tell against the basic reliability of the mind when it functions discursively. We cannot defend nor do we even need to examine every application in this matter. Our task is more modest: it is to discover the meaning of the reasoning processes, and to indicate why these are basically trustworthy, and why, therefore, anyone who carefully proceeds in this manner is following a very natural and a very fruitful way of acquiring new truths.

Since the matter in question here is reasoning itself, what we say cannot be established by ratiocination but must be the fruit of analysis.

II • THE ANALYSIS

(a) *Reasoning and Logical Causality.* By ratiocination is meant a process of the mind whereby it proceeds from what it already knows to the knowledge of some new truth. If this process is from a universal truth to a less universal or a particular one, it is called deduction. If it proceeds from particulars to a universal truth it is called induction.

Thus from the knowledge that any spiritual being is naturally immortal, I can learn that the human soul, being spiritual, is immortal. Or from observing the expansion of many metals when heated, I may be able to conclude that any metal at all will expand when heated.

In either procedure the supposition is that I do not have true knowledge of that to which I conclude, but want to know whether this predicate, for example, "immortal," or "expansion when heated," is to be affirmed or denied of this subject, for example, the human soul or metal. In the process of ratiocination the motive or the reason for asserting or denying the identity of subject and predicate is made manifest, and precisely because of the premises, I arrive at the conclusion.

This means that the premises have much more than a spatial or temporal priority in respect to the conclusion. In them is found the very *reason, motive,* or *cause* of my assent to the conclusion. Presupposing the truth of the premises and the presence of the correct form of reasoning, the mind is moved toward the conclusion in such a way that it is *not logically free.* Accepting the premises, it must assent to the conclusion, precisely because of the motive found in them.

In this sense reasoning is said to be a process which involves *logical causality.* In the order of knowledge it is one and the same thing to exercise causality and to function as a motive.

(b) *Function of the Medium.* Moreover, this causality is exercised precisely by the third thing, the medium, to which are referred the subject and predicate of the conclusion. That is to say, what moves the mind to assent to the conclusion is something *extrinsic* to both subject and predicate, else there would be no *process* at all in this operation, and the one extrinsic thing involved is the medium, the third "thing" which mediates between and links the subject and the predicate so that their relation to one another becomes known.

Thus when there is no immediate knowledge of the relation between a subject and a predicate, their relationship may be found by comparing them to the same third "thing," whose relation to the subject, S, and to the predicate, P, is already known. As a result the relationship of S and P is affirmed or denied precisely because something else mediates between them and provides the *motive* for the assent and thus *causes* the conclusion to be drawn.

As will be seen, this medium is a middle term in deduction and it is experience in induction.

III • SUMMARY OF THE ANALYSIS

(a) Reasoning involves logical causality. This is seen in the fact that before the reasoning process takes place the conclusion is not known; during the process a reason or motive for it is found which demands intellectual assent or dissent. And reflection on the process shows that the entire reason or cause of the conclusion is found in the premises. Cf. *Analysis (a)*.

(b) This logical causality is exercised precisely by the *medium* or mediating third "thing" which links the subject and predicate. It is because of this medium, extrinsic both to S and to P, that the mind is necessitated, since there is no immediate grasp of their relationship. Cf. *Analysis (b)*.

IV • THE STATEMENT IN BRIEF

Reasoning proceeds by logical causality, exercised concretely by the medium.

V • DEFINITIONS

Ratiocination: an act whereby the mind proceeds from what it already knows to some new knowledge.

Deduction: the advance from a universal truth to a less universal or particular truth.

Induction: the advance from particular truths to a general truth.

Logical Causality: the influence, in the order of knowledge, of known truths such that they necessitate or motivate the acceptance of a new truth (the conclusion).

Medium: that which serves as an intermediary.

Extrinsic Medium: one which is formally identical with neither the subject nor the predicate, and yet is already known to be materially identical with each of them.

CHAPTER ELEVEN

DEDUCTION

I • THE SITUATION

(a) *Two Questions.* The chief question raised in regard to deduction is whether it can possibly result in new knowledge. It would seem that if the conclusion is drawn from the premises, it must have already existed in them; and if it is already known, how can one be said to acquire *new* truth by this process? Once that point is examined, there remains this further question: since the truth of the conclusion presupposes the truth of the premises, where is the ultimate foundation for the truth of the premises to be discovered?

(b) *Source of New Mediate Truths.* The purpose of this chapter should be clearly defined and kept in mind. It is not to examine every case of deduction, since that is impossible and unnecessary, nor is it meant to defend the frequent invalid uses made of this process. Rather, it is to show that the process of deduction is or can be the source of genuine new truth. When the human intellect functions in this way, it is employing a native power which is *basically* fruitful and reliable, so that we are not limited to the immediately experienced, but have the power to use present truths for the acquisition of new ones.

(c) In this analysis it is presupposed that we are dealing with *bona fide* cases of deduction, in which the premises are granted to be true and the correct form of reasoning observed.

II • THE ANALYSIS

(a) *Illustration.* The deductive process amounts to this: I am confronted with a subject, S, and a predicate, P, and I do not

know whether it is true to say that S is P or that S is not P. However, I find out that there is a common something, M, whose relation to each of them I do know, and which I proceed to use as a
mediating term. Thus knowing that every S is M and that P is
not M, I now come to know that S is not P.

For example, I do not *know* whether the human soul is immortal
or not. We can suppose, however, that I do know that anything
spiritual is immortal. If I also know that the human soul is
spiritual, I can conclude that it is immortal.

(b) *How the Conclusion Is Contained in Premises.* Without
being unduly distracted by the content of the example given, we
can examine what the process means. For one thing, we have
already seen that to derive a conclusion from two premises means
that it was already somehow contained in them. Moreover, we
have seen that the conclusion follows with logical causality from
the premises.

Now the question is: Just how are the premises to be said to
contain the conclusion? Theoretically, three possibilities can be
conceived: The conclusion may be said to be either actually, merely
potentially, or virtually in the premises.

If it is *actually* in the premises, then it is formally or really
known in the premises, even though it may be only implicitly.
Thus in knowing "animal" I actually though implicitly know "body."

If it is *merely potentially* contained in the premises, then it may
be or can be derived from them, but no note of necessity is included.

This latter note is precisely what is meant by the designation
virtually. If the conclusion is virtually contained in the premises,
then it not only can but should be admitted, if the premises are
granted. That is to say, the note of a *necessary* sequence is admitted. Thus the mind would be necessitated, in the logical order,
to admit the conclusion.

(o) *Not Merely Potentially.* Keeping in mind what was seen in
the preceding chapter, we can now see that the *merely potential*
presence of the conclusion in the premises is not enough. The
added characteristic of logical necessity and the fact that the
premises are causes of the conclusion demand more than such
mere potentiality.

(d) *Not Actually.* Nor can it be said to be actually, that is, implicitly, known in the premises. If it were thus present, it would be found and known either in knowing the major premise or the minor. However, the major premise *taken alone* does not refer in any determinate way to the individuals, since they are in no way actually known in knowing the major. Thus if I know that everything spiritual is immortal by nature, and if that is all I know, I can in no way make a statement about the natural immortality of the human soul. The generality and the indeterminacy of the major must be applied to a definite and determinate individual or individuals before any new knowledge can be had. Thus knowing that M is P leaves me without knowledge of S or of its relation to M or to P.

Nor does the minor premise *taken alone* actually contain the conclusion, since it offers no knowledge of the predicate in question. All I know here is that S is M, and thus in this judgment alone I do not know that M is P. Hence from the minor I cannot arrive at the knowledge of the relation between S and P.

(e) *Conclusion Virtually in Premises.* Moreover, since the conclusion is actually contained neither in the major nor in the minor, the mere "spatial" or "temporal" juxtaposition of them will result in no new knowledge at all. In that case I would merely possess two separated and isolated judgments. All that is required, however, is that these two now be seen in their interrelationship. Once that is seen, the linking function of the medium is grasped, the two judgments are thereby known as forming a unity, the necessary determination is provided, and the conclusion is thereby apprehended and affirmed.

This means that one can best designate the way in which the conclusion is found in the premises as *virtual,* that is, not-actually and not-merely-potentially, but potentially with the added characteristic that once the premises function precisely *as premises*—not as isolated propositions but as interrelated through a common medium—I am logically *necessitated* to admit the conclusion.

(f) *New Knowledge.* That this process begets *new* knowledge becomes clear when it is realized that the conclusion is in no way *actually* (explicitly or implicitly) known in either the major or the minor. To say that it is virtually known means that I can, should, and must draw the conclusion from these premises. Thus what is *indeterminate* and hence unknown becomes determinate knowledge through this process. In other words, *new* knowledge is derived from deduction.

(g) *Foundation in Analytic Principles.* From what has preceded we have seen that deduction proceeds by logical causality exercised through a middle term and that this form of reasoning can be said to provide new knowledge. Now there arises a question relative to *the truth of the conclusion* and its relation to the truth of the premises, particularly of the major premise, since that is usually the universal statement which indeterminately contains the conclusion and which is determined and applied by the minor premise.

To make clear what we are seeking, this must be now said: Deduction is a fact of our intellectual experience, and we naturally arrive at new propositions through its use. In line with what has already been said on the intellect as a faculty for truth, it must be admitted that truth is acquired or can be acquired in this way. Were this not so, the intellect in its natural operation would be erroneous and thus radically not a faculty for truth.

Hence the question here is this: If the conclusion is accepted as true because of the premises, and precisely because of the truth of the major, why is the major premise accepted as true? If the major is the conclusion of a previous deduction, it would be accepted as true because of the major premise on which it rests. Why is that premise then accepted as true? If it is accepted because of a previous deduction, just how far backward is the process to proceed? If it proceeds *in infinitum* or indefinitely, there would never be an intelligent reason for accepting any conclusion. Since that cannot be admitted, we must admit that ultimately deduction rests on some truths which are not themselves conclusions or mediately known truths. That is to say, to avoid an infinite process and to account for the possibility of truth through deduction, we must realize that the process ultimately is founded on and gets its probative force from some truths which are immediately known through a mere analysis and comparison of the subject and predicate.

Such truths are called *analytic principles.* They are called *principles* because from them or in the light of them something else is known. They are *analytic* because the mere analysis of subject and predicate is enough to manifest their relation—that is, no strict proof or deduction is required in order to know them. Hence, too, they are not mediate truths, but truths naturally and immediately known.

(h) *The Principle of Contradiction.* It should be noted that we have in no way determined just what these analytic principles are; all that has been seen is that there must be some such principles, or else deduction is impossible and no proved conclusion can be accepted.

However, to single out one such principle for special consideration and explicitation we can turn to the principle of contradiction (or of noncontradiction, as it is frequently called). This, in so far as it is said to be a *law of reality* (ontological principle), can be enunciated in the following terms: A thing cannot simultaneously and under the same aspect be and not be. As a *law of the mind* (logical principle) it can be formulated thus: A predicate cannot be simultaneously affirmed and denied of any one subject.

That this principle is a law of the mind is manifest from the fact that every judgment is a determinate statement about a determined subject, and hence to have any meaning it must involve the rejection of its opposite as false. Thus if I say that the atom can be split, I reject the truth of the statement that the atom cannot be split. That is to say, by every judgment or affirmation I say *what is* concretely, I state something definite about something, and imply that it cannot simultaneously not be what I affirm. Hence, as Aristotle says, by no matter what is said the principle of contradiction is affirmed.

Were this not so, the principle itself could be denied without contradiction. But the very denial would be offered as a statement having a determinate meaning and meant to convey a definite truth. And in so far as that is true, the very attempt to deny this principle necessarily involves its affirmation. Thus the principle of contradiction is a law of the mind.

(i) *A Law of Being.* In line, moreover, with previous analyses it must be recognized that the fundamental reason why this principle governs the mind is because it is the law of *being itself*. The mind knows and judges what is and affirms something determinate because what is is determined and at any given moment is what it is. By knowing, we do not make things to be or to be such and

such, but rather judge them to be because they are. Thus the
principle of contradiction means ontologically nothing more than
this fact of each being being what it is; being self-identical, it
cannot simultaneously be what it is not.

(j) *Most Fundamental Principle.* Furthermore, it can be seen
that this principle is the first, the most basic and the most known
of all principles. For one thing, the terms in its enunciation are
equivalently *being* and *non-being,* the simplest of all possible
terms; hence nothing simpler can be found to mediate between
them. Thus being is immediately known.

Also, I naturally know this or that being, and that there are many
beings with which I am in contact. That I know *many beings* involves the recognition of many *unities,* each distinct or separated
from all others. Knowledge of this distinction or division of one
thing from others means knowledge of negation. Hence, my immediate contact with the plurality of things means the immediate
knowledge of both *being* and *negation.* These are the terms of the
principle of contradiction, so even in respect to its terms, the principle is immediately grasped by me.

That it is most known of all principles, no matter what they are,
is seen by reflecting on the fact that every other principle must
presuppose it, whereas it itself presupposes no other principle.
Every valid principle must be concerned with some being or some
group of beings; it therefore presupposes being and the basic principle of being which holds true wherever there is being. The
principle of contradiction actually presupposes no other principle as
such, because apart from it there can be nothing but nonmeaning
or self-contradiction, which actually cannot be.

Since this principle is coextensive with being, it is possessed or
known in every act of knowledge. It is present in every deduction,
and were it to be violated in any process, that process would be
invalid. Thus, whereas we saw that deduction must ultimately rest
on some analytic principles to have validity, it is now clear that
among those principles the principle of contradiction holds first
place. Every deduction must ultimately rest on this principle.

(k) *Common and Proper Principles.* Although it is impossible
to go into an analysis of other analytic principles at this point
(some of them will be analyzed in other tracts), it can be mentioned that other such principles are finality, causality, sufficient

reason, comparative identity, good is to be done and evil avoided, and so on. The principle of identity—what is is—or of the excluded middle—anything either is or is not—are the same in content as the principle of contradiction, and so should not be too sharply distinguished from it. They are really only different formulations of one and the same truth.

The principles just mentioned can be called *common* analytic principles, since they apply to all being or to various sorts of being.

There are also *proper* analytic principles which are formed by an easy, intuitive *application* of these principles to some definite subject matter. Examples are the divisibility of extension, the equality of quantities, the stability of nature, the equality of intensities, and so on.

(1) *No Deduction from Common Principles.* When it is said that deduction must ultimately rest on some common analytic principles, it is not meant that conclusions are ultimately *deduced* from these principles, but rather that they can be resolved into them. So the correct meaning of this *resolution* must be grasped. First, from *common* principles nothing at all can be deduced. Pertaining as they do to being itself, they can be said to be intuitively or directly applied to any particular being or sphere of being. The "inferiors" are actually contained in and known in them, hence there can be explicitation but no new deductive knowledge in the application to individuals. So in questioning the truth value of any particular conclusion and by saying that it ultimately rests on analytic principles, we do *not mean* that we can show how it is deduced from these principles. We do mean that the conclusion in question can be *resolved into some analytic principle or principles.* Thus, conclusion D is based on a major premise C; C is true because of B; B is deduced from A. This process cannot go on indefinitely. Hence, some premise is accepted because its denial would involve the denial of the principle of sufficient reason, or of the principle of causality, or of the principle of contradiction. In this way the conclusion is resolved into first analytic principles.

Thus, extreme Rationalism is avoided by recognizing that there is no such thing as a common principle from which all knowledge can be deduced—in fact nothing at all is deduced from common analytic principles. The resolutive function of the first principles consists merely in the recognition that in the deductive process the

initial premise is accepted as true because its denial would involve
the denial of being itself and of its principles.

(m) *Résumé.* By this analysis, then, all we have seen is that
the human mind has the power to function reliably in a deductive
manner, that in this process it acquires new knowledge, and that
new knowledge can be shown to be true by the resolution of the
conclusion into first analytic principles. Nor does this mean that
such resolution has to be carried out for every case of deduction;
that would be endless. What is meant is that in valid deductions
this resolution is possible and is the basic reason why the deductive
procedure is a basically valid and reliable one.

III • SUMMARY OF THE ANALYSIS

(a) In deduction, since the conclusion is derived from the
premises, it must be somehow contained in them. To say that it
is contained in them *merely potentially* would not say enough, since
this would not include the notion of logical causality.

Nor is it *actually* known either in the major or minor premise.
The former contains the subject only in an indeterminate way, not
sufficient of itself to give formal knowledge. The latter would provide no knowledge of the predicate. Hence it is not actually found
even implicitly in either premise.

However, it must still be said to be contained in the premises in
such a way that the conclusion *can* and with logical necessity *must*
be admitted once the premises are granted. In this sense we can
say that the conclusion is *virtually* present in the premises. Cf.
Analysis (a), (b), (c), (d), (e).

(b) That the conclusion drawn does provide *new* knowledge is
clear from the fact that what was not known and what was only
indeterminately contained in the premises becomes now deter-
minately and actually known. Cf. *Analysis (f).*

(c) The truth of the conclusion depends on the truth of the
premises, especially of the major premise. To establish the truth
of this major there cannot be an infinite process. Hence deduction
must ultimately rest on some analytic principles whose truth is self-
evident. Cf. *Analysis (g).*

(d) The principle of contradiction is a law of the mind, since
every judgment says something determinate about a determined

subject, and hence implicitly affirms the principle. Cf. *Analysis (h)*.

(e) The principle of contradiction is primarily a law of being, since the mind in knowing conforms to what is. Cf. *Analysis (i)*.

(f) The principle of contradiction is the first, the most fundamental, and the most known of all principles. This becomes clear from reflecting on the fact that its terms are the simplest possible; that knowledge of any being and of a plurality of beings means knowledge of *being* and of *negation;* and that this principle is presupposed by all other truths, whereas it itself does not presuppose any other particular truth. Cf. *Analysis (j)*.

(g) There are both common and proper analytic principles. Nothing can be deduced from the common principles, but in their light all knowledge proceeds. Hence they have a resolutive function with regard to deduction, in the sense that ultimately it can be shown that an initial premise is to be accepted or else some analytic principle would be violated. Cf. *Analysis (k), (l), (m)*.

IV · THE STATEMENT IN BRIEF

Through deduction new knowledge is acquired in the sense that what was known merely virtually becomes actual knowledge. Ultimately this process rests on valid analytic principles.

V · DEFINITIONS

Deduction: the process whereby the mind goes from a universal truth to a less universal or particular truth.

Formally Known: what is actually known, either implicitly or explicitly.

Potentially Known: that which is not actually known, but is capable of being known.

Virtually Known: that which not only can be known, but which must be admitted if the premises are granted.

Principle: that from which anything proceeds.

Logical Principle: that from which anything is known.

Analytic Principle: one in which the relation between the subject and predicate is evident from the mere consideration of the terms.

VI • SUPPLEMENTARY NOTES

The main objections against this position on deduction and on first principles have been foreseen in the text, but since some of them are constantly recurring, it might be profitable to isolate them for special consideration, even at the risk of undue repetition.

(a) There is no new knowledge got through deduction, since the conclusion is already known in the premises.

REPLY: Until the meaning of being *virtually present* in the premises is grasped, this objection will appear to be valid and unanswered. The fact is that in knowing the major the conclusion is present only in an indeterminate way not sufficient for actual knowledge, and hence is not known. Through the determination provided by the minor, sufficient specification is provided for the intellect now to know the conclusion. Thus if the major and minor are both granted and their interrelation through the middle term is recognized, the conclusion is thereby drawn with logical necessity and new, determined knowledge is attained.

(b) The function of first principles seems to indicate we hold a sort of extreme Rationalism whereby we are able to acquire all knowledge by deduction from all-inclusive principles such as the principle of contradiction.

REPLY: This difficulty totally misunderstands the function of first principles. How we acquire them will be discussed in the next chapter. For the present we need only indicate that knowledge of them has been shown to be necessary. Moreover, nothing is deduced from common analytic principles. Rather, they can be called the *light* shining on all data presented, or the natural bases on which all knowledge rests. In their light and on them as a foundation, the intellect proceeds in its acquisition of truth. Their function is one of resolution, not one of deduction; no truth can ever violate them, nor can any proposition be denied when its denial would deny a first principle. Hence, there is no extreme Rationalism; knowledge still proceeds from sensible experience, and one of its procedures is by way of deduction in the light of first principles. Moreover, deduction is not the only way of acquiring truth, as will be seen in the next chapter.

(c) The principle of contradiction needs to be rationally established, and it cannot hold in certain areas, where indeterminacy prevails, especially in respect to becoming.

REPLY: To be rationally established in this context cannot mean demonstrated, as has already been seen. This principle can be explicitated by analysis only and needs no further establishing. The reference to certain *obscure areas* where the principle does not seem to hold merely means that there are areas where we do not know and may never know just how things operate. To go from saying that we cannot determine a mode of operation to the statement that the reality in question does not operate in a determined way is a gratuitous leap. We cannot afford to deny what is obvious for the sake of what is admittedly obscure or even unknown. Not determinable by us and not determined in itself are two quite different states.

Finally, many difficulties are found in the fact of becoming, but nothing which tells against the principle of contradiction. We never experience *pure becoming*, and in fact it is an impossibility. Rather what is had is a *being which becomes*, and that involves no contradiction.

CHAPTER TWELVE

INDUCTION

I • THE SITUATION

(a) *To Examine Bases and Meaning of Induction.* Deduction is not the only sort of ratiocination employed by man as he grows in the possession of truth. It is a fact that if he knows some general truth proper to some subject matter, he can deduce new knowledge relative to some previously unknown "inferior"; and it is a fact that he can form some such universal proper analytic judgments by the immediate application of some common analytic truth to a determined subject matter. But experience shows that these are by no means the only sort of general or universal judgments which we claim to have.

Most of the great scientific advances have been brought about by the use of induction, and it is the basis and meaning of that process which we now propose to investigate. The twofold exaggeration of rejecting induction as a means of ever getting to certain truth, and of accepting too readily what merely appears to be induction, must be avoided. This caution must be observed by those who use and apply induction; our function is to justify the process itself as a basically reliable and fruitful method for the acquisition of truth. From reliable sources we can today find out the melting points of various metals, the boiling points of liquids, the results of certain elements' combining in various ways, and many other facts about the way many things operate. Even in very ordinary knowledge, we can know much about the obvious operations of some things. The question here is how such knowledge is acquired, whether it is reliable, and why.

(b) *Natures and Operations.* In our terminology we can say that what we are discussing is the connection between a *nature* and its *operations.* Thus when it is said that Na plus Cl properly combined will give sodium chloride, or that gold will melt at such and such a temperature, or that water will freeze at a definite temperature under certain circumstances, these are not proposed as mere guesses or as merely probable, but as factual. The scientist may deny that he knows anything about "natures"; from his point of view he may do so, and we need have no quarrel with him over his denial. However, it is up to the philosopher to speak of *natures* as such, to show that there is an underlying "principle" of operation present in each; and much of his knowledge about these natures he must accept from the empirical scientists. That there are natures is a philosophic problem; how things operate is a scientific question; and the relation between the two is again a philosophic question. Thus speaking scientifically, there are no "natures"; but the scientist who would say "There are no natures such as the philosophers discuss," would be overreaching himself and speaking now not as a scientist but as a philosopher—and at that point we would have to disagree and reject his philosophic position.

(c) *Relation to Empirical Sciences.* This discussion, then, touches very sharply on the empirical sciences and on the general lines of the inductive method or methods so characteristic of those sciences. Rather than conflict, our analyses will equivalently show that the general line of investigation followed by men who engage in scientific experiment is good and reliable because it expresses one of the characteristic ways in which the mind can come to know truths. This does not mean to say that it has always given certitude, or even that it can in many difficult areas ever hope to arrive at certain knowledge.

It does mean that in some cases we have acquired certain truths through induction, that the process itself is basically reliable, and that even where there is little or no hope of success, the investigation is good because underlying it there is a reliable human method.

(d) So we do not want to start evaluating every use of induction that has been made, or to ask the scientist as such to accept our position, but we shall provide a humanly intelligent background for his experiments and for his inductions. There is a community of interests in this question, at least to the extent that whatever the

field of knowledge may be, man must use *human* knowledge, the
various aspects and means of which we are analyzing.

II • THE ANALYSIS

(a) *Synthetic Judgments.* Having once acquired the ideas of
whole and *part,* I can immediately know in general that a whole is
greater than any of its parts; having learned the meaning of
spiritual and *natural immortality,* I can come to recognize that any
spiritual being is naturally immortal.

But having learned what *gold* is and what *melting at 467° C.*
means, I can think about them indefinitely and I will not discover
whether the one is to be affirmed or denied of the other. So, too,
in a deep dungeon I cannot tell whether it is raining or snowing
or sunny outside, or whether the grass is green or burnt. These
and countless similar judgments of fact require experience, and
they enunciate a passing state of affairs. Judgments affirming individual passing experiences are called *particular synthetic judgments: particular* because they have to do with an individual,
passing event; *synthetic* because they are the fruit of experience.

However, our judgments based on experience are not all limited
to particular, transitory events. When I say that hydrogen and
oxygen in a definite context will result in water, that heavy bodies
will fall to the ground, or that fire burns, my statement means more
than the expression of something transitory. It really affirms not
only that this has happened, but that it is what regularly happens
and even that this is what will continue to happen regularly.

Such judgments are called *synthetic principles: principles* because
they express a universal truth, and *synthetic* because they derive
from experience.

(b) *Some Knowledge of Natures Implied in Scientific Studies.*
Now the question is: Are these synthetic principles *merely a totalization* of what has happened, do they merely *summarize* previous
experiences, or do they indicate the ability of the mind to go much
deeper in its knowledge of things?

That something deeper is involved is *implied* by *scientific procedures* as well as by *what has already been seen* in our analyses
of human knowledge.

Thus the *empirical scientist* in any given experiment, and especially in a complex one, takes many things for granted in respect to the materials he uses. He knows just how some of the elements and compounds employed will act and interact, and he does not delay over them, but proceeds to the *new* element which he is investigating. In other words there are many things of which he is already so sure that they merit no more verification. We can say that he is certain about them. Were one to suggest that in given circumstances sodium plus chlorine might not result in sodium chloride, or that it might suddenly change and result in potassium chloride or in sulphuric acid, the suggestion would be ignored. These and similar easy examples from physics, chemistry, and other empirical sciences imply more than the scientist *qua* scientist need admit or even consider; they indicate that some *necessary* characteristic of the material used is known, that something is definitely known about the natural actions and reactions of things, and that these are so well established already that they can be taken as the bases for further investigations. Philosophically we can say that this means that something is already known about the very *nature* of the thing in question. Furthermore, it is known not only that the thing actually *operated* that way, but that the concrete uses made of it in experiments indicate a conviction and a realization that it must and will always operate that same way in similar circumstances.

Thus in scientific procedures we can see that there is *implied* a recognition of the ability to know something about the naturally necessary, determined operations of material things; which means that something can be known in universal fashion about the *natures* which so operate.

(c) *Relation to Abstraction.* Moreover, *in what we have already seen* in respect to the human mind and its ability to abstract the universal from the singular and the concrete is also implied man's ability to know something about natures or essences themselves in the strict sense of the word. That is to say, when treating of universals we indicated that through abstraction we can come to know a thing under one or another aspect, through its perceived characteristics and operations. Much of what is thus conceived may be and is quite transitory and accidental to the *nature* or essential being in question, but some of the characteristics abstracted may be, and actually are, essential to or of the very nature of the being which is known. By way of example we might say that there at least appears to be a vast difference between the relation of S and P in the statements that "a piece of paper is white in color," and that "a piece of paper will burn when thrown on a bed of red hot coals," or "a stone thrown from the top of a tower will fall to the ground."

Yet granted that through precision we form all sorts of concepts, and granted that some of them seem to be more "important" for the being in question than others, just how do we or can we ever determine that this or that characteristic not only is true of the subject, but is so necessarily true of it that we must admit that it pertains to the thing's very nature or essence? And if we do select some characteristics as necessary or essential, by what right do we do so?

(d) *The Inductive Procedure.* Actually, the procedure goes somewhat like this: I experience this metal expanding when heated; again I experience another metal expanding when heated. The same experience is repeated time and again with various metals and under different circumstances. Finally, I conclude that *metal* expands under heat, meaning thereby that it is of the very nature or essence of metal to so expand. The same is done in regard to falling heavy bodies, to boiling points, and to many easily determined operations. Thus from several or many individual judgments of what has been experienced, I conclude to a general judgment, to a law of nature, that is, a synthetic principle.

Such procedure, however, would seem to be a bit arbitrary. Twenty, forty or even one thousand cases examined are not *all;* in fact no pretense is made to examine *all* possible cases. Hence by what right can we say that from relatively few cases examined I can legitimately infer a universal statement or law?

(e) *Twofold Foundation for the Procedure.* To answer this question it must be realized that there is a *twofold* foundation for the process here in question, that is, for induction, which is the process whereby the intellect proceeds from individual truths to universal truths. By a *foundation* we by no means intend a principle from which these laws can be deduced. Rather it refers to certain *basic conditions* which are fulfilled either in the knower or in what is known.

(1) Thus there is *a logical foundation* on the side of the intellect in the fact that the human mind is capable of distinguishing be-

tween what does and what does not pertain to the nature or essence
of the thing. This is due fundamentally to its power of abstraction, and to the recognition that some sufficient reason must exist
for the constant recurrence of some same characteristic under varying circumstances. There must be a sufficient reason for the repeated, unvarying experience of a given phenomenon or operation
found wherever this or that reality is found. We are capable
intellectually of seeing that for certain operations the merely accidental is not a sufficient explanation, and the only way to explain
what happens is to admit that the operation is proper to the nature
of the being in question. Thus something is known of the nature.

Thus it is of the nature of ordinary fire to burn, of heavy bodies
to fall to the ground, and of H_2O and SO_3 to result in H_2SO_4, and
so on.

(2) There is also a *real* or *ontological* foundation for induction,
found in things themselves. That is to say, corporeal things exist
and act in a definite, determined way, unless something else inter-
feres with them. They cannot suddenly "decide" to act in an
arbitrary or novel fashion, since they lack freedom, which alone
could introduce such novelty. Corporeal beings are what they
are, and necessarily act according to their natures. Sodium, for
example, properly compounded with chlorine cannot elect to become gold. The very natures involved preclude any such happen-
ing.

The entire edifice of scientific research is reared on this at least
implicit conviction, and without it no natural certitudes would be
held, nor would so much scientific progress have been made.
Progress is made only when previous work does not have to be
constantly redone, but when it can be accepted as duly established
and reliable for future use.

Nor does this mean that we already know all there is to know
about things, nor that every attempt at the knowledge of natures
has been successful. In fact there is very much in human knowledge that is quite undetermined—provided the indeterminacy is
put in the right place. We do *not* hold *that material things are
undetermined;* they are totally and completely determined, and
cannot possibly act in a chaotic, disorganized, or arbitrary way.
What is undetermined is our knowledge of things: owing to difficulties of experimentation, to the elusive nature of some particles
being examined, or to the crudity of even the refined instruments

used, it is not easy and may not even be possible to detect, measure, or in any way ascertain with complete accuracy just how a given nature operates.

All this can be granted, and the fact still remains that corporeal natures in themselves are determined in their being and in their operations (this is known as the *analytic principle of the stability of nature*), and that we can and do know with certainty many things about those natures through their operations.

(f) *No Unjustified Universalizing.* In induction, therefore, we rise from "inferiors" to a general law.

The *motive* or the medium is experience.

Moreover, the process of induction has a firm foundation both in the logical and in the ontological orders. The *logical foundation* is the ability of the mind to determine what does pertain to the nature in question. The ontological or *real foundation* is the fact that natures are determined in themselves and in their operations, so that they act in a given way unless there is outside interference.

Hence there is no gratuitous leap to the universal. What are had are various experiences. These are enunciated in individual judgments of experience. When enough of these are had, the mind can recognize, in the light of the principle of the stability of nature and of its own ability to know, that the only sufficient reason for what has been experienced is that this or that operation or characteristic pertains to the very nature of the thing and hence can be enunciated as being true wherever that nature is found. Hence it is true even of similar individuals not actually experienced.

So induction is more than a mere generalization or summation of previous experiences.

(g) This fulfills all that was originally said about reasoning in general. Induction proceeds from what is already known (immediate judgments of experience) to new knowledge (the general law) by means of an extrinsic medium (experience). It is therefore a valid source of new truth.

(h) *Moral Laws.* With the necessary modifications the same general development can be provided for the acquisition of moral laws; that is, the general enunciations of how free natures (men) will act. Although there is freedom here, and hence man can deliberately choose to act contrary to his nature, still what is natural for him can be determined through induction. In fact a great deal of our actual living relies upon our knowledge of how men will act in given conditions. Some examples are: mothers love their children; no one tells lies for nothing. Without going into the details of these moral laws, we can mention an interesting point touching on the example of lying. The use of the lie detector is an experimental confirmation of the position that lying is contrary to man's nature. The way this instrument works indicates that it takes an effort on the part of a human being to tell a lie.

(i) *The Number of Instances.* One further question can now be asked: How many instances must be examined before a synthetic principle can be enunciated? To that there is simply no determined answer possible. In practice it will depend on the nature of the thing being studied, on the instruments available, on the acumen of the examiner, on the ease or difficulty of the problem, and on other factors. Nor is it necessary for a theory of knowledge to provide more than the justification for the general process and some indication of how it is to be applied. The application must be made by others. Hence, in general it can be said that as many instances must be examined as are necessary to render evident to a capable investigator the fact that this or that operation or characteristic really does pertain to the very nature of the being in question. Only when that point is reached does the mind feel that it has fulfilled its function and rest satisfied with the results.

Finally, it must be stated that induction is a very varied process; and we have aimed at justifying it basically only, leaving the various techniques for the actual application to individual cases.

(j) *Acquiring Analytic Principles.* Quite different from the inductive process is the way in which analytic principles are acquired. Here, it is true, experience is also required, since we do not derive human knowledge in some *a priori* fashion quite independent of our experiences in this world. But in the formation of analytic principles experience is necessary only to provide the *matter* for the universal judgment, that is, from experience we abstract the subject and the predicate. It does not function as the medium or the motive for the general affirmation, and in that sense no particular judgments of experience are strictly necessary before the universal is affirmed.

Thus I can acquire the ideas of "whole" and "part" only through abstraction, working on what is sensibly experienced. Once these are had, however, I can immediately see that no limitation to this one being is involved, but that it is rather of the very nature of *any and every whole* to be greater than *any of its parts,* or to be equal to the sum of all its parts. The same can be said of other analytic principles.

Hence, here experience provides the *matter,* that is, the subject and the predicate. It does not provide the motive for linking them in a universal way.

The motive is the immediate intelligibility of the relation between the subject and predicate, once they are grasped. Thus it is said that analytic principles are those in which the relation between subject and predicate is manifest from the mere consideration of the *terms.*

III • SUMMARY OF THE ANALYSIS

(a) There are *particular synthetic judgments,* such as "It is raining," or "This fire is hot."

There are also *synthetic principles,* such as laws of nature or moral laws. Cf. *Analysis (a).*

(b) Synthetic principles are more than mere totalizations of previous experiences. They give some knowledge of the natures of things.

This is *implied* in the procedures of scientists who accept previously established characteristics as holding good generally.

It is also *implied* in our previous analyses of our power to abstract the universal from the particular. Cf. *Analysis (b), (c).*

(c) The validity of this process is recognized by reflection on its twofold foundation: logical and ontological.

The process itself involves sense experiences, immediate judgments of experience, and finally the universal or synthetic principle.

Its logical foundation is the intellectual power both to abstract and to recognize the difference between what is accidentally and what is essentially true of the thing in question.

The ontological foundation refers to the "principle of the stability of nature," which recognizes that corporeal natures are determined in their being and in their way of operating.

Hence we can know things as they are, and among their operations we can come to know some which are true of the nature it-

self, and so will be true of that nature wherever it is found. This
is, then, not a mere totalization of cases observed, but a well-
founded universalizing, and it results in new knowledge. Cf.
Analysis (d), (e), (f), (g), (i).

(d) Moral laws, relating to the actions of free natures, can also
be found by induction, in the same general way. Cf. *Analysis
(h).*

(e) Analytic principles are not acquired through induction, but
from experience the subject and predicate are got through abstraction. Their relationship is immediately intelligible, and hence all
that is required is a consideration of the terms of the relation—
that is, of S and P. Cf. *Analysis (j).*

IV • THE STATEMENT IN BRIEF

*The truth of synthetic principles is made manifest through in-
duction. The truth of analytic principles is made manifest
by a mere consideration of the terms.*

V • DEFINITIONS

Synthetic Principle: a general truth, expressing a law of nature or a
moral law, having experience as its motive.

Law of Nature: a universal judgment expressing how corporeal
natures operate.

Moral Law: a universal judgment expressing how free natures
operate.

Induction: the process whereby the intellect proceeds from individual truths to universal truths.

Foundation for induction: a condition which is required for
induction to be possible.

Logical Foundation: a necessary condition which must be possessed by the mind. This necessary condition is the ability
to know natures and what belongs to them.

Real Foundation: a necessary condition found in that which is
known. It is enunciated in the principle of the stability of
nature.

The Inductive Process concretely includes abstraction, immediate judgments of experience, application of analytic principles, and the universal judgment.

Analytic Principle: a universal principle the truth of which is known from a mere consideration of the terms.

VI • SUPPLEMENTARY NOTES

(a) To go from particular truths to a universal truth is not legitimate, because in any such process the effect is superior to the cause.

REPLY: If the *total cause* is kept in mind, the difficulty vanishes. The cause is not merely the particular judgments of experience, but it is those judgments seen in the light of the analytic principle of the stability of nature in addition to the intellect's ability to know universals and to determine what belongs to the nature and what does not—that is, the merely accidental. Thus when the point is reached where it is recognized that some trait is of the very nature of the being, one no longer is dealing with an individual *qua individual,* but rather with the individual *qua having such a nature.*

(b) If induction proceeds from analytic principles, then it would seem to be really deduction.

REPLY: There is no question of deducing anything from a common analytic principle, but rather a sort of intuitive application of the principle to the matter in question, and through that application a proper synthetic principle is formed from what has been experienced.

Having seen that there are various sources from which truths are acquired, it can now be asked if there is one general and final norm which underlies all of them. This is seen to be objective evidence (Chapter Thirteen). Moreover, there remains to be discussed the quality of the assent (doubt, opinion, certitude, and so forth) (Chapter Fourteen), and the reliability of human testimony as a source of knowledge (Chapter Fifteen).

• *Suggested Readings For Chapters Thirteen Through Fifteen*

Carraghan, Gilbert, *A Guide to Historical Method,* pp. 70–80, 232–320. New York: Fordham University Press, 1946.

Newman, J. H. Cardinal, *A Grammar of Assent,* pp. 240–50. New York: Longmans, Green & Co., Inc., 1947.

Vier, Peter C., O.F.M., *Evidence and Its Function According to John Duns Scotus,* pp. 31–62. St. Bonaventure, New York: Franciscan Institute, 1951.

Wilhelmsen, Frederick D., *Man's Knowledge of Reality,* ch. 14, pp. 165–180. Englewood Cliffs, N. J.: Prentice-Hall, Inc., 1956.

CHAPTER THIRTEEN

THE ULTIMATE NORM
FOR TRUTH

I · THE SITUATION

What we are to consider in this chapter, the universal and ultimate norm for truth, has in a sense already been settled from the first reflex recognition of the existence of truth. However, for purposes of clarification it needs to be treated here in some detail.

By recognizing and admitting the fact that I can know truth I also admit that I can distinguish truth from falsity, and that there exists some speculative norm according to which I thus distinguish between them. Since I can know various particular truths derived from various sources, there must be some special or particular norms which apply to my recognition of those particular truths. Moreover, since my intellect is *one* faculty whose function is to seek and to find truth, there must be one general or universal norm which guides it and beyond which there can be no appeal in its task of recognizing truth and rejecting falsity, no matter what the particular source may be.

It is in view of these brief considerations that we can say already that there must actually exist some universal and ultimate norm in the light of which the intellect recognizes and affirms all particular truths. Hence our main question is not to establish its existence, but rather to determine more precisely just what that norm is and how it functions.

II • THE ANALYSIS

(a) *The Need for Objective Evidence.* Reflecting on individual judgments we see that they can come from various sources: from immediate sense experience, deduction, induction, immediate analysis of the relation between subject and predicate, or from testimony, by which we rely on the knowledge and the truthfulness of some other person. In all these cases, moreover, certain conditions proper to each source of truth must be fulfilled before the "truth" is accepted. However, reflection shows that in every case the object known, or that which is affirmed, is known and affirmed only because it becomes *evident* to the intellect.

That is to say, what is known is *what it is,* and *its being what it is* is brought home to the mind either by induction or by deduction or by sense experience, or in some other way. Until the mind sees what is, it is not determined to judge. When it begins to see what is, it inclines towards making a judgment, and when it clearly sees what is *as it is,* it is determined to judge. To say that "seeing what is as it is" is what determines the intellect is just to admit in other words that the all-pervading or universal norm for truth is objective evidence.

(b) *Further Meaning of Objective Evidence.* By objective evidence, therefore, is meant *a characteristic of reality whereby its intelligibility is presented to and recognized by the mind.* It therefore involves two factors. Primarily objective evidence is a characteristic of the object, and is one with the intelligibility of the object. In this sense we might say that everything is *in itself* evident, since in theory all that is is knowable. But for anything to be actually known this light of reality must be made to shine on the mind, so that objective evidence secondarily involves a relation to the person knowing—it must be presented to and recognized by the mind. Until that is done, we cannot speak of truth as being possessed, and so a norm of truth would have no place.

Hence, objective evidence can be said to be the self-revelation of reality presenting itself to the mind.

Knowledge can be looked upon as a unique sort of dialogue between the known and the knower. Reality speaks or reveals itself—because it is basically intelligible; at times the attentive mind "hears" what is said and, hearing it, affirms what is said precisely because reality itself has first spoken or presented itself. Thus in judging, the mind says that *this is so* because it knows that this is so. In other words, what is affirmed is affirmed because it is *evident* to the knowing mind.

(c) *Kinds of Objective Evidence.* Since we acquire truth about reality in various ways, it is clear that various sorts of evidence can be distinguished. Thus I can accept as true the statement that "artificial satellites have been put in orbit around the earth," or that a given complicated mathematical computation is correct. Yet what I accept in these and similar cases is surely not evident to me. What is evident is the fact that the authority for such statements is reliable. The statements themselves can be said to have merely *external evidence;* there must be *internal evidence* for the reliability of the authority.

Other truths may be accepted because they have been demonstrated either deductively or inductively. Such conclusions are made evident by the processes used and thus can be and are accepted as evident. However, their evidence is *mediate.* As all demonstration derives its strength from the immediate analytic principles on which it rests, so, too, all mediate evidence is ultimately accepted because of the immediate evidence of those basic principles.

Finally, therefore, there is the *immediate evidence* of immediate judgments of sense experience and of universal analytic principles. For the statement that "it is now raining" or that "I am writing on this paper," no further appeal is necessary beyond the reality which immediately presents itself to my senses. For the statement that "a whole is greater than any of its parts," or that "a circle is not a square," no appeal is necessary beyond the reality which immediately presents itself to my intellect in understanding the subject and predicate.

(d) *Relation Between These Kinds of Evidence.* Beyond this sort of evidence one cannot and need not go in seeking a norm for truth. External evidence relies on internal evidence. Internal evidence which is mediate ultimately rests on evidence which is immediate. Immediate evidence of sensation or intellection is the ultimate reason one can have for accepting any truth, and all truth eventually rests on it.

This can be illustrated by a slightly different reflection on the intellect and the meaning of its judgment. The mind is a "seeing" faculty which does not create but recognizes and affirms reality. For this to happen, reality must become "visible" or manifest itself as it is to the mind. When it does this through some intermediary, the latter must be immediately knowable or rest on something which is so knowable, else there would be an infinite process backward. When what-is manifests itself as immediately knowable, the mind clearly "sees" and judges. Thus immediate evidence is a sufficient and ultimate norm for truth.

As was seen before in discussing deduction, the fact that each case of valid deduction can be "resolved" into first principles does not mean that such resolution need be actually carried out in every instance. Therefore even though mediate evidence rests on immediate evidence, it does not mean that in every instance one must trace it back in order to have true certitude. It does mean that the mediate evidence accepted does have a foundation in the evidence which moved one to accept the analytic principles on which the processes rest. However, assent is never justified without some objective evidence, and hence we can say in general that the ultimate and universal norm for truth is objective evidence.

III • SUMMARY OF THE ANALYSIS

(a) Although truths are acquired in various ways, the real function of each way is to present reality just as it is to the mind. Until the intellect sees reality as it is, it should refrain from judging; when it grasps reality as it is, it affirms it in a judgment. This manifestation to the mind of reality as it is, is objective evidence. Cf. *Analysis (a)*.

(b) Objective evidence is primarily a characteristic of the object, but it also involves a relation to the mind knowing. It is a self-revelation of reality to the knower. Cf. *Analysis (b)*.

(c) External evidence is accepted because of some internal evidence. Mediate evidence is accepted because of some immediate evidence. Immediate evidence, in respect to senses or to the intellect, is accepted because it is itself sufficient to determine the intellect. This can be recognized by reflection on the fact that the intellect is not a faculty for choosing but for seeing what is and affirming what it sees. When some reality is immediately present, nothing further is needed. Cf. *Analysis (c), (d)*.

IV • THE STATEMENT IN BRIEF

The ultimate and universal norm for truth is objective evidence.

V • DEFINITIONS

Norm: that according to which one judges.
> *Speculative Norm:* a norm for distinguishing between truth and falsity.

Universal Norm: one which holds for all truths.

Ultimate Norm: one beyond which there is no appeal, and in the light of which proximate norms function.

Objective Evidence: a characteristic of reality whereby its intelligibility is presented to and recognized by the mind.
> *External Evidence:* evidence which is not found in what is known, but in something else (testimony).

> *Internal Evidence:* evidence which is found in the thing which is known. It is either:
>> *Mediate:* that which is had by virtue of demonstration (conclusions) or

>> *Immediate:* that which is had when the reality known is immediately present either to the senses or to the intellect (analytic principles and immediate judgments of experience).

VI • SUPPLEMENTARY NOTES

(a) The explanation given in this chapter seems to mean that objective evidence is the source or principle from which all truth is derived.

REPLY: First of all, there is no such source or principle for human knowledge. It is acquired, rather, from various sources and in ways that correspond to the various sources; for example, immediate sense experience, deduction, induction, immediate analysis of S and P, and authority. First analytic principles serve for purposes of resolution, not of deduction. Moreover, objective evidence is not a principle at all; it is rather the *precise reason* why each and every truth is ultimately accepted. It is a characteristic of what is known, not a source of what is known. It adds nothing to the reality of what is, but simply means that what-is has, in particular circumstances, the characteristic of being manifest to the knower.

(b) Since objective evidence requires that any given reality be manifest to the mind, it would seem that in the last analysis the ultimate norm is actually *subjective*, that is, my clear perception ultimately determines my affirmation or negation.

REPLY: It is true that what-is must be manifest to the mind, and objective evidence does include this relation to the mind as one of its elements; however, that does not suffice to make it a merely subjective norm. Rather, the clarity of my perception is due to the clarity of the object manifesting itself, not vice versa. The object must *be* clearly visible or intelligible in order to beget clear vision or knowledge. Confirmation of this position is had in all studies and research, which are all so many efforts to let what-is speak for itself and answer various questions which are put to it, and so to render itself evident to the inquirer.

CHAPTER FOURTEEN

TRIPLE CERTITUDE

I • THE SITUATION

(a) *Quality of the Assent.* Having thus far considered the possibility, sources, nature, and ulitmate norm for human truth, we can now reflect on the quality of the assent to truth. In the face of a given problem or situation many intellectual reactions are possible. Thus I am capable of *error,* which is a positive difformity between the mind and reality, or of *ignorance,* which is a lack of knowledge. In a given case I may have a slight *suspicion* of where the truth lies, or *positive doubt,* when there are apparently equal reasons for two opposite solutions. Finally, I may express an *opinion,* which means a weak sort of assent accompanied by the realization that I may be wrong.

(b) *Certitude.* Over and above these various intellectual reactions, I can at times realize that I actually do know what judgment is to be made, and I make it knowing for sure that I am right. In this case I have *certitude,* which means a firm assent of the mind to truth with no prudent fear of error. It is this quality of the assent to truth, this certitude, which we are to examine in order to understand its nature and to see if it admits of degrees.

(c) *Formal Certitude.* That the practical living of our lives implies many practical certitudes of various sorts is easily seen in our use of fire, electricity, and various instruments, as well as in our trust in other people both for information and for other help. However, our interest is in the nature and possibility of *formal certitude* where the motive for the firmness of the assent is known, and where I therefore know that the certain assent is well-founded.

It will help in this matter to keep some concrete examples in mind, such as:

This man is telling the truth.
Mrs. Doe loves her son John.
This driver will not deliberately seek a collision.
If I step off the roof, I will fall to the ground.
That fire will burn my hand.
This water will boil at about 100° C.
The melting point of gold is about 1063° C.
God exists.
The human soul is spiritual.
The principle of contradiction, or of causality, and so forth.

In such cases, am I *certain* that they are true, and if so, with what sort of certitude?

II • THE ANALYSIS

(a) *The Motive for Certitude.* Considering the meaning of certitude in general as the firmness of an intellectual assent to truth with no prudent fear of error, it can be seen that any such intellectual response perfectly fulfills the function of the intellect. Other intellectual reactions such as doubt, suspicion, opinion, and so on, are not perfect acts of the mind. As a faculty for knowing truth, its ability is most perfectly actuated when it knows what is, and knows that it knows it with no prudent fear of being mistaken. Nothing better can be asked of it.

Concretely, this means that the given reality has revealed itself and the mind has clearly seen what is. In other words, formal certitude can be had only when there is objective evidence for what is known. As was seen before, the human mind is a faculty which *sees* what is, and seeing it must affirm it; as such it can be moved to *perfect* intellectual actuation only when what-is is presented to it with objective evidence.

Hence it must be said that there can be no formal certitude without objective evidence, and that objective evidence is the essential motive for every formal certitude.

Formal certitude, therefore, includes these elements:

(1) negatively, absence of all prudent fear of error;

(2) positively, determination of the mind to one position;

(3) positively, some objective evidence which is the motive for the determination of the mind.

(b) *Metaphysical Certitude.* In regard to the list of examples given earlier, one can see that he may be more certain about some than others; hence we now want to consider the various possible types of certitude. Some of the examples given refer to what a free agent will do in definite circumstances (John will not kill me); others refer to how a corporeal being will operate in given circumstances (Hydrogen and oxygen will combine to give water); still others refer to the very essences of things (the Principle of Contradiction).

ANALYTIC PRINCIPLES: If I reflect on my assent to the Principle of Contradiction, I can see that a denial of it is impossible. I am absolutely sure that an exception to it is impossible. And in this case my certitude can be called *absolute* or *metaphysical*.

The same can be said about other first analytic principles, since they refer to the very essence of a being, and an exception is completely impossible. Thus no part can be greater than its whole, every contingent being has a cause, and so on. Any exception to these would violate the very essences of things and hence would ultimately violate the principle of contradiction. The certain assent here is founded on being presenting itself evidently to the intellect, and the necessity of its existing according to these principles is an absolute necessity.

(c) *Mediate Analytic Judgments.* The same sort of certitude can also be recognized as belonging to mediate analytic judgments. If these are derived from metaphysically certain analytic premises, then the conclusion itself is metaphysically certain. For example, let us presuppose that I have absolute certitude that whatever is spiritual is naturally immortal, and that the human soul is spiritual. In this case I would have metaphysical certitude that the human soul is naturally immortal.

(d) *Synthetic Principles.* In the chapter on induction we saw that if that process is pursued carefully, it can at times give us knowledge of the natures of things. We saw that this is actually done in some easy cases. Now to say that we acquire some knowledge of the very nature or essence of a being means that the characteristic so ascertained will always and everywhere belong to that sort of being. For it not to be true of the being would involve a contradiction, since something essential to it would be missing. In fact, its absence would mean an essentially different being, rather than the one originally in question.

Thus synthetic principles, provided they have been rightly acquired, are metaphysically certain, since they are knowledge about the essences or natures of things. Every such principle equivalently says, "*It is of the nature* of x to do y."

(e) *Immediate Judgments of Experience.* The same can also be said about these. If I actually feel pleasure or pain, if I am actually writing, or talking and shaking hands with someone, if I actually see and hear rain falling, I cannot possibly deny these immediately experienced sensations and events. To do so would actually be an overt denial of the principle of contradiction. When such judgments merely enunciate *what is actually experienced,* there is no room at all for error or doubt and I have metaphysical certitude about them.

(f) *The Meaning of Physical Certitude.* When I turn, however, to such judgments as: "This fire will burn," "If I step from the boat I will sink into the water," "This metal will expand when heated," or to many other examples of how corporeal things will actually operate, the situation is not quite the same. Here it is not precisely a question of natures or essences, but of the actual operations of those natures.

Since (as will be shown in another tract) things are totally contingent, they require God's activity for the production of their effects or operations. So while leaving the thing *what it is,* God can suspend its proper operation, and in that case we have what is called a miracle. Thus it is of the nature of a heavy body to fall to the ground or to sink in water, yet it can be sustained so that it does not fall or sink. Or it is of the nature of ordinary fire to burn, but God can work a miracle. When that happens the *nature or essence is not changed,* only the operation of the thing does not take place. If a man walks on the water, it still remains of the nature of his body and of water that he sink into it, yet the effect is suspended miraculously.

When dealing, therefore, with corporeal agents and with their natural operations, the opposite is *not* excluded as *impossible* or as contradictory.

Can I then ever have formal certitude as to how these things will act? In answering this question care must be taken to avoid the two extremes of saying that we regularly have such certitude or that we never have it. Since the first requisite for physical certitude is knowledge of a law of nature, and since certain knowledge of such laws is definitely limited, formal certitude in this area is limited. Yet, since some aspects of reality are easily known and some laws of bodies have been fully established, there can be cases of formal certitude.

(g) *Conditions and Application.* The conditions to be fulfilled are simple enough. There must be:

(1) knowledge of the law of nature (synthetic principle), and

(2) the clear absence of any reason for suspecting an exception. If something in the surrounding circumstances suggests the positive intervention of God, it would be imprudent to say that certitude has been reached.

It should be kept clearly in mind that it is to some individual, concrete examples of what will happen that physical certitude applies; and it is had only when the law of nature which applies is known with metaphysical certitude. When the law has not yet been so established, there is only hypothesis and theory and probability. I do not put my hand into a hot fire, because I know that my hand will burn. I do not step off a roof, because I know that I will fall. Many other examples from daily life can be cited.

In these cases I can have *certitude:* a firm assent with no fear of error. It is *categorical,* because I am sure of what will actually happen, not just of what ought to happen. It is *not metaphysical* certitude, since the opposite is not impossible, nor is there any positive motive over and above the law of nature. This type of certitude is called *physical certitude.*

(h) *Hypothetical Certitude.* As distinguished from absolute certitude, physical certitude is called hypothetical, in the sense that the opposite is not impossible, but can be brought about through God's intervention. So the firmness of the assent varies from the firmness had in metaphysical certitude. In the latter the firmness is founded on the absolute necessity of something being essentially what it is; in the former, the firmness is founded on an objective necessity also, and a necessity which is well-founded in its own order, but is hypothetical in the sense that it can be suspended by God.

(i) *Reductively Metaphysical Certitude.* At the same time we can affirm that God will not *generally* or very frequently work miracles in respect to any given nature. This is true because having founded natures, He would be acting contrary to His divine Wisdom and Goodness were he generally to work exceptions. For this new positive reason, the statement that "fire will generally burn" is certain with reductively metaphysical certitude.

SUMMARY: What has been said can be summarized in the following examples, and can then easily be applied to others. Taking one expression such as "fire burns," we can qualify it as follows:

(1) "It is *of the nature* of fire to burn."
  This is a law of nature, a synthetic principle; hence we have metaphysical certitude in making this statement.

(2) *"This fire* will burn that paper."
  This is an individual application of the law; here we have physical certitude.

(3) "Fire will *always* burn."
  This says that the effect will always concretely follow and that there will never be a miracle in this matter; here we have no certitude at all.

(4) "Fire will *generally* burn."
  This relies on the Wisdom of God who will not generally work exceptions; here we have reductively metaphysical certitude.

(j) *The Meaning of Moral Certitude.* Keeping in mind what has already been seen, we can make the analysis of moral certitude a short one. Here it is a question of the operations of a free nature.

It is presumed that by the process of induction some moral laws (that is, how it is natural for free natures to act) have been acquired. These laws, too, are metaphysically certain since they affirm that this or that mode of operation is *natural* to free agents. Moreover, certitude will have to do with applications of these laws to the concrete, singular activity of some individual man.

Thus the statement "No one tells lies without reason," means that "it is of the very nature of man that he does not lie without reason." This does not mean that no man ever has or ever will lie without reason; it does mean that when this is done, a man acts contrary to his nature. Here we are again dealing with hypothetical certitude, and the exception here is due to an abuse of free will on the part of an individual.

(k) *Conditions and Application.* To have moral certitude there must be a *known moral law,* found by induction. There must also be a *clear absence of reasons* for suspecting an exception here and now.

Confidence in the reliability of our knowledge of the way other people act or will act is basic to much of our ordinary living and it plays a big part in the progress of scientific studies. Men have even been put to death on the reliable witness and truthfulness of others. Hence, although the application of formal certitude is much more difficult in this area than it is in physical certitude, it is nonetheless very important.

Countless examples may be proposed. Let us just say that at times I am certain that "this witness is not lying," "this scientist has honestly recorded his findings," "this driver will not deliberately seek a collision," and so on.

In such cases I have *certitude,* which is *categorical,* which is *not metaphysical,* and which is called *moral certitude.*

(l) *Reductively Metaphysical Certitude.* In a way similar to that given above (i), moral certitude can be *reduced* to metaphysical certitude by the addition of a new positive motive to the general law. Thus I can reflect that it is not possible that men generally or as a usual practice abuse their liberty in a given way, and hence I can conclude that moral laws will *generally* be actually followed by men.

SUMMARY: We can briefly exemplify what has been said on moral certitude with the example of men lying gratis, thus:

(1) "It is of man's nature that he does not tell lies without reason." This is a moral law and a synthetic principle; we have metaphysical certitude.

(2) "This man is not now lying." This is an application of a moral law; we have moral certitude.

(3) "No man will ever tell lies without reason." This means there will never be an abuse of freedom in this matter; we have no certitude at all.

(4) "Men will not generally tell lies without reason." This means there will not generally be such an abuse of freedom; we have reductively metaphysical certitude.

(m) *Certitude Is Analogous.* Clearly in this matter of moral certitude the firmness of the assent is again special and different from the other certitudes. There can be a firmness, founded on the moral law and the objective necessity which this implies, but since an exception can be had owing to the deliberate abuse of freedom, great care is needed in its use.

It can thus be seen that the motive of certitude is applied analogously to the three varieties discussed; since certitude is the firmness of the assent and this relies on the objective necessity manifested to the mind, and since this necessity is itself different in each case, it must be admitted that certitude is really three-fold. In all cases it can be *formal* certitude, yet the element of firmness is both present in each type and yet differs according to the necessity.

(n) *Certitude Is Not Statistical.* What has been said on physical and moral certitudes must not be confused with "statistical certitude" which has to do with summaries and statistical predictions and which can have many valuable uses. However, it is not necessarily concerned with a strictly *natural* tendency and it cannot be applied to this or that individual with certitude.

III • SUMMARY OF THE ANALYSIS

(a) Since the intellect is a faculty for knowing reality, it is *perfectly* actuated only when what-is clearly manifests itself to the mind. Thus objective evidence is the essential motive for all certitude. Cf. *Analysis (a).*

(b) *Metaphysical* or *absolute certitude* is had in regard to immediate analytic principles, mediate analytic judgments, synthetic principles, and immediate judgments of experience, since in all these the opposite is excluded as *impossible* in that it would violate the very nature of being. Cf. *Analysis (b), (c), (d), (e).*

(c) *Physical certitude* can be had in regard to some concrete operations of corporeal beings; for example, this fire will burn, this heavy body will fall if released. The motive is (a) some known law of nature, (b) the clear absence of reasons for suspecting an exception.

Although this certitude is hypothetical rather than absolute, since an exception is not absolutely impossible, it is nonetheless formal certitude and categorical, in that it states what will actually happen; and so the opposite is excluded as incompossible, that is, not simultaneously possible.

Reductively metaphysical certitude refers to the fact that natural
agents generally will produce their natural operations, since miracles
will remain rare exceptions owing to God's Wisdom and Goodness.
Cf. *Analysis (f), (g), (h)*.

(d) *Moral certitude* is had in regard to some concrete operations
of free beings; for example, this man is not lying; this driver will
not deliberately seek a collision.

The motive is (a) some known moral law, (b) the clear absence
of reasons for suspecting an exception.

Although this certitude is hypothetical rather than absolute (an
exception is not excluded as impossible), it is nonetheless formal
and categorical certitude.

If it is stated that this or that moral law will not be violated
generally, this can be called reductively metaphysical certitude,
since men will not generally act contrary to their nature. Cf.
Analysis (i), (j), (k).

(e) Certitude as applied to the three types discussed is analogous,
since the same elements (positive and negative) are present in
each case according to the necessity on which it is founded. Cf.
Analysis (l).

IV • THE STATEMENT IN BRIEF

*There are three sorts of formal certitude: absolute, physical
and moral, each of which rests on objective evidence.*

V • DEFINITIONS

Formal Certitude: a well founded firm assent to the truth with no
    prudent fear of error.
Metaphysical Certitude: firm assent based on the very essences or
    natures of things. This is called *absolute,* since the opposite is
    excluded as impossible.
Physical Certitude: firm assent based on the relation between nat-
    ural causes and their effects. Its motive is (a) a law of nature,
    (b) the clear absence of reasons for the contrary. This is
    hypothetical in the sense that the opposite is not impossible.
    But where physical certitude is had, what will actually happen
    is known, and so the opposite is excluded as incompossible,

Triple Certitude

Triple Certitude 111

that is, as not being simultaneously possible. So it is stated
categorically, This fire will burn.

Moral Certitude: firm assent based on the relation between free
causes and their acts.

Its motive is (a) a moral (not ethical) law, (b) the clear absence of reasons for suspecting an exception.

This also is hypothetical, and yet its opposite is excluded as
incompossible.

Rests on: that is, is supported by the *motive* of objective evidence.

Motive: the ground or reason which determines a firm assent.

Objective Evidence: a characteristic of reality whereby its intelligibility is presented to and recognized by the mind.

CHAPTER FIFTEEN

HUMAN TESTIMONY

I • THE SITUATION

(a) *Introduction.* We have considered various ways in which one can acquire truths from personal experience. In all of them, whether immediate judgments of experience or demonstrated truths, whether analytic or synthetic principles, I affirm that something is true because in one way or another it is rendered *evident.* In all of these cases, when the proper requisites have been met, I can be said to have *intrinsic objective evidence.* That is to say, the object itself manifests itself to me and is known by me either mediately or immediately. However, there is a whole realm of truths not included under those considerations.

(b) *Social Aspects of Truth.* We not only live with other men, but also effect a constant social communication of truth. Hence, the question arises as to the precise value of truths received on the testimony of others. Not much reflection is needed to see the very practical importance of this question, in regard to the communication both of events and facts and also of doctrinal matters.

Thus I may not have witnessed a certain event (a hurricane or a fire), but some friend was there and he tells me about it. Or I was never at the south pole, but I do rely on the account given by someone who was there. Again, I learn of the French Revolution and accept it as factual on the authority of historians.

Of great importance, too, is the use made of testimony in doctrinal matters. Take for example a scientific field such as biology. It is true that each student must experientially learn much of the basic matter in this or any other field. However as time goes on

112

he accepts more and more on the word of competent men in his field. It would be simply impossible for any one man to repeat every experiment required in order to know experientially the truth of all the experiments and conclusions that have been established even in his own field of specialization. The same thing applies to politics, news, economics, to dealing with the members of one's family and with business associates, and to all other branches of human knowledge and hundreds of life situations.

(c) *Progress in Truth.* This, then, is a matter of vast ramifications and of very necessary and important usage in the social and intellectual development of man. We do inherit much; each of us does not exactly have to begin *ab ovo* in our intellectual life. One generation passes on much that it has acquired, and so human knowledge can and does make great strides forward. Were that not so, were there no communication of knowledge, we might all be intellectually still back in prehistoric times.

(d) *Our Aim.* The application of this matter is so varied in both practical daily life as well as in scientific and cultural life, that we can be expected to do no more here than indicate the general characteristics which must be enjoyed by testimony before it can be accepted as a source of certain truth. Moreover, it must be said that before one admits true certitude in any given case, he should be very careful and cautious in his application.

Our concern here, as it has been in other considerations, is to indicate that there does exist this other source for truth, and to point out in general the conditions required for it to be valid and reliable.

II • THE ANALYSIS

(a) *Testimony of an Immediate Witness.* Where someone is reporting to me about an event he himself has witnessed, then at times I can accept what he says as certainly true, if (1) the fact of his testimony is clear, (2) he knows what he is talking about, and (3) he is telling the truth. In such circumstances, he is recognized as being in possession of the truth, and if he tells what he knows, then he puts me in possession of the same truth.

Naturally, the application of these conditions is not always easy, but when and if they are fulfilled, I have with certainty acquired a new truth. The fact of his testimony is the easiest to verify, since I am either hearing it from him directly, or reading what he has written. So, presuming that I can understand him, I know what he is communicating. Secondly, there are times when I can know that he has knowledge of the event in question; especially when that event was something outstanding or very easy to grasp, and he was in a position to know it. Thirdly, his truthfulness might require more careful consideration, but at times I can be sure from my knowledge of his probity, prudence, and good name, and from the circumstances which hold here and now, that he is actually telling the truth.

At times these conditions can be and are fulfilled, and when they are it means that the witness has knowledge which he truthfully communicates to me, so that I now know some truth which was not the object of my own experience.

(b) *Many Witnesses.* Where the same event was witnessed by many people, and where I receive many *independent* accounts of what actually happened, the degree of my certitude in regard to that event is even intensified. Although in this case there will be and should be a certain amount of divergence in respect to minor details, the fact of substantial agreement in reporting the event will itself require a sufficient reason. And here the only sufficient reason, granting the independence of the various testimonies, must be the truth of the event.

(c) *Remote Events.* In the case of events which happened long, long ago, and in which I do not have immediate contact with the personal witness or witnesses, I am confronted with a situation which requires special study and the special techniques of the historian. All these, including knowledge of sources, the nature of documentation, and so on, are too specialized and difficult to handle here, so we need merely mention that what has to be determined even in these cases is fundamentally the same as what was mentioned above.

(d) *Doctrinal Testimony.* In cases in which what is proposed for my belief is not some particular event but some teaching or doctrine, the conditions to be met are still fundamentally the same. Thus I must know (1) that some doctrine is proposed as being true, (2) that it is proposed by someone who knows what he is talking about, and (3) that it is truthfully communicated.

That some teaching or doctrine is proposed can easily be established. That the person proposing it is competent and has certain personal knowledge of it may not always be easy to verify, but at times I can know it through knowledge of the position of respect which he occupies in his own field, his already proved competence, his modesty of character and his dedication to truth, the importance of what has been proposed, the coherence of what is now taught by him with other truths already known, and so forth. Finally, his truthfullness here and now may be known by me either directly or from knowledge of his dedication to his work, or from knowing his good reputation in his field and the fact that any lying would be ruinous to his authority.

At times these conditions can be and are fulfilled, and so I can accept as true the doctrine or teaching which has been proposed; in so doing I believe the truth of a doctrine on the statement of this doctrinal witness and have acquired a new truth through human testimony.

This sort of testimony is, of course, the sort we mentioned earlier in stating that progress in all sorts of scientific studies would be impossible were there not social communication and acceptance of doctrinal matter. Obviously, doctrinal testimony is not to be accepted lightly, but the growth and the progress of human learning very definitely relies on its trustworthiness. It is true also that this sort of testimony is more easily accepted by learned men than by the incompetent.

(e) *Evidence.* Consideration of what has been said reveals that the truth of the event or doctrine does not, by means of testimony, become intrinsically evident to me; the supposition in this matter is that what I accept is not itself experienced by me. What I myself must really *know* (that is, what must be intrinsically evident to me) is the fact that testimony has been given by one who knows and who tells me the truth. But the truth which I accept on authority remains only extrinsically evident; and so it is by an act of human faith that I accept it. I *know that* it is credible; I do not positively *see what* is believed. Moreover, in line with what was said in connection with moral certitude, it is not always necessary for us to establish with positive proofs the truthfulness of the witness; at times the general truth that men do not as a rule lie without reason, together with the absence of any positive reason for the witness to be lying here and now, can suffice for moral certitude in testimony.

(f) *Caution Needed.* Naturally, in matters of this sort, one must be very careful to distinguish the emotional reactions or the personal evaluations and commentaries of the witness from the substance of what is presented by him. This is particularly necessary in the evaluation of news accounts, editorials, political matters, and the like. This amounts to saying that although this field of testimony is a fruitful and necessary source of truths which we cannot or have not directly experienced ourselves, it is still a source which must be used with careful discrimination.

III • SUMMARY OF THE ANALYSIS

(a) The testimony of a direct witness to some event can provide me with new truth, if I know: (1) the fact of his giving testimony, (2) the fact that he knows the event in question, and (3) the fact of his truthfulness here and now. I can sometimes know these facts. When many independent witnesses report the same substantial event, my certitude can be even higher. Concerning historically remote events, special techniques must be used. Cf. *Analysis (a), (b), (c).*

(b) When what is proposed for my acceptance is not an event but some doctrine or teaching, I can be justified in accepting it on authority if I know: (1) the fact of its being proposed, (2) that the one proposing it is competent in the matter, and (3) that he is telling the truth. These conditions can sometimes be met, and so I can at times accept such truths on authority. Cf. *Analysis (d).*

(c) The evidence possessed for the new truth which is accepted is and remains extrinsic evidence; however, I must have intrinsic evidence for the knowledge and truthfulness of the witness. Cf. *Analysis (e).*

IV • THE STATEMENT IN BRIEF

Under certain conditions historical and doctrinal testimony
can be valid sources of new truths.

V • DEFINITIONS

Testimony: the act whereby a witness transmits to another some knowledge which he has. This can be either historical or doctrinal.

Historical Testimony: that testimony which deals with some fact or event.

Doctrinal Testimony: that testimony which presents some doctrine or teaching.

A *Witness:* one who transmits some truth to another.

An *Immediate Witness:* one who has himself had direct experience of the truth in question.

A *Mediate Witness:* one who himself received from another the truth which he is now transmitting.

Credibility: the sum of those characteristics which render the truth proposed actually believable. It is made up of both the fact of testimony having been given, and the authority of the witness.

Authority of the Witness: a characteristic of the witness which leads to his being believed. It comprises his knowledge (that he was not mistaken), and his truthfulness (that he does not wish to deceive).

CONCLUSION TO

PART I

One who has carefully followed the analysis given in these chapters should find himself in possession of a realistic position in respect to knowledge which is neither naive nor over-simplified, but which tries to take into account all the factors of human knowledge and explain them in an objective and coherent fashion.

We began by analyzing the only data available in this sort of investigation, that is, concrete acts of judgment, and from consideration of these acts we acquired some knowledge of the cognitive potency. Speaking in a most general way, we saw that the very making of any act of judgment involves not only a self-awareness, but also a recognition of certain basic truths; this involved a recognition of the truth-capacity of my mind. More specifically we saw that we are capable of knowing objective truth, the object of which is real being, independent of the mind for its existence, and fundamentally intelligible.

In more detail we saw that under limited conditions the senses are reliable sources of knowledge about sensible reality, that universal ideas are acquired by abstraction and yet they represent a nature which is in reality, that even the universality of the concept has a foundation in reality, that truth formally exists in the judgment, that the process of ratiocination involves logical causality, that deduction provides new truths, that induction leads to the knowledge of synthetic principles, that truth can also be acquired from human testimony, when what is known is not itself accessible to me, that the ultimate norm for truth is objective evidence, and, finally, that there are three kinds of certitude, each of which differs from the others, yet each of which is formal certitude.

Briefly, then, we have examined sensation, conception, induction, deduction, and testimony for their functions as sources of truth.

Although each of these could, of course, provide the matter for long, detailed studies, what has been presented offers a logically coherent, basically satisfactory and experientially well-founded theory of knowledge.

With these general lines sketched and both the power and the limitations of the human mind indicated, the student who has reflected on these matters should have advanced from a naive scepticism or a naive confidence in his knowing powers to a reflective recognition of the basic capacities of the mind in its various ways of acquiring truth, and to the confidence that the human pursuit for truth in the many fields of study open to man is not a futile or unrewarding effort, but one that can and should be engaged in with diligence and caution. He should now have the assurance that truth is neither the arbitrary fiction of a diseased mind nor the unattainable goal of a hopeless search, but a reliable grasp of reality which unfolds gradually and reveals step by step the profound riches of being.

Part Two ——————————————————

OTHER THEORIES
OF KNOWLEDGE

An Introductory Note

on *Pure Positions*

in Philosophy

The present section aims at presenting briefly the explanation and evaluation of human knowledge proposed by other philosophers or philosophic schools. In this connection, it must be remembered that it is impossible, in a work of limited scope, to discuss these positions with all the nuances and individuating characteristics proper to each philosopher. Such work belongs rather to a history of philosophy.

For our purposes it will be quite enough if we can succeed in grasping the fundamental tenet, the basic attitude and the central inspiration inherent in each different solution offered. A comparison with medical procedure may help to make this point clear. Just as no two individual cases of tuberculosis are exactly the same, since each has slightly varying characteristics depending on the organism which is affected, and the doctor is perfectly justified in saying that both patients have radically the same disease, so, too, in respect to theories of knowledge, we can admit individual differences of development and emphasis and still detect and indicate a fundamental oneness of insight and approach on the part of several philosophers.

What can thus be isolated and presented as the underlying common explanation or insight in matters philosophical is known as a "pure position." That the consideration of pure positions is a necessity in philosophy becomes clear when one reflects on the fact that without them we would be involved in an endless and overwhelming mass of details and individuals. That it is possible has been indicated in the analysis of the mind's power to abstract. That this is a customary procedure can be empirically confirmed by turning to any history of philosophy whatsoever.

We shall proceed, then, to present very briefly various pure positions relative to human knowledge. In each case we shall try to understand the fundamental insight, the reasons which led to such a view, the success of the doctrine in explaining some of the data of cognition, and, when it is required, the failure of the doctrine to explain other data which should have been taken into account.

CHAPTER SIXTEEN

SCEPTICISM

I • DESCRIPTION

(a) *Pyrrho's Doctrine.* Diogenes Laertius says in his famous *Lives* that Pyrrho adopted

> a most noble philosophy, . . . taking the form of agnosticism and suspension of judgment. He denied that anything was honorable or dishonorable, just or unjust. And so, universally, he held that there is nothing really existent, but custom and convention govern human action; for no single thing is in itself any more this than that.
> . . . But Aenesidemus says that it was only his philosophy that was based upon suspension of judgment, and that he did not lack foresight in his everyday acts. He lived to be nearly ninety.[1]

The doctrine touched upon here is that of Universal Scepticism which teaches that the human mind can never attain to any truth with certitude. Hence, the proper attitude for which one should strive is a total suspension of judgment. The state of mind, then, for which we are urged to strive, is one of universal doubt.

(b) *Not So Extreme in Most Things.* Although it is true that this extreme position has been adopted by relatively few philosophers and by none who can be considered truly outstanding in the history of thought, yet it deserves our attention not only because some few have held it, but also because very many other doctrines are more or less infected with the same basic view. *Insofar* as they share this view, they are subject to the same evaluation.

---

[1] R. D. Hicks (trans.), Diogenes Laertius' *Lives of Eminent Philosophers,* Loeb Classical Library (Cambridge, Massachusetts: Harvard University Press, 1925), II, 475.

(c) *Basically a Respect for Truth.* Although this may sound paradoxical, it does seem that we must admit that the reason for the existence of this doctrine of Universal Scepticism is to be found precisely in the Sceptics' profound respect for truth. They look on truth as so important and valuable that they are unwilling to accept lightly anything which is offered by others or experienced by themselves.

Many of them were struck by the prevalence of error in human knowledge, by the number of positions once dogmatically proposed and later disproved, and by men's readiness to accept as true anything that attracts them. Added to all that is the necessity man is under to establish the reliability of the mind by the use of the mind itself, which seemed to them to be a begging of the question. Moreover, even those who defend the power of the human mind are forced to admit that human knowledge is only partial, and that it only gradually comes to any knowledge of things.

From these and similar considerations they conclude that one can never rest secure in the possession of any truth at all, and that the proper attitude is a total suspension of judgment and an admission of doubt as the most balanced state of mind.

(d) *Inspires Caution.* Although admitting that utter Scepticism is an extreme position, we can also admit that contact with it can help to introduce a healthy note of caution into our own studies and reflections, and can put us on our guard against a too-ready acceptance of "truths." Moreover, in the history of thought it has occasioned profound analyses on the part of men such as Aristotle and St. Augustine relative to some very fundamental truths. In fact, the Scepticism of his day partly moved Descartes to attempt a reappraisal of metaphysics in the hope of finding irrefutable grounds for philosophy.

(e) *As a State of Mind.* What preceded has to do with Scepticism as a doctrine which is proposed for our acceptance. It can also be considered as *a fact*, that is, as an actual state of mind. In this sense a Sceptic is a man who actually doubts everything, or who keeps his mind in a state of suspension from all judging either about speculative or about practical matters. This, then, is nothing more than putting the doctrine into practice.

(f) *Gets at Foundations of Knowledge.* It must be noted that what is at issue here is exactly what was treated in Part One, Chapter One, concerning the general ability of man to know truth. Hence it is not a limited question concerning sensation or conceptualization or demonstration. We must admit that the acquisition of truth is often difficult and in some particular cases it may have to be admitted that success is unforeseeable. Yet we have seen that we do know some truths and do have reliable means for acquiring others. In contradistinction to this view, Scepticism is totally negative; it is radical despair about man's intellectual abilities. As such it can never yield fruitful results in the domain of knowledge.

It is true that some—Pyrrho, for example—felt that not despair but utter tranquillity of mind should be the attitude adopted. But this must still be the tranquillity of emptiness or despair. At any rate even the Sceptics admit that in actual fact they have to live with practical certitude. So their denial centers around speculative truths, and this denial they propose as a doctrine.

II • THE ANALYSIS

(a) *Speculative Certitudes Implied.* Turning first to Scepticism as *a fact* (that is, an actual mental state of universal doubting) we can start by accepting an admission made by the Sceptics that even they have to live with "practical" certitudes. By this is meant that in ordinary daily living and acting they must proceed "as if" there were some human truth and certitude, even though there actually is none.

However, this very living and acting with "practical" certitudes does necessarily involve some "speculative" certitudes. Thus failure to walk off a cliff or into a ditch, keeping out of the path of a train or an auto, eating food and avoiding poisons, all indicate a general knowledge of and distinction between things to be sought after and things to be avoided. Hence, at least a speculative value judgment of general implications lies at the roots of these practical decisions. They imply the speculative certitudes that harmful things are to be avoided and helpful things are to be desired. The same can easily be illustrated by other "practical" certitudes which involve speculative certitudes.

(b) *Total Scepticism Impossible.* Again, if we limit our consideration to doubt in the speculative order only, this proposal of doubt implies a recognition not only of the fact of doubt but also of the existence of serious reasons for embracing this attitude. It also involves knowledge of the difference between the states of certitude and doubt, and of the conditions which must be met before a firm assent is given. Thus there is an admission that the intellect should seek conformity with reality and that it can be moved to assent only by objectively valid motives.

(c) *A Self-Defeating Doctrine.* Looking now at Scepticism as a general doctrine to be taught or as a general theory of knowledge, we can see that it proposes a definite teaching on the nature of human knowledge. Its stand is that nothing is certain, and this doctrine it proposes as a tenet we should all adopt.

In other words, a clearly determined doctrine with a clearly determined meaning is proposed. Thus stated it means that Scepticism is certainly the true explanation of human knowledge, and hence one thing is exempt from the total incertitude of our knowledge— Scepticism itself.

However, if this is the case, not everything is uncertain; we are freed from the need of doubting about Scepticism itself, and hence Scepticism overcomes its own doctrine.

In this sense, Scepticism is self-defeating and self-destructive, since its assertion implies its own negation.

Nor is the situation saved for Scepticism by the assertion that it is proposed only as a probability or as a mere conjecture, for even in that case, as Aristotle insists, something having a determined meaning is proposed, and the very determinateness of the meaning involves some certitude, that is, what is proposed as merely probable or conjectural has *this* or *that* determined and, hence, certain meaning. Without a determined meaning no position at all is taken and no doctrine is proposed.

(d) *Overcoming Scepticism.* What has been done, then, is to approach this matter from the point of view of the Sceptics themselves, to accept, for the sake of discussion, the points which they themselves make, and then to point out how their own assertions actually mean that their doctrine is self-destructive and that universal adherence to this state of mind is impossible. In this sense it can be said that at least in a theoretical way the Sceptic can be brought to see that the human mind is really capable of truth.

III • SUMMARY OF THE ANALYSIS

(a) Since Sceptics admit that they live with practical certitudes, and since these imply some speculative certitudes, it must be admitted that Sceptics, too, implicitly admit some speculative certitudes. Cf. *Analysis (a)*.

(b) Speculative doubt necessarily involves some speculative certitudes and some knowledge of the nature of the human intellect. Hence, Sceptics who profess such a doubt do have some speculative certitudes and some knowledge of the nature of the human intellect. In other words, total Scepticism as a state of mind is impossible. Cf. *Analysis (b)*.

(c) Scepticism as a doctrine teaches that nothing is certain. This position, however, is proposed with a determinate meaning, and hence is self-contradictory and self-destructive. Cf. *Analysis (c)*.

IV • THE STATEMENT IN BRIEF

Universal scepticism viewed as a state of mind is impossible; as a doctrine it is self-contradictory.

V • DEFINITIONS

Scepticism: the profession of doubt.

Universal Scepticism: the profession of doubt in respect to all human truth.

As a State of Mind: that is, as actual doubting. This may refer to doubting either in the practical or in the speculative order.

Is Impossible: no one can actually doubt everything; or no one can *be* a universal sceptic.

As a Doctrine: as the teaching or theory that universal doubt is to be entertained.

Self-Contradictory: implying its own negation and the implicit affirmation of its contradictory.

VI • SUPPLEMENTARY NOTES

The core of all objections against the position just proposed can be summarized in two brief statements. For reasons of clarity these will be mentioned and answered in a few words.

(a) The senses have frequently led us into error, and so we should put no confidence in any of their presentations.

REPLY: There is clearly no denying the fact of error, nor the fact that the senses themselves have often been the occasion for our errors. In fact it seems necessary to admit that the practical living of our lives makes it nigh impossible to avoid all errors. We must make prudential decisions, and often enough without the time or opportunity to get all the facts. If we always waited for sufficient evidence to justify formal certitude, living would be a very heavy burden.

However, the meaning of our position is not that we must always wait for formal certitude, nor that we can in every case achieve such certitude. Our point is that we can and do at times attain a firm possession of the truth. The very admission of error is an admission of this fact.

Regarding the senses and error this can be said: The error has not really been on the part of the senses, but on the part of the intellect judging beyond the evidence. The senses are themselves material and very limited faculties, each of which can operate reliably only within a very small area. They cannot and should not be expected to do more than their nature allows, and one of the functions of the intellect is to recognize these limitations and judge accordingly. Of themselves the senses are highly specialized, and, unless they are diseased, they are naturally competent within their limits. Hence error in these matters is due to my willingly going beyond the evidence in my judgments. The solution, then, is not one of despair in respect to the senses, but one of intelligent control. We are never physically forced to error.

(b) Reason itself is used to justify reason, hence the justification is not reliable.

REPLY: In line with the analyses made here and elsewhere we can now say that no begging of the question is involved in my intellectual recognition of my own intellectual abilities. It is true that we cannot appeal to something irrational (outside reason itself) for reason's justification, since in that case the nonrational justification would itself need further justification, and no solution would ever be possible. Rather, the procedure must proceed from intellectual acts to capacities, as the data themselves demand.

VII • SUGGESTED READINGS

Aristotle, *The Basic Works of Aristotle*, ed. Richard McKeon. Cf. *Metaphysics*, Bk. IV, ch. 4; Bk. XI, ch. 6, pp. 737–43, 858–60. New York: Random House, Inc., 1941.

Laertius, Diogenes, *Lives of Eminent Philosophers*, translated by R. D. Hicks (Loeb Classical Library) II, 475 ff. Cambridge, Massachusetts: Harvard University Press, 1925.

Santayana, George, *Scepticism and Animal Faith*. New York: Charles Scribner's Sons, 1929.

Vaihinger, H., *The Philosophy of "As If,"* translated by C. Ogden. New York: Harcourt Brace & Company, Inc., 1925.

CHAPTER SEVENTEEN

METHODIC DOUBT

A. Methodic Doubt in General

I • THE SITUATION

(a) *Nonsceptical Doubt.* Having considered the nature and meaning of the doubt proposed by the Sceptics and having decided that such an approach finally implies the negation of what is explicitly stated, we can now turn to consider the method proposed by others whose avowed purpose was to overcome the defeatism involved in the sceptical attitude, and who, nevertheless, wished to approach the problem from the very grounds of Scepticism in the sense that they proposed an initial doubt, methodic, however, in nature, rather than sceptical, as the way to establish the mind's ability for truth.

(b) *Ignorance, Doubt and Problems.* It must first be admitted that in the ordinary sort of problem which arises either in the sciences or in philosophy, the most intelligent approach is by way of admitting either that one is ignorant of the answer, or at least that one has only an honest doubt of what the answer really is. Hence, the fruit of the study and investigation will be the removal of the ignorance or the doubt, and the providing of the correct answer. What we find in many philosophers of the modern era is the desire to approach the question of knowledge in this same way.

(c) *Doubt and Problem of Knowledge.* Thus, without going into the various characteristics whereby we can distinguish one of these men from another, we can first of all consider the very general question of the possibility and validity of approaching the question in this way. Later we shall consider more carefully the doubt of Descartes and of Cardinal Mercier.

In general then, the approach of methodic doubt means that I must deliberately convince myself that in the matter of the mind's aptitude for truth the best approach is to say that I really do not know even implicitly whether or not I can know truth. I can know only what is in my mind and can know neither its origin in "outside" sources nor its fundamental conformity to that "outside." I thus deliberately choose to doubt about the mind as an instrument for truth, either because there are good reasons for and against it, or because I see no reasons at all for either side.

(d) *Purpose Is Liberation from Error.* It must be borne in mind that this approach does not intend to make us the slaves of error and doubt, but rather to use doubt as an instrument which is to lead us to confidence in our intellect. At the same time it must also be borne in mind that this method is to be used faithfully, methodically, continuously and logically. Since it is the only method allowed, it will not be permitted to smuggle in some other procedure along the line, else one will have been unfaithful to the method.

All this means that a consistent and logically pursued doubt is expected to eventuate in the rejection of Scepticism and in the establishing of my intellectual ability for truth on a firm and lasting basis. Can these two be conjoined? That is the question to which we will seek an answer in respect to Descartes and Mercier later on; but that is also the question to which we now want an answer as it is taken in its general formulation, omitting for the moment the qualifications attached by those authors.

II · THE ANALYSIS

(a) *Begins and Ends in Doubt.* All that we are given to work on here is that my knowledge, not only in some particular applications, but even more in its general capacity for truth, is not to be trusted. In other words, although using the mind to arrive at respect for the mind—and methodic doubt entertains the sincere hope that this respect will be reached—this position states that I should entertain a serious and sustained doubt about the noetic nature of the mind itself. Hence in whatever process I follow,

I must honestly doubt each step taken. Wherever I call a halt to this process, I must still entertain that doubt, because it must logically apply to the last as well as the first step, or otherwise I have somehow ceased to employ the method, and that will call for an examination of what new method I have introduced at that point.

To put it briefly, the method of doubt can lead only to a conclusion of doubt, provided it be logically employed.

(b) *Wants Demonstration.* This procedure is quite different, of course, from what we did in the previous chapter. There we considered the doubt and the doctrine of doubt proposed by the Sceptics and drew our inferences; here, however, we are personally asked to entertain the doubt itself and thus overcome it. Again, we are not saying precisely that this methodic doubt cannot really be entertained in this matter, although that has already been seen to be the case as a result of our previous studies, but we are saying that the method of doubt, once applied to the fundamental capacity of the mind as an instrument for truth and once logically pursued, will lead to no resolution of the initial question.

Hence, although on the surface the method of doubt may seem a more open-minded and frank and honest way to approach this or any other question, it must be admitted that the subject matter being investigated here does not and cannot lend itself to this sort of approach.

This position in its very formulation seeks a strict demonstration to remove the initial doubt. To satisfy that demand it would be necessary to find the proof *outside* the intellect itself, or else there would be a complete begging of the question. That would require some nonrational and nonintellectual instrument of proof. The only alternative is to speak of demonstrating the very grounds of demonstration.

III · SUMMARY OF THE ANALYSIS

Methodic doubt, logically employed in solving the "problem" of knowledge, can only succeed in rendering the solution impossible, since the very method demands that it apply to every step in the process, the last as well as the first.

B. The Cartesian Methodic Doubt

I • THE DESCRIPTION

(a) *Intellectual Chaos.* Descartes was a religious man, a capable mathematician, and a serious thinker who was appalled by the Scepticism and Atheism of his contemporaries. Hence, he sought to bring all the powers at his command into a vigorous effort to set a few fundamental truths firmly on a lasting and unshatterable foundation. He was convinced that philosophy had become almost bankrupt because every position imaginable had been already proposed, and every affirmation of a position had been accompanied by its denial. Proofs proposed by one were rejected by another.

(b) *A Fresh Beginning.* He decided that by his day truth and error were so mixed up that it was hopeless to try to disentangle them, and so the best thing was to make an entirely fresh beginning. In his philosophy he proposed to do only three things: to show that the soul is immaterial, that God exists, and that there is an extended universe. What he sought were proofs so clear and so cogent that none would ever again be able to deny these, and that Scepticism and Atheism would thus be forever ousted from the world.

(c) *The Way Out.* As a prelude to his works, Descartes pointed out that he would continue to observe his usual ethical, religious and civil duties so as to continue a peaceful life among men. But apart from those, he would now proceed to doubt everything which he could find a reason for doubting. Hence, he methodically doubted the validity of all sense presentations, all truths got by demonstration, memory, imagination, mathematics, and all other sources.

Most characteristic of his methodic doubt was his decision to treat the probable as false, and to distrust the senses as sources of knowledge. He even went so far as to suppose the existence of some powerful malignant spirit who constantly deceived him in all that he thought.

(d) *One Firm Truth.* This method led him into a very turmoil of doubt, which continued until he found one unshatterable truth: his own existence. With that he felt that he had overcome the doubt, and could now proceed on that basis to erect his whole philosophy.

His doubt, then, can be described as real, positive, exaggerated, methodic and possibly universal.

II • THE ANALYSIS

(a) *A Reawakening.* Descartes did begin a new trend in philosophy by focusing attention on the importance of analyzing and evaluating human knowledge. Moreover in doing this he stressed the role of the subject in any such study. He also reawakened interest in metaphysics and put a new emphasis on the value of method in philosophy.

(b) *The Fruit of Doubt.* On the other hand, it can be said that the analysis made above on the method of doubt is fully applicable to Descartes' methodic doubt. That this is true is amply confirmed both by the failure of Descartes' philosophy itself and by the fate it met when his devoted followers explicitated its logical implications.

(c) *Doubting Existents.* More specifically it must be said that Descartes' method of doubt was applied to *all existence,* and thus one is left with only the ideas of essences and their interrelationships. Hence, in respect to existence, the best one can have is an essentialized existence which may possibly have some meaning in the ideal order, but cannot give contact with the real *existential* self or with any other existing thing. He speaks of the self as a *res cogitans,* and in his context this thinking thing is either nothing more than various isolated ideas or "acts of thought," or, if it pretends to be more and to include the substantial self having these thoughts, it is not justified by his method.[1]

(d) *Aware Only of Thought.* Moreover, this thought-thinking fails to explain human knowledge as it is concretely experienced, since human knowledge involves a concrete existing subject *thinking about something:* knowledge is knowledge *of* an object, not just isolated thought. The knowledge-experience is basically relational, including *knower, act* of knowing, and *object* known, and

---

[1] Cf. F. Copleston, *A History of Philosophy* (London: Burns, Oates & Washburn, Ltd., 1946–59) IV, p. 96.

this is avoided rather than explained by the pure thought postulated by Descartes.

Hence, his type of double *divorced* thought from reality, and instead of explaining it, created a new and an erroneous problem which has plagued modern philosophy. It must be remembered that our analysis has tried to show what the implications and logical results of the Cartesian method really are. Descartes himself clearly intended to arrive at real existents and clearly felt that thought is about something (Meditation III). Whether his method is logically consistent with such desires and claims is another question.

III • SUMMARY OF THE ANALYSIS

(a) Although Descartes fostered interest in philosophic thinking, his doctrine of methodic doubt is open to the same difficulties as were mentioned in general on pages 133–134. Cf. *Analysis (a), (b).*

(b) Descartes' methodic doubt leaves one isolated in the mental order, and hence cannot gain contact with any existent, either the self or any other reality, and hence does not explain what is concretely experienced as human knowing. Cf. *Analysis (c).*

C. The Negative Doubt of Cardinal Mercier

I • THE DESCRIPTION

(a) *Avoid Positive Doubt.* Cardinal Mercier, famous for his efforts at revitalizing Scholasticism, in order both to meet the challenge of Scepticism and to solve the problems of knowledge which had developed from the Cartesian method, proposed that the proper approach to Epistemology was by way of a methodic doubt, but a negative one. This, he felt, would provide the answer sought and would still be worlds apart from the positive doubt of Descartes. He clarifies his position by stating the initial attitude we should have in respect to: (1) our cognoscitive faculties, (2) mediate judgments, (3) immediate judgments;

> (1) The philosopher as he begins the work of reflection, which he undertakes in order to evaluate the nature of certain knowl-

edge, *has not the right either to deny or to declare unreliable, any more than he has the right to affirm, the aptitude of the cognitive faculties for certitude in knowing truth.*

Consequently, the initial state of reflection, with respect to the *power* of our cognitive *faculties*, should be *voluntary ignorance, abstention.*[2]

(2) Since every demonstration supposes doubt (either real or methodic) about the conclusion to be proved, reflecting reason begins with *doubt* about all *mediate* judgments.[3]

. . . we believe that, in facing truth in general, all truth, it is natural and legitimate to *attempt a universal doubt.*[4]

(3) What is to be the attitude of the spirit in the presence of these *immediate, indemonstrable* propositions? . . . It is legitimate, it is in keeping with the demands of a rigorous critique to *try* to put them in doubt.[5]

In connection with this attempt to doubt immediate truths, he claims that subsequent parts of his work will show that the effort is in vain—they cannot be doubted.

(b) *From Ideal to Real.* To complete our brief look at Mercier's position it must be added that a sketch of his procedure indicates what he has in mind. He begins by stating the problem to be handled (Book I). Next he outlines the conditions needed in order to be able to examine the questions raised (Book II) and points out the two basic problems which he will consider: (a) the objectivity of knowledge and (b) the objective reality of the terms of judgment. He arrives at the objectivity of our knowledge *in the ideal order* (Book III) and finally comes to the objective reality of our concepts (Book IV). (Cf. *Critériologie Générale,* pp. 383–385.)

II • SUMMARY

Mercier proposes that we begin with an elected state of *ignorance,* negative doubt about our aptitude for truth, then that we proceed to justify judgments in the ideal order, and finally that we get to the conformity of our knowledge to reality.

His doubt is methodic, universal, negative, and maybe real.

---

[2] D. Mercier, *Critériologie Générale,* 8th ed. (Louvain: Institut Supérieur, 1923), pp. 115–116. Translation mine.

[3] *Ibid.,* p. 117.

[4] *Ibid.,* p. 119.

[5] *Ibid.,* p. 120.

III • THE ANALYSIS

(a) *Same Fruit of Doubt.* This negative doubt as an approach to Epistemology is subject to the same analysis given on pages 133–134, namely: if persistently followed, it offers no logically consistent solution.

(b) *Initial Doubt Unnecessary.* If his position could be interpreted to mean that certain immediate truths in all their implications *fully* withstand the doubt, then it might be acceptable. However, in that case there would have been no need to insist on beginning with the method of doubt, since it would be an admission that it would be not only an unnecessary but even an impossible way to begin.

(c) *Demonstration Impossible.* Furthermore, there still remains the difficulty that merely ideal objectivity must first be shown, and only after that can the correspondence of knowledge be shown through "critical reflection." This means that a real doubt must be entertained as to the intellect's capacity, and that this can be removed only by demonstration. In answer we can say that if the intellect is not made to know truth, and if we are not naturally aware of this power, no amount of demonstration could ever avail to uncover it. If it is a natural datum of cognitive experience, it needs no strict demonstration, and if it is not such a datum, no amount of demonstrating will uncover it.

IV • SUMMARY OF THE ANALYSIS

(a) Mercier's position is open to the same difficulties as were mentioned in general on pages 133–134. Cf. *Analysis (a).*

(b) No initial doubt is necessary or admissible. Cf. *Analysis (b).*

(c) The process from ideal to real requires demonstration, and no demonstration is necessary or even possible in this matter. Cf. *Analysis (c).*

V • THE STATEMENT IN BRIEF

Methodic doubt renders logically impossible any solution to the question of the intellect's capacity for truth.

VI • DEFINITIONS

Doubt: That state of mind in which a suspended judgment ensues, because of the inability to decide whether the judgment is true or false. Doubt is said to be:
Real: if the mind is actually in a state of indecision.
Fictitious: if it is merely pretended.
Positive: if there are apparently equal reasons for both sides.
Negative: if there are reasons for neither side.   (Ignorance)
Universal: if it applies to all truths, or to truth in general.
Particular: if it is limited to some few truths.
Methodic: if it is entertained as a way of getting truth.
Definitive: if it is entertained as a permanent state of mind.

VII • SUGGESTED READINGS

Bahm, Archie J., *Philosophy: An Introduction,* ch. 13, pp. 168–76.   New York: John Wiley & Sons, Inc., 1953.

Collins, James, *A History of Modern European Philosophy,* pp. 154–64. Milwaukee: The Bruce Publishing Company, 1954.

Copleston, Frederick, S. J., *A History of Philosophy,* IV, ch. 3, 90–115. London: Burns, Oates & Washburn, Ltd., 1946–1959.

Descartes, René, *A Discourse on Method,* translated by J. Veitch.   Chicago: Open Court Publishing Company, 1946.

——, *The Meditations,* translated by J. Veitch. Chicago: Open Court Publishing Company, 1946.

Gibson, A. Boyce, *The Philosophy of Descartes,* ch. 3, pp. 74–103.   London: Methuen & Co., Ltd., 1932.

Mercier, D. Cardinal, *Critériologie Générale* (8th ed.), Pt. II, Bk. 2. Louvain: Institut Supérieur, 1923.

CHAPTER EIGHTEEN

RELATIVITY OF TRUTH

I • THE SITUATION

The rejection of despair as a justified position and the consequent admission of man's ability to know truth does not yet exhaust the possibilities for new and divergent theories of knowledge. Thus, the very admission that truth is possible evokes the question of what is meant by *truth* in this context. Since on the one hand truth is a very personal possession proper to each knower, and on the other the world in which we live and which we know is constantly changing, is it possible to say in any meaningful way that two men can share the *same truth*, or that one man can communicate the truth he knows to another? Or is each one's truth utterly personal and different and incommunicable? Or should it be said that truth is neither simply "absolute" nor simply "relative," but that further qualifications are necessary?

In facing these questions, and keeping the consideration still very general, we shall first recall some statements made by a few philosophers, and then proceed to analyze the pure position known as Relativism.

II • DESCRIPTION OF THE THEORY

(a) *The Greeks.* Turning to the ancient Greeks we find that according to Protagoras "Man is the measure of all things, of the things that are, that they are, of the things that are not, that they are not." [1]

---

[1] Kathleen Freeman, *The Pre-Socratic Philosophers* (Oxford: Basil Blackwell & Mott, Ltd., 1949), p. 350.

This is generally taken to mean that each individual's perceptions are immediately true for him at any given moment, and that there is no means of deciding which of several opinions about the same thing is the true one; there is no such thing as "truer," though there is such a thing as "better." He was led to this conclusion by observing that material objects are continuously changing, and that the person perceiving them also changes according to his own age, bodily condition and so on.[2]

Gorgias, another Greek sophist, expressed his philosophical views in his essay "On Being." They can be summarized in the three statements: "I. Nothing exists . . . II. If anything exists, it is incomprehensible. III. If it is comprehensible, it is incommunicable."[3]

It is notable that Plato nowhere refers to Gorgias' nihilist views, though he devoted serious attention to those of Protagoras; . . . Aristotle, however, took Gorgias' views more seriously, and wrote a monograph (not extant) against them . . .[4]

However one may evaluate the efforts of Gorgias at philosophy, it would at least seem clear that the burden of his essay supports the complete relativity of all truth. For his position seems to be either that we can know nothing as it is, or if we come to know something, our knowledge is so personalized and so proper to each one that it cannot be shared with another.

(b) *An Historical Gap.* Historically speaking it is interesting to note that whereas Relativism exerted strong influence in the ancient world and aroused powerful reactions on the part of Plato and Aristotle, it suffered a rapid decline with the advent of Christianity, and by the middle of the fourth century "it practically becomes *quantité négligeable* and stays underground for some nine hundred years . . . In the fourteenth century relativism reappears . . . but since 1840 it has steadily risen, up to the present time, and it reached one of its peaks in 1900–1920 . . . This means that our ethical (and other mentality) is predominantly relativistic and atomistic."[5] Although this study presented by Sorokin is mainly

---

2 *Ibid.* Cf. also *The Basic Works of Aristotle*, ed. McKeon, *Metaphysics*, 1005a–1011b, pp. 735–749. Cf. also Plato, *Theaetetus*.

3 Freeman, *op. cit.*, p. 359.

4 *Ibid.*, p. 362.

5 Pitirim A. Sorokin, *Social and Cultural Dynamics* (New York: American Book Co., 1937), II, p. 517.

ethical, what he says about Relativism as peculiar to ancient and then after a long gap to modern philosophy, applies also to the knowledge of truth itself.[6]

(c) *Modern Writers.* When we turn, therefore, to more recent philosophers in search of relativistic doctrines, we can find countless examples. Men such as J. Stuart Mill, Comte, Renouvier, and Dilthey, or movements such as Modernism, Psychologism, Historicism, and Scientific Agnosticism could be cited. However we shall limit ourselves briefly to some views expressed by the two American philosophers William James and John Dewey.

(d) *William James.* William James, in his consideration of what it means when one says that truth involves the agreement of our ideas with reality, and speaking in the name of his pragmatic philosophy, says that:

> Pragmatism . . . asks its usual question. "Grant an idea or belief to be true," it says, "what concrete difference will its being true make in any one's actual life? How will the truth be realized? What experiences will be different from those which would obtain if the belief were false? What, in short, is the truth's cash-value in experiential terms?" [7]

> The truth of an idea is not a stagnant property inherent in it. Truth *happens* to an idea. It *becomes* true, is *made* true by events.[8]

> *'The true,' to put it very briefly, is only the expedient in the way of our thinking, just as 'the right' is only the expedient in the way of our behaving.* Expedient in almost any fashion; and expedient in the long run and on the whole of course; for what meets expediently all the experience in sight won't necessarily meet all farther experiences satisfactorily.[9]

> Meanwhile we have to live to-day by what truth we can get to-day, and be ready to-morrow to call it falsehood.[10]

Speaking again of Pragmatism he points out that:

> Her only test of probable truth is what works best in the way of leading us, what fits every part of life best and combines with the

---

[6] Cf. *ibid.*, pp. 230–231. Cf. also Pitirim A. Sorokin, *The Reconstruction of Humanity* (Boston: Beacon Press, 1948), pp. 104–105.

[7] William James, *Pragmatism* (New York: Longmans, Green & Company, 1914), p. 200.

[8] *Ibid.*, p. 201.

[9] *Ibid.*, p. 222.

[10] *Ibid.*, p. 223.

collectivity of experience's demands, nothing being omitted. If theological ideas should do this, if the notion of God, in particular, should prove to do it, how could pragmatism possibly deny God's existence? She could see no meaning in treating as 'not true' a notion that was pragmatically so successful.[11]

On pragmatistic principles, if the hypothesis of God works satisfactorily in the widest sense of the word, it is true.[12]

(e) *John Dewey.* Turning now to John Dewey, we find the same general pragmatic attitude toward truth exemplified. There is no truth until "inquiry" has been made or until the idea has been tested by its consequences. Hence all ideas or statements, even of facts, must be considered as mere hypotheses, and in the light of this explanation he works out his theory of "truth."

Here it is enough to note that notions, theories, systems, no matter how elaborate and self-consistent they are, must be regarded as hypotheses. They are to be accepted as bases of actions which test them, not as finalities . . . They are tools. As in the case of all tools, their value resides not in themselves but in their capacity to work shown in the consequences of their use.[13]

If ideas, meanings, conceptions, notions, theories, systems are instrumental to an active reorganization of the given environment, to a removal of some specific trouble and perplexity, then the test of their validity and value lies in accomplishing this work. If they succeed in their office, they are reliable, sound, valid, good, true. If they fail to clear up confusion, to eliminate defects, if they increase confusion, uncertainty and evil when they are acted upon, then are they false. Confirmation, corroboration, verification lie in works, consequences . . . That which guides us truly is true— demonstrated capacity for such guidance is precisely what is meant by truth.[14]

(f) *Two Reactions.* Before going on to our own analysis and evaluation of this relativistic position it will be worthwhile to record the reactions of Bertrand Russell and Pitirim Sorokin. In

---

[11] *Ibid.,* p. 80.

[12] *Ibid.,* p. 299.

[13] John Dewey, *Reconstruction in Philosophy* (New York: Henry Holt and Company, Inc., 1937), p. 145.

[14] *Ibid.,* p. 156. For a precise, clear, and interesting presentation and analysis of the doctrines proposed by both James and Dewey, see Bertrand Russell, *A History of Western Philosophy* (New York: Simon & Schuster Inc., 1945), pp. 811–828.

concluding his chapter on Dewey's philosophy and especially referring to the doctrine on truth, Russell says:

> In all this I feel a grave danger, the danger of what might be called cosmic impiety. The concept of "truth" as something dependent upon facts largely outside human control has been one of the ways in which philosophy hitherto has inculcated the necessary element of humility. When this check upon pride is removed, a further step is taken on the road towards a certain kind of madness—the intoxication of power which invaded philosophy with Fichte, and to which modern men, whether philosophers or not, are prone. I am persuaded that this intoxication is the greatest danger of our time, and that any philosophy which, however unintentionally, contributes to it is increasing the danger of vast social disaster.[15]

Speaking of what present day philosophies need to do if they are to be successfully transformed into helps for cooperation rather than incitements to hatred and enmity, Sorokin says:

> They must abandon other forms of ideological degredation of man and sociocultural reality. They must cease the illogical relativizing of truth and other values that leads to the obliteration of the boundary line between truth and error, goodness and evil, beauty and ugliness. The relative has meaning only when contrasted with the absolute. Without any absolute the very concept of the relative becomes empty and meaningless.[16]

III • ANALYSIS

(a) *A Common Theme.* Although there are as many variations in the theme of Relativism as there are philosophers who propose it, and the relativity of knowledge is said to be due to reality itself, on the score that it is not intelligible as it is in itself, or to the human mind, on the score that it is not capable of knowing things as they are, and what is said to be true is considered to be dependent on utility, the consequences which will ensue, one's social conditions, or on the personal characteristics of the individual knower, still amid all these variations a common fundamental theme can be detected, and Relativism can be generally described as the doctrine

---

15 Russell, *op. cit.*, p. 828.
16 Sorokin, *The Reconstruction of Humanity*, pp. 116–17.

which teaches that the object of our knowledge cannot be known as it is in itself, but only as it is modified by us, or merely as it appears to us. Hence *what is known* is dependent on the knowing subject.

(b) *Special Emphases.* It must be admitted that this theory of knowledge does grasp and express with emphasis an aspect of the situation which is experienced by everyone in the course of his intellectual development. This can be expressed in various ways:

(1) *It Opposes Rationalism.* The views proposed by Relativists are opposed to an extreme rationalistic mentality which advocates an exaggerated glorification of the human intellect. In this view the mind is supposed to have some sort of direct intuition of essences, to be able to arrive at a comprehensive knowledge of reality in all its relations, and to be capable of developing all-inclusive systems which explain everything. This is to postulate almost divine knowledge for men, and in its opposition to such extreme claims Relativism offers a valuable word of caution.

(2) *It Stresses The Role of Experience.* Moreover, in the growth of Relativism one can detect the influence of modern philosophical Empiricism and of the spirit developed by the rapidly expanding natural sciences. This involves the recognition of the partiality of human knowledge and of the consequent need for constant experiment and unremitting recurrence to experience for the testing and confirmation of positions taken. This also is a point well made.

(3) *It Stresses The Influence of Environment.* Finally, the Relativists have seen that religious upbringing, varying political, economic and educational backgrounds, social conditions, and countless other factors all concretely influence men in their acceptance or rejection of certain positions. These also influence one in how much or little he grasps about various subjects.

(4) *In a Word, It Emphasizes Human Limitations.* By way of summary, we can say that any adequate theory of knowledge must face the fact that: (a) human knowledge is not absolute in the sense of being divine or comprehensive, (b) it must be in constant contact with experience and not devise dreamings of what things would be or try to make everything fit into predetermined categories, (c) it is much dependent on one's training and environment, (d) it's truth is a very personal and intimate possession, so much

so that the notion of possession takes on a very special meaning in this context.

(c) *A Necessary Distinction.* However, to be truly adequate a theory of knowledge must also face certain other facts and factors in the human truth-situation. All that has been said pertains only to *the way in which* each one grasps truth; it does not touch on *that which* is grasped. That is to say, the truth-situation includes not only an intellectual act of knowledge, but also a something which is known and which is affirmed *to be as it actually is.* Can that which is known and which is affirmed be mind-dependent, or does truth require an objectivity and an independence of what is affirmed, which would mean that the mind in grasping *what is* conforms itself to it because it is? If truth means this conformity to *what is,* then some objectively stable or "absolute" character must be acknowledged. Thus, if I know that the Principle of Contradiction is true, and another person knows it also, then *what is known* is the same in both cases, although our ways of knowing it may differ somewhat. Again, we affirm that Scepticism is wrong, because it is wrong, that is, *because being is what it is;* and whether I affirm it now or later, or whether it is affirmed by me or others, *that which is affirmed* is objectively and immutably true, and it must be affirmed not because of my way of grasping it, but because of *what is.*

For illustration we can take the statement of a transitory event such as, "There was a World War II." The truth of that judgment in no wise depends on my subjective modifications or on the personality of the person making it, that is, in regard to the truth of *what is affirmed,* nor does it depend on the pleasure it gives me or anyone else, nor on its utility. Rather, what is affirmed truly depends on *what is.*

A full theory of truth, therefore, must keep in mind this distinction between *what is known* and the *way in which it is known.* A certain relativity can be admitted on the subjective side touching the way one knows, but truth must be seen to be objectively absolute in respect to what is known. This need will become clearer in what follows.

(d) *Phenomenalism.* The theory of Relativism means that truth, even in respect to what is known, is dependent on the subject knowing. Hence, the object is never known as it is, but merely as it appears to the knower and as it is modified by him; and in that sense truth or falsity is never determined by what is, but by the activity of the knower; and this same activity rather than *what is* serves as the norm for truth.

Taken seriously this can only mean that we can never get to know anything definite or determinate about reality, because we are limited only to appearances and our knowledge at best is of phenomena. This total inability to know reality as it is and the consequent lack of determination and of certitude in human knowledge can be said quite logically to share generously in the sceptical despair over human knowledge already discussed. If we can know nothing *as it is*, then we really know nothing and human knowledge ceases to be a cognitive experience.

Moreover, in this context the diligent work of empirical scientists and the studies and reflections of thinking men are equally inexplicable. For what reason do men concentrate their powers on data presented and why do they develop ever more precise and complicated instruments for detecting and recording the data found in reality? It can only be done because reality discloses itself to us, and the more attentive we are and the better our instruments the more it can tell us about itself. We are not satisfied with a first hasty appearance, but apply ourselves attentively to be sure that we really get at *what is*. And even then we can admit that what we have got is far from all that can be learned. Still, the little that we learn is true—it is a characteristic of the object.

(e) *Internally Inconsistent.* Again it can be said that Relativism is an internally inconsistent doctrine. On the one hand it insists that we can never know anything as it is in itself; on the other hand it itself is proposed as the only objectively valid and absolutely true explanation of the truth-experience of man, and as one to which I should conform by admitting its validity. Were the Relativists' position to be adopted, it would mean that any theory of knowledge which is found useful or helpful or any one which appears even momentarily true should be accepted as true. And if that is so, there is no logical reason why Relativism itself should be proposed for our acceptance.

(f) *No Norm for Truth.* Moreover, since there is no objective norm for distinguishing between what is true and what is false, it would follow that there is no discernible basis for distinguishing a true from a false affirmation; hence, one and the same thing could, with equal truth-value, be at once affirmed and denied. In terms of human knowledge that means intellectual chaos.

This would mean that whatever knowledge a person might presume to have is useless and foundationless. It means also that the efforts made to arrive at scientific knowledge are in vain; and finally, it means that any attempt to communicate knowledge to another is not worth trying. With human knowledge utterly dependent on the knower even in respect to what the object is, human truth is quite literally "dethroned," [17] and there remains nothing of any value to communicate to others—in fact such communication is meaningless.

IV • SUMMARY OF THE ANALYSIS

(a) Relativism is opposed to an extreme Rationalism, encourages constant recurrence to experience, and emphasizes the concrete influences which come to bear on man's knowledge. In this sense it rejects a false and exaggerated sort of "absolute" in human truth. Cf. *Analysis (b)*.

(b) However, as a theory of knowledge it is one-sided and hence exaggerated in its own way. It fails to take fully into account the experience of knowing as the self-knowing-something. If this has any meaning at all, it means the self becoming conformed to *what is*. Hence, knowledge is simultaneously subjective-objective. Relativism does not first distinguish between *what is known* and the *way in which it is known,* and then retain both sides of the distinction. Cf. *Analysis (c)*.

(c) Relativism means that human knowledge is purely phenomenal, and hence lacks the determination and certitude necessary to overcome a total Scepticism. Since nothing can be known as it is, study and experimentation lose all meaning. Cf. *Analysis (d)*.

(d) Relativism is internally inconsistent, since it is proposed as the objectively valid and true doctrine of knowledge although it simultaneously teaches that there is no objective truth. Cf. *Analysis (e)*.

(e) Since there is no objective norm for truth according to Relativism, the simultaneous affirmation and negation of the same

---

[17] Cf. Dietrich von Hildebrand, *The New Tower of Babel* (New York: P. J. Kenedy & Sons, 1953), Chap. II, "The Dethronement of Truth," pp. 57–102.

thing could both be proposed with equal truth-value. Hence, knowledge can never be reliable, and communication of knowledge becomes meaningless. Cf. *Analysis (f)*.

V • THE STATEMENT IN BRIEF

Relativism as a theory of knowledge is inadequate and internally inconsistent: ultimately it renders human truth meaningless.

VI • DEFINITIONS

Relativism: a theory of knowledge according to which human truth is *totally* dependent on the subjective conditions of the knower.
Totally: not only in respect to *the way in which one knows,* but also in respect to *what is known.* This may be said to depend on utility, effectiveness, pleasure, suitability, "cash-value" or on other factors, but all these are in relation to the knower. In this context, what is known is merely an appearance; the norm for truth is purely subjective.
Inadequate: not taking into consideration all the factors of the experience.
Inconsistent: implicitly denying what it explicitly affirms.
Meaningless: of no real or permanent or coherent value.

VII • SUGGESTED READINGS

von Hildebrand, Dietrich, *The New Tower of Babel,* ch. 2, pp. 57–102. New York: P. J. Kenedy & Sons, 1953.

James, William, *Pragmatism.* New York: Longmans, Green & Co., Inc., 1914.

Russell, Bertrand, *A History of Western Philosophy,* pp. 811–28. New York: Simon & Schuster, Inc., 1945.

Sorokin, Pitirim, *Social and Cultural Dynamics,* II, 515–18. New York: American Book Company, 1937.

CHAPTER NINETEEN

IDEALISM

I • THE SITUATION

Another very influential theory of knowledge in modern philosophy is Idealism. In terms of what has already been said, it can be said that Idealism rejects the despair of Scepticism and the arbitrary and inconsistent meaning of truth found in Relativism. Hence, it admits that man can and does possess truth, and that he knows objects as they are in themselves. However, the question now arises about the nature of those objects in themselves—that is, what is the ontological status of the *something* which is known? The precise interpretation of this question centers around the intramental or the extramental reality of what is known. Put briefly this amounts to asking, can the mind "go out" to a something other than itself and know it as it is, or can that something other "enter into" the mind so as to become known, or *is* there any extramental reality at all?

II • DESCRIPTION

(a) *Illustration of The Position.* Although admitting that Idealism has been a very popular doctrine in modern philosophy, and that there have been as many variations in its development as there have been Idealists, we shall attempt to grasp and evaluate the fundamental insight which underlies and inspires all the varieties. Moreover, although countless examples are available in the works of men like Berkeley, Fichte, Schelling, Hegel, Green, Bradley, Cassirer, Brunschvicg, Croce, Gentile, and so on, we shall content

ourselves with only a few illustrations from the writings of Bishop
George Berkeley:

> That neither our thoughts, nor passions, nor ideas formed by the
> imagination, exist without the mind, is what everybody will allow.
> And it seems no less evident that the various sensations or ideas
> imprinted on the sense, however blended or combined together
> (that is, whatever objects they compose) cannot exist otherwise
> than in a mind perceiving them. I think an intuitive knowledge
> may be obtained of this, by any one that shall attend to what is
> meant by the term *exist* when applied to sensible things. The
> table I write on, I say, exists, that is, I see and feel it; and if I
> were out of my study I should say it existed, meaning thereby that if
> I was in my study I might perceive it, or that some other spirit
> actually does perceive it. There was an odour, that is, it was
> smelled; there was a sound, that is to say, it was heard; a colour
> or figure, and it was perceived by sight or touch. This is all that I
> can understand by these and the like expressions. For as to what
> is said of the absolute existence of unthinking things without any
> relation to their being perceived, that seems perfectly unintelligible.
> Their *esse* is *percipi*, nor is it possible they should have any existence,
> out of the minds or thinking things which perceive them.
> It is indeed an opinion strangely prevailing amongst men, that
> houses, mountains, rivers, and in a word all sensible objects have
> an existence natural or real, distinct from their being perceived by
> the understanding . . . What are the forementioned objects but
> the things we perceive by sense, and what do we perceive besides
> our own ideas or sensations; and is it not plainly repugnant that any
> one of these or any combination of them should exist unperceived? [1]

> . . . it is evident from what we have already shown, that extension,
> figure and motion are only ideas existing in the mind, and that an
> idea can be like nothing but another idea, and that consequently
> neither they nor their archetypes can exist in an unperceiving sub-
> stance. Hence it is plain, that the very notion of what is called
> *matter* or *corporeal substance,* involves a contradiction in it. [2]

It is clear that the emphasis in these quotations bears chiefly on
the denial of the extramental existence of material substance and
on the consequent intramental existence of all "sensibles." This
precise point we shall consider a bit later on, turning for the present
to the central theme and basic meaning of Idealism.

---

[1] George Berkeley, "A Treatise Concerning the Principles of Human Knowledge," *The Works of George Berkeley, Bishop of Cloyne,* editors A. A. Luce
and T. E. Jessop (London: Thomas Nelson, 1949), Vol. II, p. 42.

[2] *Ibid.,* p. 45.

(b) *The Principle of Immanence.* In approaching his study of human knowledge the Idealist places great emphasis on the apparently total heterogeneity of mind itself and what is commonly considered to be material reality which is known. He is convinced that since knowledge necessarily means an identity between the mind and the object known, and since he sees no possibility of identity between spirit and matter, the original, naive attitude of men in regard to the extramental and independent existence of the object must be abandoned, and it must be held that what is known is something totally internal to the mind. So, when we know something, the mind need not be said to "go out" to meet and know that object; in fact there is no such "outside" object at all. More technically there is no *transcendent* element in knowledge (and *transcendent* here simply means independent of or not produced by mind).

As a pure position, then, Idealism is committed to the basic Principle of Immanence according to which the only object knowable is one whose total "being" is its "to be known" and which is consequently internal to the mind itself.

(c) *Idealism and the Ego.* What has thus far been said is the guiding idea of all Idealists, even when one distinguishes them into the two classes of *empirical* and *transcendental* Idealists. The only difference between these positions centers about the interpretation of the knowing *ego.* Empirical Idealism admits the existence of only the individual, actually experienced *ego.* Transcendental Idealism postulates over and above the experienced individual *ego* the existence of a more basic, nonexperienced, underlying *ego,* the source of unity, universality, and necessity. However, the point common to both positions, and the one which interests us, is the denial of any mind-independent thing-in-itself.

(d) *Reasons for the Position.* Among the reasons offered for adopting this position, we can mention the following:

(1) Insistence on the fact that an object in order to be known must be present within the mind. In the early part of our study we saw that part of the initial meaning of what it is to know does involve the need of an "identity" between knower and known.

(2) In the knowledge of truth Idealists stress the important place occupied by the knower; that is, were there no knower, nothing would be known. This is expanded to mean that as

the object known is *made known* by the activity of the subject, so its *being* is identical with its *being known*.

(3) They see a contradiction in the notion of a transcendent or
mind-independent being. That is to say, any such being
would have to be simultaneously external to the mind and
internal to the mind. As transcendent, it would be "outside"; but as known, it would have to be "within" the mind.
This dual existence the Idealists consider to be impossible.

(e) *Applications of the Doctrine.* For these and similar reasons
the philosophy of Idealism developed, and in keeping with its
fundamental principle there emerged explanations and corrections
touching on the knowledge situation which can be summarized
briefly as follows:

(1) Although it must be admitted, even by Idealists, that men
naturally, spontaneously, and thus, inevitably, tend to be
Realists—that is, they tend to look upon what is known as
something independent of mind, and something to which
mind must conform itself, this realistic tendency must be
corrected and recognized as erroneous in the light of idealistic
philosophy.

(2) Hence the ontological situation is not that a thing is known
because it *is;* rather it is, because it is known. The act of
knowing, then, precedes the being of what is known—in fact,
knowledge is an act creative of being.

(3) Knowledge in this context can be only the pure coming to
be of the knowing act itself.

(4) Truth itself can in no sense find a norm or rule in what is,
but it must be measured solely by the activity of knowing.

(f) *Bearing on Ontology.* The preceding is a brief presentation
of the basic meaning and implications of Idealism taken as a pure
epistemological position. That this theory involves also an Ontology
or explanation of the nature of being is perfectly obvious, and since
the two are so intimately united, we shall consider several illustrations, accompanying them with brief comments.

With the general assertion that whatever is must be mind-
produced, a theistic Realism can express general agreement. This
Realism will grant with the Artistotelico-Thomistic tradition that
the first Mover (or Creator) must be *nous* or mind, and that all
limited or changing beings owe their existence to this mind. How-

ever, the Realist will insist that this does not mean that things exist merely intramentally, and all the more will he deny that things are the product of the individual human knower. In fact he will point out that what we first know are the material, nonmental beings of our experience, and only from a study of what they are can we arrive at the knowledge of an immaterial being to account for their existence.

Another important emphasis in Idealism is the polemic against matter and the insistence that matter is totally unintelligible or even that it does not exist. The Realist will admit that matter as interpreted by the Idealist is unintelligible and moreover that the latter is right in denying the existence of matter as so viewed, for matter, in that sense, would mean mere passivity, a totally inert and static something which would be impenetrable by mind. But the Realist points out that that sort of matter as such does not exist precisely because material existents are not pure matter but material *things*. That is, a material thing is a composite, made up of both act and potency; it is not merely passive or static (a prime matter, in Aristotelian terminology) but in its very nature includes the dynamic act of existing, and of existing and operating in such and such a manner. Hence, material beings are knowable in virtue of their activating and actual perfections, and so from this point of view also, the Idealists have given an interpretation of reality which is only partially true.

(g) *Some Positive Aspects.* In concluding this description it should be pointed out that the Idealists have done well in stressing with great emphasis the need for acknowledging the identity between knower and known, if true knowledge is to be had. Were subject and object totally discrete and separated, were they totally heterogeneous, there could be no such thing as knowledge. Moreover, the Idealists were impressed with the fact that elements such as harmony, beauty, unity, universality, and necessity play a large part in human knowledge, and they are elements which must be accounted for. In their understanding of matter as something totally passive and neutral they not only could not find those elements, but even found them incompatible. Hence they turned to the mind itself for the required explanation, and in doing so, they once again grasped and graphically expressed their views of the dynamic aspects of human knowledge. In doing so they again did a service to philosophy by opposing the view of mind as a purely passive, static reception of images from "outside," and stressed the activity of mind in its act of knowing.

III • THE ANALYSIS

(a) *A Rejection of Diversity.* Although admitting the value of the emphases just mentioned, and granting that Idealism has done even more than that to further the advance of human thought in the difficult field of philosophy, we must nonetheless examine it carefully to determine whether or not it does justice to the human knowledge experience, and, thus, to decide whether or not it is an adequate theory of knowledge.

The first point to be made is this: Idealism admits the fact that men are naturally Realists; they spontaneously recognize that the object known is *other* than themselves knowing, and they accept as true that which is conformed to what is.   However, to explain how the object can be *other*, that is, to admit the diversity between knower and known, is difficult to reconcile with the identity which must exist between subject and object.   Because it is *difficult* to explain knowledge realistically, Idealists conclude that it is *impossible.*

This can be put another way: The experienced facts of human knowledge are clear enough, and these include the double factor of identity and diversity between knower and known.   These precisely are the factors which have to be explained, and in the explanation it must be indicated just *how* they can be reconciled. To explain the *how* in this context is admittedly not easy.   However, explain them both we must, if we are to propose an adequate theory which will take into account the concrete experience along with all its evident data.   To say that the realistic explanation is difficult and hence impossible is like saying that to shoot a rocket around the moon is difficult and therefore impossible.

Rather, it must be concluded that human knowledge will never be adequately explained by any theory which suppresses important data and then explains what is left over.

(b) *An Oversimplified Explanation.* It is precisely because of this suppression of certain elements of the knowledge situation that Idealism can be called a naive, over simplified and reductionist theory of knowledge. Human knowledge is obviously a complex process and a difficult one to explain, precisely because it is a unique experience not paralleled anywhere in the material world. Hence it can be expected to have aspects which are hard to grasp and harder to express. To be honest, we can only take all those elements into consideration and try to find their explanation in a coherent and logical theory. To say that "diversity" introduces an embarrassing element and consequently must be suppressed, is not to explain human knowledge as it is concretely experienced by me or by other men; it is rather to explain it away. Hence, by denying the extramentality of the object known, and by reducing everything to mind itself, one can arrive at a very coherent, logical and convincing theory, but it will not explain human knowledge *as it is concretely experienced.*

(c) *An Inconsistency.* Moreover, in line with what has already been seen in previous chapters, even the Idealists have already admitted a basically realistic position, and to that extent have admitted the untenability of Idealism. What we mean is this: Our analyses have already shown that both Scepticism and Relativism are to be rejected since they are inadequate as theories of knowledge. The Idealist admits this, and in admitting it he admits two things:

(1) Those doctrines are objectively wrong, and because they *are* wrong in themselves, we must admit they are wrong, that is, the mind must *conform* to *what is* the objective fact.

(2) Contrary to the relativistic position, it is not enough for me to affirm something which appears to me in order for it to be true. In that minimal sense at least I must submit to what is and thus recognize some transcendence of what is affirmed. Hence, at this point, without going explicitly into the question of sensible or material reality as such, I can still see that insofar as the Idealist admits an objective difference between affirming Scepticism or Dogmatism, Idealism or Realism, so that as a result he admits some of those affirmations to be objectively right and others wrong, he admits that we must conform to *what is* and thus recognize some transcendence of what is. Put another way, we can say that were Idealism right, it would not matter what I affirmed, since my affirmation would make it true, in that knowledge is creative of its object.

(d) *Empirical Idealism Inadequate.* Considering more particularly the position of Empirical Idealism, a brief reflection will show that it is not self-consistent and that it is relativistic.

(1) Its inconsistency consists in the fact that it wants to teach that there are many individual knowers; however, since everything known is the projection of the individual knower, it follows that even the assertion that there are other persons is itself a projection of the knower. Hence not only are material objects made to be by my thought, but the same is true of other knowers. The result is that the only existent is myself thinking, all else is but a moment in my thought, and I have arrived at a position of utter Solipsism,[3] which is a radical denial rather than a coherent explanation of human knowledge.

(2) It need only be mentioned that since in this idealistic position everything is the projection of my ego, what *I* know is true, and we are faced with a totally relativistic position on truth.

(e) *Transcendental Idealism Inconsistent.* Transcendental Idealism likewise manifests many internal inconsistencies and contradictions, mainly because in its very formulation it presupposes and proposes a realistic basis of knowledge. This can be seen by the following considerations:

(1) In regard to *objective experience* Idealism holds that our knowledge is creative of its object, that is, only that of which I am conscious actually exists. Now *the fact of my creation of the object known* is something of which I was in no way conscious until the Idealists appeared; hence, according to their system, this creation by mind was objectively true, independent of my knowledge of it, and this, systematically speaking, means the admission of a realistic basis of knowledge by Idealism itself.

(2) In respect to *the knower* Idealism teaches that there is a common, transcendental knower, and of this, too, I was most certainly not conscious. Hence, this also existed prior to and independent of my knowledge of it. Here again we

---

[3] *Solipsism* can be defined as a doctrine according to which the individual *ego* and its modifications constitute the whole of reality, nothing else having any independent existence.

have a realistic presupposition in an idealistic explanation—
hence an internal inconsistency.

(3) Moreover, the *notion of truth* loses all meaning in this system, since the only reality at any given moment is what is
grasped at that moment; hence all previous "truths" are
submerged or rejected by the one now in possession. This
process, however, is a continuous, never-ending one, so that
at no moment can I ever say that what I know is definitely
true. There is thus no objective truth at all, despite the
verbal rejection of Scepticism and Relativism.

(4) *Knowledge itself* ceases, in the last analysis, to be a relational act whereby a subject knows something about an
object. To be a subject, in an idealistic context, means "to
think," and to be an object means only "to be thought."
Taken seriously this position leaves us no ground for suppos-
ing that there is anything at all beside the *act* of thinking;
hence, there is no justification within Idealism for a subject
which knows or an independent object which is known.
Knowledge thereby ceases to be relational, since the very
terms of the relation have no reality.

(f) *Need for Demonstration.* As a final point in respect to
Idealism in general, it can be said that it is a position which needs
formal demonstration and which cannot be proved except by begging the question. It needs to be demonstrated, since, even as
Idealists themselves admit, human knowledge as it is concretely
experienced is realistic. Hence, we are by no means immediately
aware of the truth of Idealism, and so it must be arrived at mediately, or by demonstration. However, that is impossible, since the
only way that it can be explained or proved to be idealistic, is by
presupposing that it is such, and by describing it as though it were
idealistic—in other words, by begging the question. Were it an
adequate and objective theory of knowledge it would, like any valid
reflective process, describe the concrete experience as it is, then
seek an explanation of the very facts experiencd which would be
consistent with the data, and finally solve difficulties in the light
of what is.

IV · SUMMARY OF THE ANALYSIS

(a) Human knowledge is concretely and realistically experienced as involving both the identity and the diversity of knower and known. Since these elements are difficult to reconcile, Idealism rejects the realistic theory of knowledge. Cf. *Analysis (a)*.

(b) Idealism is a naive and reductionist theory, since it oversimplifies by neglecting part of the data, and explains away the object by reducing everything to mind. Cf. *Analysis (b)*.

(c) In the rejection of Scepticism and Relativism, the Idealist admits some objective truth to which mind must conform, but this admission implies a basic Realism. Hence to that extent Idealism is not self-consistent. Cf. *Analysis (c)*.

(d) Empirical Idealism is basically solipsistic and relativistic, and hence not an adequate theory of knowledge. Cf. *Analysis (d)*.

(e) As a doctrine Transcendental Idealism contains many internally inconsistent elements:

(1) It teaches that knowledge is creative of its object, although this very creation of the object is said to be a fact which was *true*, even though I did not know it.

(2) The transcendental, underlying *ego* which it postulates must also have existed independent of its being known, since the doctrine proposed urges us to accept it as *a fact*.

(3) Truth itself can have no permanent value, since it is only what is known at any given moment and must yield to the next moment.

(4) Since all that exists is the act of thinking, knowledge ceases to be in any sense relational; hence the admission that truth is a conformity between knowledge and object known becomes meaningless. Cf. *Analysis (e)*.

(f) Since the total immanence of the object is not an immediate datum of experience, we must come to know it mediately through demonstration. However, it cannot be demonstrated unless experience is first described in terms of immanence. Hence it cannot be proved without begging the question. Cf. *Analysis (f)*.

V • THE STATEMENT IN BRIEF

Integral idealism, whether empirical or transcendental, is self-inconsistent and inadequate as a theory of knowledge.

VI • DEFINITIONS

Idealism: a theory of knowledge which teaches that the object known is dependent for its reality on the activity of the mind knowing.

Integral: when used in connection with Idealism, positively denying the existence of any thing-in-itself. This is the solipsistic position towards which any Idealism, no matter how qualified, actually tends.

Empirical Idealism: the theory which admits the existence of only the individual, experienced *ego*.

Transcendental Idealism: the theory which admits, over and above the individual experienced *ego*, the existence of a more basic, nonexperienced, underlying *ego*, the source of unity, universality, and necessity.

Self-Inconsistent: implying in its explanation elements which it explicitly rejects.

Inadequate: failing to take into account all the data concerning what is to be explained.

VII • SUGGESTED READINGS

Bahm, Archie J., *Philosophy: An Introduction,* chs. 4, 6, 7, pp. 57–65, 76–103. New York: John Wiley & Sons, Inc., 1953.

Blau, J., *Men and Movements in American Philosophy,* ch. 6, pp. 187–227. Englewood Cliffs, N. J.: Prentice-Hall, Inc., 1952.

Hoernle, R. F. Alfred, *Idealism as a Philosophical Doctrine.* London: Hodder & Stoughton, Ltd., 1924.

Werkmeister, W. H., *A History of Philosophical Ideas in America,* ch. 14, pp. 289–316. New York: The Ronald Press Co., 1949.

CHAPTER TWENTY

ACOSMISM

I • THE SITUATION

The consideration of integral Idealism has resulted in the explicitation of the fact that *some* objectivity of knowledge must be admitted, that is, to the extent that the objective falsity of Scepticism and of Universal Relativism, and the objective truth of their denial must be granted. Hence in this minimal sense the truth of some sort of Realism has been recognized. However, the history of modern philosophy has been replete with either a very studied or an out-of-hand rejection of the reliability of sensation as a source of knowledge, and hence, our analysis must be brought to bear on this question. For the present our study will be limited to the question of whether or not sensible reality is recognized as existing. Later we shall consider the question of the nature and objectivity of the sense qualities.

II • THE DESCRIPTION

(a) *Berkeley's Principle.* As was indicated in the previous chapter, the outstanding exponent of Acosmism, the doctrine which denies the extramental existence of a material world, is George Berkeley. Although he himself insists that no one has more respect than he for sense experience and claims to do nothing more than explicitate what every man feels to be true about sensation, it must be admitted that his doctrine has generally been judged to be acosmistic, and it is difficult to gather any other impression from the reading of his works. His new principle "Existence is *percipi* or *percipere* or *velle,* that is, *agere,"* in view of his own explanations and exemplifications, makes his meaning quite clear.

(b) *Only Spiritual Substances and Ideas.* The basic thesis of Acosmism is that there is no such thing as material reality. All that can be known and all that can be are spiritual substances and ideas. The theory does not claim that man's ordinary experience is an illusion; the illusion enters only when men tend to project their ideas outward and to speak in terms of an independently existing material universe.

How such a position could have logically developed in the history of thought will become clearer after the considerations in the next chapter, but for the present we shall say only that when ideas are made the only objects directly known and are thus said to terminate our knowledge act, it is an easy step to the denial of any material reality beyond the ideas themselves.

(c) *Regulation of Sequence of Ideas.* Acosmism, then, at least theoretically allows that there is a plurality of spiritual, intellectual substances, and that these know only their ideas. It adds that the sequence of ideas is not purely arbitrary, but follows a regular order, usually said to be regulated and determined by the Supreme Spirit, so that we seem to live in an ordered, progressive, developing world. But the only "things" progressing are ideas.

One of the reasons for this position is the claim that "an idea can be like nothing but an idea," so that an immaterial idea can never represent material reality. Hence, the only real things are minds and ideas.

III • THE ANALYSIS

(a) *Truth Means Conformity.* We have already seen that we are capable of a nonrelativistic knowledge of truth respecting something independent of our knowing it, and that knowledge, to be meaningful, has to mean a conformity with what is. Insofar as it does not mean such conformity, it is open to the difficulties already discussed.

(b) *My Knowledge and Sensation.* Without going into the myriad details which can arise on the question of sensation, we can now say that the same thing is true of our knowledge of sensi-

ble reality.   If my knowledge purports to be of sensible reality,
and if there is no such reality, then to that extent my knowledge
has no meaning.

That I do naturally and spontaneously claim to know such
sensible reality is one of the most evident facts of my intellectual
experience.   In fact, my regular experience is such that what I
directly and explicitly know with the greatest ease are the material
things which surround me.   In all my knowledge I find myself in
need of some sort of material image to accompany my intellectual
knowledge.

Moreover, when I come to know something about "immaterial"
or "nonsensible" being, I do so by some sort of comparison to or
negation of things sensed.   Thus, for example, negative words such
as infinite, immaterial, immortal, invisible, inaudible, manifest one
way in which I can proceed from the material or sensible to the
formation of other ideas.   Still others are formed by relation or
analogy to sensible things.   Finally, even knowledge of our own
acts and existence is intimately linked to our awareness of ourselves realized and grasped through our own acts of sensation.

(c) *Judgments About Sensible Existents.*   It is a fact of experience, admitted even by Acosmists, that I naturally, constantly
and inevitably judge that sensible things do exist in reality as I
judge them to be.   Moreover, the recognition of error and the
existence of doubt in these matters and all the studies of the empirical scientists are so many confirmations of this natural and
human conviction.

In these cases, when error is suspected or when greater knowledge is desired, a man does not sit back in the seclusion of his
room, close his eyes and shield his other senses; what he does is to
apply them ever more attentively to the sensibly experienced
reality.   He invents instruments such as the telescope or micro-
scope, and he devises measuring tools such as the spectrometer.
In doing all that, he is doing nothing more than prolonging and
enlarging the range of the senses, since any such instrument would
be meaningless without the senses themselves to read and interpret
the new data.

(d) *Undermining of Mind.*   Again, it can be said that were the
Acosmists right in this matter, the senses and man's judgments about
sensible reality would be regularly, constantly, and incorrigibly

wrong. We really do experience things as sensibly existent; in fact, that is one of the most basic characteristics of human knowledge as it concretely progresses. Error as to the objective existence of sensible things would mean that human knowledge in its very bases and of its very nature is erroneous. The mind could never be said to know anything, because it could not be trusted. In this sense, Scepticism in respect to the reality of sensible things ultimately involves a Scepticism in respect to the mind's very capacity for truth. It begins as a partial Scepticism, but cannot avoid leading logically to a total Scepticism.

(e) *Data Not to Be Suppressed.* Finally we can point out that behind this idealistic trend in modern philosophy is the conviction that any sort of Realism has to be naive, credulous and unscientific. However, the charge is often made without a sufficient grasp of the nuances of a balanced realistic explanation. It is not naive to affirm that I have the experience of perceiving things as sensibly existent, or of intellectually affirming their reality. That actually is the human experience, and that precisely is the datum which any theory of knowledge must analyze and explain. Hence naiveté cannot properly be said to enter into that admission; it can only be introduced in respect to the *explanation* of the *meaning* of that experience and of *how* it is had.

That explanation, as we have seen, is not easy when all the elements are preserved intact. It becomes relatively simple when some of the data are suppressed, as for example, when the experienced otherness of what is known or the object itself is reduced to the status of a mental phenomenon. Then a coherent explanation can be developed, but it only explains what knowledge would be were it limited to mental beings, whereas the situation to be explained is the awareness of the other as other. A developed Realism seeks to explain this experience, preserving both the immediacy of knowledge and the complexity of the way in which the human person actually knows reality.

IV • SUMMARY OF THE ANALYSIS

(a) We have already seen that knowledge means the non-relativistic conformity with what is. Cf. *Analysis (a).*

(b) I naturally and constantly have the experience of knowing sensible reality. Material images seem to accompany all my knowledge. I form ideas of "nonsensible" things by relation to or analogy to the sensible. Cf. *Analysis (b)*.

(c) I spontaneously and inevitably judge that sensible things exist in reality. This is further confirmed by the fact that when error is suspected, the reaction is to be more attentive in the use of the senses. Scientific studies and instruments also illustrate the same situation. Cf. *Analysis (c)*.

(d) That being granted as the concrete human experience, it must be said that were the intellect to judge falsely in this matter, the error would be due to the very nature of the mind itself. Basic distrust of the mind in respect to the existence of sensible reality logically means basic distrust of it in its capacity to know what is, so that an initial Scepticism in respect to sensation involves finally a total Scepticism in regard to truth. Cf. *Analysis (d)*.

(e) The fact of experiencing existent sensible reality is part of the data in the noetic experience. It is basic and originative in respect to human knowledge; hence the function of a theory of knowledge should be to explain *how* we can have knowledge of this *other*, not to oversimplify the given situation by suppressing one of the experienced elements. Cf. *Analysis (e)*.

V • THE STATEMENT IN BRIEF

Acosmism contradicts concrete experience, implies total Scepticism, and oversimplifies the data.

VI • DEFINITIONS

Acosmism: a theory which denies the existence of material reality, and says that all existents are spiritual substances and ideas.

Experience: both sensitive and intellectual awareness of the self and of the other.

Implies: logically leads to if accepted.

Oversimplifies: omits one or other of the elements of the very situation which is to be explained.

VII • SUGGESTED READINGS

Bahm, Archie J., *Philosophy: An Introduction*, ch. 4, pp. 57–65. New York: John Wiley & Sons, Inc., 1953.

Berkeley, George, *Three Dialogues Between Hylas and Philonus.* Chicago: Open Court Publishing Company, 1946.

————, *A Treatise Concerning the Principles of Human Knowledge.* Chicago: Open Court Publishing Company, 1946.

Luce, A. A., *Berkeley's Immaterialism.* London: Thomas Nelson & Sons, Ltd., 1950.

CHAPTER TWENTY-ONE

RATIONALISM

I • THE SITUATION

(a) *Descartes' Division of Man.* Whereas Descartes' anthropology will require more detailed consideration in the field of rational psychology, it must at least be pointed out here that one of the outstanding elements of his philosophy is his division of man into two complete and almost unrelated substances. On the one hand he held the *self* to be a spiritual substance whose whole nature is to think; on the other hand he held that the body is a corporeal substance whose essence is extension. Since Descartes looked on these two substances as two complete beings, and since he nonetheless recognized the fact that man experiences himself as a unity, there arose for him the problem of relating the one to the other. Descartes' efforts to solve this problem cannot be called successful. He did insist on the substantial unity of the *self*, the spiritual substance. Moreover, he insisted on some unity of the *man*, and this he called a composite unity. However, as to just what he meant by this "composite unity" he never made quite clear: The unity of man could not be a substantial unity, since it is the unity of two "things" or "beings"; yet he did not want to assert that it was a mere accidental unity, so he invented his composite unity. Actually this was more an evasion than a solution, and it was not long before the fruits of his division were reaped.

(b) *The Sequel: Idealism and Materialism.* Apart from his immediate successors, and in particular Fr. Malebranche, whose Occasionalism showed some of the logical implications of the Cartesian duality in man, there arose in the course of time the two philosophic positions known as Idealism and Materialism. These can be looked

168

upon as clear expressions of the logical implications of Descartes' doctrine. Whereas he taught that man is two beings instead of one, they reasserted the naturally experienced unity of man's being. However, they did this against the background of Descartes' distinction, and so their reassertion of unity took the form of opting for one or other of the "beings" which he proposed. Thus, according to the Idealists, all that exists is "pure thought" and the experience of a material body is an illusion. For the Materialists all that exists is the body, and the spirit is at most an epiphenomenon of matter.

(c) *Rationalism and Empiricism.* There are still other positions which are possible against the same background, and it is in some of them that we are now interested. While Rationalism and Empiricism do not necessarily require the negation of either soul or body, they do mean for all practical purposes the suppression of the one or the other. Thus in general, Empiricism so stresses the uniqueness of "sense" experience that it rejects the system-building and generalizing functions of the mind, and really reduces all knowledge to what is immediately sensed, although sensation may be variously explained. Rationalism, on the other hand, so emphasizes the intellectual powers of man that it effectively rejects the role of sensation in man's acquisition of knowledge. We shall here be occupied with Rationalism, leaving Empiricism to the next chapter.

(d) *Roots of Rationalism.* For the moment we can say that by Rationalism we mean a doctrine which sees knowledge merely as the unfolding of the mind's innate powers, in such a way that from one or some few self-evident principles all knowledge can be derived without recourse to experience. The roots of this position can be detected in Descartes' statement:

> . . . I have taken the existence of this thought [that is, mind or the thinking self] for the first principle, from which I very clearly deduced the following truths, namely, that there is a God who is the author of all that is in the world, and who, being the source of all truth, cannot have created our understanding of such a nature as to be deceived in the judgments it forms of the things of which it possesses a very clear and distinct perception. Those are all the principles of which I avail myself touching immaterial or metaphysical objects, from which I most clearly deduce these other principles of physical or corporeal things, namely, that there are bodies extended in length, breadth, and depth, which are of diverse

figures and are moved in a variety of ways. Such are in sum the principles from which I deduce all other truths.[1]

However, the classic expression of Rationalism is to be found in the works of Spinoza and Leibnitz, so we shall turn to them.

II • DESCRIPTION

Benedict de Spinoza (1632–1677)

(a) *Four Ways of Knowing.* Undoubtedly influenced by Descartes' equation of the self with the thinking principle or soul, Spinoza set out to explain reality and man's knowledge of it. The first step in his explanation is to determine the various sorts of knowledge of which man is capable. Next, he evaluates them and determines that the grade of knowledge which perceives a thing through its essence is the best. He goes on to insist that the attainment of this sort of knowledge is within man's competence. Finally, in his *Ethics* he proceeds to apply this highest type of knowledge in his explanation of all reality.

But, before all things, a means must be devised for improving the understanding and purifying it, as far as may be at the outset, so that it may apprehend things without error, and in the best possible way.[2]

In order to bring this about, the natural order demands that I should here recapitulate all the modes of perception, which I have hitherto employed for affirming or denying anything with certainty, so that I may choose the best, and at the same time begin to know my own powers and the nature which I wish to perfect.

Reflection shows that all modes of perception or knowledge may be reduced to four:

(1) Perception arising from hearsay . . .
(2) Perception arising from mere experience . . .
(3) Perception arising when the essence of one thing is inferred from another thing, but not adequately; this comes when from some effect we gather its cause . . .

---

[1] John Veitch (trans.), René Descartes, *Meditations and Selections* (Chicago: Open Court Publishing Co., 1952), p. 115.
[2] R. Elwes (trans.), Benedict de Spinoza, *Improvement of the Understanding, Ethics and Correspondence* (New York: Tudor Publishing Co., 1936), p. 5.

(4) Lastly, there is the perception arising when a thing is perceived solely through its essence, or through the knowledge of its proximate cause.[3]

As can be seen, then, he pointed out that we have four kinds of knowledge: that of hearsay, that of daily experience, that of scientific proof, and that of an intuitive knowledge of essences. This last he felt was the only truly secure and reliable knowledge, and he proposed to lead men to the acquisition of it. He insisted that we are capable of arriving at such perfect knowledge. His work *On the Improvement of the Understanding* was written in order to show men that their intellects were "sick" due to preoccupation with mere probabilities and the lower types of knowledge and needed to be "healed" by following his method and seeing reality as he saw it.

> The fourth mode alone apprehends the adequate essence of a thing without danger of error. This mode, therefore, must be the one which we chiefly employ.[4]

(b) *Intuition of Essences.* The intuitive knowledge of essences to which he would lead us would mean that we would be able to see what things are in themselves and in all their relationships. In fact, a basic principle in his thinking is this: "The order and connection of ideas is the same as the order and connection of things."[5] That is to say, we should not try to come to the knowledge of causes from the experience of their effects; rather the reverse is the correct way: We can and should proceed from the knowledge of causes to effects. Thus since God is ontologically the First Cause and the One from Whom all else proceeds, knowledge must begin with a knowledge of Him, and from knowing Him all else can be known.

> We may add that the idea in the world of thought is in the same case as its correlate in the world of reality. . . . Further, from what has just been said—namely, that an idea must, in all respects, correspond to its correlate in the world of reality—it is evident that, in order to reproduce in every respect the faithful image of nature,

---

[3] *Ibid.*, p. 6.
[4] *Ibid.*, p. 9.
[5] *Ibid.*, *Ethics* II, prop. 7, p. 82.

our mind must deduce all its ideas from the idea which represents
the origin and source of the whole of nature, so that it may itself
become the source of other ideas.[6]

The nature of God, which should be reflected on first, inasmuch as
it is prior both in the order of knowledge and the order of nature,
they [others whom he considers to be confused in their thinking]
have taken to be last in the order of knowledge, and have put into
the first place what they call the objects of sensation.[7]

This method he actually employs in *The Ethics*, where he begins with the study of God and proceeds to interpret all else in the
light of what has been learned about Him. Spinoza's method is
called by him a geometric one, and this appellation refers both to
the external formulation and progress of his argument, which uses
definitions, axioms, proofs, scholia, lemmata as in geometry, and
also to his conviction that his proofs are as convincing and air-tight
as any geometric proof.

(c) *Knowing That All Is Part of God.* More concretely, what
Spinoza discovers is that there is but one substance, God. "Besides
God no substance can be granted or conceived." [8] The divine
substance has certain attributes of which we know only two:
thought and extension.[9] All other "things" are only modes of this
one divine substance, falling under one or other of these attributes.

> Individual things are nothing but modifications of the attributes of
> God, or modes by which the attributes of God are expressed in a
> fixed and definite manner.[10]

Thus the soul is a mode under the attribute of thought; all material bodies are modes under the attribute of extension.[11] He goes
on from this to a most detailed consideration of human life, its

---

6 Elwoo, *op. oit.*, p. 10.

7 *Ibid., Ethics* II, prop. 10, note 2, p. 86.

8 *Ibid., Ethics* I, prop. 14, p. 49.

9 *Ibid., Ethics* II, "Prop. I: Thought is an attribute of God, or God is a
thinking thing." "Prop. II: Extension is an attribute of God, or God is an
extended thing." pp. 79–80.

10 *Ibid., Ethics* I, prop. 25, coroll., p. 61.

11 *Ibid., Ethics* II, def. 1. "By body I mean a mode which expresses in a
certain determinate manner the essence of God, insofar as he is considered as
an extended thing." p. 78.

bondage and liberation; all of his considerations being deduced systematically from his initial premises. Without going into further detail, it can be seen that this position means that there is really only one being—the one substance, God. All else is a "part" of God, and even "the human mind is part of the infinite intellect of God." [12] This doctrine has been called both a Pantheism and a Theopanism (all is God versus God is all).

Realizing how strange his doctrine must sound, Spinoza makes this plea:

> Here, I doubt not, readers will come to a stand, and will call to mind many things which will cause them to hesitate; I therefore beg them to accompany me slowly, step by step, and not to pronounce on my statements, till they have read to the end.[13]

Gottfried Wilhelm Leibnitz (1646–1716)

(d) *Monads and Substances.* Unlike Spinoza, Leibnitz insisted on the substantiality of individual things in this world. God is not the only substance. However, in his aprioristic explanation of the nature of substance Leibnitz insisted that everything composite must be made up of simples, and everything divisible must be made up eventually of indivisibles. Hence, substances must be simple and indivisible beings, which are also in some sense spiritual in nature. These substances he called monads, and so everything we meet is made up of a combination or aggregation of unextended, simple monads which are dynamic in character. Composites (mere aggregates of monads) are due to the combination of various monads under the direction of one principle (or entelechy) monad for the purpose of clarifying the perceptions of that monad.

> The Monad, of which we will speak here, is nothing else than a simple substance, which goes to make up composites; by simple, we mean without parts.
>
> There must be simple substances because there are composites; for a composite is nothing else than a collection or *aggregatum* of simple substances.

12 *Ibid., Ethics* II, prop. 11, coroll., p. 87.
13 *Ibid.,* prop. 10, note, p. 88.

Now, where there are no constituent parts there is possible neither extension, nor form, nor divisibility. These Monads are the true Atoms of nature, and, in fact, the Elements of things.[14]

(1) *Change Means Growth in Perceptions.* In Leibnitz' view the only "change" which takes place in the monad and the only progress made is to proceed from one perception now had to another and clearer one, and each new perception is but the unfolding and flowering of the previous one.

The action of the internal principle which brings about the change or the passing from one perception to another may be called Appetition.[15]

Nor does all this mean that he admits any interaction between the monads. Each is a self-enclosed world, so built by God that it mirrors all that is true in its surroundings, yet without receiving any influence or impulse from what is outside itself. Still it is "as if" the others did influence it.[16] Leibnitz admits the influence of only one Monad on others, and that is God, the Supreme Monad, who made all as they are, who gave them their natures, and who thus gave them initially all that will ever happen to them. Their development is just the gradual unfolding or perception of what is already within them. Thus, each substance already contains all that has ever happened to it and all that will ever happen to it.

Furthermore every substance is like an entire world which it portrays, each one in its own fashion . . . It can indeed be said that every substance bears in some sort the character of God's infinite wisdom and omnipotence, and imitates him as much as it is able to; for it expresses, although confusedly, all that happens in the universe, past, present and future, deriving thus a certain resemblance to an infinite perception or power of knowing.[17]

As the individual concept of each person includes once and for all everything which can ever happen to him, in it can be seen, *a*

---

14 G. Montgomery (trans.), Leibnitz, *Discourse on Metaphysics; Monadology* (Chicago: Open Court Publishing Co., 1957), p. 251.

15 *Ibid.*, p. 253.

16 *Ibid.*, p. 23.

17 *Ibid.*, p. 14.

priori the evidences or the reasons for the reality of each event, and why one happened sooner than the other.[18]

(2) *Human Knowledge.* What has been said on the nature of substance (monad) in the philosophy of Leibnitz implies what is to be taken as at least the ideal in human knowledge. My knowledge proceeds independent of any influence from things, it does not originate with experience, but rather I already know, at least virtually, all that I will ever know. All that is needed is the unfolding and clarification of my present concepts. Thus all knowledge is really *a priori* and innate. So, too, in knowing any individual I can, at least theoretically, know all that has ever happened to *it* and all that will ever happen to *it*. This is based upon his view that all true propositions are analytic in nature, since the predicate is contained in the subject. Of course Leibnitz admits that such explicit knowledge is hard for us to come by, and he somewhat reluctantly admits that we must often accept something only because we see that it happens or has happened; still, the ideal sort of knowledge remains *a priori* and innate.

(e) *Basic Principles.* Behind all that has just been said, and important for understanding the Rationalism of Leibnitz, is the role which he ascribes to certain basic principles. The two first are the Principles of Contradiction and of Sufficient Reason:

> Our reasoning is based upon two great principles: first that of Contradiction, by means of which we decide that to be false which involves contradiction and that to be true which contradicts or is opposed to the false.

> And second, the principle of Sufficient Reason, in virtue of which we believe that no fact can be real or existing and no statement true unless it has a sufficient reason why it should be thus and not otherwise. Most frequently, however, these reasons cannot be known by us.[19]

This same attitude is expressed in another way when he says:

> Thus the content of the subject must always include that of the predicate in such a way that if one understands perfectly the con-

---

[18] *Ibid.*, p. 19.
[19] *Ibid.*, p. 258.

cept of the subject, he will know that the predicate appertains to it also.[20]

The ideal of knowledge, then, is to be in the position of being able to contemplate essences themselves in the light of these basic principles, and to be able to deduce from them all the predicates which can ever be applied to them. Naturally, the application of these to concrete existents is not easy, but he feels that some help can be found by the use of another principle, that of perfection. This principle amounts to saying that God always chooses the greatest amount of perfection and produces the best possible world. Thus, if I can know that this or that contingent fact will actually promote perfection, then that fact is true.

> Hence it is seen to be most evident that out of the infinite combination of possibles, and the infinite possible series, that one exists by whose means the greatest possible amount of essence or possibility is brought into existence.[21]

Hence, in order to decide whether a given essence does actually exist or will actually exist, all one needs to know is whether its actual existence does or will render the universe more perfect. Theoretically, such knowledge is possible; however, in practice only God has such a grasp of things, and man must fall short of the ideal. So here Leibnitz yields a bit in his Rationalism and appeals, maybe reluctantly, to the need for experience. He admits that in practice we frequently know that the greater perfection of the universe is to be realized by the actual existence of this or that being only from its actual existence. The trouble is that it is too hard for us to know how it will fit in with the universal order of things. So experience tells us that it actually exists; then by principle we know that the reason for that is the fact that its existence does promote universal harmony and perfection.

III • THE ANALYSIS

(a) *The Logic of Ideas.* Although what has been said above is minimal in respect to the total philosophy of these two men, still

---

20 *Ibid.,* p. 13.

21 M. Morris (trans.), Leibnitz, "On the Ultimate Origination of Things," *The Philosophical Writings of Leibnitz* (London: J. M. Dent, 1934), p. 34.

it is sufficient to illustrate what we mean by their rationalistic attitudes and positions. It also illustrates another general point worth mentioning here. In the dialogue of minds which we see in the history of thought, it is interesting and informative to see certain ideas being developed and carried to their logical conclusions. There seems to be an inner logic of ideas which, once they are proposed, gradually makes itself felt, for even great minds cannot completely divorce themselves from their intellectual environment and background. That is to say, the history of thought is not just the recording of isolated thoughts of isolated men; rather, each inherits much from his predecessors, and each does his thinking under the influence of problems and solutions which are current in his day.

Thus in Rationalism we see the logical fruition of notions activated by Descartes and his successors. If the soul and body are really two complete beings, then it becomes an easy step to insist on their functioning with almost total independence of each other, and so sense experience can logically be said to be relatively unimportant for the mind. If there is a body, its function is not noetic, nor is it related to the noetic. Then one can say that all knowledge is from, of, and through the mind alone. But what can knowledge mean in that context? Since it does not derive from experience, it can be only the unfolding from within of the mind's content, and so the mind is capable of devising intellectual systems, principles, and explanations in a totally *a priori* fashion.

Moreover, in this context it is easy to see with Spinoza that the truest and best and richest idea is the one with which we should begin our reasonings, and we should seek to deduce all else from that perfect idea.

(b) *System and Certitude.* One of the chief insights which these men seem to have had is the need we are under for systematization in knowledge. That man has a natural drive in that direction is a fact of intellectual experience, admitted even by those who for doctrinal reasons reject all system-building. Such work is but the fruition of man's power to abstract and of his ability to gain some knowledge of essences or natures. Lest we get lost in myriad and unrelated singular details, Rationalism is an invitation to recognize and draw value from the power of the mind to construct, to rationalize, to explain, to know essences.

So the laudable goal of Rationalism is to construct a rational,

objective, certain system of philosophical truths, free from all subjective and emotional elements, and one which would be comparable
to the objectivity and certainty of mathematics. The wranglings
and sophistical arguments and scepticism of much previous phi-
losophizing drove them to desire and to work for the reinstating of
philosophy in a preeminent position of respect and honor.

(c) *Neglect of Experience.* However, if we take a general view
of the position under discussion and reflect on its implications for
knowledge, we are immediately faced with an apparently total
neglect of experience. Knowledge is said to be innate knowledge,
at least virtually, of self-evident principles from which truths about
reality can be deduced. Even when knowledge is said to be occasioned by experience, it is never said to be derived therefrom;
rather, knowledge is only the unfolding of reason itself. Hence,
Rationalism urges us to seek to view all reality from an *a priori*
viewpoint, with the hope of first preparing and developing a system to which reality must then conform. However, that sort of
procedure cannot be maintained in the face of a careful analysis of
the facts of knowledge.

(d) *Too Exclusively Deductive.* Moreover, the rationalistic
mentality is an exaggeration of the human intellectual power.
Imbued with the ideal of *a priori* demonstration, it would seek to
deduce all truth from certain basic principles and to make the
order of knowledge the same as the order of being. In another
study it will be shown that the ontological order is one in which
things somehow proceed from a First Cause, so that in the order
of being God first exists, and only secondly do created beings begin
to exist. However, man's knowledge does not proceed in this way;
we do not first know God and then through Him come to know
things. Rather, all that we know, even the most basic principles, is
based on and originates in our experience. Only with difficulty
do we come to know God, and only by constant labor do we succeed in building up any sort of systematic and coherent explanation
of reality. Even the most extreme Rationalists, although retaining
the ideal of the *a priori* deduction of all truths, are forced at times
to grant some grudging place to experience, since otherwise they
are at a loss to explain certain factors in their theories. The embarrassing elements are the contingent facts of daily experience.

These can in no wise be deduced from principles; they must be learned from actual contact. Then, too, it is a fact that we learn many new truths from sources other than deduction, such as induction or testimony.

(e) *Theory vs. Practice.* On the relation between a philosophic theory and reality this can be said: The theory or doctrine proposed must be one which fits the facts as they are actually experienced, and where there is a real conflict, it is the theory which must yield. We are not in a position to concoct theories and then demand that reality conform to them. Rationalism has hit on and developed a partial truth in respect to the dynamic aspects of the mind, but it has exaggerated these to the point where the human conditions for knowledge and its clearly experienced reliance on sense experience have been lost from sight. Thus, Rationalism is a powerful and logically developed theory of what human knowledge would be were it really innate and were we capable of deducing all knowledge from certain basic principles.

IV • SUMMARY OF THE ANALYSIS

(a) Rationalism is a logical fruition of the cleavage introduced by Descartes between soul and body. As a complete being, the soul does not need the body, and knowledge becomes the unfolding from within of the mind's own content. Cf. *Analysis (a).*

(b) Moreover, Rationalism exemplifies the intellectual power to know natures and to build systems of thought, and in doing so it sought the sort of certainty which would reinstate philosophy in a place of honor. Cf. *Analysis (b).*

(c) However, Rationalism neglects the role of sense experience, postulates innate knowledge, and develops *a priori* theories to which reality must conform. Cf. *Analysis (c).*

(d) Rationalism overestimates the role of deduction. Its ideal of the *a priori* demonstration of everything fails to consider other sources of truth, such as sense experience, testimony and induction. Cf. *Analysis (d).*

(e) Rationalism exaggerates the power of the human mind. Theories must be based on experience, not vice versa, and where there is clear conflict with facts, the theory must yield. In the last analysis, then, Rationalism shows us logically what human knowledge would be were it innate, rather than what it actually is. Cf. *Analysis (e).*

V • THE STATEMENT IN BRIEF

Rationalism stresses the active power of the intellect to build systems and to deduce truths; however, it exaggerates the a priori aspects of knowledge, neglects the role of experience, is too exclusively deductive and so cannot be an adequate theory of knowledge.

VI • DEFINITIONS

Rationalism: a theory of knowledge which proposes the *a priori* demonstration of all truths as the ideal of human knowledge.
Deduction: demonstration which proceeds from the universal to the less universal or particular.
A Priori: prior to and independent of experience.

VII • SUGGESTED READINGS

Collins, James, *A History of Modern European Philosophy,* ch. 6, pp. 201–10; ch. 7, pp. 262–71. Milwaukee: The Bruce Publishing Company, 1954.

Copleston, Frederick, S.J., *A History of Philosophy,* IV, pp. 15–24. London: Burns, Oates & Washburn, Ltd., 1946–1959.

Leibnitz, Gottfried, *Discourse on Metaphysics and Monadology,* translated by G. Montgomery. Chicago: Open Court Publishing Company, 1957.

Morris, C. R., *Locke, Berkeley, Hume.* London: Oxford University Press, 1949.

Spinoza, Benedict de, "On the Improvement of the Understanding," *Philosophy of Benedict de Spinoza,* translated by R. Elwes. New York: Tudor Publishing Company, 1936.

CHAPTER TWENTY-TWO

EMPIRICISM

I • THE SITUATION

Having seen how the rationalistic viewpoint tends to exaggerate the power of the human intellect to know truth and to concoct systems explicative of all reality to which reality itself must conform, we now turn to examine a reaction against that mentality in the name of experience. The classical period expressive of this empirical stress is to be found in seventeenth and eighteenth century England, typified by the works of John Locke, George Berkeley and David Hume. However, the influence of that viewpoint was by no means exhausted with their death; rather it has ever since profoundly influenced the scientific and philosophic mentality of the modern era. In fact it is true to say that it is largely due to the inspiration of Empiricism that modern scientific endeavors have been so diligently pursued, just as it is true to say that the sciences themselves have been one of the factors influencing the empirical mentality in philosophy. The keynote of this outlook is emphasis on experience, but on experience taken in one very definite meaning.

II • THE DESCRIPTION

A. *John Locke* (1632–1704)

(a) *His Purpose.* According to Locke too much time and effort had been spent on the elaboration of systems grandiose in conception and brilliant in execution but lamentably unfaithful both to reality itself and to the power of the human intellect to know truth. Hence, he proposes that we follow a more humble and honest path when he points out the aim of his work:

> This therefore being my purpose, to inquire into the original, certainty, and extent of human knowledge, together with the grounds and degrees of belief, opinion, and assent . . .[1]

This sort of inquisition, he feels, is most necessary, since, before we can intelligently apply the mind to the elaboration of systems we should determine the limitations and the capacities of the instrument which is to be used. He feels that failure to do this is what accounts for so many of the exaggerations to be found in the history of human thought.

> If by this inquiry into the nature of the understanding I can discover the powers thereof, how far they reach, to what things they are in any degree proportionate, and where they fail us, I suppose it may be of use to prevail with the busy mind of man to be more cautious in meddling with things exceeding its comprehension; to stop when it is at the utmost extent of its tether; and to sit down in a quiet ignorance of those things which, upon examination, are found to be beyond the reach of our capacities.[2]

At the other extreme from Rationalism, he teaches that all knowledge originates in sense experience, and is limited to a knowledge of ideas themselves. So, we shall here consider the object of knowledge, the objective reality of what is known, and the explanation of general ideas.

(b) *The Object of Knowledge.* That of which we are immediately aware and that which constitutes the immediate object of our knowledge is, according to Locke, an *idea* and only an idea. Thus, when it is said that we must be satisfied with what we experience, he means that we know only our ideas, whether of *sensation* (drawn from the external world) or of *reflection* (drawn from the internal operations of the mind).

---

[1] J. A. St. John (ed.), "An Essay Concerning Human Understanding," *The Philosophical Works of John Locke* (London: George Virtue, 1843) Introd. ii, p. 80. Most of the texts quoted can also be found according to chapter and number in Edwin A. Burtt (ed.) *The English Philosophers from Bacon to Mill* (New York: Modern Library, 1939), pp. 238–402; here p. 244.

[2] St. John, *op. cit.,* Introd. iv, p. 80; also Burtt, *op. cit.,* p. 245.

Since the mind, in all its thoughts and reasonings, hath no other immediate object but its own ideas which it alone does or can contemplate, it is evident that our knowledge is only conversant about them.

Knowledge, then, seems to me to be nothing but the perception of the connexion and agreement, or disagreement and repugnancy of any of our ideas. In this alone it consists. . . . For when we know that white is not black, what do we else but perceive that these two ideas do not agree?[3]

Knowledge, as has been said, lying in the perception of the agreement or disagreement of any of our ideas, it follows from hence, that,

First, We can have knowledge no further than we have ideas.

Secondly, That we have no knowledge further than we can have perception of their agreement or disagreement.[4]

So far the position is clear: Empiricism stresses experience, but experience is not immediately of objective reality; rather the object immediately experienced is the idea.

(c) *Objective Reality.* What has been said does not mean that for Locke the only things in existence are ideas; some ideas, he says, do truly represent objective reality. To indicate what he means, he distinguishes between our ideas of primary and our ideas of secondary qualities (an echo of Descartes). The ideas which we have of primary qualities (solidity, extension, figure, motion or rest, and number) do correspond to and represent reality as it is, and these qualities do exist in reality as we experience them.

However, secondary qualities (color, taste, sound, for example) do not exist as such in reality. Instead, he holds that there exist in reality some "powers" capable of causing these ideas in us, but such ideas do not correspond to or represent what actually is. Hence, these ideas are subjective.

From whence I think it easy to draw this observation, that the ideas of primary qualities of bodies are resemblances of them, and their patterns do really exist in the bodies themselves; but the ideas produced in us by these secondary qualities have no resemblance of them at all. There is nothing like our ideas existing in the bodies themselves.[5]

---

[3] *Ibid.*, part IV, chap. i, nos. 1, 2, p. 383; *ibid.*, p. 317.

[4] *Ibid.*, part IV, chap. iii, nos. 1, 2, p. 391; *ibid.*, p. 327.

[5] *Ibid.*, part II, chap. viii, no. 15, p. 145; *ibid.*, p. 267.

(d) *Idea of Substance.* Locke does not deny the existence of substances, whether spiritual or material, but he equivalently says that we can have no knowledge of them as they are. In this matter he distinguishes between "real" and "nominal" essences or substances; the former referring to what they are in themselves (unknowable by us), and the latter referring to the mere grouping together of accidents or qualities or operations. Such groups are all that we can know as substances. In this explanation the Phenomenalism of Locke becomes clear, since all we can know are the mere appearances or accidents, and these in no way reveal, even in a minimal sense, anything about the principle of operation or the substance itself. In this context, then, accidents are looked on as a sort of cloak which hides rather than manifests the nature of the being, and with these accidents our knowledge terminates.

> So that if anyone will examine himself concerning his notion of pure substance in general, he will find he has no other idea of it at all, but only a supposition of he knows not what support of such qualities which are capable of producting simple ideas in us; which qualities are commonly called accidents . . .
>
> An obscure and relative idea of substance in general being thus made, we come to have the ideas of particular sorts of substances, by collecting such combinations of simple ideas as are, by experience and observation of men's senses, taken notice of to exist together, and are therefore supposed to flow from the particular internal constitution, or unknown essence of that substance.
>
> . . . our specific ideas of substances are nothing else but a collection of a certain number of simple ideas, considered as united in one thing.[6]
>
> . . . it is not to be wondered that we have very imperfect ideas of substances, and that the real essences, on which depend their properties and operations, are unknown to us. . . . This consideration alone is enough to put an end to all our hopes of ever having the ideas of their real essences; which whilst we want, the nominal essences we make use of instead of them, will be able to furnish us but very sparingly with any general knowledge or universal propositions capable of real certainty.[7]

(e) *Generalizations.* What has been seen concerning Locke's views on substances enters intimately into his study of general

---

[6] *Ibid.,* part II, chap. xxiii, nos. 2, 3, 14, pp. 245, 250; *ibid.,* pp. 294–95, 302.
[7] *Ibid.,* part IV, chap. vi, no. 12, pp. 421–22; *ibid.,* p. 358.

ideas. Since all that we know is what is concretely experienced here and now, and since experience really means sense experience, it is to be expected that Locke will have trouble in explaining universal ideas. He does find such trouble, and basically it is because of his insistence that we can know only "nominal" essences. Only singular things exist, and so our true ideas should be only particular ones. However, Locke sees that if that position is taken too literally, it would reduce science to a knowledge of the isolated and disconnected singulars, and thus universalization would be without foundation. Since he is not willing to go quite that far, and yet refuses to admit any abstraction which would in any way touch on the real nature of anything, he tries to find another way out.

He seeks to avoid the extremes of utter singularity and of universalization through abstraction by the devices of *separation* (the mental selection of those ideas which are shared by several beings) and *relating* (referring those ideas to several beings). Thereby, he feels that one forms ideas which are general in character without being universal in the traditional sense.

> The use of words, then, being to stand as outward marks of our internal ideas, and those ideas being taken from particular things, if every particular idea that we take in should have a distinct name, names must be endless. To prevent this, the mind makes the particular ideas received from particular objects to become general; which is done by considering them as they are in the mind, such appearances, separate from all other existences, and the circumstances of real existence, as time, place, or any other concomitant ideas.[8]
>
> For when we nicely reflect upon them, we shall find that general ideas are fictions and contrivances of the mind, that carry difficulty with them, and do not so easily offer themselves as we are apt to imagine. For example, does it not require some pains and skill to form the general idea of a triangle (which is yet none of the most abstract, comprehensive, and difficult), for it must be neither oblique nor rectangle, neither equilateral, equicrural, nor scalenon; but all and none of these at once. In effect, it is something imperfect, that cannot exist; an idea wherein some parts of several different and inconsistent ideas are put together.[9]

---

[8] *Ibid.*, part II, chap. xi, no. 9, p. 161; *ibid.*, p. 281.
[9] *Ibid.*, part IV, chap. vii, no. 9, p. 426.

His explanation is at least an admission that merely discrete, singular and isolated experiences can not provide a solid foundation for scientific work. He also admits that in some unspecified way "nature" provides the foundation for our generalizing, but this remains linked with his view that all we can know are sense qualities and that these actually constitute the only essence knowable. This very shallow foundation for universal ideas will provide a point for Hume to attack.

B. George Berkeley (1685–1753)

(f) *Empiricism Continued.* We have already examined the acosmistic position of Bishop Berkeley and seen that in his view the existence of material reality is rejected, whereas the existence of ideas and spiritual substances is admitted. Here we are concerned with him more precisely as an Empiricist, as one who extols the all-important role of experience. In this matter, he continues the position inaugurated by Locke in his insistence that the immediate object of knowledge is and can be only the idea in the mind. Moreover, in his polemic against matter, he quite agrees with Locke on the subjectivity of the secondary qualities, but goes on to say that the very same arguments also tell against the objectivity of the primary qualities; hence he propounds the subjectivity of both.

(g) *The Object of Knowledge.* It is clear that Berkeley can in no wise be called a Realist. We know immediately only our ideas, but they do not represent reality at all; in fact they themselves, and spiritual substances, are the only realities. So there is no meaning in the assertion of their correspondence with a reality apart from the knower; their truth consists in their correspondence with each other. Thus not only are ideas the immediate objects known, but they are the only reality, and so the experience which we have is only one of having a succession of ideas.

It is evident to anyone who takes a survey of the objects of human knowledge, that they are either ideas (1) actually imprinted on the senses, or else such as are (2) perceived by attending to the passions and operations of the mind, or lastly (3) ideas formed by help of memory and imagination, either compounding, dividing, or barely representing those originally perceived in the aforesaid ways.[10]

Referring to Locke's distinction between primary and secondary qualities and to his views concerning their objective reality, Berkeley says:

> Some there are who make a distinction betwixt *primary* and *secondary* qualities. . . . The ideas we have of these [secondary] they acknowledge not to be the resemblances of anything existing without the mind, or unperceived, but they will have our ideas of the primary qualities to be patterns or images of things which exist without the mind, in an unthinking substance which they call *matter*. By *matter*, therefore, we are to understand an inert, senseless substance, in which extension, figure, and motion do actually subsist. But it is evident from what we have already shown, that extension, figure, and motion are only ideas existing in the mind, and that an idea can be like nothing but another idea, and that consequently neither they nor their archetypes can exist in an unperceiving substance.[11]

(h) *Generalization.* The empiricist interest in particular experiences and distrust of the universal or general is pushed one step further by Berkeley. He constantly refers to such ideas as abstract; and he insists that no such ideas can be got by abstraction, since there is no reality to provide a basis for this process. In fact, he feels that Locke conceded too much to abstraction in his explanation of the formation of general ideas. The following is his description of what this abstraction means; immediately afterwards his evaluation of it will be given.

> . . . For example, the mind having observed that Peter, James, and John resemble each other in certain common agreements of shape and other qualities, leaves out of the complex or compounded idea it has of Peter, James, and any other particular man, that which is peculiar to each, retaining only what is common to all, and so makes an abstract idea wherein all the particulars equally partake; abstracting entirely from and cutting off all those circumstances and differences which might determine it to any particular existence.

---

[10] Burtt, *op. cit.*, "A Treatise Concerning the Principles of Human Knowledge," no. 1, p. 523.
[11] *Ibid.*, no. 9, pp. 525–26.

And after this manner it is said we come by the abstract idea of man, or, if you please, humanity, or human nature . . .[12]

Whether others have this wonderful faculty of abstracting their ideas, they best can tell; for myself, I find indeed I have a faculty of imagining, or representing to myself, the ideas of those particular things I have perceived, and of variously compounding and dividing them. . . . I cannot by any effort of thought conceive the abstract idea above described.[13]

From this point on Berkeley explains our apparent possession of general or universal ideas by denying that there really are any such ideas at all; and by stating that insofar as we seem to have them, we have only some singular idea, and we make it stand for several others. Using the idea of triangle as an example, he says that all we can do is think of some one particular triangle, then we can pay no attention to the exact angles or sides which delimit this one, and thus we seem to have an idea which is a sign of and which is applicable to many.

. . . And here it is to be noted that I do not deny absolutely there are general ideas, but only that there are any *abstract* general ideas. . . . An idea which considered in itself is particular, becomes general by being made to represent or stand for all other particular ideas of the same sort.[14]

. . . Thus when I demonstrate any proposition concerning triangles, it is to be supposed that I have in view the universal idea of a triangle; which ought not to be understood as if I could frame an idea of a triangle which was neither equilateral, nor scalenon, nor equicrural; but only that the particular triangle I consider, whether of this or that sort it matters not, doth equally stand for and represent all rectilinear triangles whatsoever, and is in that sense *universal*.[15]

C. David Hume (1711–1776)

(i) *And Empiricism.* Since Hume continues the tradition we have been considering and brings it to its logical fruition, we shall

---

12 *Ibid.*, Introd., no. 10, p. 513.
13 *Ibid.*, Introd., no. 10, p. 513.
14 *Ibid.*, Introd., no. 12, p. 515.
15 *Ibid.*, Introd., no. 15, p. 517.

indicate only in the briefest fashion how the same themes are handled by him. His views on knowledge as meaning nothing more than a succession of disconnected or "loose" impressions or ideas connected only by the associative power of the imagination, his insistence on the unknowability of causality and of substances, and his consequent limitation of experience to what is immediately sensed—these and other points in his doctrine have led writers to call him variously a Sceptic, an Agnostic, a Phenomenalist, and a Naturalist as well as an Empiricist. It was Hume's theory of knowledge which more than anything else alerted Kant to the necessity of re-examining human knowledge and finding the foundations for scientific studies.

(j) *The Object of Knowledge.* In line with his immediate predecessors, Hume seems to assume it as most evident that what we immediately know can be only our *perceptions.* By these he means roughly what had been meant by ideas, but Hume restricts the use of the term *ideas* to those perceptions which are somewhat vague or dim, and gives the name *impressions* to those perceptions which are vivid and clear. In respect to the existence of and conformity of perceptions to outside things, Hume is not exactly clear. At times he seems to deny their existence; at other times he prefers to say only that we can not know them, but are led to believe that they exist, and very often a sort of common sense Realism breaks forth in his way of talking and indicates that he looks on knowledge as conformity to objective reality. The following quotations give some indication of these attitudes.

> We may observe, that 'tis universally allow'd by philosophers, and is besides pretty obvious of itself, that nothing is ever really present with the mind but its perceptions or impressions and ideas, and that external objects become known to us only by those perceptions they occasion.[16]
> For as to the notion of external existence, when taken for something specifically different from our perceptions, we have already shewn its absurdity.[17]

(k) *Generalizations.* Since experience can mean only my perception in the immediate present of this or that particular event,

---

[16] L. A. Selby-Bigge (ed.), David Hume, *A Treatise of Human Nature* (Oxford: Clarendon Press, 1888, reprinted 1958), Bk. I, Part II, Sect. VI, p. 67.
[17] *Ibid.,* Bk. I, Part IV, Sect. II, p. 188.

knowledge becomes only the perception of disparate, unconnected and individual perceptions with no objective bond between them. That I seem to grasp them as connected, is only the function of the imagination, which in various ways tries to associate various perceptions together.

> 'Tis still true, that every distinct perception, which enters into the composition of the mind, is a distinct existence, and is different, and distinguishable, and separable from every other perception, either contemporary or successive. . . . The understanding never observes any real connexion among objects, and . . . even the union of cause and effect, when strictly examin'd, resolves itself into a customary association of ideas.[18]

This position has been referred to as the "loosening" of ideas, in the sense that each is complete in itself and not really connected with another; there is mere succession. In this context it is easy to see that there can be no generalization based upon an objectively grasped similarity of natures, but only particular ideas. Since, however, there do seem to be universal ideas, Hume explains them as particular ideas joined to some common term.

> A particular idea becomes general by being annex'd to a general term; that is, to a term, which from a customary conjunction has a relation to many other particular ideas, and readily recalls them in the imagination.[19]

> Let men be once fully perswaded of these two principles, *That there is nothing in any object, consider'd in itself, which can afford us a reason for drawing a conclusion beyond it;* and, *That even after the observation of the frequent or constant conjunction of objects, we have no reason to draw any inference concerning any object beyond those of which we have had experience.*[20]

The two principles just quoted are symptomatic of the empirical mentality in that they are a warning against "going beyond" what is immediately experienced by the senses. It is in the light of these principles that any attempt to acquire knowledge of substances, of spiritual principles, and of causality, is doomed to failure, since these cannot be "experienced."

---

[18] *Ibid.*, Sect. VI, pp. 259–60.
[19] *Ibid.*, Part I, Sect. VII, p. 23.
[20] *Ibid.*, Part III, Sect. XII, p. 139.

III • THE ANALYSIS

(a) *Reaffirmation of Experience.* Seen against the background of Rationalism and of the Cambridge Platonists, this empirical reaffirmation of the need for experience, together with the uncompromising rejection of Innatism and the attack upon the excess of *a priori* system-building, introduced a much needed corrective into the field of modern philosophy. Once again the humble origins of man's knowledge as well as the fact that it is only by slow and laborious efforts that he progresses in his acquisition of truth were affirmed. His most developed systems, his most grandiose conceptions, even the most basic principles of his knowledge, all must somehow be discovered by him through a gradual analysis of the reality which he experiences. When the classical Empiricists wrote, these things badly needed saying.

Moreover, the Empiricist's insistence that only individuals exist was a point well made, although such insistence may appear to many today to be unnecessary. And in connection with it went the recognition of the fact that in the conception of the universal concept the intellect has a very definite and active role to play.

(b) *The Object of Knowledge.* However, reflection on the value of Empiricism cannot stop there, since the Empiricists in their laudable stress on the role of experience cannot be said to have arrived at a fully satisfactory view of knowledge, and the logical implications of the position can be seen by following the movement from Locke through Hume.

The basic assumption of Empiricism in respect to the object of knowledge is that the only thing immediately knowable is an idea, so we never have immediate contact with what is. We can examine this first in Locke's Representative Realism. His doctrine is called a *Realism* because he admits the extramental reality of sensible things, and it is called *Representative* because he admits that at least our ideas of primary qualities do actually represent what is in reality. In this sense his views on cognition constitute a copy or representative theory of knowledge. For Locke, then, ideas are a strictly instrumental means in knowing; they are the immediate objects grasped, and only mediately do we sometimes come to know some sensible qualities.

This position is, however, logically inconsistent. For one thing, if we have no immediate contact with what is, by what right can we ever say that any idea truly corresponds to what is as it is? The very possibility of such an assertion would presuppose that a comparison be made between the idea and the reality of which an idea is had; and to make such a comparison it would be necessary to have (1) the idea, and (2) some direct contact with the other term of the comparison, that is, with the reality in question. Without this, I can never say that my idea is an exact representation of what is.

Likewise, it is in keeping with what has just been said that the word "postulate" can be accurately used in respect to Locke's position. If all that I immediately know is the idea itself, then that is all that I ever will know immediately, and beyond that I can have faith in the reliability of my knowledge, I can hope that it knows what is; in a word, I can *postulate* that my knowledge accurately represents what is, but in the context I can never know that this is the case, because there is no way of ever making the required comparison.

(c) *Reification of Ideas.* The empirical position, then, actually means the reification of ideas. That is to say, it does not recognize that ideas are a unique sort of means whereby man knows what is, or that their sole function is to link the knower and the known; rather it makes of them the object of knowledge, by describing them as means which are themselves the things known. This is to make ideas into things known, or to reify ideas.

This admission gives rise to a further reflection which the logic of the situation seems to demand. If ideas are raised to the status of things known and if all that we can grasp are ideas, then it must logically be admitted that we cannot immediately know even our ideas. Between the object known (the *idea* itself) and our knowledge of it there would have to intervene an idea of that idea; moreover, that same process would have to be repeated indefinitely. Although this may sound a bit extreme as stated, still it is logically unavoidable, if Representative Realism is taken seriously. For what logically coherent reason can be adduced for stopping with the immediate knowledge of an idea when the initial position taken is that what is known can be known only mediately through knowledge of an idea of it?

Finally, the idealistic implications of this position have been abundantly illustrated by Berkeley's development. Once the mind is affirmed to know only its ideas, it is logically simple to question whether there *is* anything more, and to reach the point of denying that there is any extramental reality. As Berkeley said, the same reasons for denying objectivity to secondary qualities can be brought to bear on primary qualities, and thereby on all material existents. When that point is reached, one is no longer explaining the entirety of the elements experienced by the human person in knowing, but is suppressing some of the relevant data.

Hume's Scepticism and Agnosticism in regard to all substances and to the reality of causality is another indication of the logical results of Mediate Realism as a theory of knowledge.

(d) *Knowledge of Substance.* Although the explanation of substance and accident will be more carefully provided in the study of general metaphysics, we can point out here that the empirical view as to their relation is unsatisfactory. The Phenomenalism characteristic of Empiricism sees in "accidents" and operations a sort of cloak or shell which surrounds the underlying substance and which blocks it from view. This is a very naive interpretation, or misinterpretation. It fails to see that accidents are modifications of and manifestations of substance itself, that as such they reveal something and tell us about reality itself, since it is the substantial reality which is thus accidentally modified. The operations do proceed from and thus do tell us something about the very nature which is operating. This may often be a minimal knowledge, but it is something positive, and could not be had if accidents were somehow little "beings" which cloaked what is from our view. Thus the choice offered by Empiricism between an intuitive grasp of essences, which they reject, and a complete ignorance of natures is not one which we need make.

(e) *Generalization.* Empiricism weakly admits that some sort of universalization is required for scientific knowledge. Locke even goes so far as to say that the reason we can generalize is somehow connected with similarities between things. Verbally, at least, this appears to be an admission of the foundation for abstraction. However, his basic Phenomenalism prevents him from going deep enough, and so he could not actually provide a solid foundation for scientific knowledge. If all we can really know are singular things as such, then generalization loses all effective meaning, and human knowledge must be content with isolated contacts with reality.

This same attitude prevents Berkeley from finding a solution in the case he presents of the triangle. On the level of his discussion he is perfectly right in saying that we cannot arrive at knowledge of a triangle which is not one determined kind. What he means is that we cannot have a sense image of triangle as such. He is right, but he is not right in denying the possibility of an intellectual conception of triangle.

Finally, Hume's use of association in this matter cannot provide more than a contingent and passing connection, and it results in a purely mental construct, with no solid foundation in reality. This too cannot be a sufficient basis for scientific knowledge. It was considerations of this sort which aroused Kant to a reconsideration of this whole position and of the grounds for the possibility of science.

(f) *Being Empirical.* To sum this all up briefly, the chief difficulty with Empiricism is that it is not sufficiently empirical. The position that all we can know are sense presentations and that experience is limited in meaning to sense experience is an oversimplification of the data. It is understandable in the light of historic reasons, but it cannot hope to provide an adequate explanation of human knowledge. There is more to our knowledge than meets the eye or any other sense, precisely because the senses are not the only faculties we have for knowing. Experience is indeed the sole appeal we have, but the experience of the Empiricists is not the totality of human experience; it is but part of it. Our experience is both sensible and intellectual, and these two are not identical, despite the intimate connection which exists between them. So in what I sensibly experience my intellect is capable of grasping what is only intellectually perceptible, and that is not limited to the mere singular, isolated, and disparate, but can touch on the foundation of unity, universality, and necessity.

IV · SUMMARY OF THE ANALYSIS

(a) The rejection of Innatism and of *a priori* systems together with renewed stress on the importance of experience as a source of knowledge are strong points in the empirical philosophies. Cf. *Analysis (a)*.

(b) However, in making ideas the immediate object of human knowledge, the Empiricists have rendered it logically impossible to know whether there is conformity to reality. Knowing immediately only the idea itself, I can make no comparison to determine its reliability as representative. Cf. *Analysis (b)*.

(c) Their position raises ideas to the status of "things" known. This involves the further inconvenience that as "things" known, ideas themselves must be only mediately known, and so knowledge becomes either logically impossible or else an act of faith. Cf. *Analysis (c)*.

(d) Idealism, Scepticism, or Agnosticism are the natural fruits of this theory of knowledge. Cf. *Analysis (c)*.

(e) Although rejecting an intuitive view of essences, Empiricism also sees accidents and operations as hiding rather than revealing the substance or nature of the thing. Cf. *Analysis (d)*.

(f) The limitation of experience to mean singular sense experiences results in the inability of Empiricism to provide a solid foundation for universal ideas or for scientific knowledge. Mere phenomenal similarity or the imaginative association of ideas is not profound enough to uphold any necessity or universality. Cf. *Analysis (e)*.

(g) In the sense that the Empiricists limit altogether too stringently the meaning of human experience, we can say that their basic difficulty is a failure to be sufficiently empirical. All the elements of human experience must be accepted and explained, without arbitrary limitations being imposed. Cf. *Analysis (f)*.

V • THE STATEMENT IN BRIEF

Empiricism introduces a much needed emphasis on the role of experience; however, its views that only ideas can be immediate objects of knowledge, that even a minimal knowledge of substances is impossible, and that universal ideas are not founded on the natures of things, result in a theory of knowledge which is inadequate.

VI • SUGGESTED READINGS

Bahm, Archie, *Philosophy, An Introduction,* chaps. 3–5, pp. 50–75. New York: John Wiley and Sons, Inc., 1953.

Burtt, Edwin (ed.), *The English Philosophers from Bacon to Mill.* New York: The Modern Library, 1939.

Collins, James, *A History of Modern European Philosophy,* ch. 8, pp. 328–43; ch. 9, pp. 368–78; ch. 10, pp. 410–18. Milwaukee: Bruce Publishing Co., 1954.

Copleston, Frederick, *A History of Philosophy* (5 vols.), IV, pp. 24–29. London: Burns, Oates, and Washbourne, 1946–1959.

Morris, C. R., *Locke, Berkeley, Hume.* London: Oxford University Press, 1949.

Note on Nominalism and Conceptualism

I • THE SITUATION

(a) *Why This Note.* It may seem strange that at this point a note should be introduced touching on what appears to be an obsolete medieval question. However, for one thing it was lively discussion of the question of universals by the various medieval schools which helped develop the solution proposed in the first part of this work, and so it has both doctrinal and historical interest here. For another thing, it will be easily seen that between these schools and the Empiricism we have just considered there is a kindred spirit.

Moreover, in speaking of contemporary philosophy the term "Nominalism" is frequently used, and so some indication of its meaning will be useful. It involves a mentality which does appeal to "practical" minds, who are more interested in "facts" and in what works than in general theorizing.

The perpetuity of interest in this medieval question is indicated by this reference to the legal profession:

> It may seem a bold and reckless statement to assert that an adequate discussion of cases [in a court of law] involves the whole mediaeval controversy over the reality of universals. And yet, the confident assertion of immutable principles of justice inhering in the very idea of free government made by the writers of these decisions,

and the equally confident assertion of their critics that there are no such principles, show how impossible it is to keep out of metaphysics.[1]

(b) *Exaggerated Realism.* Boethius (480–524) adumbrated a general theory of Moderate Realism; but in the discussion which then began the first stage was dominated by exaggerated Realism. According to this position there is such a complete and precise correspondence between our concepts and reality that when one has an idea, generic or specific, there exists some substantial reality exactly corresponding to it. Thus, when I have the idea "man" and apply it to Peter and James and John, there must exist a unitary substance corresponding to my idea, and the individuals just share in or participate in it; or it can be said that they are just accidental modifications of that one substance.

Hence, between the way I know and the way things are, there is complete correspondence, which would mean, for example, that two men would be substantially one and only accidentally different, and that when a new man is born all that happens is this: A new property is acquired by an already existing substance. This theory was held by William of Champeaux (1070–1120), although he later abandoned it owing to the sustained attack of Abelard.

CRITIQUE: We mention this doctrine only in passing, so by way of critique we need only say that it is an extremely naive position, and one which destroys individuality. It is naive because it supposes a complete correspondence between the way I know and the way things are, and it destroys individuality because if logically pursued, it would result in complete Monism.

II · DESCRIPTION

A. *Nominalism*

(a) *Basic Insight.* The central tenet of Nominalism is that all that exists is the singular, individuated, isolated thing; hence all that we can really know and all about which we can form concepts are these individuals; hence no universal exists in reality, and no

---

[1] Morris R. Cohen, "The Place of Logic in the Law," *Harvard Law Review,* XXIX (1916), 628.

universal ideas exist in the mind. We do, of course, use one word (*nomen*) to designate several things, but this is only a sort of shorthand in terminology and has no counterpart either in the mind or in reality.

(b) *Many Applications.* It should not be thought that Nominalism was narrowly limited to the question of universals. Actually, as a movement it was vitally concerned with metaphysical questions such as the proofs for the existence of God, the nature of substance and accident, the nature of causality, the relation of philosophy to theology. And it was the Agnosticism implied in their position when applied in these spheres which gave rise to the violent opposition they met once they stepped outside the field of logic. It is also true that they did enjoy a very wide popularity in the universities of Europe from the end of the fourteenth until well into the sixteenth century.

(c) *Logical and Real Function.* Here we are most interested in the nominalistic attitude on universals, and on that matter it is true to say that it taught that universality pertains only to the logical function of terms (to their use in a proposition), and does not pertain at all to their real function (to their use as standing for things). This actually means that there is ultimately no real foundation in things for the universal, since there are simply and solely individual things in reality. It should, of course, be remembered that those who held this doctrine were in an intellectual environment where it was most important to counter the tenets of exaggerated Realism. Apart from that, one valuable result of their work was their development of studies in logic, but even that eventually degenerated into "logic-chopping."

William of Ockham is frequently associated with this movement; however, he can be interpreted as having taught a form of Moderate Realism, but without the metaphysical background of exemplary ideas in God. Still, he did in great part inspire the "Ockhamist Movement" which was quite openly nominalistic in our present sense of the word.[2]

---

[2] Cf. Copleston, *op. cit.*, III, pp. 122–27.

B. Conceptualism

(d) *The Doctrine.* As a pure position Conceptualism teaches that there are not only universal words but also universal ideas, but that these lack objective reference, and so have no foundation in reality. This position, also, is a reaction against exaggerated Realism.

Although as pure positions Nominalism and Conceptualism are easily distinguished, in practice it is very difficult to classify individual medieval philosophers in respect to these positions. In each case there are certain necessary qualifications which make a too definitive classification a bit unjustified. Actually most medieval Nominalists are called such simply because they make no explicit mention of the role of the universal concept, and were they so questioned it might be that they would admit such a concept. Of course the same cannot exactly be said of more recent Empiricists or their followers. For these thinkers, the concept is only a sort of vague or undifferentiated grasp of a singular "image."

Still, among the medievals, it can be said that Peter Aureoleus and Henry of Harclay (fourteenth century) did have conceptualistic tendencies, in that they looked upon the universal as being formed by a sort of impression of the individual.[3]

This spirit is much more evident in a philosophy such as Kant's, where universality is definitely and solely due to the imposition of the categories of the mind.

III • THE ANALYSIS

(a) *Singular Individuals.* The basic fact recognized by both of these positions is, as in Empiricism, that things as they exist in reality are individual and singular. This had to be emphasized against an exaggerated Realism which insisted that *genera* and *species* are found in the extramental order in exactly the same way as they are found in the mind. In addition to this, Conceptualism recognized that universals as such have formal existence only in the mind, and thus in some sense at least they show the dynamic

---

[3] *Ibid.*, pp. 40–42.

and active power of the mind, rather than its mere passivity in knowing.

(b) *On Nominalism.* If this position were to be taken seriously as it is proposed, it would mean the end of significant human speech and of all scientific knowledge. This sounds like an extreme statement to make. However, if speech is a material function whose purpose is to communicate knowledge and manifest what we know, then it must represent what is in the mind. Hence words cannot have a meaning which is nonexistent in the mind, and if there are different types of words, they must represent different types of ideas. As we saw in Part One, Chapter Seven, we use universal words which are applied univocally and distributively to many individuals. Hence universal ideas must exist in the mind.

(c) *On Conceptualism.* Here again we find one element of knowledge exaggerated. If universal ideas are conceived to be mere mental fabrications or purely subjective forms imposed on reality, then the objectivity of knowledge can hardly be defended with any logical consistency. What must be kept in mind is that although the formal extramental existence of universals as such cannot be granted, it must be admitted that there is and there is seen to be a foundation in reality for the intellectual generalization which takes place. Things are objectively similar in varying degrees, and it is this recognized similarity which provides the objective justification for the mind's way of conceiving a nature. We must beware of the naive position which demands for objectivity that what is known exist in reality in the same way as it exists in the mind. What is known does exist in reality and it also exists in the mind; however, the way it exists differs in both cases, even though there is a foundation in reality for the mind conceiving the nature as it actually does.

IV · SUMMARY OF THE ANALYSIS

(a) Stress on the singularity of things was needed in the face of exaggerated Realism. Cf. *Analysis (a).*

(b) Nominalism would mean a lack of significance for universal terms, and so for scientific knowledge. However, various types of words indicate various types of ideas in the mind, and so universal words manifest universal ideas. Cf. *Analysis (b).*

(c) Conceptualism is inconsistent with the truth and objectivity of knowledge. Although the universal does not formally exist in things, it is founded on the objective similarities found in reality. Cf. *Analysis (c)*.

V • THE STATEMENT IN BRIEF

Nominalism cannot justify verbal predication; Conceptualism cannot justify the objectivity of judgments; and both positions exaggerate the singularity of things.

VI • DEFINITIONS

Nominalism: a doctrine which teaches that only words are universal, that there are no universal concepts, and that reality is utterly singular.

Conceptualism: a doctrine which teaches that both words and concepts can be universal, but that reality is utterly singular.

Verbal Predication: the use of some one term to designate many things.

Objectivity of Judgments: the correspondence to reality of affirmations or negations made by the mind.

To Exaggerate Singularity: to find not even the foundation for universality in things, but to see them all as isolated and disconnected facts.

VII • SUGGESTED READINGS

Copleston, Frederick, *A History of Philosophy* (5 vols.), II, ch. 14, pp. 136–55; III, 56–60, 122–27, 148–52.   London: Burns, Oates, and Washbourne, 1946–1959.

Gilson, Etienne, *The Unity of Philosophical Experience*, ch. 3, pp. 61–91. New York: Charles Scribner's Sons, 1937.

CHAPTER TWENTY-THREE.

THE KANTIAN CRITIQUE

I • THE SITUATION

One of the most prominent and certainly one of the most impressive studies of human knowledge in the history of philosophy is Kantian Criticism. Most modern philosophers have been at least indirectly influenced by Kant's attempt to reconcile Rationalism and Empiricism, and to find and explain the conditions necessary for scientific knowledge. His is a vast system which cannot be mastered without long and diligent study, and hence it cannot be expected that this brief consideration will do more than touch some of the main points of his theory of knowledge, omitting its application to other branches of philosophy. Moreover, since Kant's explanations are quite extensive, it will be impractical to give quotations; so the student is urged to turn directly to the *Critique of Pure Reason* or else to some history of modern philosophy for corroboration or more detailed explanations of the description given below.

II • THE DESCRIPTION

(a) *The Background.* Historically preceding Emmanuel Kant, and profoundly influencing his thought, were the Rationalists (Leibnitz, Spinoza, Wolff) and the Empiricists (Locke, Berkeley, and especially Hume). In general it can be said that the Rationalists exaggerated the power of the human intellect and insisted on its ability for *a priori* knowledge and grandiose conceptual systems, whereas the Empiricists so exaggerated the role of individual and isolated experiences that their doctrine provided no logical foundation for general or universal knowledge of any sort, with the result that scientific knowledge and the communication of any general truths becomes logically impossible.

Kant sought to eliminate the exaggerations and to reconcile the two positions and thus to explain *how* scientific knowledge is possible.

(b) *Types of Judgment.* In doing so he began by proposing certain definitions which were to influence his own explanations and which are necessary for us in order to understand him:

Analytical Judgments are those in which the predicate is already known in knowing the subject. These he considers to be merely tautological; hence they do not contribute to the advance of knowledge.

Synthetic Judgments are those in which the predicate is not contained in the subject. These do increase our knowledge.

Empirical Judgments have to do with the particular facts of sense experience.

A Priori Judgments express factors independent of sense experience.

Kant accepted scientific knowledge in the fields of mathematics and the physical sciences as an indisputable fact and sought only to explain *how* it was possible. So, since such knowledge actually contains the characteristics of necessity and universality, and since experienced facts are totally particular, isolated, and contingent, as the Empiricists had said, Kant concluded that there must be an *a priori* element in all scientific knowledge. Hence scientific judgments are at once *synthetic* (they increase our knowledge) and *a priori* (being necessary and universal).

(c) *A Priori Categories.* Just *how* precisely is this possible? Kant's reply is that such judgments both rest on experience, in the sense that the *matter* for knowledge must be derived from sense experience, and also include an *a priori* element which is contributed by the mind of the subject knowing, because universality and necessity are not found in the sensible reality. This is to say that what is got from sense experience is a chaotic, disorganized, disunited, unformed matter, and that must be unified, formed or shaped, and organized by the knowing mind. Hence of its nature the mind has a tendency to unify sense presentations. Moreover, this is not just a general, undifferentiated tendency to unity in general, for it must be admitted that the mind regularly produces various sorts of unities, and hence must be endowed naturally with various different principles or categories according to which it produces the various unifications of sense presentations. These are the famous Kantian *a priori categories* of the understanding, for example, cause-effect, substance-accident, possibility-impossibility, necessity-contingency.

(d) *Noumenon and Phenomenon.* All this means that according to Kant, when I say that I know something, the object which I actually know is not the thing as it is in itself, but rather the object known is the combination of a "matter" coming from outside and a "form" which is imposed on it by my understanding. He calls the thing-in-itself a *noumenon*, and what is actually known, a *phenomenon*. Hence we never know *what is* but only *what appears*, and what appears is at least partially a projection or creation of my understanding. True reality is forever inaccessible to the human understanding.

(e) *The Imagination.* The question naturally arises as to why some one definite category rather than any one at random comes into action for this or that particular sense presentation, or why particular sense presentations are subsumed under some definite category. What determines the selections involved? To answer this Kant postulates what he calls the "schemata of the imagination," the function of which is to organize the sense presentations for a definite category, and to adapt the categories to the sense presentations in question.

(f) *Kantian Elements of Knowledge.* We can summarize the preceding by pointing out briefly the elements of human scientific knowledge as Kant proposes them:

(1) The *matter*, which is a chaotic, disorganized something coming from the unknowable thing-in-itself, or noumenon.

(2) Sense *forms* of space and time which inform the matter on the sense level. These are mentioned here just for purposes of completion.

(3) The *schemata* of the imagination which organize the sense presentations and prepare the categories.

(4) The *categories* of the understanding which bring about various unities and thus constitute or create what is actually known.

As can be seen, only the first element has any objectivity, the other three being imposed by the knower. All four together make up the object known, the phenomenon, the noumenon remaining totally unknown in itself.

III • THE ANALYSIS

(a) *Passivity, Activity, and Knowledge.* It should be easily seen that what Kant sought to accomplish in his own intellectual environment was something badly needed and valuable. The exaggerations of Rationalism and Empiricism had to be countered, whereas what was valid in their insights had to be preserved. Thus we can appreciate Kant's insistence on the importance of experience, since no true philosophy can afford to shut itself up in the empyrean towers of pure isolated thought to concoct its theories apart from the realities in which we live and move. If knowledge is relational, then it must be knowledge *of something,* and so man's experiences of being are of prime importance. However, it is also true that in knowing, man is not purely passive or receptive; the intellect does have its active and vital role to play, and here Kant is also right in seeking to grasp and express the dynamic aspects of the mind in its operations. He is right, too, in finding a clue to or an indication of this dynamism closely allied to the universality and necessity characteristic of scientific knowledge. Sensible things are surely individual and contingent, whereas science is of the universal and the necessary—at least as an ideal.

(b) *The Unknown Noumenon.* Although that is true, the question remains as to whether Kant was entirely successful in his ambitious project. It is one thing to detect and reject the exaggerations of other positions; it is another to offer a positive doctrine which is itself free of exaggerations.

One of the first things which strike us in considering Kant's theory of knowledge is that he admits the existence of extramental reality, of the thing-in-itself, the noumenon, and at the same time he says that it is entirely unknowable by man. In one who sought to deliver knowledge from the Scepticism of Hume this is difficult to understand, since it is really equivalent to a total Agnosticism. It means that we simply do not know reality, and our experience of knowledge as relational loses all meaning. Such an explanation of human knowledge is in fact not an explanation of what we concretely experience, but rather a denial of some of the relevant data. The knowledge act is experienced as relational, that is the fact to be explained; to deny access to or to suppress the term of that relation is not to explain what is experienced, but to explain what knowledge might be were it not relational. Since by knowledge we do not know the other as it is, we might just as well say that by knowledge we do not know.

(c) *Phenomena and Objectivity.* In regard also to what is actually said to be the object known, the phenomenon, it must be said that since this is constituted by the categories of the understanding, it exists only when it is thought. In this sense it is a creation or projection of the mind, and hence also there is no meaningful objectivity to knowledge.

(d) *An Idealistic Doctrine.* Moreover, in this context *truth* would seem to be deprived of all meaning, or at best it would be equivalent to what truth means in a relativistic and an idealistic theory of knowledge. If noumena are unknowable, it is *as if* they did not exist at all, and so far as knowledge is concerned they might just as well not exist. In fact since we can know nothing about them, we cannot logically admit that they really do exist; their existence is at least irrelevant and maybe even deniable. Thus we are faced with a theory which is idealistic, and hence subject to the criticisms previously given.

(e) *Failure of the Schemata.* Kant himself saw a real difficulty in the question of how the various categories and the various groups of sense presentations ever become coordinated. He sought to answer it by means of the schemata of the imagination. However, this is not a very effective answer, since it merely puts the same difficulty in a different location, and the question still remains as to how and why the schemata function as they do. If it is in purely arbitrary fashion, knowledge itself becomes arbitrary. If it is ultimately due to some psychological association, after the fashion of Hume, then at the very center of scientific knowledge there is found a merely contingent and temporal factor (the imagination and its associations being outside and inferior to the understanding) and hence the universality and necessity of knowledge is lost. Finally, if the cognoscitive faculties work as they do because of some structure objectively true of reality and according to which the mind knows and reacts, then the fundamental position of Kantianism has been abandoned, and some grasp of the thing in itself and in its structure has been admitted. Kant cannot consistently admit this last alternative, and so in all rigor, he should admit that even the necessary and universal elements, which he sought so hard to justify, cannot be justified in his system.

(f) *An Oversimplification.* Finally, there is an oversimplification in the view that because sense realities are particular and contingent, there is no foundation at all in reality for the characteristics of universality and necessity. Any such conclusion is based on the empiricist assertion of the "loosening of ideas" (Hume), according to which each thing is totally different from all others, and there is no relation between any two ideas. However, in our experience we do find countless similarities between things, and it is on the basis of those similarities that the mind forms its universal ideas. In other words, certain structures are not limited to any one individual, but are repeated or are capable of being repeated again and again. We are capable of grasping this common element, and although it exists in each case individualized, it can nonetheless be known as common to many. Hence there is a foundation in the very individuals themselves for the universal.

Moreover there is some foundation for necessity for the twofold reason: (1) that while a thing is what it is, it cannot simultaneously be otherwise, and in that minimal sense it now is necessarily what it is, and (2) that we can gradually come to know something about the natures of things; this tells us what anything having that nature necessarily is.

IV • SUMMARY OF THE ANALYSIS

(a) Kant sought to avoid the exaggerations of both Rationalism and Empiricism, and to stress the need of experience as well as the active, dynamic aspects of human knowledge. Cf. *Analysis (a)*.

(b) Still, since the noumenon remains totally unknowable, the experience of knowledge as relational is not explained, and the theory itself becomes equivalently a total Agnosticism. Cf. *Analysis (b)*.

(c) The phenomena exist only when they are thought, and so are mere creations of the mind; hence there is no objectivity to knowledge. Cf. *Analysis (c)*.

(d) Truth itself in this context is relativistic and idealistic; truth is totally dependent on the subject knowing, since it is just *as if* things did not exist at all. Cf. *Analysis (d)*.

(e) Since the schemata of the imagination are given a determining role in knowledge, and since these are temporal and contingent, neither universality nor necessity are really justified by Kant. Cf. *Analysis (e)*.

(f) The denial of a foundation for universality and necessity in reality is really an oversimplification. True, things are individual and singular; but they are also objectively similar, and their similarities can be known and recognized as the foundation sought. Cf. *Analysis (f)*.

V • THE STATEMENT IN BRIEF

Kant's theory of knowledge is not an adequate explanation of knowledge in respect to the object known, the nature of truth itself, or the necessity and universality which it seeks to justify.

VI • DEFINITIONS

Kant's Theory of Knowledge: a doctrine which teaches that *what is known* is merely the phenomenon, and that it results from the imposition of the categories of the understanding on some "matter" which comes from an unknowable noumenon.

Not Adequate: not taking all the factors of the act of knowledge into consideration.

Object Known: what is known, as has been seen, cannot be totally dependent on the creation of the knower.

Truth: absolute and concerned with real being

Necessary: that which must be what it is.

Universal: that which is or is conceived to be common to many.

VII • SUGGESTED READINGS

Collins, James, *A History of Modern European Philosophy*, ch. 11, pp. 455–512. Milwaukee: Bruce Publishing Co., 1954.

Ewing, A. C., *A Short Commentary on Kant's Critique of Pure Reason.* London: Methuen, 1938.

Gilson, Etienne, *Being and Some Philosophers,* ch. 4, pp. 113–32. Toronto: Pontifical Institute of Mediaeval Studies, 1949.

Kant, Emmanuel, *Critique of Pure Reason,* translated by Norman K. Smith. New York: The Humanities Press, 1950.

CHAPTER TWENTY-FOUR.

EXISTENTIALISM

I • THE SITUATION

Like most classifications used to designate several different individuals in the history of human thought, the term "Existentialism" is one which covers a tremendous variety of views, many of them in violent conflict with one another. It is important, however, that we take a look at this very recent emphasis in philosophy, keeping an eye out particularly for those elements which have to do with human knowledge.

It is quite generally conceded that the Father of Modern Existentialism is Søren Kierkegaard (1813–1855) the Danish thinker. Two who have in recent decades attracted attention are Jean-Paul Sartre, who is an avowed Atheist, and Gabriel Marcel, an avowed Catholic. Under the same heading are generally included a host of others, among whom we can mention Albert Camus, Martin Heidegger, Karl Jaspers, Simone de Beauvoir, Maurice Merleau-Ponty and Nicolas Berdyaev.

II • THE DESCRIPTION

(a) *Background.* The wide disparity of doctrines presented by these writers makes it extremely difficult to present even a generally accurate summary of their positions. However, the following somewhat negative statement can be made as an initial step towards understanding them: Existentialists embody a strong reaction against the apparent meaninglessness of man on the modern scene—and that meaninglessness is generally ascribed by them to the influence of philosophical Idealism and of the scientific or technocratic interests of the era in which we live. This can stand a bit of clarification.

(b) *Systems.* Philosophically, these writers feel, the modern era has been dominated by Idealism and particularly by rationalistic exaggerations. Men sought to erect grandiose systems of thought, of pure thought, and to fit reality into their schemes. In this way the all-important place was accorded to the logical generalized system as such, and there seemed to be a rather total disrespect for and disregard of the individual as such, particularly of the human person. The person became just another fact which might prove embarrassing for the system, or at best was looked on as just another moment in the progress of pure thought. In the realm of speculation this might be looked upon as a tyranny of the universal or of the system. One of the outstanding examples of this mentality is found by Existentialists in the philosophy of Georg W. Hegel, who sees in reality only the unfolding of absolute thought thinking itself; man, like all other "things" is just a moment in the self-realization of the absolute mind. In this sort of context, man's existence seems to be reduced to a significance very nearly approximating zero.

(c) *Technocracy.* On the other hand, technological advances in this machine age have tended to unseat man from any position of dignity in the world. He is made to serve the machine, instead of having the machine serve him. At least this is how the Existentialists would have us look at the matter. Man has been degraded; he has become a mere cog in a grand machine; he is subordinated to technocratic advances; where he can be replaced by a more efficient machine he is eliminated, and the elimination does not in practice imply his enhancement. Each one can reflect for himself on how much validity is to be accorded such indictments.

In order to present a slightly more detailed analysis of these views we shall examine a few of the individuals in this "movement," limiting ourselves to Søren Kierkegaard, Jean-Paul Sartre and Gabriel Marcel.

A. Søren Kierkegaard

(d) *Hegelianism and Kierkegaard.* There are two outstanding factors contributing to the position at which Kierkegaard finally

arrived: (1) the philosophy of Hegel, and (2) Kierkegaard's understanding of the message and meaning of Christianity.

All that has been so summarily said about Idealism above can
be applied here to Kierkegaard's understanding of Hegel. He began with unfeigned admiration for the grandiose scheme and plans
of Hegel; he sought answers to the meaning of life and of man in
Hegel; he waited patiently for answers to his queries. Finally, he
became convinced that none were forthcoming and that none would
ever be provided by any such philosophy. The principal reason,
he felt, was that Hegelianism omitted the one thing that was important: existence.[1]   Hence he caricatured the Hegelian system by
comparing it to an enormous castle in which no one can live:

> In relation to their systems most systematisers are like a man who
> builds an enormous castle and lives in a shack close by; they do not
> live in their own enormous systematic buildings.[2]

Moreover, he says that the youth went to Hegel for guidance and
for answers to problems, only to be deceived by him. In short,
Kierkegaard sees in Hegelianism only the depersonalization of the
human being.

(e) *Christianity and Kierkegaard.*   On the other hand, Kierkegaard came to see in Christianity the expression of a diametrically
opposite view of man—of the supreme value of the individual person, and of each one's personal dialogue with a personal God.
This meant the irreplaceable dignity and the immeasurable importance of each human person. One should not become lost in
the crowd or be overwhelmed by the importance of numbers.
In fact,

> a crowd in its very concept is the untruth, by reason of the fact
> that it renders the individual completely impenitent and irresponsi-
> ble . . .[3]

---

[1] Cf. David Swenson (trans.), S. Kierkegaard, *Concluding Unscientific
Postscript* (Princeton: Princeton University Press, 1941), pp. 99–113.

[2] Dru, Alexander (ed. and trans.), *The Journals of Søren Kierkegaard*
(London: Oxford University Press, 1938), entry no. 583. p. 156.

[3] Walter Lowrie (trans.), S. Kierkegaard, *The Point of View* (London:
Oxford University Press, 1939), p. 114. Cf. the entire section treating of
"The Individual," pp. 111–138.

For Kierkegaard each one is an isolated and singular individual; he must do all he can to develop his God-relationship, and thus become more truly human.

(f) *Other Themes.* Kierkegaard distinguishes three "stages" or levels of existence on which men live, and these he calls the aesthetic, the ethical and the religious. He means to say that some devote their lives to the pursuit of pleasure (sensual or intellectual), others are dominated by a sense of duty, and finally the few try to realize the "freedom of the sons of God."[4] It is to the realization of this total personal freedom, unhampered by authority as such or by doctrine as such, that Kierkegaard invites all, if they are to live the truly Christian life. His analyses of the various types of human living are profound psychological studies and manifest both a penetrating knowledge of what motivates people and a masterful power of expression on his part.

SUMMARY: In general we can say that in Kierkegaard there is a violent and even extreme reaction to speculative thought and to systems in general, since he identified all philosophical system-building with Hegelianism. There is an extreme reaction to doctrinal aspects of Christianity, to authority or intermediaries between God and the individual, to any sort of mob or crowd thinking, and to any sort of generalizing or interest in essences. On the other hand, there is insistence on total human freedom for the development of the personality, although this, he feels, can never be realized except by total dedication to Christ; on the supreme value of the individual as distinguished from the "crowd" or the "race"; on the identity between truth and "subjectivity"; and on the importance of "existence."

B. *Jean-Paul Sartre*

(g) *Man and Atheism.* In Sartre's thinking, too, the central place is held by the human person and all else is interpreted in

---

[4] This theme is found repeatedly in Kierkegaard's works, but especially in these two translated by Walter Lowrie: *Either/Or* (Princeton: Princeton University Press, 1946) and *Stages on Life's Way* (Princeton: Princeton University Press, 1945).

the light of man.   However, the climate is very different from that
of Kierkegaard's thought, since Sartre's Atheism is avowedly at the
root of his interpretations and of his analyses, so much so that it
even determines his selection of the data to be considered.   He
postulates Atheism as basic to his own personal thinking, although
he says that the whole question doesn't merit much attention,
since even if there is a God it makes no practical difference at all.
Still, he does proceed to give his "proofs" for the nonexistence of
God, and these really amount to a Sartrian account of how the
notion of God ever arose in man's mind.[5]

(h) *In-Itself and For-Itself.*   As Sartre considers things, we see
that around us are things which exist for me or for others, but do
not exist for themselves.   Thus, this piece of paper exists for me,
not for itself, nor do I exist for it.   Yet it does exist, and so he
calls it an "in-itself" since it exists there, in itself.   I, however,
recognize that I not only exist, as an in-itself, but I also exist for-
myself, for my own development and self-realization.   Hence he
divides all beings into "in-itself" and "for-itself."   The in-itself is
merely *there;* it has no meaning or perfection or order as it is; it
is chaotic and can be described as dense and viscous or gooey.
Actually these are purely psychological terms which he transfers
to reality to describe it as he reacts to it.

Among things one finds moments of consciousness, that is, men,
the self.   How I come to be, I do not know, but the fact that I
shall die and become part of the in-itself shows that my origin was
just a momentary fracture of the meaningless in-itself; thus for
Sartre man amounts only to an unstable "bubble" of consciousness
which erupts without reason on the crust of the in-itself.   This
means to say that man somehow or other just appears without
explanation and unpredictably.   Then man projects meaning and
order and whatever there is of beauty onto the world itself.[6]

---

[5] Hazel Barnes (trans.), J. P. Sartre, *Being and Nothingness* (New York:
Philosophical Library Inc., 1956), pp. 80–82, 89–90.   Most of Sartre's philosophical ideas are developed in this difficult and cumbersome work (*L'être
et le néant* is the original text).   More briefly he presents some of his leading
ideas in *Existentialism,* translated by B. Frechtman (New York: Philosophical
Library Inc., 1947).

[6] These ideas are difficult to grasp in the works of Sartre, for example in
his *La Nausée* or in *Being,* but some help might be found in seeing them in
J. Collins, *The Existentialists* (Chicago: Regnery, 1952), pp. 51–63.

(i) *Freedom and Truth.* Man's curse and his privilege is to be free; in fact, for Sartre, man is total freedom and nothing else. His whole life is one of choice, and even if he decides not to choose, that very decision is choice.[7] His life is meaningless, it is a "useless passion," since he must live vigorously for nothing. There is no God, no rewarder, and man will revert to the nothingness of the in-itself in the end. However, in the meantime he must choose, and he is totally responsible for every choice. Not that this makes any difference, since there is no one to whom he is responsible except himself, and whatever he chooses is right, since there really is no objective right or wrong. Neither is there any such thing as objective truth, since what I choose is my truth, and since I project the values and the meanings which things are to have.

C. Gabriel Marcel

(j) *Man and Subjectivity.* Marcel is so opposed to Sartre's way of thinking that he does not even want to be called an Existentialist lest he be somehow associated with Sartre. However, he has generally been numbered among these writers, so we shall briefly touch on his views, keeping in mind that what particularly interests us is anything related to a theory of knowledge.

Like Kierkegaard, Marcel is an avowed Christian, who feels that in his interpretation of "being" the reality of his faith must play its part. Again we find a thinker who is not very much concerned with analyzing the intrinsic principles of being or with finding "essences" in the traditional sense of that word. Like most Existentialists, he finds man to be the center of his thinking and its point of departure; he seeks the meaning of man, of human life, and of all other things in their relation to man. So again the stress is on subjectivity and on the meaning of the person. Again there is a distrust of the general or the universal, and some profound analyses of psychological states and reactions of the human person. He adds many carefully elaborated studies on what it means to be a man, to *possess* a body, to have *sensation*, to know *being*, to mention only a few topics.

---

[7] Frechtman, *op. cit.*, p. 48. Other ideas expressed in this paragraph are also found scattered throughout this same work.

(k) *Mystery and Problem.* In order to discuss what sort of knowledge I have of being, Marcel makes his famous distinction between "mystery" and "problem." I am facing a "problem" whenever something lies before and independent of me for my dissection or study or measurement, such as holds true in any chemical or physical study. In such investigations I can proceed in a very objective manner and arrive at an accurate formulation of the solution. However, there are some questions which can be asked, where what is being discussed cannot be treated with such complete objectivity, since the questioner cannot in any way separate himself from the data under consideration. The outstanding example of this is "being"; I can ask questions about being, but I can in no way step outside being in order to consider it; hence I can never get a purely "objective" view of being. In fact, the most rewarding way to study being, according to Marcel, is precisely to refuse to "objectify" it, but rather to begin with the self as being. Here then, there is "encroachment" upon the data itself by the examiner, in the sense that what is being examined is as much within me as outside me. As can be seen, Marcel is not here using the term "mystery" in any theological sense, but only to indicate a very special sort of problem. Neither does mystery mean that what is considered is unknowable; in fact the opposite is said to be true: it is a mystery, for example, *being, existence,* and *freedom,* because it is so close to and personal to me; it is the light of the intellect knowing, and it pervades all problems and studies. However, our knowledge of being must not be confused with the sort of conceptual knowledge had in our understanding of other things, of problems. Marcel is opposed to the overstress on conceptualization which he feels has been prevalent in philosophy. Mysteries such as *being* and *existence* cannot be conceptualized.

(1) *Participation.* As was said, Marcel's type of analysis stresses the uniqueness of the individual person, and does not so much seek the metaphysical structure of man as the meaning of man and of his activities. Thus for example, in his analysis of sensation, he rejects the interpretation of it as a *message* emitted by a "transmitter" and picked up by a sort of "receiving set," where the senses are looked upon as a sort of receiving station to gather in and interpret messages. Rather, for him sensation means a being-with and a being-in-continuity-with things, or, as he puts it, it is under-

standable only as a way of participation in the world. This view involves the larger one, quite opposed to Sartre, that man is at-home in the world, that his relation should be one of communication with and union with all things, whether they be inanimate or animate, human or divine. The same is seen in respect to the world when you consider that man is body-soul in a unique unity, that he does not "possess" his body in any way similar to the way we "possess" money or an auto. Rather, we are "one" incarnate-spirit, and that oneness involving the body itself is our being at-one with the material world. There is then no hostility or opposition between man and his world, there is rather union and participation and at-homeness.[8]

III · THE ANALYSIS

(a) *Special Emphasis on Man.* Although it must be admitted that each Existentialist differs from the other, and that it is difficult to evaluate them together, still certain basic tendencies or ideas can be pointed out and their implications explicitated.

The first thing to be said is that this strong emphasis on the meaning and role of the individual person is something good and extremely profitable in itself, all the more so as a corrective for the over-conceptualization and the over-systematization which characterizes so much of modern philosophy. The human person does occupy a privileged place in the world in which he lives; his freedom and his creative abilities are the unique source of novelty on earth, and as such he has deserved special treatment.

It is not so much that he had not received such special consideration prior to the Existentialists (the history of thought is full of such special considerations for man), but it must be admitted that more often than not he was studied in the light of categories devised for external reality, and in that sense he was effectively treated very frequently as one "thing" among the entirety of things.

---

[8] Most of what has been said is based on Marcel's French works, but one who is interested can find many of the themes in the following works which have been translated into English:

M. Harari (trans.), G. Marcel, *The Philosophy of Existence* (New York: Philosophical Library Inc., 1959); K. Farrar (trans.), G. Marcel, *Being and Having* (Boston: Beacon Press, 1951); R. Hague (trans.), G. Marcel, *The Mystery of Being,* 2 vols. (Chicago: Henry Regnery Company, 1951).

Any such statement, of course, needs much qualification, since a too bald statement can be unfair to so many from Socrates to the present day who sought to study and express the role of man and of his special ethical life. However, what these more recent writers have chosen to call "objectivity" as distinguished from "subjectivity" can generally be said to characterize the over-all treatment previously given.

Moreover, the reaction of the Existentialists against the depersonalization of man wrought by the influence of technocracy, and against the "objective" mentality characteristic of the modern era, which stems from the tremendous advances of the sciences, is a well-founded reaction. This is all the more true because of the influence of those "objective" sciences which treat of man, such as psychology, psychiatry, sociology, and even biochemistry and biophysics. Existentialism recognizes basically that there is something special and unique in man's freedom, and it most clearly explicitates what Kant meant when he said that the human person must never be treated as a mere means. So, omitting for the moment any of the exaggerations which may be detectable in their doctrines, we can say that the Existentialists have made a valid protest at least to the extent that they are a warning against the depersonalization of man and a protest against over-conceptualization and exaggerated objectification in the study of man.

(b) *Danger of Subjectivism.* At the same time, so far as knowledge is concerned, there are many dangers which are deserving of consideration in the existentialistic mentality, and they must be honestly and seriously considered. From the point of view of knowledge itself, there is great danger that the notion of "subjectivity" may lead imperceptibly to an outright Subjectivism and to Relativism. If each person is so unique in every way, if, as Sartre insists, he makes his own values, then he makes his own truth. What he decides on for himself and what he freely chooses is good for him, without respect to others; so, too, what he decides has truth-value, and must have that value for him. In this way we have come full circle, if it is taken literally, and we have in contemporary clothing an old exaggeration, which can be associated with the Idealism against which these men so violently reacted. This becomes all the more clear in the philosophy of Sartre when he insists that man projects all meaning and order onto chaotic matter. So, although his thought does not produce the being of the world from nothing, still it does clothe reality with a meaning which it does not already have even fundamentally.

(c) *Irrationalism.* To put this another way, it is true that for Sartre reality is irrational and meaningless apart from man. In a way he is logically forced to admit this, since there is no God, nor exemplary ideas in the divine Mind, and no source of intelligibility, since this ultimately requires a relation to the divine Mind, and there is no mind but man's. This should mean that being is basically unintelligible, and that sort of thinking has long been typed as Irrationalism. In that context truth cannot mean a correspondence of the mind to what is, since there can be no intelligible reception from outside.

(d) *Knowledge of Natures.* What has been said so far is particularly relevant to Sartre, but taking a synoptic view of those positions indicated, and of others not mentioned, we can see that there is a definite attitude toward human knowledge and its powers implied in their rejection of systems or of any attempt at a system, an attitude that characterizes all of the Existentialists. This is basically a dissatisfaction with the claim that something can be known of human nature in itself and in its fundamental relations; its implication is that what we can know is the individual person only, and what can be said must be limited to the singular, the subjective, the immediately personal. Even those, like Marcel, who praise the work done by the natural sciences and who seem to imply that on some other level there is a validity to be conceded to them seem to say that the same can in no wise be said of man's nature or of our knowledge of it.

Now, it is true that each human person is irrepeatable and irreplaceable, and in that sense each human person is unique. However, there is more to the picture than that. Each of us is a man, endowed with a human nature, which is fundamentally the same in all of us. As we can know something about other natures, so we can know truths about human nature, truths which will hold for men wherever and whenever they are found. As can be seen, the negation just now ascribed to Existentialism and the affirmation of knowledge of human nature we are here insisting on, has an intimate relation to the power of the human mind to abstract and to enunciate universal truths for various natures. That power we have already vindicated for the human mind; the negation of it is an exaggeration. Recognition of the following truths, which would bring about a balanced attitude, is needed here: (1) Much of my knowledge of self is so intimate and personal that it is not conceptualizable, and yet is true knowledge. (2) Through conceptualization I can and do know much about my human nature and the nature of all men.

(e) *Nominalistic.* So the impatience manifested with systematization in matters touching the human is a modern expression of a nominalistic frame of mind. It would say that each one is so unique that general statements are only so many words which do not touch the reality of the individual person. We must in that case be content with individual bits of knowledge about my individual self, with no possibility of generalizing. Of course it must be admitted that I cannot *be* anything except my individual self; however I can *know* more than that. This anti-system and anti-general tendency and its consequent nominalistic implications seem to be true of every Existentialist, even of those who might explicitly reject them.

(f) *Overstress of Will.* Moreover, there is such emphasis on the role of will and of freedom, even to the belittling of intellectual knowledge itself, that there is also a tone of Voluntarism detectable. That is to say, the role of the will seems to be divorced entirely from knowledge and to be even substituted for knowledge. This is of course intimately linked to the Irrationalism mentioned earlier; what I know is unimportant, and what I freely choose is what is true and valuable and good.

(g) *Meaning Added to Objective Structures.* What has been said here can easily be supplemented by considerations from the points of view of ethics and theodicy, but our concern has been with the implications for a theory of knowledge. If what has been said seems a bit harsh or even exaggerated, we might reflect on the meaning of a remark made by a close friend of Gabriel Marcel, and a man who knows Marcel's thinking as well as anyone else. Fr. Troisfontaines says that:

> It would be just as much a pity to neglect rational structure and objective solidity as to neglect subjective analyses and the existential atmosphere. Of the two functions of language, that of suggesting and that of making precise, which culminate in pure poetry or mathematics, existentialism is inclined to confine itself to evocation,

scholasticism excels in definition.   The two functions are indispensible to the life of the spirit.   What do you get when you separate flesh from bone?   A skeleton and a formless mass: two dead things. This is a comparison which should be pondered as much by those who ridicule objectivity as by those who distrust subjectivity.[9]

In a sense it can be said that what is here implied is that no one philosophy can ever claim to have exhausted the real or to have provided all the definitive answers to all questions.   The truth is always being discovered, and we should welcome it no matter what source happens to present it—or to present even one aspect of it. Hence, Fr. Troisfontaines says that in Scholasticism we are provided with a basic core of truths, that we are led to a gradual knowledge of natures and of relationships, that through it we come to know basic structures and constituitive principles, and much more of everlasting value.   However, there is a tone of cold-blooded objectivity to it all, there is a scientific and rational analysis and proof, there is an outline which while valid can stand much continual filling in.   And that is one of the things which Existentialism can do with its insights and its probing descriptions and analyses of human experience and meaning.   He means that it can put a bit of flesh on the bones of the principles and basic truths.   However, those analyses must be done in the light of those basic truths. Truth cannot contradict truth; even though it can be viewed from many sides.   Scholasticism establishes the fact of freedom; Existentialism can show what freedom means in concrete situations. Scholasticism can show the intrinsic principles of matter and form in things, and of the composite nature of man, including his spiritual soul; Existentialism can describe what this means in particular situations, or meditate on what it means to be "incarnate," or to be in touch with things by sensation or intellection.   There can be a complementarity in the treatment, and when there is opposition, the source of error or exaggeration must be discovered.

IV · SUMMARY OF THE ANALYSIS

(a) Existentialism's emphasis on the special place occupied by man in the world, its stress on subjectivity as distinguished from

---

[9] M. Jarrett-Kerr, C. R. (trans.), Roger Troisfontaines, S. J., *Existentialism and Christian Thought* (London: A. & C. Black, 1949), p. 39.

objectivity in the study of man, and its opposition to the depersonalization of man in modern times are points of value worth
stressing. Cf. *Analysis (a)*.

(b) There is, however, a danger of falling imperceptibly from
subjectivity into Subjectivism; and where each person is seen as so
unique that he makes or creates his own values, there is danger of
a sort of Idealism. Cf. *Analysis (b)*.

(c) For Sartre, since there is no Divine Mind, there is no ground
for intelligibility, and hence there is present a sort of Irrationalism.
Cf. *Analysis (c)*.

(d) In the general rejection of systems, there is implied a negation of the mind's power to know something about natures. There
is lacking a balance between the uniqueness of each one's personal
knowledge of self and the possibility of some general knowledge of
natures, even of human nature. Cf. *Analysis (d)*.

(e) This exclusion of general knowledge, which is found in all
the Existentialists, points to a nominalistic view of knowledge. Cf.
Analysis (e).

(f) The role of will and of human freedom is at times made to
be the all-important point. What I freely choose is good and true.
This also is a sort of Irrationalism. Cf. *Analysis (f)*.

(g) Granted these dangers as existing in Existentialism as it has
concretely been presented, it may still be admitted that the descriptions and analyses of human nature and of its meaning in
respect to the world in which man lives can play a complementary
role in philosophy. They can add a bit of fullness to the consideration of fundamental principles and of the intrinsic principles which
constitute composite beings. In particular, their emphases on the
uniqueness of man's role can be welcomed by and given a solid
foundation in keeping with the views of a *philosophia perennis*.
Cf. *Analysis (g)*.

V • THE STATEMENT IN BRIEF

*Existentialism stresses the unique role of man and of his free-
dom; however, it tends to overstress singularity and choice, and
so leans towards nominalistic and irrationalistic views.*

VI • DEFINITIONS

Existentialism: From Fr. Troisfontaines we can borrow the following definition: Existentialism is a passionate return of the individual to his own freedom, in order in the unfolding of its processes to extract the significance of his being.[10]

---

[10] Jarrett-Kerr, *op. cit.,* p. 25.

VII • SUGGESTED READINGS

Although the literature in this field is vast, the following are suggested because they are brief yet instructive:

Copleston, Frederick, S.J., *Contemporary Philosophy,* pp. 125–227. London: Burns, Oates & Washburn, Ltd., 1956.

Collins, James, *The Existentialists,* pp. 188–224. Chicago: Henry Regnery Company, 1952.

Marcel, Gabriel, *The Philosophy of Existence,* translated by M. Harari. New York: Philosophical Library Inc., 1949.

Sartre, Jean-Paul, *Existentialism,* translated by B. Frechtman. New York: Philosophical Library Inc., 1947.

Troisfontaines, Roger, S.J., *Existentialism and Christian Thought,* translated by M. Jarrett-Kerr, C.R. London: A. & C. Black, 1949.

Selected Bibliography

Allers, Rudolf, "On Intellectual Operations," *The New Scholasticism* (January, 1952).

Aquinas, St. Thomas, *Basic Writings of Saint Thomas Aquinas*, ed. Anton C. Pegis, 2 vols. New York: Random House, Inc., 1945.

———, *Introduction to Saint Thomas Aquinas*, ed. Anton C. Pegis, ch. 3, pp. 280–429. New York: Random House, Inc., Modern Library Edition, 1948.

———, *On Being and Essence*, trans. A. Maurer, C.S.B. Toronto: Pontifical Institute of Mediaeval Studies, 1949.

———, *On Truth*, 3 vols., trans. Mulligan, McGlynn, Schmidt. Chicago: Henry Regnery Company, 1952–1954.

Aristotle, *The Basic Works of Aristotle*, ed. Richard McKeon. New York: Random House, Inc., 1941.

Bahm, Archie J., *Philosophy: An Introduction*. New York: John Wiley & Sons, Inc., 1953.

Berkeley, George, *Three Dialogues Between Hylas and Philonous*. Chicago: Open Court Publishing Company, 1946.

———, *A Treatise Concerning the Principles of Human Knowledge*. Chicago: Open Court Publishing Company, 1946.

Blau, Joseph L., *Men and Movements in American Philosophy*. Englewood Cliffs, N. J.: Prentice-Hall, Inc., 1952.

Bochenski, I. M., *Contemporary European Philosophy*, trans. D. Nicholl and K. Aschenbrenner. Los Angeles: University of California Press, 1956.

Burtt, Edwin (ed.), *The English Philosophers from Bacon to Mill*, New York: Random House, Inc., The Modern Library Edition, 1939.

Collins, James, *The Existentialists*. Chicago: Henry Regnery Company, 1952.

———, *A History of Modern European Philosophy*. Milwaukee: The Bruce Publishing Company, 1954.

Copleston, Frederick, S.J., *Contemporary Philosophy*. London: Burns, Oates, & Washburn, Ltd., 1956.

———, *Existentialism and Modern Man*. London: Blackfriars Publications, 1948.

———, *A History of Philosophy*, 5 vols. London: Burns, Oates, & Washburn, Ltd., 1946–1959.

Descartes, René, *A Discourse on Method*, trans. J. Veitch. Chicago: Open Court Publishing Company, 1946.

———, *The Meditations* and *The Principles of Philosophy*, trans. J. Veitch. Chicago: Open Court Publishing Company, 1946.

Donceel, Joseph, S.J., "A Thomistic Misapprehension," *Thought* XXXII (1957).

Dewey, John, *Reconstruction in Philosophy*. New York: Henry Holt & Company, Inc., 1937.

———, *John Dewey's Philosophy*, ed. Joseph Ratner. New York: Random House, Inc., Modern Library Edition, 1939.

Ewing, A. C., *A Short Commentary on Kant's Critique of Pure Reason*. London: Methuen & Co., Ltd., 1938.

Freeman, Kathleen, *The Pre-Socratic Philosophers*. Oxford: Basil Blackwell & Mott, Ltd., 1949.

Garraghan, Gilbert, *A Guide to Historical Method*. New York: Fordham University Press, 1946.

Gibson, A. Boyce, *The Philosophy of Descartes*. London: Methuen & Co., Ltd., 1932.

Gilson, Etienne, *Being and Some Philosophers*. Toronto: Pontifical Institute of Mediaeval Studies, 1949.

———, *Le Réalisme Méthodique*. Paris: P. Téqui, 1935.

———, *Réalisme Thomiste et Critique de la Connaissance*. Paris: J. Vrin, 1947.

———, *The Spirit of Mediaeval Philosophy*, trans. A. H. Downes. New York: Sheed & Ward, Inc., 1936.

———, *The Unity of Philosophical Experience*. New York: Charles Scribner's Sons, 1937.

Henle, Robert, S.J., *Method in Metaphysics*. Milwaukee: Marquette University Press, 1951.

von Hildebrand, Dietrich, *The New Tower of Babel*. New York: P. J. Kenedy & Sons, 1953.

Hoenen, Peter, S.J., *Reality and Judgment According to St. Thomas Aquinas*. Chicago: Henry Regnery Company, 1952.

Hoernlé, R. F. Alfred, *Idealism as a Philosophical Doctrine*. London: Hodder & Stoughton, Ltd., 1924.

Holloway, M., "Abstraction from Matter in Human Cognition," *The Modern Schoolman* XXIII (1946), 120–130.

James, William, *Pragmatism.* New York: Longmans, Green & Co., Inc., 1914.

Joyce, George H., S.J., *Principles of Logic.* London: Longmans, Green & Co., Inc., 1926.

Kant, Emmanuel, *Critique of Pure Reason,* trans. Norman K. Smith. New York: Humanities Press, 1950.

Keller, Leo W., S.J., *The Problem of Error from Plato to Kant.* Rome: Gregorian University Press, 1934.

Laertius, Diogenes, *Lives of Eminent Philosophers,* trans. R. D. Hicks (Loeb Classical Library, vol. 2). Cambridge, Mass.: Harvard University Press, 1925.

Leibnitz, Gottfried, *Discourse on Metaphysics* and *Monadology,* trans. G. Montgomery. Chicago: Open Court Publishing Company, 1957.

Lonergan, Bernard, S.J., "The Concept of *Verbum* in the Writings of St. Thomas Aquinas," *Theological Studies,* VII, nn. 1–4.

————, *Insight.* London: Longmans, Green & Co., Ltd., 1957.

Luce, A. A., *Berkeley's Immaterialism.* London: Thomas Nelson & Sons, Ltd., 1950.

Marcel, G., *The Philosophy of Existence,* trans. M. Harari. London: The Harvill Press, Ltd., 1948.

Maritain, Jacques, *The Degrees of Knowledge,* trans. G. B. Phelan, *et al.* New York: Charles Scribner's Sons, 1959.

————, *The Dream of Descartes,* trans. M. Andison. New York: Philosophical Library, Inc., 1944.

————, *Existence and the Existent,* trans. L. Galantiere and G. Phelan. New York: Pantheon Books, 1949.

————, *The Range of Reason.* New York: Charles Scribner's Sons, 1952.

————, *Science and Wisdom,* trans. B. Wall. New York: Charles Scribner's Sons, 1940.

Mascall, E. L., *Existence and Analogy.* London: Longmans, Green & Co., Ltd., 1949.

Mercier, D. Cardinal, *Critériologie Générale,* 8th ed. Louvain: Institut Supérieur, 1923.

Morandini, Franciscus, S.J., *Logica Maior,* 2nd ed. Rome: Gregorian University Press, 1951.

Morris, C. R., *Locke, Berkeley, Hume.* London: Oxford University Press, 1949.

Mouroux, J., *The Meaning of Man,* trans. A. H. Downes. New York: Sheed & Ward, Inc., 1948.

Neill, T. P., *Makers of the Modern Mind.* Milwaukee: The Bruce Publishing Company, 1949.

Newman, J. H. Cardinal, *A Grammar of Assent.* New York: Longmans, Green & Co., Ltd., 1947.

Nicoll, Donald, *Recent Thought in Focus.* London: Sheed & Ward, Ltd., 1952.

Pieper, J., *Leisure, the Basis of Culture,* trans. A. Dru. New York: Pantheon Books, 1952.

Plato, *The Dialogues of Plato,* 2 vols., trans. B. Jowett. New York: Random House, Inc., 1937.

———, *Plato's Theory of Knowledge* (The *Theaetetus* and the *Sophist* of Plato), trans. F. M. Cornford. London: Routledge & Kegan Paul, Ltd., 1951.

de Raeymaeker, Louis, *The Philosophy of Being,* trans. E. Ziegelmeyer. St. Louis: B. Herder Book Company, 1954.

Régis, L. M., O.P., *St. Thomas and Epistemology,* The Aquinas Lecture, 1945. Milwaukee: Marquette University Press, 1946.

———, *Epistemology,* trans. I. C. Byrne. New York: The Macmillan Company, 1959.

Renard, Henri, S.J., "The Metaphysics of the Existential Judgment," *New Scholasticism,* XXIII (1949), 387–394.

Ross, W. D., *Plato's Theory of Ideas.* Oxford: Clarendon Press, 1951.

Russell, Bertrand, *A History of Western Philosophy.* New York: Simon & Schuster, Inc., 1945.

Santayana, George, *Scepticism and Animal Faith.* New York: Charles Scribner's Sons, 1929.

Sartre, Jean-Paul, *Existentialism,* trans. B. Frechtman. New York: Philosophical Library, Inc., 1947.

Smith, Gerald, "A Date in the History of Epistemology," *The Maritain Volume of the Thomist.* New York: Sheed & Ward, Inc., 1943.

Smith, Vincent E., *Idea-Men of Today.* Milwaukee: The Bruce Publishing Company, 1950.

Spinoza, Benedict de, *Philosophy of Benedict de Spinoza,* trans. R. Elwes. New York: Tudor Publishing Company, 1936.

Stallknecht, N. P., and R. S. Brumbaugh, *The Spirit of Western Philosophy.* New York: Longmans, Green & Co., Inc., 1950.

Thilly, F., *History of Philosophy,* revised by L. Wood. New York: Henry Holt & Company, 1951.

Troisfontaines, Roger, S.J., *Existentialism and Christian Thought,* trans. M. Jarrett-Kerr. London: A & C Black, Ltd., 1949.

Tyrell, Francis M., "Nature and Function of the Act of Judgment," *New Scholasticism,* XXVI (1952), 393–423.

Vaihinger, H., *The Philosophy of "As If,"* trans. C. Ogden. New York: Harcourt, Brace & Company, Inc., 1925.

Vier, Peter C., O.F.M., *Evidence and Its Function According to John Duns Scotus.* St. Bonaventure, New York: Franciscan Institute, 1951.

Weaver, R. M., *Ideas Have Consequences.* Chicago: University of Chicago Press, 1948.

Werkmeister, W. H., *A History of Philosophical Ideas in America.* New York: The Ronald Press Co., 1949.

Wild, John, *Introduction to Realistic Philosophy.* New York: Harper and Brothers, 1948.

—— (ed.), *The Return to Reason.* Chicago: Henry Regnery Company, 1953.

Wilhelmsen, Frederick D., *Man's Knowledge of Reality.* Englewood Cliffs, N. J.: Prentice-Hall, Inc., 1956.

INDEX

Index 235

Judgment (*Cont.*):
formal truth found only in, 63–64, 65
general analysis of, 3–7
procedure in, 4–5
selection of act of knowledge for, 3–4
summary of, 6
individual, sources of, evidence and, 97, 99
initial act of, truths known from, 4–5, 6, 7
Kant's types of, 203
mediate analytic, absolute certainty of, 104
objectivity of, defined, 201
as one term of relation, 64–65
particular synthetic, 86, 92
reflex, 10, 11
relationship to truth, 62–68
transcendence involved in, 28
true:
and elements, 48
objectivity of content of, 26
transcendence and, 28, 32

K

Kant, Immanuel, 189, 194, 202, 218
Kantian Critique, 202–9
a priori categories and, 203–4
as attempt to reconcile Rationalism and Empiricism, 202–3, 205, 207
background of, 202–3
defined, 208
elements of knowledge proposed by, 204–5
failure of schemata of the imagination, 206–7, 208
idealism of, 206, 208
imagination in, 204
impossibility of knowing noumenon, 205–6, 207
influence of, 202
noumenon and phenomenon in, 204, 205
objectivity to knowledge lacking in, 206, 208
oversimplification in denial of foundation for universality and necessity, 207, 208

Kantian Critique (*Cont.*):
types of judgments proposed by, 203
Kierkegaard, Søren, Existentialism of, 210, 211–13, 214, 215
Christianity and, 212–13
Hegelianism and, 211–12, 213
levels of existence in, 213
Knowing, defined, 6
Knowledge:
active and passive aspects of, 35
elements and sources of, 36–94 (*see also specific types*)
concepts, 48–61
deduction, 73–83
induction, 84–94
judging, 62–68
reasoning, 69–72
sensation, 38–47
existence of sensible objects as evidence of, 31–32, 33
explicit and implicit, defined, 7
intellectual, 6
invariability as aspect of, 21–24
Kantian elements of, 204–5
methodical, 6
natural, 6, 11, 12
need for systematization in, Rationalism and, 177–78
object of, according to Empiricists, 182–83, 186–87, 189, 191–92
of objects, nature of, 26–35
sensitive, 6
theory of:
Kantian, defined, 208
method of approaching, 13–17
must have both subjective and objective elements, 147, 149
variability of, according to knower, 20–21, 23, 24
Known, the:
defined, 24
formally, 75, 81
potentially, 81
virtually, 75, 81

L

Law, moral, as motive for moral certitude, 108, 110, 111
Law of being, 77–78
Law of nature:
defined, 93

Index